CONCEPTUAL
概念·建筑 ARCHITECTURE

高迪国际出版（香港）有限公司 编

史强 译

大连理工大学出版社
Dalian University of Technology Press

图书在版编目(CIP)数据

概念·建筑：汉英对照 / 高迪国际出版（香港）有限公司编；史强译. — 大连：大连理工大学出版社，2011.9
　　ISBN 978-7-5611-6494-5

　　Ⅰ. ①概… Ⅱ. ①高… ②史… Ⅲ. ①建筑设计—作品集—世界—现代 Ⅳ. ①TU206

中国版本图书馆CIP数据核字（2011）第174713号

出版发行：大连理工大学出版社
　　　　　（地址：大连市软件园路80号　邮编：116023）
印　　刷：深圳市精彩印联合印务有限公司
幅面尺寸：246mm×290mm
印　　张：21
插　　页：4
出版时间：2011年9月第1版
印刷时间：2011年9月第1次印刷
责任编辑：袁　斌　张　泓
责任校对：王秀媛
封面设计：蔡馨瑶

ISBN 978-7-5611-6494-5
定　　价：298.00元

电　话：0411-84708842
传　真：0411-84701466
邮　购：0411-84703636
E-mail：designbooks_dutp@yahoo.cn
URL：http://www.dutp.cn

如有质量问题请联系出版中心：（0411）84709043　　84709246

CONCEPTUAL
CONCEPTUAL
CONCEPTUAL
CONCEPTUAL

FOREWORD 序言

Ivo Buda 伊沃•布达

Architecture is a visionary and magical game that requires a particular calling that approaches the religious, like the role of a priest. The process of architectural design is almost manic: rules are learned by everyday existence, watching life on the streets, people, nature, details, construction. The architect looks at the world around him to learn and understand from what has evolved in nature and been built by humans. This requires a great love for all that surrounds us, to be able to grasp such beauty and distil it into architectural design.

The architect works closely with the land and nature, with his actions he creates modifications and integrations above or below the land surface. In fact, when the architect begins a project, he begins an adventure, a journey through the places and the history of the earth. Making architecture is going through these places with poetic sensibility to convey the excitement of light, materials, and space. The game of architecture is a bridge between man and nature.

The turning point for an architectural work is the idea. The preparation of the idea is based on research involving patient introspection, aesthetics, technology and history. The idea is like the fruit of a tree, is caught in a moment but it needs to grow into a tree, and the quality of the fruit is directly proportional to the properties of the tree.

Although there are many external forces, the design is primarily shaped by personal choice. This process involves both intuition, which at times is completely unrelated to conventional logic, and the precise analysis of data and facts. In all the powers of the human mind, intuition is perhaps the most fascinating, mysterious and least understood. It involves insight, which can be attained at any time of the day or even while dreaming, suddenly revealing the meaning of a complex concept, a deep understanding of a situation, or simply a solution to a problem. It's like a mysterious light that shatters the darkness, a revelation that seems to come from a unknown part of us that we do not control. Even science, thought to be the most rational application of the human mind, is strongly shaped by intuition and inspiration. For example, the famous chemist Kekule von Stradonitz (1829—1896) said he found the structure of benzene in the figurative symbols of a dream, and upon awakening had to work hard to check its validity. And it is well-known that Picasso used to say: "I first find and then I search."

Insight and the development of new ideas are made possible by those who doubt the obvious, distrust the known and embrace the unknown, uncertainty and doubt. It is possible to find innovative solutions, if you approach a situation through the eyes of a child, without reference to what has been already established and instead create the original and the audacious.

Ivo Buda

　　建筑是一场梦幻神奇的比赛，它要求一种特别的使命感来完成一种信仰，就像牧师的职责一样。建筑设计的过程几乎是狂热的，从日常生活中学习规则、在街道上观察生活，人群、自然、细节、建造。建筑师环顾他周围的世界，通过自然界的演变和人类的建造来学习和理解。这就要求他们对周围一切的热爱，并能够掌控这种美，将这种美浓缩到建筑设计中。

　　建筑师与土地和自然的关系密不可分，通过他的行动，创造出地表之上或之下的改变和整合。实际上，当一个建筑师开始设计一个项目时，他便也开始了一段冒险，一段不同地点和地球历史的旅程。建造建筑就是带着诗意的感性，通过这些地方来传达对光、材料与空间的感动。建筑是人与自然之间的桥梁。

　　对于一个建筑作品来说，构思是一个转折点。构思的准备是建立在耐心的内省、美学、技术和历史等研究的基础上的。构思就像是果树的果实，摘下果实只需片刻，但是果树的成长需要很长一段时间，而且果实的质量与果树的品质成正比。

　　尽管存在着很多的外力，但设计主要还是由个人选择所塑造。这个过程同时牵涉到有时与传统逻辑完全无关的直觉以及对数据和事实的珍贵分析。在所有的人类精神力量之中，直觉恐怕是最富有吸引力、最神秘和最不好理解的。它涉及到洞察力，在白天的任何时候甚至是在梦中都能获得。直觉会突然揭示一个复杂概念的含义，对一种情况的深刻理解或者简单地解决一个问题。直觉就像能打破黑暗的神秘之光，就像是来自我们无法控制的世界的一个启示。即使是科学，被认为是人类头脑最理性的应用，也受到直觉和灵感的强烈影响。例如，著名的化学家凯库勒（1829—1896）曾经说过，他就是在梦中发现了苯的结构图，而醒来之后需要努力地工作以验证它是否正确。众所周知，毕加索曾经说过："我首先发现，然后寻找"。

　　洞察力和新构思的发展是通过那些有明显疑问的人而产生的，他们怀疑已知的并乐意采纳未知的、不确定的和有疑问的。如果以一个儿童的眼光来观察一个环境，对已经建立的一切都一无所知，相反的便能创造出独特而大胆的东西，这样也许就会找寻到富有创意的设计构思。

伊沃•布达

Elena Giussani 艾伦纳•吉萨尼

Architectural concept and its meanings

Successful buildings capture the spirit of their surroundings, even as they assert their own identity. They're visually appealing, healthy, comfortable, flexible, secure and efficient, a pleasure to be in. And they pay their way, adding real value for their owners and users. Architecture reflects our knowledge of how people and organizations use and experience place and space. Uncovering the emotional and behavioral needs of the people who experience space is critical. Since real life is never linear, a strong vision and a flexible approach give plans the resilience they need to guide development over time. Whether they're for a city, a community or an individual, the successful plans are robust enough to overcome the push and pull of the unforeseen, while creating added value at every stage of implementation.

When we design a concept we attempt to decipher the profound meaning of context's present state and to imagine its potential future. We wish to create a memorable, marvellous fragment of the perceptible universe we are demanded to complete with the project. We also try to imagine the new building we are about to create as a receptive cavity for everyone's voice to resound and this can be possible only if we will be able to express what the context most deeply is actually missing.

The theory of forms refuses any a priori solutions and any uncritical application of academic styles, proposing instead an individual and personal creation who is always highly aware of the demands of his own age. The aim of architecture is not to show technological ability or functional perfection but to express with clarity the wonderful story that is enclosed and also not yet expressed in every theme. That story will be told with few essential forms. Those necessary become a symbol.

The value of architecture is in presenting a metaphor, in its being a symbol of the universal that overcomes the formalisms of stylistic language. It takes as its starting point the contingencies of history and of the morphology of its own particular context, to proceed on to a successive abstraction of a significance that is higher, more open, more extraordinary. Every creation is a metaphor of concrete reality and of its historical depth.

Form in architecture contains functionality, economy, memory and tension towards the future; it expresses the symbol with essentiality, it is the synthesis of the poetics of the visionary with that of the possible and openness to unusual solutions. Geometry is not just platonic and mathematical solids. It is not the foundation of invention, not an instrument for the control of inspiration, but for its liberation.

Architecture is a form of art and communication that transcends the particular and aims to open itself up to many different meanings by stepping over the bounds of various disciplines: science, philosophy, metaphysics. The architectural form denies static fixity and tries to overcome the laws of gravity and a rigid horizontal-vertical imposition. It is an a-tectonic architecture, sloping and suspended, on which a continuity of tensions impose an apparent motion. Both through curving forms and broken lines, the overall effect of the volumes is that of a juxtaposition that achieves unstable equilibriums. The pure volumes are distorted and distribute their weight at ground level in an unusual way; the undulating surfaces are ambiguous and inclining, open towards an imaginary infinite continuity into the surrounding space; the inner space flows through the volumes without clear interruptions and the bounder lines between internal and external are fleeting, but unequivocally defined.

The project is a process of knowing. As such, it does not have a definite end but can expand itself indefinitely in time and in its contents. The aim of every project is to prefigure a possible future. Architecture realises in concrete form a piece of the complexity: that does not mean that every building constitutes a complex entity, but that it is part of a situation of complexity to which it contributes with its own form.

If the architectural external volume is the expressive and metaphorical element that determines a single, unique, imaginative form, the inner space represents a cognitive experience set in time that transports the user into a visionary world. Inside those spaces elevated pathways are layered over each other at different levels which permit horizontal crossing of the buildings and which open dramatic views below. The pathways give in the architectonic concept the idea of a fourth dimension, time, by overcoming the limits set by the fixity of space. The building lives therefore by its own function, even though not conditioned by it, and changes according to the spontaneous movement of its users and through the passing of time which is revealed by means of the entry of the light into the open internal spaces so as to define them always in different ways according to the seasons.

Elena Giussani
Studio Nicoletti Associati

建筑概念与其含义

一个成功的建筑，即使本身着重表现自己的特点，也能够抓住周围环境的精神所在。成功的建筑拥有吸引人的外形，洁净舒适又不乏灵活多变，结构牢固并高效。人处于其中将是非常愉快的事情。这些建筑物有所值，能够为其所有者和使用者带来实际的价值。建筑学所反映的正是人们和团体是如何使用并体会地点和空间的，而揭示使用空间的人们的感情和行为需求则是重点所在。由于现实生活从不是线性的，优良的远见和灵活的方法可以让设计方案在建筑本身跨越时间的发展中拥有所需要的弹性。无论设计是为了一个城市、一个社区，还是为了个人，成功的建筑设计在人们使用它创造价值的每一个阶段都能够克服不可预见的影响。

当我们设计一个概念的时候，我们试图解读现阶段环境的重大意义并想象这个概念未来的形象。我们希望能够在完成设计项目的同时，创造出的一个难忘的属于可认知宇宙的美妙片段。我们同时尝试将即将创造出的建筑想象成一种易于让人接受并能让每一个人的声音回响的洞穴。只有在我们能够表达出环境中所缺乏的最深刻的含义时，这种愿望才会变成可能。

形式理论提出了个人或个体创造的风格，而拒绝任何先验的解决方式以及对学术风格不加批判的应用，这是因为每个个体是对其所处阶段的需求最清楚的。建筑的目标不是展示技术能力或功能的完美性，而是清晰地表现每一个主题所附带的而未被表现出来的奇妙的故事。很少有主要的外在形式可以讲述那种故事，而那些产生这种效果的则成为了一种象征。

建筑的价值在于提出一个隐喻，在于它作为克服形式主义的风格语言的一个普遍的象征。它以历史的偶然性和自身的具体独特形态作为出发点，遵照一个连续的更高、更开放、更卓越的抽象意义。每一个创造都是实体存在和它历史性价值的隐喻。

建筑中的形式包括了功能性、经济性、对过去的回忆与对未来的焦虑，建筑形式表现出了象征符号与其重要性质。建筑形式融合了富于诗意的梦幻景象与诗化可能性及对不寻常解决方式的接受。几何外形也并非仅是理想化的和纯粹的数学形式。它并非是创新的基础，也不是一种控制灵感的仪器，而是为了建筑形式本身所释放出的自由精神。

建筑学是一种形式和交流的艺术，它超越了特定的事物，而着力于跨越科学、哲学、玄学等不同领域之间的限制从而使其自身包含多种不同的意义。建筑形式拒绝一成不变的静止形式，并试图超越重力定律和死板的垂直一水平的结构布局。建筑外形设计需要符合地形构造，随着地形的变化或陡或缓，而不断的对比变化则将外在的动态赋予了建筑之上。通过弧形外形与破碎的线条，建筑体量的整体效果创造出一种并列的效果，产生了不稳定的动态平衡。最纯粹的体量被加以变形并将其重量通过一种不寻常的方式分散了到地面上，起伏的表面倾斜而带有朦胧感，内部空间没有明显干扰地穿过建筑体，而内部与外部的分界线虽然在眼前一闪而过，但仍明确而清晰。

设计一个项目是一个认知的过程。因此，它本身并没有明显的终点并可以在空间和含义上无限地扩展。每一个项目的目标是预想一个可能的未来。建筑学在水泥外形中表现的是复杂的综合体的一小部分，这并不是说每一个建筑构成了一个复杂的实体，而是说建筑本身是其自身外形参与构成的复杂环境的一部分。

如果建筑的外在体量是决定了一种简单、独特又充满想象的外形的富于表现力与寓意的元素，建筑的内部空间则代表了带领使用者在特定时间里进入幻想世界的一种认知体验。在这些空间内部，提升的步道在不同高度层叠布置，使得人们可以水平穿过建筑并可以看到下面引人注目的开放景色。步道的设计克服了固定空间范围的限制，从而在建筑概念中引入了第四个维度，即时间的概念。因此，即使并非以此为条件，建筑依然可以通过其自身的功能而存在。通过这种设计，建筑可以随着其使用者的运动而改变。同时，通过将外部光线引入开放的内部空间以使建筑因季节不同而具有不同的特点，设计师也使得建筑能够随着时间流逝而不断变化。

尼克莱蒂建筑事务所
艾伦纳•吉萨尼

CONTENTS 目录

010 Court of Justice Madrid
马德里法院

018 Dubai Camel
迪拜骆驼塔

026 Taipei Performing Arts Center
台北表演艺术中心

034 The White Peacock
白色孔雀

042 Car Experience
汽车体验

050 The Helix Hotel
螺旋酒店

058 Movie Theater and Media Center of
Saint-malo 圣马洛电影院和媒体中心

064 PJCC Development
PJCC开发区

068 Taipei Performing Arts Center/Studio
Nicoletti Associati 台北表演艺术中心

072 Centers for Disease Control Complex
疾病控制中心大楼

078 Swietokrzyska Tower
Swietokrzyska塔楼

082 TEN: Campus Tenova
十：特诺华园区

086 BNT: Designing in Teheran
BNT: 在德黑兰的设计

090 LCF Farnesina High School
法尔内西纳LCF高中

094 Putrajaya Waterfront Residential
Development 普特拉贾亚滨水区住宅开发

100 Edil Tomarchio Commercial Park
Edil Tomarchio商业公园

104 Ex Fonderie Riunite
Ex Fonderie Riunite新规划

108 Seoul Performing Arts Center-
Nodeul Island 首尔表演艺术中心

114 Stadium Kajzerica
Kajzerica体育场

120 Maribor Art Gallery
马里博尔美术馆

124 The Crowd in the Cloud
云中的人群

128 Ordos UNVEIL
鄂尔多斯UNVEIL办公楼

138 Innsbruck
因斯布鲁克

144 The New Earth
新的地球

148 Aachen
亚琛

152 The Danish Pavilion EXPO
2010 丹麦2010世界博览会展馆

156 Urban Interlace(Landsbanki
Headquarters in Reykjavik,
Iceland) 城市编织（冰岛雷克
雅维克的冰岛国家银行总部）

160 Mestia Airport
Mestia机场

162 Museum of Polish History/
Design Initiatives 波兰历史博物馆

009

164 MOCA Wroclaw
弗洛茨瓦夫当代艺术馆

166 Airbaltic Terminal
波罗的海航空航站楼

172 Norwegian Wood
挪威的森林

176 THE APEIRON
Apeiron岛大楼

182 Waalse Krook
Waalse krook大楼

192 Cybertecture Egg
网络建筑：卵形大厦

198 TEK- Technology Entertainment
Knowledge Building 科技娱乐
与知识中心

206 Bella Sky
贝拉天空

210 Centro Comercial Pedregal
佩德雷加尔商业中心

218 Tornado Tower
飓风塔

224 In Remembrance of the Sinan Great
Mosque Design 纪念希南大清真寺设计

228 Izmir Opera House
伊兹密尔歌剧院

234 Rotterdam City Tower
鹿特丹城市之塔

242 The Lantern Pavilion
灯笼亭

248 City Municipality Ljubljana
卢布尔雅那市政厅

256 Four-Leaf Clover Kindergarten
四叶草幼儿园

262 Flowing Gardens
流水花园

272 Auditorium and Library for the
University of Amiens 亚眠大学礼堂与图书馆

280 STEALTH
STEALTH博物馆

288 Golf Dots – Golf Resort at Herning
赫宁高尔夫度假村

292 The Iceberg – Isbjerget
冰山

298 Museum of Polish History/
Zerafa Architecture Studio 波兰
历史博物馆

304 Quebec Museum
魁北克博物馆

308 Mercedes Benz Tower, Yerevan
埃里温的梅赛德斯奔驰塔

316 Science Center Østfold
Østfold科学中心

318 House of Culture and Movement
文化活动中心

320 Dalarna Media Arena
达拉纳媒体剧场

324 Star Light
星光

328 Peak Series
山顶系列

Court of Justice Madrid

马德里法院

Architect: Ivo Buda
Firm: Ivo Buda architetto
Location: Madrid, Spain
Area: 7,033m^2

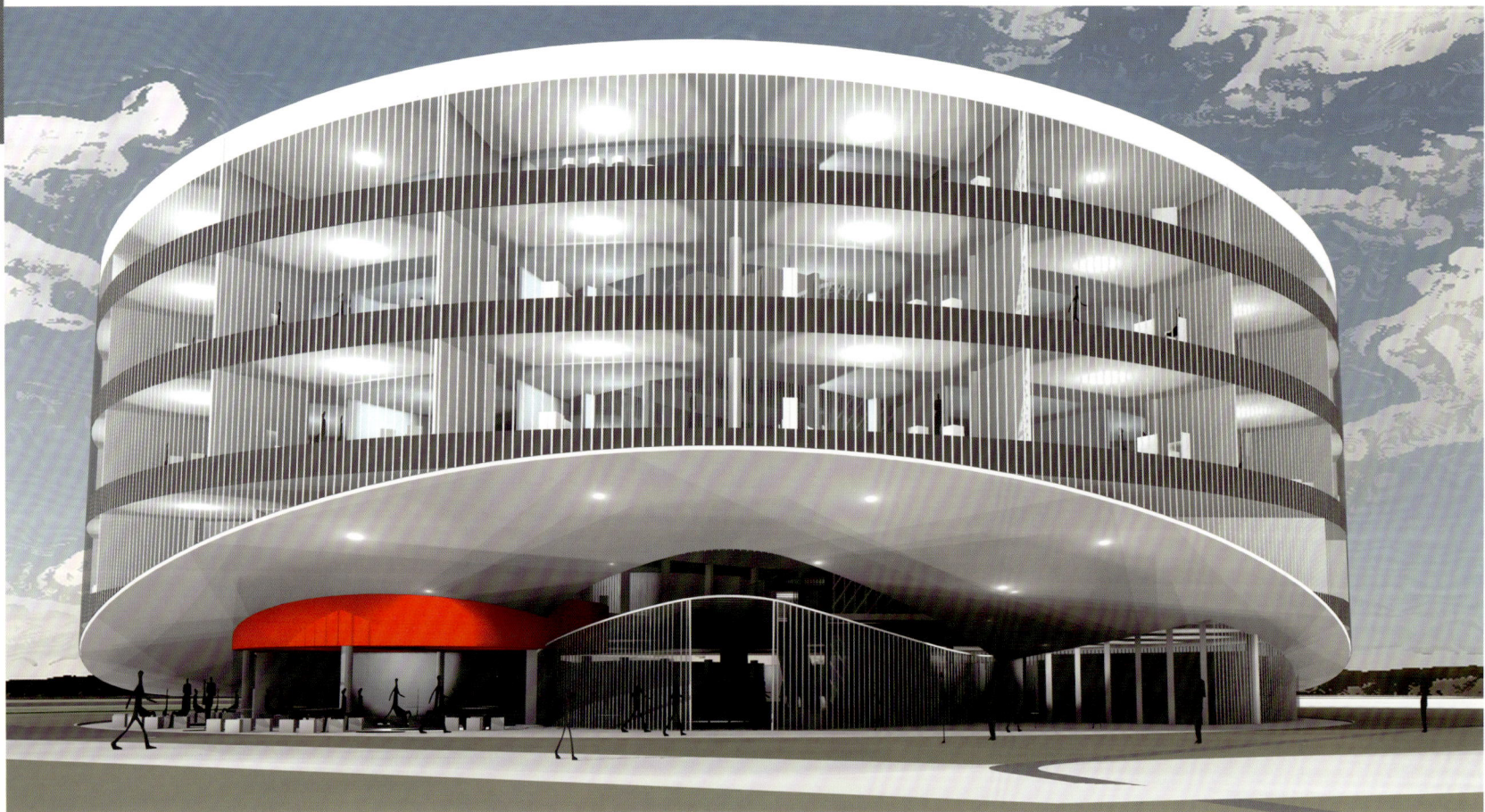

The building is connected to the ground of the campus as a shell of a turtle, with an irregular basis. In this way, it fits the plans of the campus, stands on the pedestrian path and stops people under the shell, extending their role to covering element of the public square-entrance of the building.

The spaces are created through the movement of the shell, that also allow light into the entrance (hall) and in the basement levels (rooms and bedrooms). Thanks to this movement, it is possible to use the roof surface, creating a terrace. The project is developing the spatial possibilities without altering the basic concept of the cylinder. This research aims to study the shell as symbolic form. The superposition of the overhanging shall creates the public square-entrance, resulting in a spectacular crystalline building that is transformed as people move through its surroundings. The basic concept of the cylinder maintains a dialogue with the other circular buildings and at the same time is attractive and recognizable. The facade is composed by a double skin, responsible for protecting the building.

The overall organization of the building derives from the consideration of three types of spaces linked by three separate and distinct paths: the space for the public, the space for judges and persons belonging to the administration of justice, and the space for the security and arrested. All these spaces are independent paths avoiding unwanted encounters.

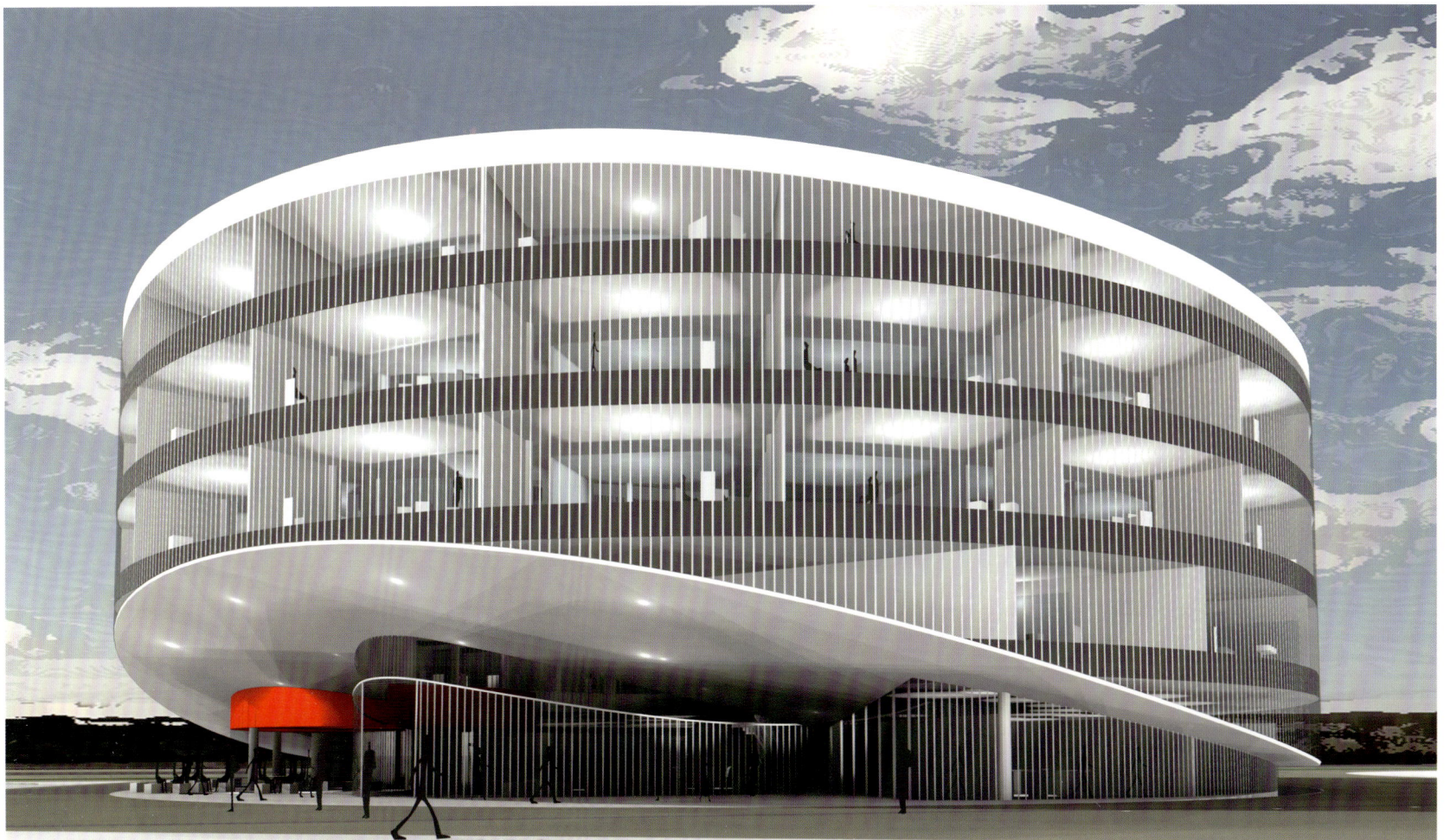

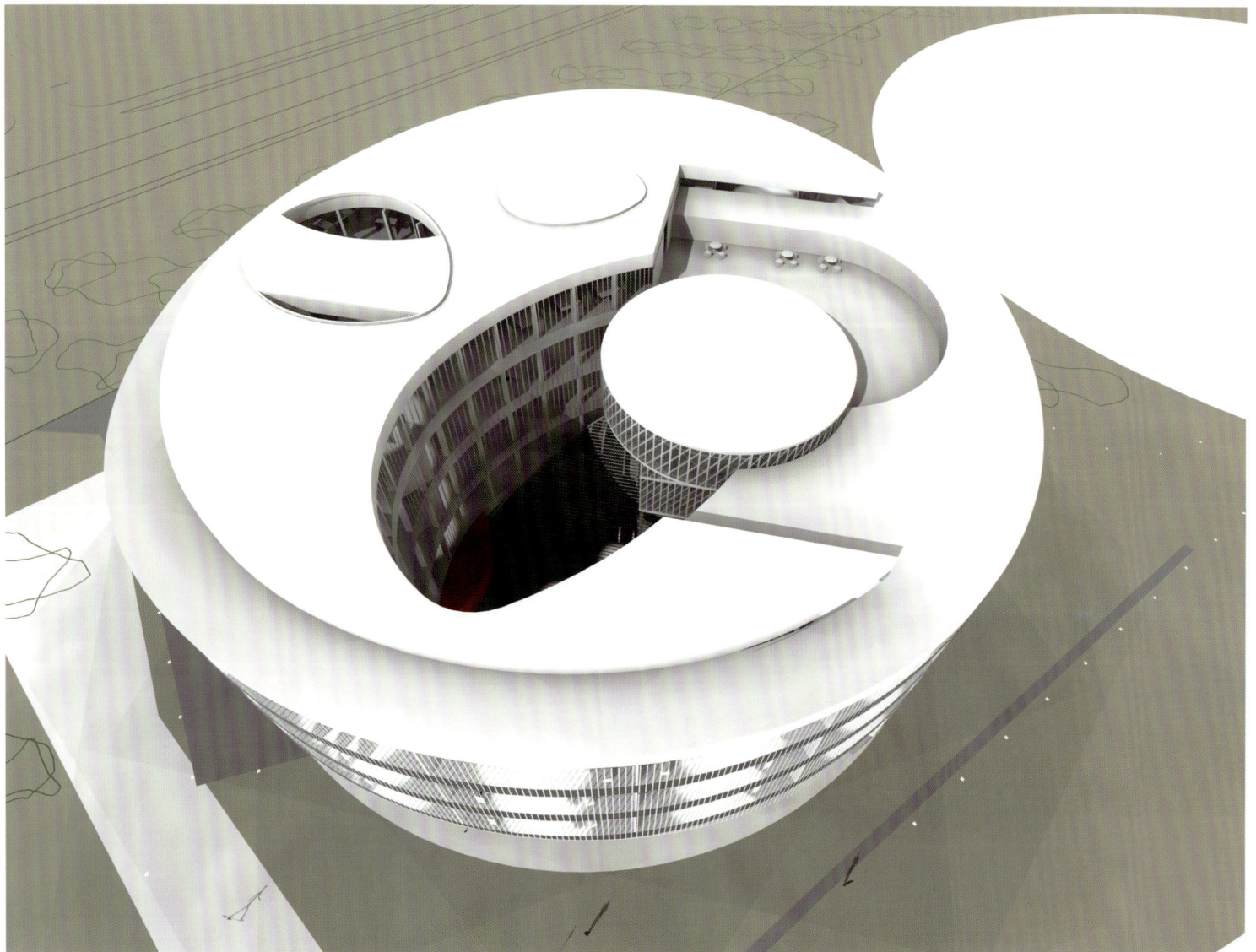

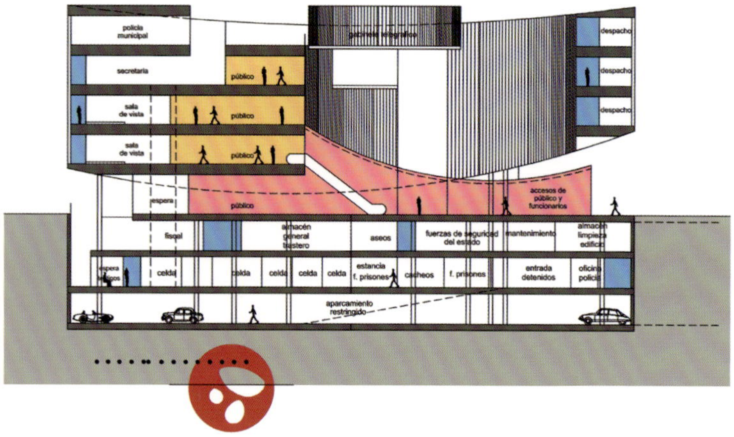

policía municipal
secretaría
público
sala de vista
público
sala de vista
público
despacho
despacho
despacho
espera
público
accesos de público y funcionarios
fiscal
almacén general trastero
aseos
fuerzas de seguridad del estado
mantenimiento
celda
celda
celda
celda
celda
estancia f. prisones
cacheos
f. prisones
entrada detenidos
aparcamiento restringido

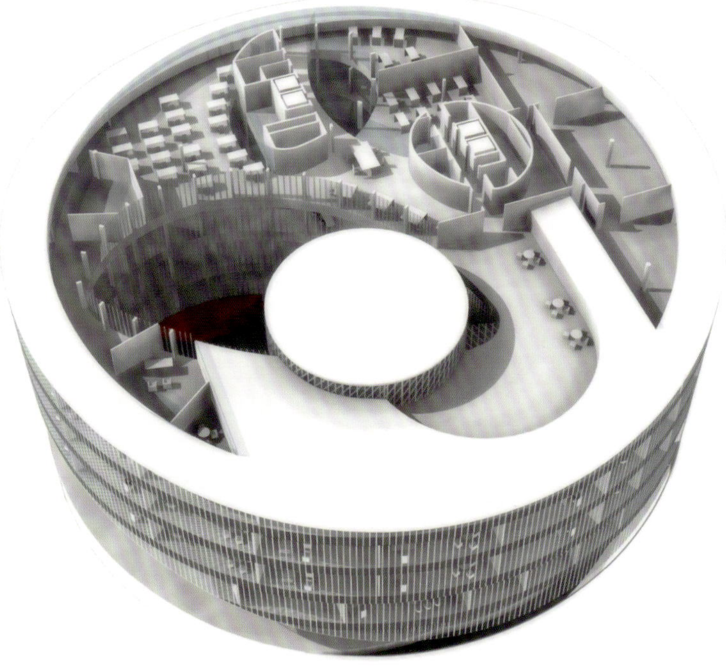

P-3

instalaciones

aparcamiento restringido

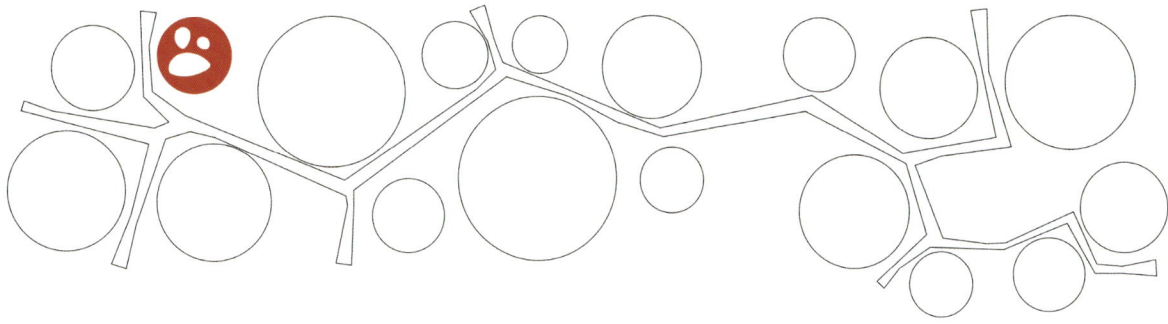

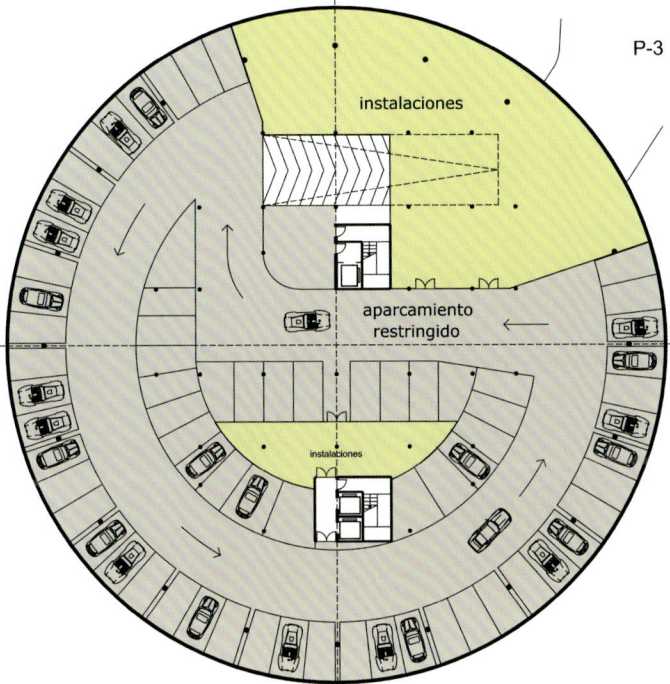

NIVEL -3 aparcamiento restringido
instalaciones
escala 1:300
0 5 10 20

P-2

vestuario y almacen limpieza - residuos

catering presos

despacho individual

despacho individual

secretaria adm.

luz natural

archivo

celda

instalaciones

aparcamiento autobus

almacén efectos

comedor

aparcamiento furgones

oficina policia

estancia f. prisiones

escalera

cacheos

f. prisiones

entrada detenidos

rec. f.

rec. f.

espera detenidos

espera detenidos

rec. f.

espera abogados

c. de limpieza

sala de declaraciones

sala de declaraciones

sala de declaraciones

rueda de reconoscimiento

locutorios

locutorios

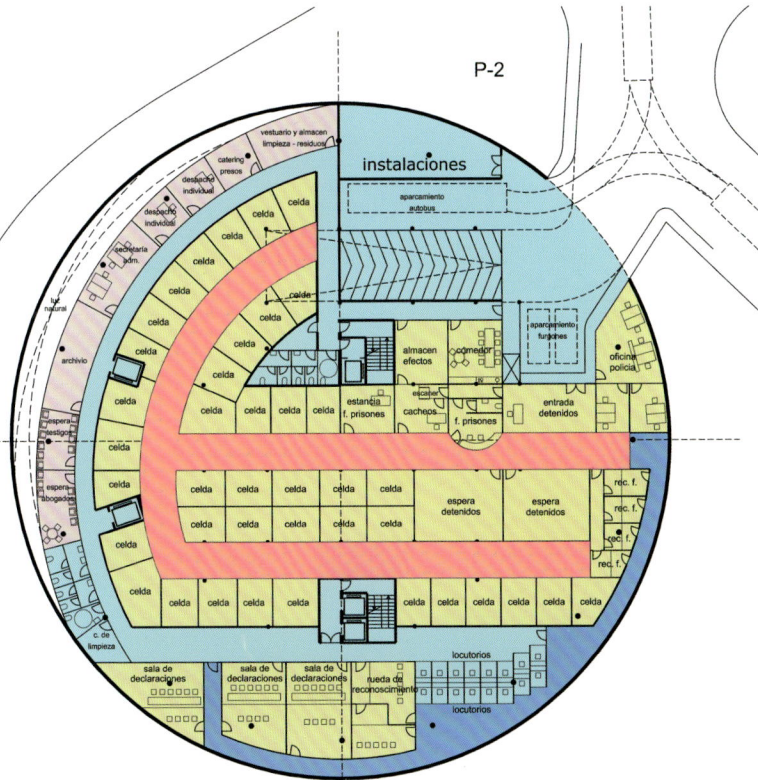

NIVEL -2 detenidos, funcionarios de prisiones
fuerzas de seguridad del estado
escala 1:300
0 5 10 20

secretario judicial

fiscal

forense

juez

apoyo al juez

descanso oficina

archivos generales

almacén general trastero

fuerza de seguridad del estado

mantenimiento

vestuario hombres

almacén limpieza edificio

despacho abogados

fiscal

vest. mujeres

mujeres

vestuario de limpieza

vestuarios hombres

despacho a. social

descanso dormitorio

seguridad edificio limpieza

cuarto taller mantenim.

vestuarios

funcionarios de prisiones

comunicaciones

vestuario mujeres

vestuario hombres

vestuario hombres

almacén trastero

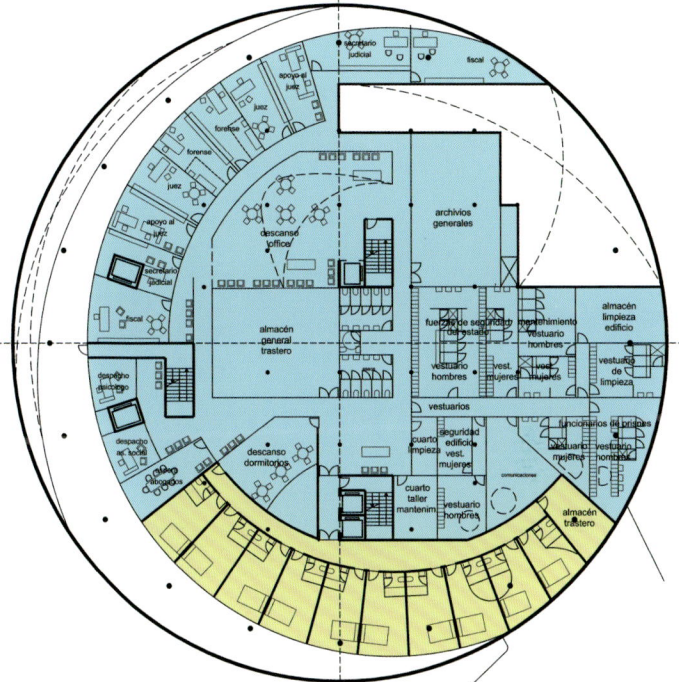

NIVEL -1 juzgado de guardia d., dormitorios,
vestuarios
escala 1:300
0 5 10 20

secretarias de juzgado de vigencia domestica

secretarias de juzgado de guardia de

RADAR

medos de p. comunic.

secretarias de juzgado de guardia diligencias

at. público

espera

seguridad control accesos

despacho psicólogo

espera

espera

juzgado de público y funcionarios

despacho a. social

at. público

salida de público y funcionarios

aseos h.

forense

nodos de p. comunic.

archivo vivo

policia judicial

cafeteria 24 h

secretario judicial

juez

policia judicial

vacio

aseos

diligencias

policia judicial

juzgado de funcionarios

aseos

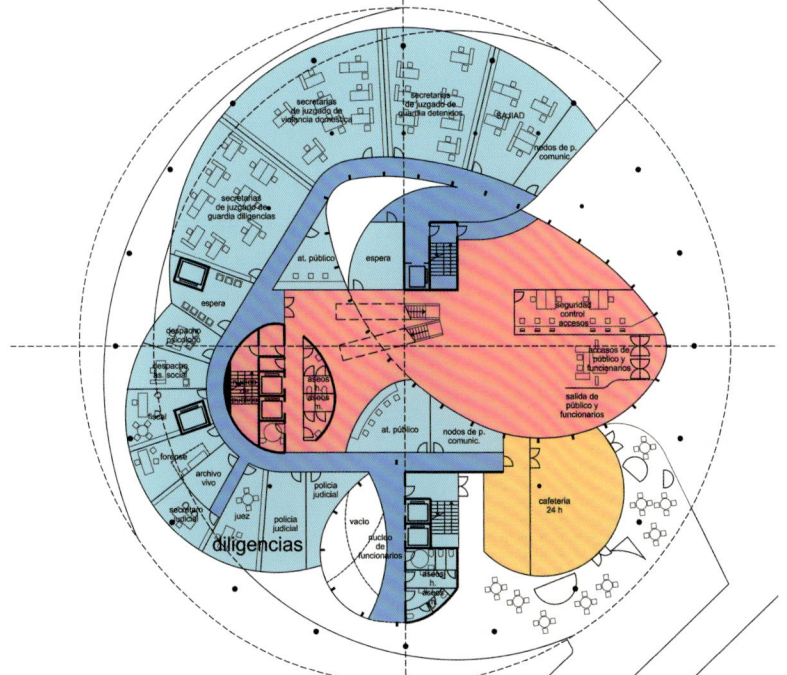

NIVEL 0 aceso a edificio, oficinas secretarias
juzgado de guardia de diligencias
escala 1:300
0 5 10 20

这个建筑与校园区域的地面相连接，仿佛是一个不规则的乌龟壳。这样的设计方法使该设计与校园的规划相适应。整个建筑坐落在人行道上，人们可以在这个壳形建筑下面驻足，使人们的角色扩展成为了这座公共建筑入口的外部元素。

建筑的空间根据壳形结构的运动趋势而设计，从而使得光线能够进入到入口大厅与底层区域（房间和卧室）。得益于这些动态趋势的设计，设计师可以利用屋顶的表面建造一个平面。这座建筑在不改变圆柱体基本概念的前提下发展了空间上的可能性，目的在于将壳形结构作为象征性的形式加以研究。叠加的突出物形成了一个公共的入口广场区域。这种设计的效果就是当人们在这座水晶般的建筑周围行走的时候，会发现建筑的景观会不断地改变。这座圆柱形建筑的基本概念是与周围的建筑保持对话联系，而同时要引人注意。建筑的双层立面结构起到对建筑本身的保护作用。

整个建筑的结构来自于由三种相分离又独特的通道相连接的三种空间：大众的空间、法官与法院管理人员的空间以及保安人员与嫌疑犯的空间。所有的这些空间都有相互独立的通道以避免这些人员之间不必要的相遇。

NIVEL 1 j. de faltas de instrucion, salas de vistas, SAJIAD colegio de abogados y procuradores
escala 1:300

NIVEL 3 4 j. de juicios rapidos de instruccion
escala 1:300

NIVEL 2 salas de vistas
escala 1:300

NIVEL 4 gabinete telegrafico, policia judicial, seguridad
escala 1:300

FLOORS 2 AND 3 : 3 COURTS FOR ACCELERATED TRIALS FOR DOMESTIC VIOLENCE AND 4 MAGISTRATES COURTS FOR ACCELERATED TRIALS

EXTERIOR FAÇADE

| core for civil servants | judge 22 m² | judicial secretary 18m² | prosecutor 22 m² | police doctor 25 m² | 2 offices x 15 m² 1 psychologis 1 social worke | 2 offices x 15 m² 1 psychologist 1 social worker | police doctor 25m² | prosecutor 22 m² | judicial secretary 18m² | judge 22 m² |

restricted traffic area — dealing with victims — restricted traffic area

| public toilets | lawyers' waiting rm. 20 m² | civil servants toilets installations | victims' waiting roo | zone for | victims' waiting room | civil servants toilets installations | |

traffic area for public and professionals

| public core | | counter 2 m | counter 2 m | | |

| secretariat 11 civil servants 120 m² with current archive | secretariat 11 civil servants 120 m² with current archive |

SECOND FLOOR : SAJIAD

| to lets and core for civil servants | psychologist 15m² waiting room 10 h2 | soc. Worker 15m² | Laboratory tech. 20m² | lawyer 18 m² |

traffic area for public and professionals

| public core | double office admin assist 22m² | administrative secretariat 8 civil servants | counter 2m | |

A : access opened by card or similar. Which allows only civil servants and accompanied victims to pass through. Suggested layout

1st FLOOR: MAGISTRATES COURT FOR MISDEMEANOURS

FACADE EXTERIOR

| core for civil servants | judge 22 m² | judicial secretary 18m² | police doctor 25 m² | prosecutor 22 m² | prosecutor 22 m² |

restricted traffic

| toilets for public | current archive toilets for civil servants installations | current archive toilets for civil servants installations | |

traffic area for public and professionals

| public core | counter 2 m | counter 2 m | |
| | secretariat 3 civil servants 40m² | prosecutors' secretariat 5 civil servants 60 m² | |

A : access opened by card or similar. Which allows only civil servants and accompanied victims to pass through. Suggested layout

GROUND FLOOR, DIAGRAM OF SUGGESTED LAYOUT

| police court, proceedings/evidenc judge, secretary, prosecutor police doctor | 2 offices x 15 m² 1 psychologist 1 social workers | judicial police 2 offices | core and toilet for civil servan |

restricted traffic

police court proceeds/evider jud. police at. public	victims' waiting room	secretariats of police courts arrested persons and domestic violence police court 11 civil servants SAJIAD office
police court proceedings/evidence secretariat 11 civil servants	waiting rm closed public	
care and toilets for public	hall public waiting area	
security check point for access	Att. public	

attention to public for secretariats, Information and registry counter 6 m

| b | b |

24 h cafeteria

b independent checkpoints for public and civil servants 1 detector arch + scanner + table 0.80x0.80m and chair, for checkpoint

access for public and civil servants

secretariats for the domestic violence and detainee police courts, very close to the restricted core which gives access to the rest of the court area in basement 1

A : controlled access for civil servants or accompanied victims to the building's restricted access area. Lawyer's access to the zone for arrested persons and basement 1

BASEMENT LEVEL -1: 1 POLICE COURT FOR ARRESTED PERSONS + 1 POLICE COURT FOR DOMESTIC VIOLENCE

police court for domestic violence — police court for arrested persons

natural lighting from patios in the ground floor or openings in the façade

| prosecutor 22 m² | judicial secretary 18m² | support staff for the judge 2 adm. asst 22m² | judge 22 m² | police doctor 25 m² | core and wash rooms civil servants | lawyers' waiting room 20m² | 2 offices x 15 m²: 1 psychologist 1 social worker | police doctor 25 m² | judge 22 m² | support staff for the judge 2 adm. asst. 22m² | judicial secretary 18m² | prosecutor 22 m² |

restricted circulation, immediate connection to the corresponding secretariats on the ground floor and to the zone for arrested persons on level -2

basement floor 1 : 10 night-shift bedrooms

natural lighting and ventilation for bedrooms from patios in the ground floor openings in the façade (preferably to the exterior)

| restricted traffic | pantry lounge for bedrooms | bedroom 1 | bedroom 2 | 3 | 4 | 5 | 6 | 7 | 8 | 9 | 10 | storeroom |

corridor exclusively for bedrooms

LEVEL -2   GENERAL DIAGRAM

restricted traffic core cargo lift		
DOUBLE HEIGHT loading and unloading dock room for waste materials parking for vehicles for transferring arrested persons	ZONE FOR ACCUSED PERSONS		restricted traffic core
TUNNEL campus level -2	DOUBLE HEIGHT sally port for bus for accused persons pedestrian access to cells		
ramp to car park level -3	ZONE FOR ACCUSED PERSONS		

LEVEL -3   GENERAL DIAGRAM

restricted traffic core cargo lift	car park	restricted traffic core
campus gallery installations	installations	
ramp to car park level -3	car park	

natural light

catering for arrested pe	cloakroom and store cleaning-waste	individual office	individual office	secretary administratv 4 mesas	archive	cloakrooms for warders for warders on floor -1
dining room for warders service area pantry	store room personal effects toilet warders	body search scanner	lounge for warders 25 m2	cell	cell	cell
sally port for bus for transferring arrested persons parking for security: level -2	2* sally port for arrested person	visual control cabin 1*				minimum 3 m
view over parking and sally port						cell
police office 8 tables 2 offices individual	1 desk for police registry	police doctors' examination rooms x 4	arrestees waiting room 40 m2	arrestees waiting room 40 m2	cell	cell	cell
		accompanied arrestees traffic				
office - lounge 15 persons police cloakroom on floor -1	restricted traffic core 6*	lawyers' waiting room 5*	3 room for statements 4*	8 phone booths	identity par- ade x3	toilets	witness waiting roo	restricted traffic core 6*

restricted traffic

1* -counter for registering warders
security glass with direct vision and access to the cell corridor. Console with 3 surveillance monitors

2* -grille: motorised sliding grille

3* -controlled access door (card or similar)

4* - statement room (see diagram)

5* Lawyers' waiting room with window with security glass and hatch towards the "grille" zone

6* access to other premises on the floor -1
direct access to the police court for Arrested persons on level: -1
access to arrested persons' lifts to the courtrooms

statement room		
	4 arrestees	
	dais	
	table 6 people	
	5.50 min	

access for judges and lawyers

security glass — intercom | arrested person | table — table — lawyer

visiting booth

| | seats | |
| | arrested persons — witnesses | |

IDENTITY PARADE ROOMS

9.- CROSS-SECTION DIAGRAM OF USAGE

							floors
telegraph office		STATE SECURITY FORCES			judicial police	judicial police	judicial police	4th
accelerated trials	central strip	accelerated trials		accelerated trials	central strip	accelerated trials	3rd
accelerated trials	central strip	accelerated trials		WAITING ROOMS	COURTR OMS		2nd
court for misdemeanour	central strip	SAJIAD lawyers and barristers	crt misdemeanrs	WAITING ROOMS	COURTR OMS		1st
c. servants proceed's crt public cafeteria	attending to public	PUBLIC HALL		ground floor level	PUBLIC HALL	attending to public	secretaries police crt	GROUND
TUNNEL	loading and unloading dock supplies, waste waiting area transfer vehicles for arrested pers	vehicle sally port transfer vehicles for arrested persons double height		maintenance	cloakrooms, archive, storeroom		BASEMENT 1
	double height	arrested pers zone,	cells	evidence, police doctors identity parade	waiting roo visit booths,	BASEMENT 2
gallery installations	connections	installations		restricted car park		BASEMENT 3

↑ civil servants traffic   ↑ public traffic   ■ area directly connected

■ lift for accused persons

Dubai **Camel**

迪拜骆驼塔

Architect: Ivo Buda, Manfredo Bianchi, Tomaz Kristof
Firm: Ivo Buda architetto
Location: Dubai, UAE
Area: 2,030m²

The audacity of a cosmopolitan city of Dubai is comparable to the courage of caravans crossing the desert. The proposal is homage to this audacity, bringing the image of the caravan to the city, a camel that comes to an oasis to drink water. The tower aims to capture the essence of this journey through the desert, the will of the pioneers to explore the unknown land and the joy in an adventure in the nature. Dubai is an adventure in the desert and an oasis with fresh water at the same time. It is represented in an organic form that elegantly reaches to the sky and at the same time keeps the natural shape of the slow movement of a camel.

A strange perspective of four distorted legs invites the visitor to explore the tower from all possible angles around it, as the image of the structure is completely different seen from different sides or even distances. Therefore the approach to the tower follows a spiral line with tower in the center. It begins already on the highway - since there are not many buildings around, the tower is well-visible from a distance. The spiral continues through the local road and parking lot, then the entrance to the park, pathways, and finally the ramps that lead to an entrance podium 5m above the ground. To offer a visitor a final tour around (and under) the building before entering it, the ramps are curved around the legs of the tower, creating octopus shape with entrance podium in the center. These walkways provide a more private and intimate approach to the building and establish a spatial separation from the park and the green areas that are left slowly below.

SITE PLAN
1:1000

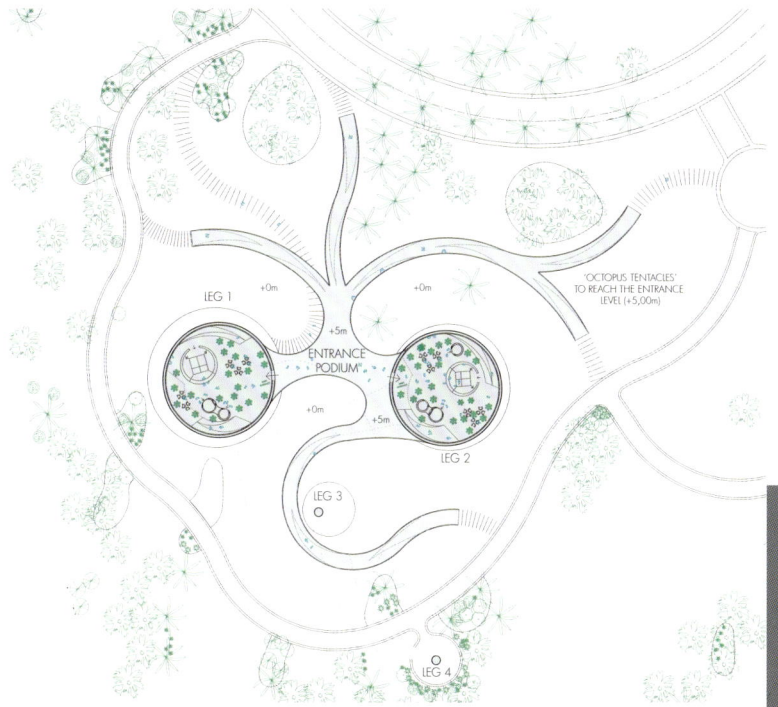

BOATING LAKE

RESTAURANT

THE CAMEL

ACCESS TO
UNDERGROUND
PARKING

LEG 1 +0m

+0m

+5m

ENTRANCE
PODIUM

'OCTOPUS TENTACLES'
TO REACH THE ENTRANCE
LEVEL (+5,00m)

+0m

+5m

LEG 2

LEG 3

LEG 4

Sections

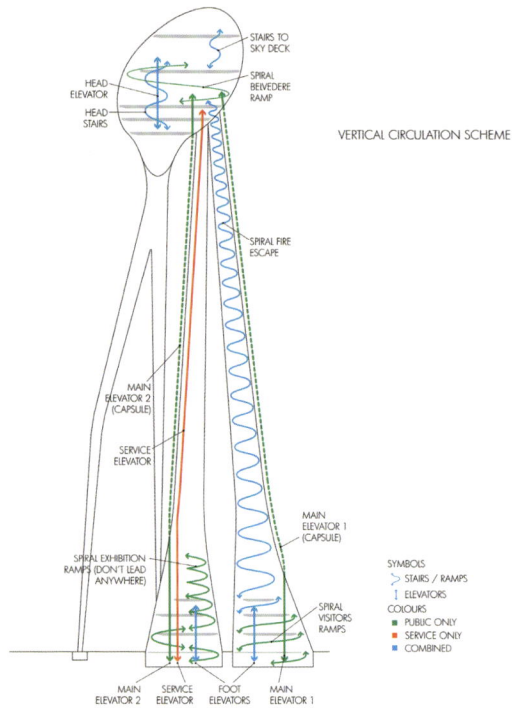

VERTICAL CIRCULATION SCHEME

STAIRS TO
SKY DECK

SPIRAL
BELVEDERE
RAMP

HEAD
ELEVATOR

HEAD
STAIRS

SPIRAL FIRE
ESCAPE

MAIN
ELEVATOR 2
(CAPSULE)

SERVICE
ELEVATOR

MAIN ELEVATOR 1
(CAPSULE)

SPIRAL EXHIBITION
RAMPS (DON'T LEAD
ANYWHERE)

SPIRAL
VISITORS
RAMPS

SYMBOLS
↗ STAIRS / RAMPS
↕ ELEVATORS
COLOURS
■ PUBLIC ONLY
■ SERVICE ONLY
■ COMBINED

MAIN
ELEVATOR 2

SERVICE
ELEVATOR

FOOT
ELEVATORS

MAIN
ELEVATOR 1

PLANTS WITH
POTS ON THE
STRUCTURE

PLANTS WITH
POTS INSIDE THE
STRUCTURE

HANGING PLANTS

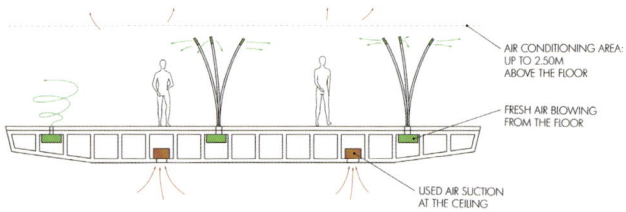

AIR CONDITIONING AREA:
UP TO 2.50M
ABOVE THE FLOOR

FRESH AIR BLOWING
FROM THE FLOOR

USED AIR SUCTION
AT THE CEILING

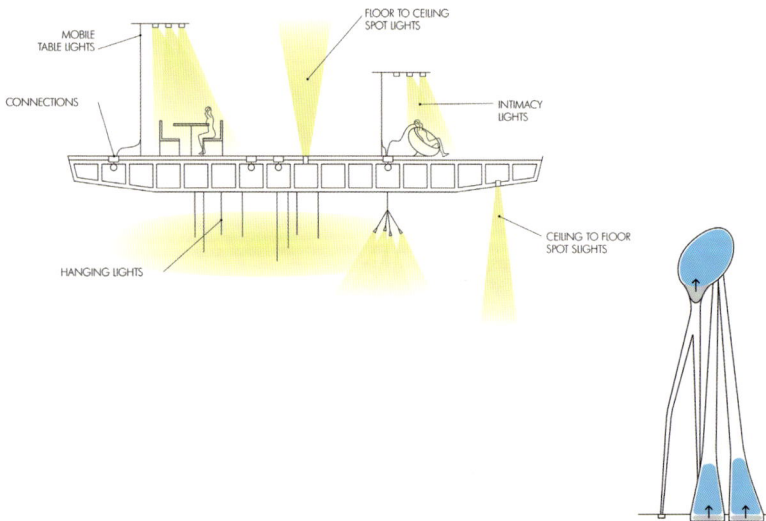

MOBILE
TABLE LIGHTS

FLOOR TO CEILING
SPOT LIGHTS

CONNECTIONS

INTIMACY
LIGHTS

CEILING TO FLOOR
SPOT SLIGHTS

HANGING LIGHTS

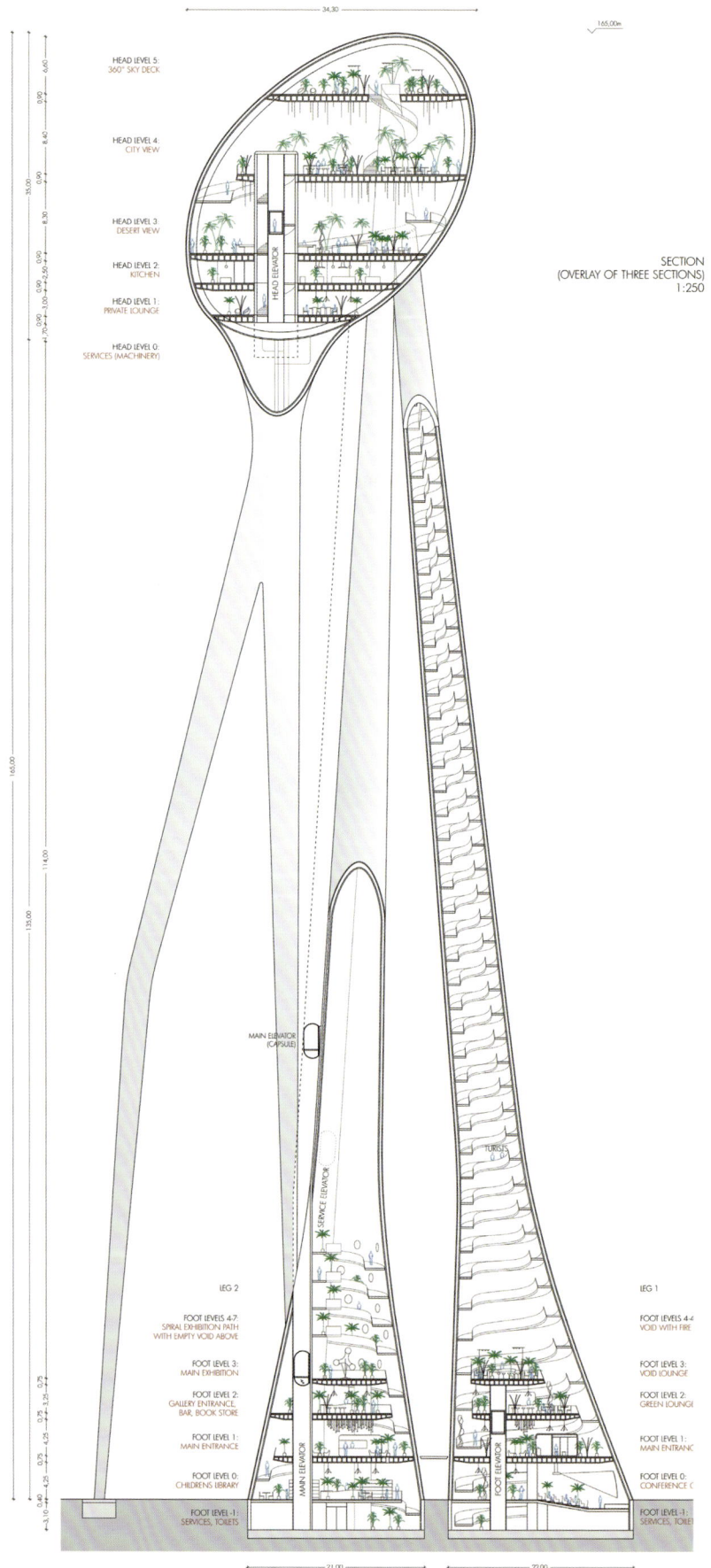

SECTION
(OVERLAY OF THREE SECTIONS)
1:250

HEAD LEVEL 5:
360° SKY DECK

HEAD LEVEL 4:
CITY VIEW

HEAD LEVEL 3:
DESERT VIEW

HEAD LEVEL 2:
KITCHEN

HEAD LEVEL 1:
PRIVATE LOUNGE

HEAD LEVEL 0:
SERVICES (MACHINERY)

HEAD ELEVATOR

MAIN ELEVATOR
(CAPSULE)

SERVICE ELEVATOR

LEG 2

FOOT LEVELS 4-7:
SPIRAL EXHIBITION PATH
WITH EMPTY VOID ABOVE

FOOT LEVEL 3:
MAIN EXHIBITION

FOOT LEVEL 2:
GALLERY ENTRANCE,
BAR, BOOK STORE

FOOT LEVEL 1:
MAIN ENTRANCE

FOOT LEVEL 0:
CHILDRENS LIBRARY

FOOT LEVEL -1:
SERVICES, TOILETS

MAIN ELEVATOR

FOOT ELEVATOR

LEG 1

FOOT LEVELS 4-4:
VOID WITH FIRE

FOOT LEVEL 3:
VOID LOUNGE

FOOT LEVEL 2:
GREEN LOUNGE

FOOT LEVEL 1:
MAIN ENTRANCE

FOOT LEVEL 0:
CONFERENCE C

FOOT LEVEL -1:
SERVICES, TOILE

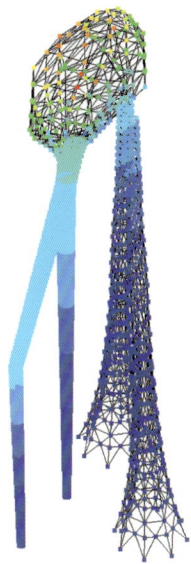

DISPLACEMENT
IN X-AXIS

4,45cm

0,32cm

MAX STRESS
VERTICAL LOADS+WIND ACTION
S.L.U.16 FOR REAR LEGS

1939 daN/cm²

155 daN/cm²

VERTICAL-SEISMIC-WIND LOADS
BUCKLING S.L.U.
GLOBAL STRUCTURE

2785 daN/cm²

214 daN/cm²

VERTICAL-SEISMIC-WIND LOADS
STRENGTH S.L.U.
GLOBAL STRUCTURE

3618 daN/cm²

270 daN/cm²

Structure

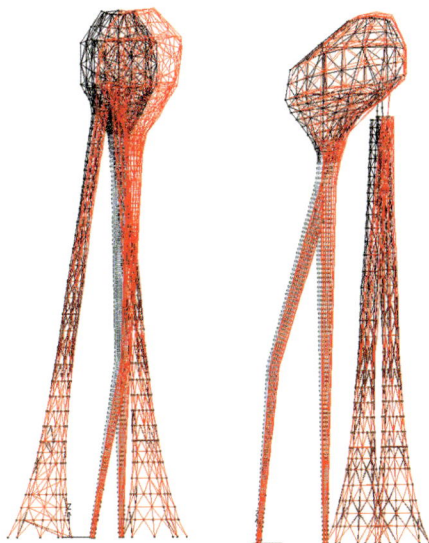

SEISMIC DEFORMATION 90°

SEISMIC DEFORMATION 0°

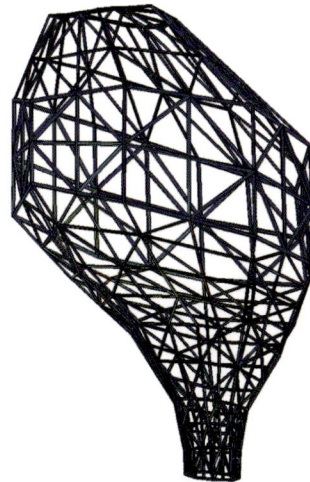

360° SKY DECK
360° SKY DECK
DOWN TO THE CITY VIEW
360° SKY DECK

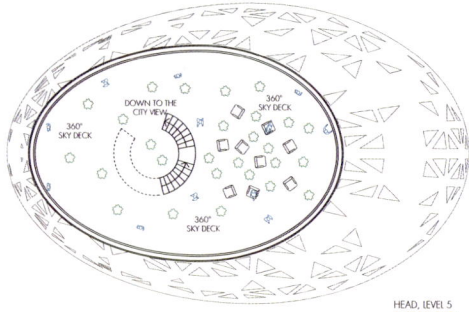

HEAD, LEVEL 5
360° SKY DECK
floor level: +158,00m
space height: 6,60m
minor axis (outer): 16,00m
major axis (outer): 23,90m

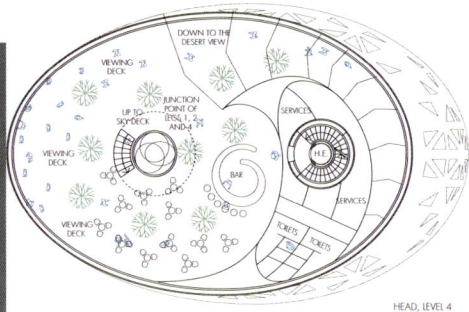

SPIRAL STAIRS/RAMPS TO THE HEAD
VOID (100M HIGH EMPTY SPACE)
S.E.
M.E.
SPIRAL STAIRS/RAMPS TO THE HEAD

LEG 1, LEVELS 4-47
VOID WITH RAMPS/STAIRS (FIRE ESCAPE)
floor level: +14,00 to +144,50m
space height: 120,00m
outer diameter: 14,10 to 4,50m

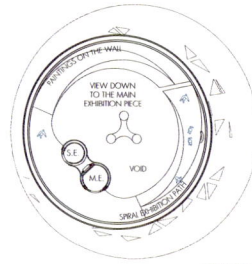

PAINTINGS ON THE WALL
VIEW DOWN TO THE MAIN EXHIBITION PIECE
S.E.
M.E.
VOID
SPIRAL EXHIBITION PATH

LEG 2, LEVEL 3
GALLERY
(SPIRAL EXHIBITION PATH)
floor level: +14,00 to +28,65m
space height: 120,00m
spiral height: 14,65m
outer diameter: 14,85m to 4,50m

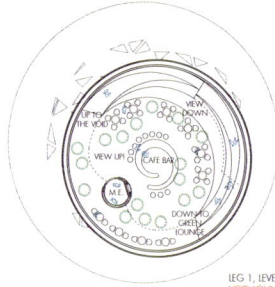

VIEWING DECK
DOWN TO THE DESERT VIEW
JUNCTION POINT OF LEGS 1, 2 AND 4
UP TO SKY DECK
SERVICES
BAR
VIEWING DECK
H.E.
VIEWING DECK
SERVICES
TICKETS
TICKETS

HEAD, LEVEL 4
CITY VIEW
floor level: +148,70m
space height: 8,40m
minor axis (outer): 21,00m
major axis (outer): 30,90m

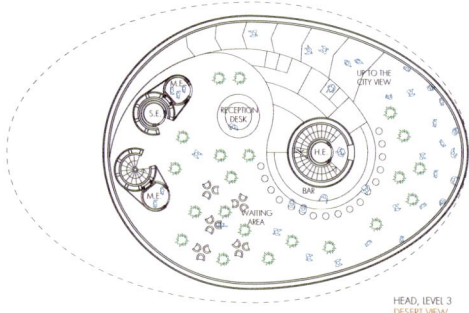

UP TO THE VOID
VIEW DOWN
CAFE BAR
DOWN TO GREEN LOUNGE
M.E.

LEG 1, LEVEL 3
VOID LOUNGE
floor level: +14,00m
space height: 120,00m
outer diameter: 14,60m

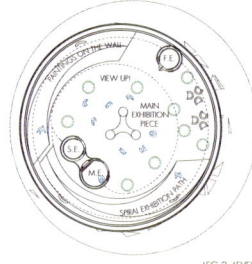

PAINTINGS ON THE WALL
VIEW UP
F.E.
MAIN EXHIBITION PIECE
S.E.
M.E.
SPIRAL EXHIBITION PATH

LEG 2, LEVEL 3
GALLERY
(MAIN EXHIBITION SPACE)
floor level: +14,00m
space height: 120,00m
outer diameter: 15,35m

M.E.
UP TO THE CITY VIEW
RECEPTION DESK
S.E.
H.E.
BAR
WAITING AREA
M.E.

HEAD, LEVEL 3
DESERT VIEW
floor level: +139,40m
space height: 8,30m
minor axis (outer): 20,15m
major axis (outer): 27,90m

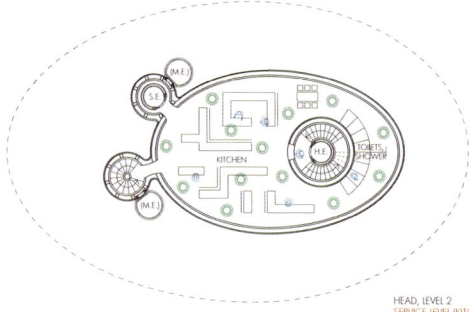

UP TO VOID LOUNGE
DOWN TO ENTRANCE HALL
TICKETS
S.E.
M.E.

LEG 1, LEVEL 2
GREEN LOUNGE
floor level: +10,00m
space height: 3,25m
outer diameter: 15,70m

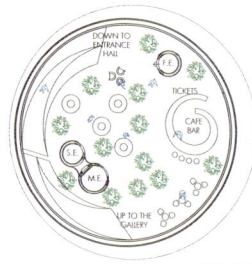

DOWN TO ENTRANCE HALL
TICKETS
CAFE BAR
S.E.
M.E.
UP TO THE GALLERY

LEG 2, LEVEL 2
GALLERY ENTRANCE
floor level: +10,00m
space height: 3,25m
outer diameter: 16,85m

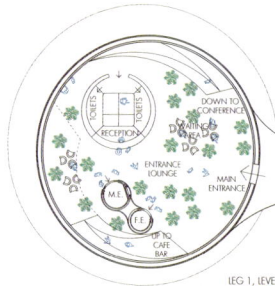

M.E.
S.E.
KITCHEN
H.E.
TICKETS
CLOAK ROOM

HEAD, LEVEL 2
SERVICE LEVEL (KITCHEN)
floor level: +136,00m
space height: 2,50m
minor axis (outer): 12,15m
major axis (outer): 19,40m

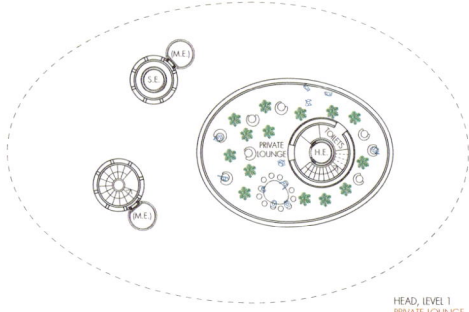

TICKETS
RECEPTION
DOWN TO CONFERENCE
ENTRANCE LOUNGE
MAIN ENTRANCE
M.E.
F.E.
UP TO CAFE BAR

LEG 1, LEVEL 1
MAIN ENTRANCE HALL
floor level: +5,00m
space height: 4,25m
outer diameter: 17,20m

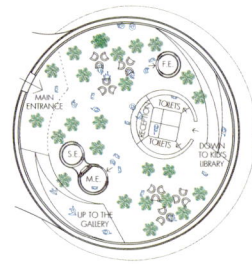

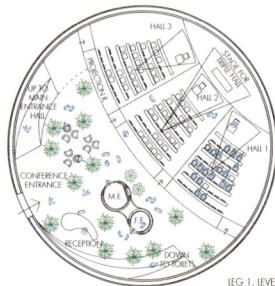

MAIN ENTRANCE
TICKETS
S.E.
M.E.
F.E.
DOWN TO KIDS LIBRARY
UP TO THE GALLERY

LEG 2, LEVEL 1
MAIN ENTRANCE HALL
floor level: +5,00m
space height: 4,25m
outer diameter: 17,20m

PRIVATE LOUNGE
H.E.
M.E.
S.E.

HEAD, LEVEL 1
PRIVATE LOUNGE
floor level: +132,10m
space height: 3,00m
minor axis (outer): 11,15m
major axis (outer): 15,15m

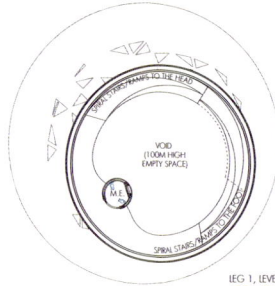

HALL 3
UP TO MAIN ENTRANCE HALL
RECEPTION
HALL 2
STORE ROOM
HALL 1
CONFERENCE ENTRANCE
M.E.
F.E.

LEG 1, LEVEL 0
CONFERENCE CENTER
floor level: +-0,00m
space height: 4,25m
outer diameter: 22,00m

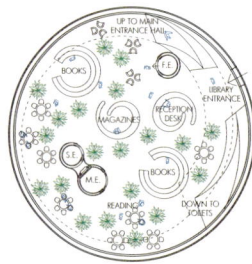

UP TO MAIN ENTRANCE HALL
BOOKS
LIBRARY ENTRANCE
MAGAZINES
RECEPTION DESK
BOOKS
S.E.
M.E.
READING
DOWN TO TOILETS

LEG 2, LEVEL 0
CHILDRENS LIBRARY
floor level: +-0,00m
space height: 4,25m
outer diameter: 21,00m

作为一个国际大都市，迪拜所体现出的大胆可以与那些穿越沙漠的商队相比。这项建筑设计方案就是这种大胆的体现。在这个方案中，设计师通过一只前往绿洲饮水的骆驼形象将沙漠商队的形象带到这座城市中。这座塔的设计意图抓住这种穿越沙漠的旅程所体现的核心精神，即那些开拓者们探索未知世界的愿望与在大自然中历险的乐趣。迪拜正代表了沙漠中的冒险生活，同时也是充满清水的绿洲。这种理念被展现为一种有机形态，优雅地伸展向天空。同时，整个设计仍保留了骆驼缓慢行走时的自然外形。

这座塔楼从不同的方向甚至不同距离来看都会形成完全不同的外观，而它的四个抽象的腿形支柱正形成了一种奇怪的角度，使得游人可以从周围通过各种可能的角度来欣赏它。因此，通往这座塔的道路也被设计成围绕塔楼中心的螺旋形。因为这座塔周围没有很多的建筑，因此整个建筑从远处就可以清楚地看到。所以，人们在高速公路上就开始了通向这座塔的旅程。螺旋形的道路沿着公路并一直延伸到停车场，然后到了公园的入口、步道，并最终到达连接至高于地面5米的入口高台的通道。为了让游人在进入建筑之前能够有机会最后一次从建筑的四周（与底部）观察它，通道在塔楼的四个支柱间曲折环绕，形成了一个包围入口高台的章鱼形状。这些步道为游人提供了一种与建筑之间更加私人化与亲密的接触，并在空间上将公园与其远处的绿化区域相分离。

Plans

Taipei Performing Arts Center

台北表演艺术中心

Architect: Ivo Buda
Firm: Ivo Buda architetto
Location: Taipei, Taiwan, China
Area: 23,100m²

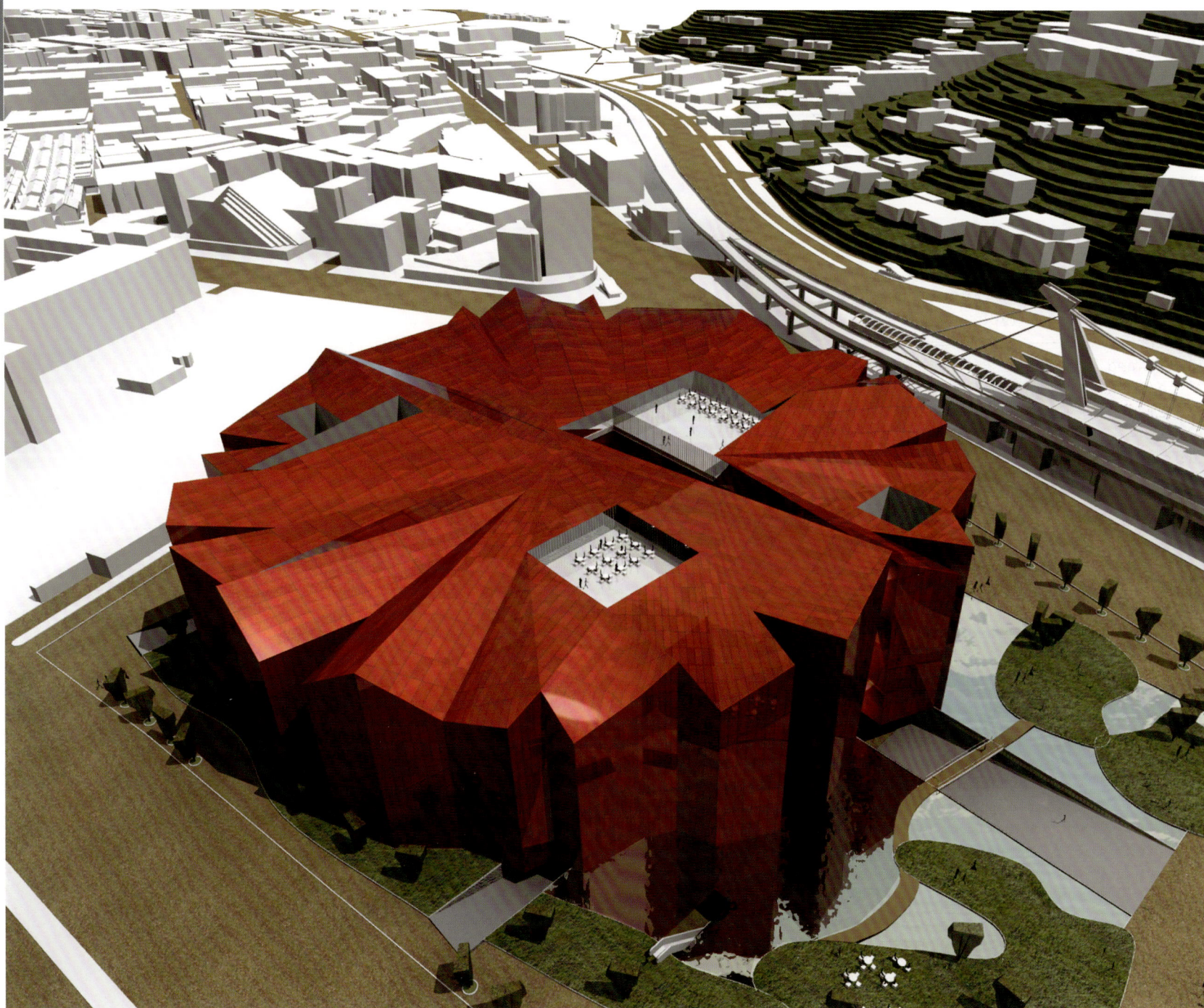

Like a big flower, a round shape, corroded little by little through a precise gesture, embraces all the diverse elements of the composition. This reference is an opportunity to research the essence of the theatre and the activity inside it. That activity has the fragrance of a flower and the intensity of a sunset. In fact the sun and the nature give people every day a repeated, but unexpected spectacle, and yet the willingness to be permeated by the unexpected often reveals new keys to the comprehension of reality. But like a flower dances in the sun and in the wind, can the architecture dance too? Yes, in the images the architecture is still, in its real life it get cross by light and people. The articulated volume in the light undertakes multiples possibilities of definition: in the sunset light it is carnal and sensual, in the midday light acute and precise. Inside the space is a continuum with the outside, compressed, secret, but luminous. The overall organization of the building derives from the connection of three types of theaters:

1. The Grand Theater
2. The Proscenium Playhouse
3. The Multiform Theater

The three theaters all have independent control and operation systems, they are connected with a common lobby with control points at the theatre entrance. The common lobby is a 24-hour open area with no requiring ticket, except only when area control is needed. The environment in the lobby is comfortable and visually is linking all the floors of the center.

The program for the Grand Theatre include classical operas, large-scale dance performances and musicals. The sitting capacity is of 1,500 seats with seating of 1,000 seats at the orchestra level and 500 seats at balcony.

The major programs for the Proscenium Playhouse and the Multiform Theatre are drama and dance performances. The sitting capacity is of 800 seats.

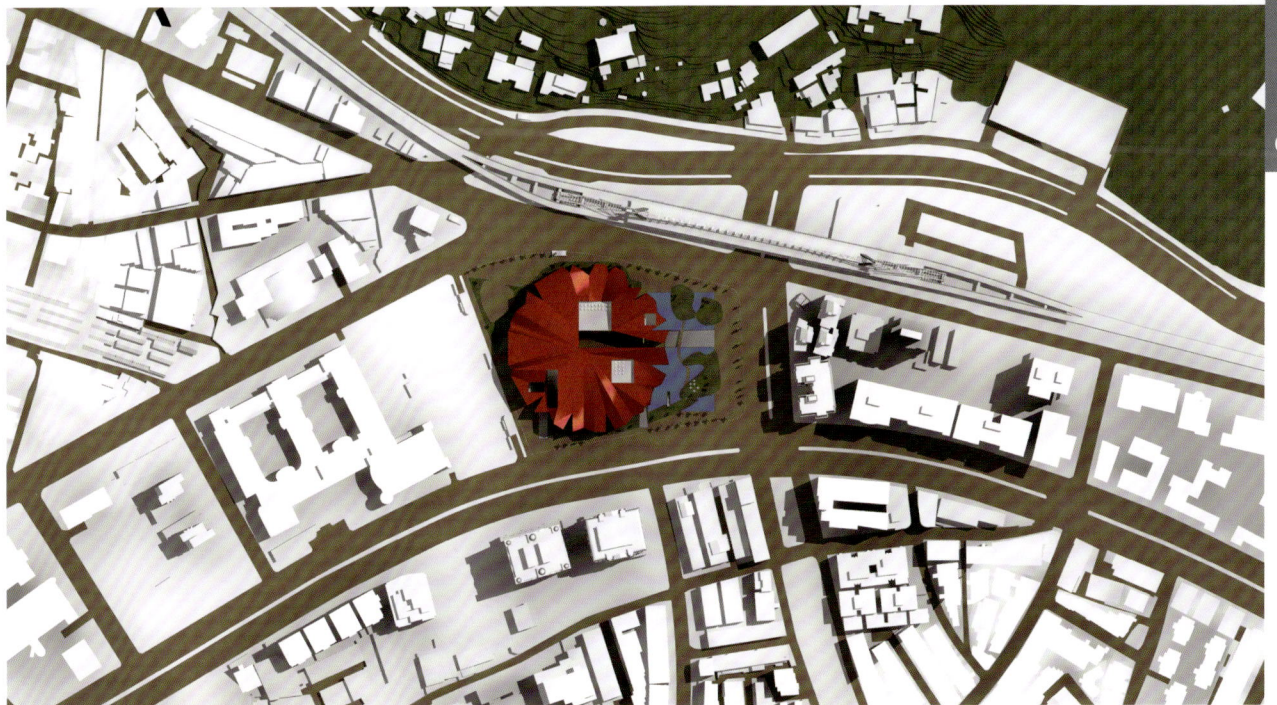

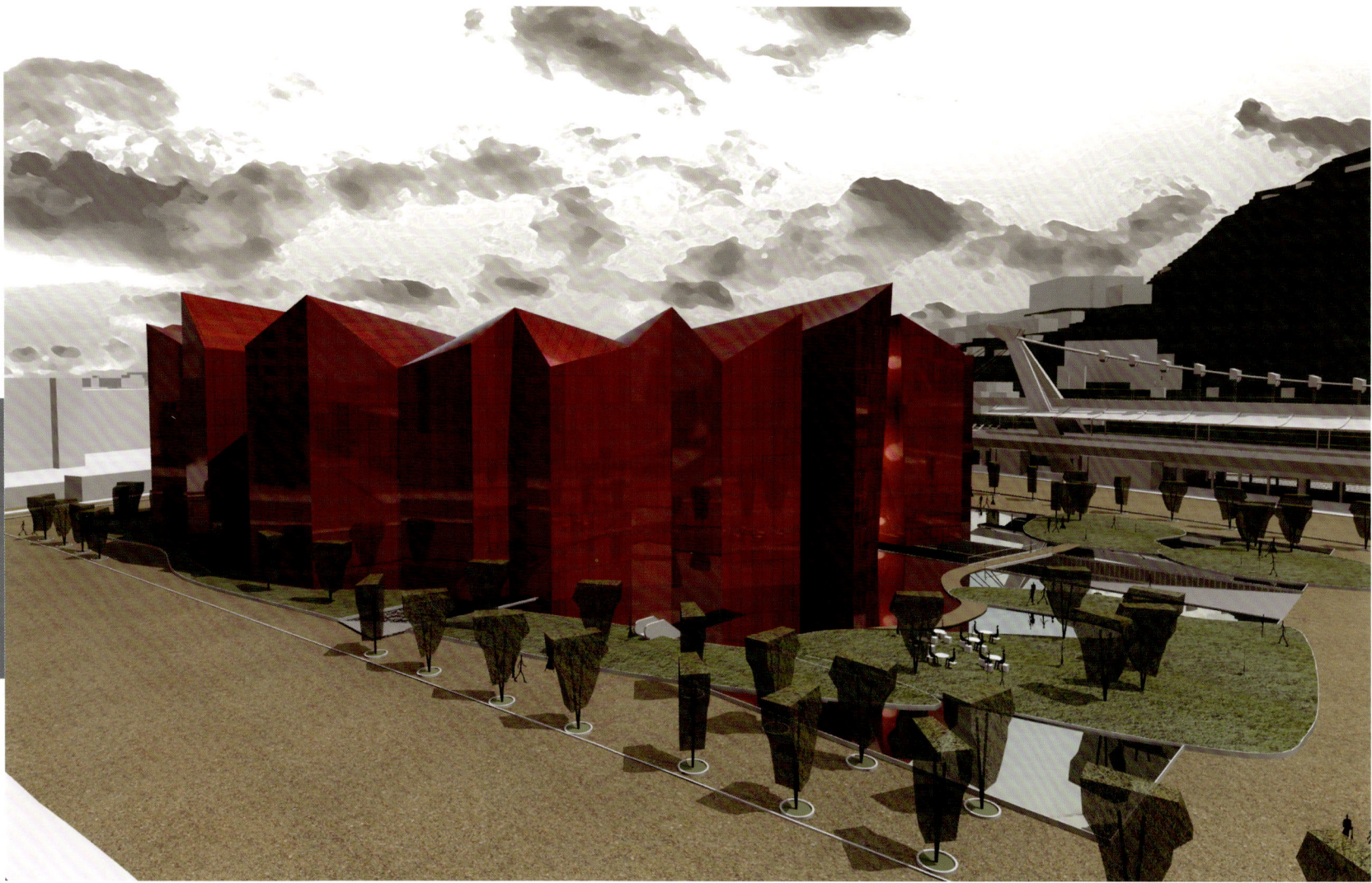

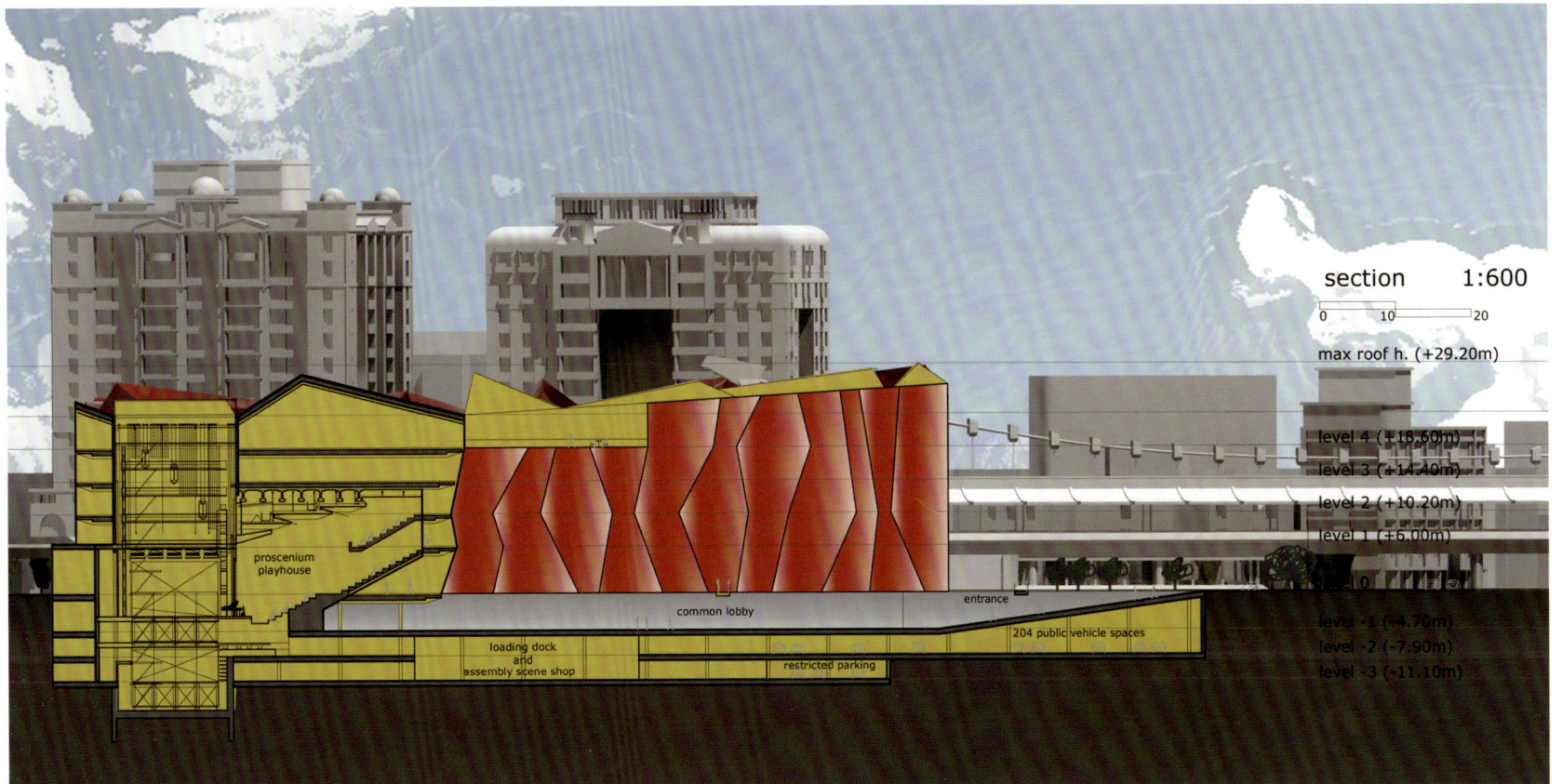

section 1:600

0 10 20

max roof h. (+29.20m)

level 4 (+18.60m)
level 3 (+14.40m)

level 2 (+10.20m)

level 1 (+6.00m)

level -1 (-4.70m)
level -2 (-7.90m)
level -3 (-11.10m)

proscenium
playhouse

common lobby

entrance

loading dock
and
assembly scene shop

restricted parking

204 public vehicle spaces

029

Areas

空间项目 SPACE IDENTIFICATION	空间名称 ITEM	面积 FLOOR AREA(m²)
一、大剧场 Grand Theater		
110	前厅 Hall	1200
120	观众区 Auditorium	1275
130	表演区 Stage	1410
140	后台准备区 Back Stage Space	1330
150	控制/设备/储藏空间 Control Room & Storage	1765
	小计 Subtotal	6980
二、镜框式中剧场 Proscenium Playhouse		
210	前厅 Hall	670
220	观众区 Auditorium	680
230	表演区 Stage	835
240	后台准备区 Back Stage Space	1090
250	控制/设备/储藏空间 Control Room & Storage	1290
	小计 Subtotal	4565
三、多形式中剧场 Multiform Theater		
310	前厅 Hall	600
320	观众区 Auditorium	1360
330	后台准备区 Back Stage Space	300
340	控制/设备/储藏空间 Control Room & Storage	745
	小计 Subtotal	3005
四、表演服务空间 Performance Service Spaces		
410	排练空间 Rehearsal Space	1830
	小计 Subtotal	1830

空间项目 SPACE IDENTIFICATION	空间名称 ITEM	面积 FLOOR AREA(m²)	备注 REMARKS
五、表演支援空间 Technical Support Area			
510	表演支援空间 Technical Space	2080	
	小计 Subtotal	2080	
六、运营管理空间 Adminstrative Space			
610	运营管理部门 Adminstrative/Managerment Offices	2095	
620	会议室 Conference Rooms	485	
630	研修咨询部门 Library	635	
	小计 Subtotal	3215	
七、公共服务空间 Public Service Area			
710	餐厅艺文空间 Restaurant and Artshop Arcade	4600	
720	停车场 Vehicle Spaces	9050	汽车200辆，机车350台 (200 parkings of car,350 parkings of moto-cycle)
730	设备机房 Electrical And Machine Room	1200	
	小计 Subtotal	14850	
八、连接空间 Linkage Space			
810	大厅 Common Lobby	2600	
	小计 Subtotal	2600	
总楼地板面积 Grand Total		39125	m²

　　台北表演艺术中心的外形如同一朵巨大的花朵，圆形的结构通过精确的设计一点一点地收缩，包容了整个设计中所有的元素。这是一次探索戏院的本质与其中活力的旅程。这座中心让人感到了鲜花的芬芳和夕阳的余辉。事实上，太阳和大自然使人们每天都能看到不断重复但又令人惊奇的景象，而充满了惊喜的意愿经常会表现出认识真实生活的关键。但建筑能够像在阳光下和风中舞蹈的花儿一样起舞吗？是的，虽然看上去建筑是静止的，但在它真正的生命里，建筑与人群和光线相互融合。在光线下的相互连接的空间可以被定义成多种多样的可能性：在夕阳下，它是世俗化和感性化的；在中午，则是敏锐与精确的。建筑的内部空间则是从外部到压缩的空间，再到隐秘但明亮的空间的连续体。

　　建筑的整体安排源自下面三种剧院的连接：

1. 大剧院
2. 舞台剧场
3. 多元化剧场

　　全部的三个剧场都有独立的控制和运转系统，并通过公共的大厅相连，在剧院的入口则设有控制点。公共大厅是免票的24小时开放区域，除非在有区域管制的情况下。大厅内的空间十分舒适，并与中心所有楼层连接。

　　大剧院可以进行古典戏曲、大型舞蹈表演和音乐剧演出，共有1500个座位，其中一层有1000个座位，二层有500个座位。

　　在舞台剧场和多元化剧场里则可以进行戏剧和舞蹈表演，共有800个座位。

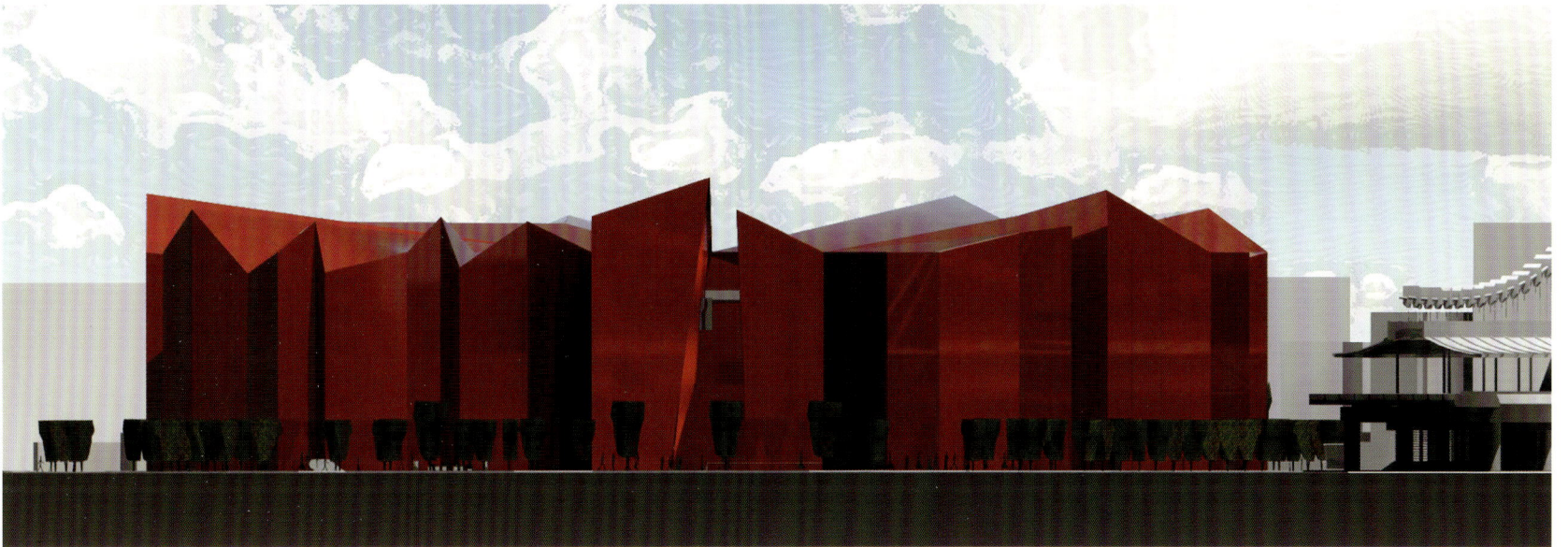

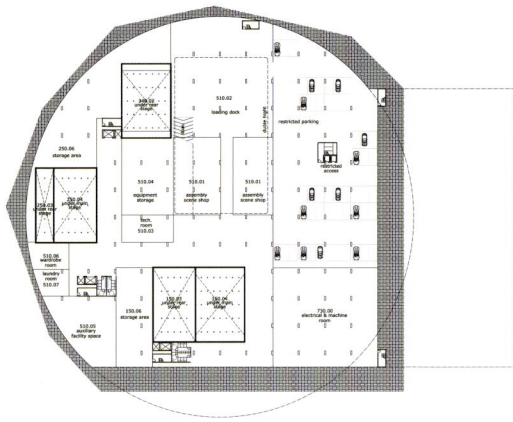

loading dock &
restricted parking
3100m²

technical space &
storage
6500m²

tot. area level -3
9600m²

level -3   1:600

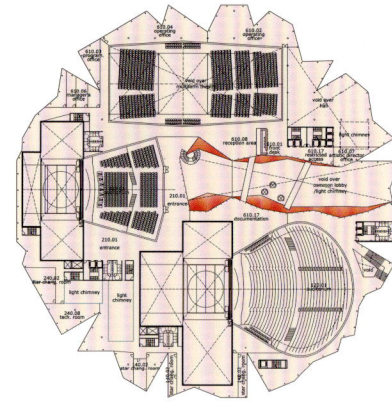

grand theatre &
back stage
870m²

p. playhouse &
back stage
740m²
administrative
space
1400m²

tot. area level 0
3100m²

level +1   1:600

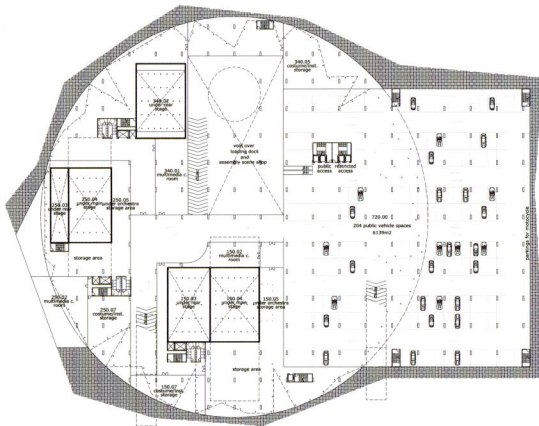

public parking
6100m²
technical space &
storage
1800m²

technical space &
storage
1050m²

technical space &
storage
1260m²

tot. area level -2
10210m²

level -2   1:600

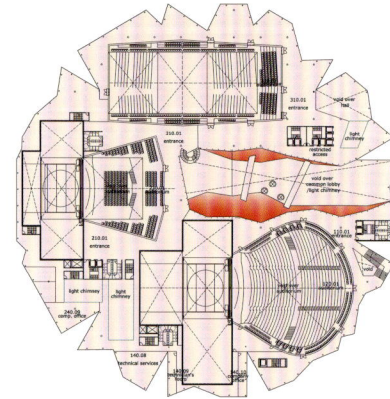

grand theatre &
back stage
1420m²

p. playhouse &
back stage
950m²

multiform theatre &
back stage
1060m²

tot. area level 0
3430m²

level +2   1:600

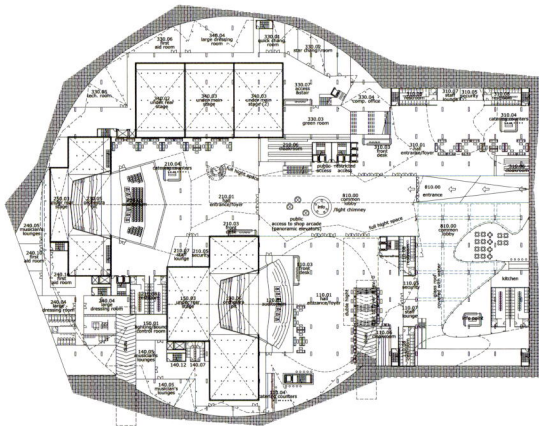

common lobby
2580m²

grand theatre &
back stage
3520m²

p. playhouse &
back stage
2860m²

multiform theatre &
back stage
3610m²

tot. area level -3
12570m²

level -1   1:600

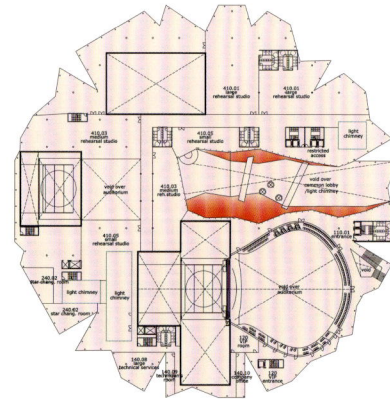

grand theatre &
back stage
840m²

p. playhouse &
back stage
350m²

rehearsal
space
3120m²

tot. area level 0
4310m²

level +3   1:600

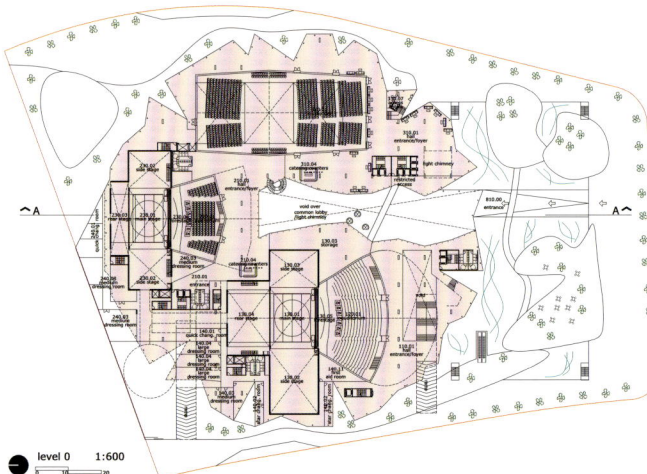

grand theatre &
back stage
1420m²

p. playhouse &
back stage
950m²

multiform theatre &
back stage
2560m²

tot. area level 0
4930m²

level 0   1:600

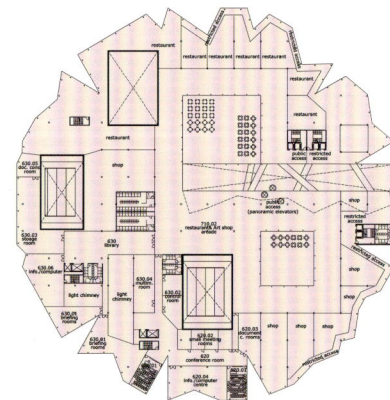

restaurant & Art shop
arcade
5480m²

library &
conference room
1850m²

tot. area level 0
7330m²

level +4   1:600

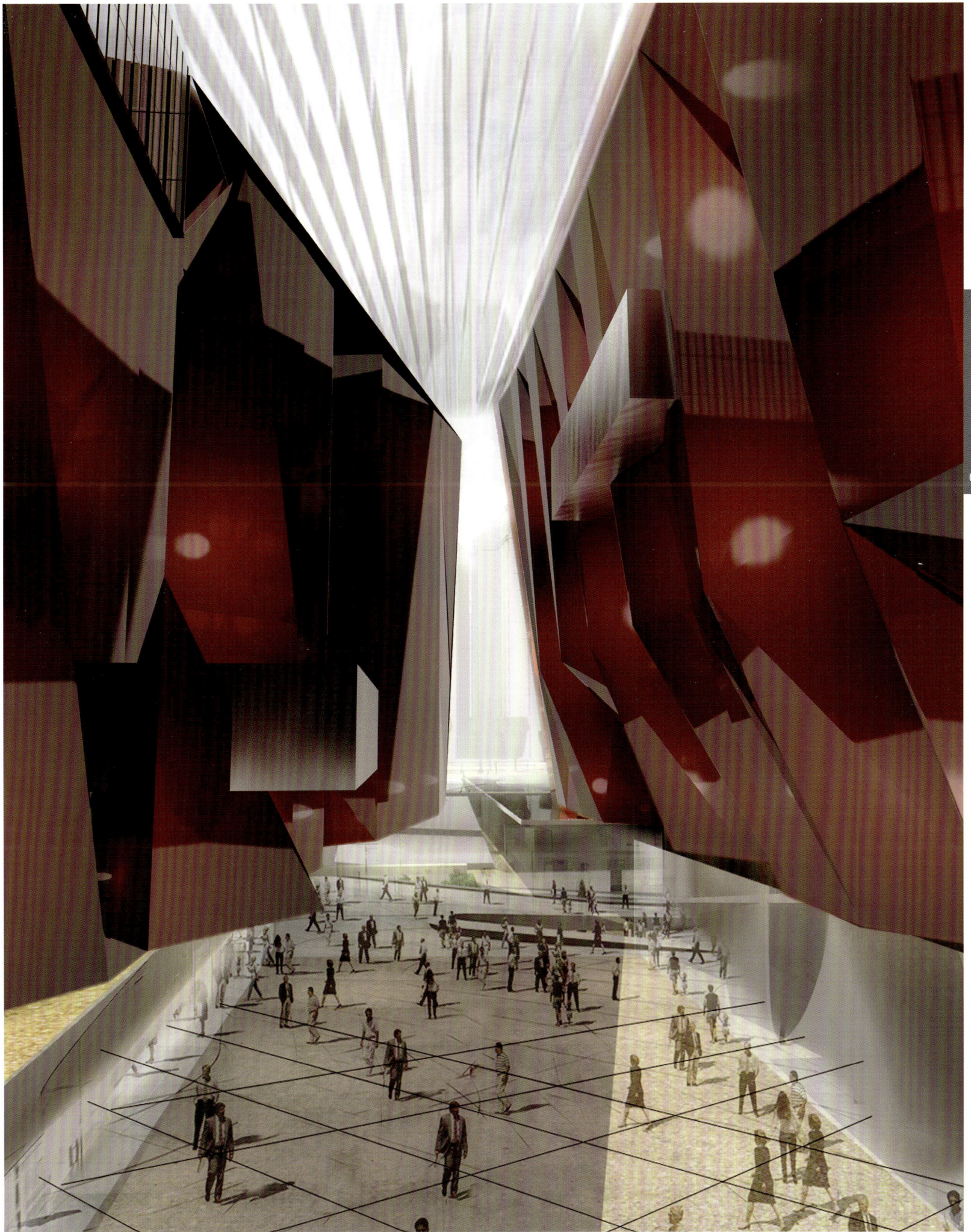

The White Peacock

白色孔雀

Architect: Ivo Buda, Manfredo Bianchi, Tomaz Kristof
Firm: Ivo Buda architetto
Location: Teheran, Iran
Area: 7,125m²

Today we are called upon to dedicate our resources and capacities towards a society built upon the richness of human diversity that echoes the beauty of the unfolded plumage of the peacock.

The Peacock is representative of glory, immortality and incorruptibility. It is a possessor of some of the most admired human characteristics, and is a symbol of integrity and beauty.

In Babylonia and Persia the peacock is seen as a guardian of royalty.

A strange perspective of the "peacock tale" made of steel diagonal elements invites the visitor to explore the building from all possible angles around it. Although its shape is very simple the building appears to have many mysteries, the corners disappears and the perception of a static volume is substituted by a dynamic grid that seems to grow and to wrap a series of transparent and translucent spaces.

The building is a flexible container for commercial, office and residential activities. Different levels are strongly connected by a system of elevators and stairs. The visitors of commercial, office and residential spaces have dedicated entrances, which are visually connected through vertical "cuts" in the building slabs. This "cuts" provides fascinating multistory commercial spaces and gives attractive views from the offices.

The structure is consisting of the "curtain" of steel diagonal elements on the perimeter, concrete stairs and lift cores, and 6 internal columns. The curtain is a three-dimensional frame with a strong horizontal strength and high capacity for vertical support loads. The presence of many stiff elements subject to cutting loads provides the dissipation of seismic energy. These elements are formed when the intersection between the diagonals are close to the slabs, which makes them eccentric with a strong horizontal strength which for seismic testing allows a high value of the coefficient of structure. The concrete core guarantees a high stiffness of the building against wind and earthquake.

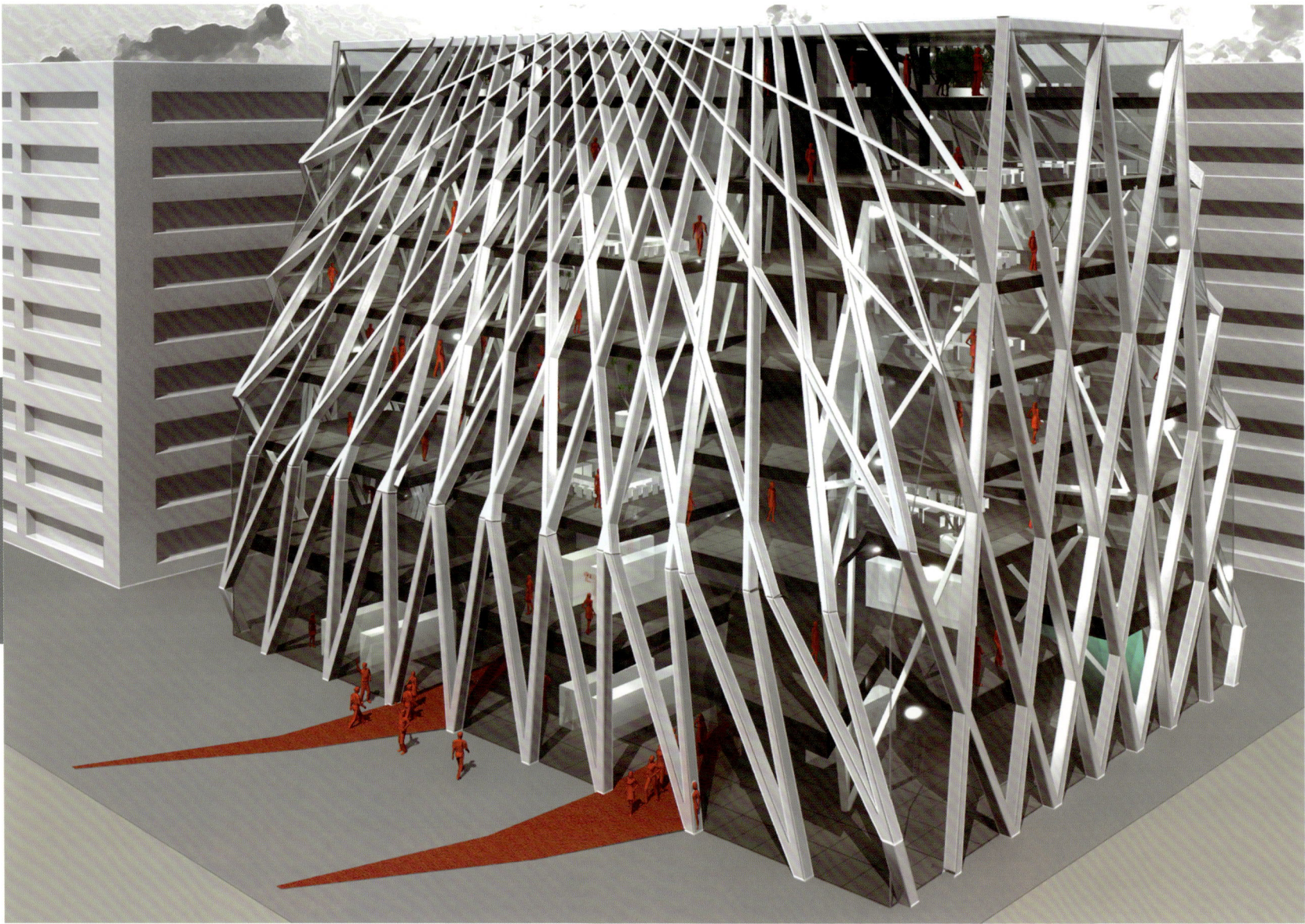

ENTRANCE FACADE

SIDE FACADE

APARTMENTS
OFFICES
OFFICES
OFFICES
OFFICES
SHOPPING
SHOPPING

PUBLIC
STAIRS

PARKING+STORA
PARKING
PARKING
PARKING

INTERNAL
STAIRS AND
ELEVATORS

CAR
RAMPS

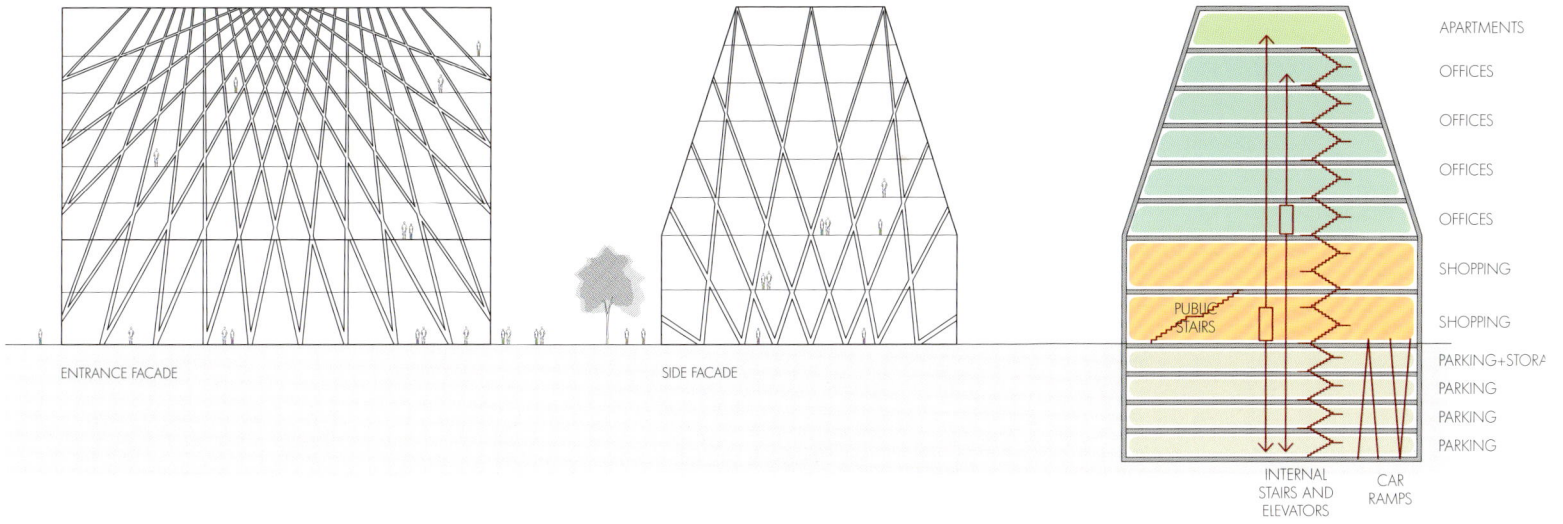

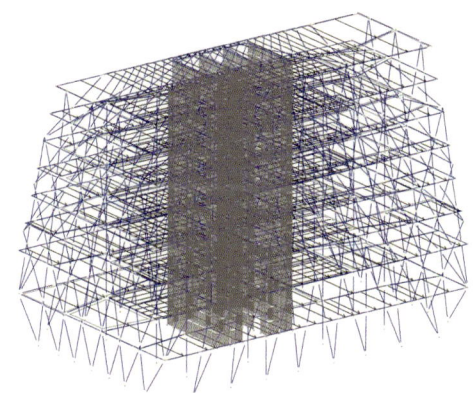

BASIC STRUCTURE DIAGRAM: CONCRETE RENFORCED CORE,
STEEL EXOSKELETON, STEEL SLABS

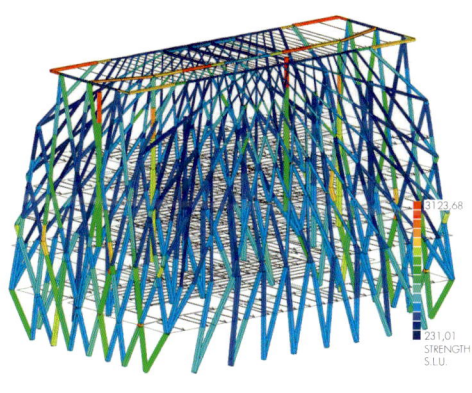

STEEL STRUCTURE, VERTICAL AND SEISMIC LOADS

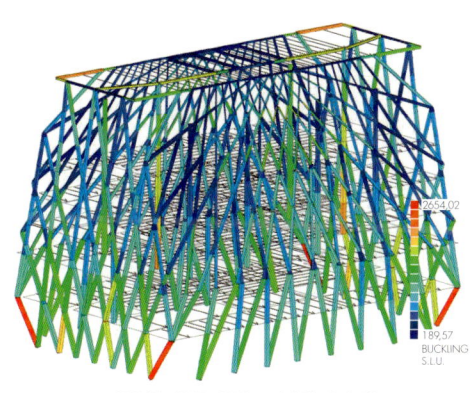

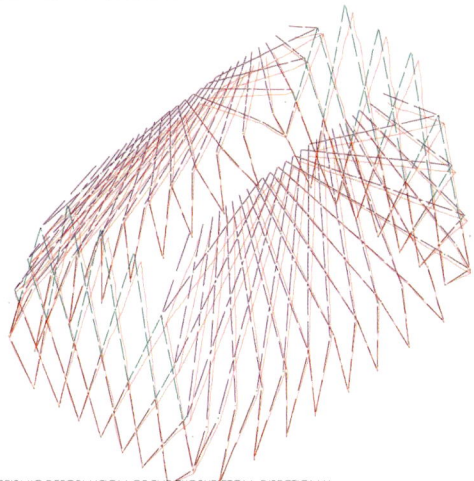

3123,68

231,01
STRENGTH
S.L.U.

STEEL STRUCTURE, VERTICAL AND SEISMIC LOADS

2654,02

189,57
BUCKLING
S.L.U.

SEISMIC DEFORMATION OF THE EXOSKELETON, DIRECTION Y

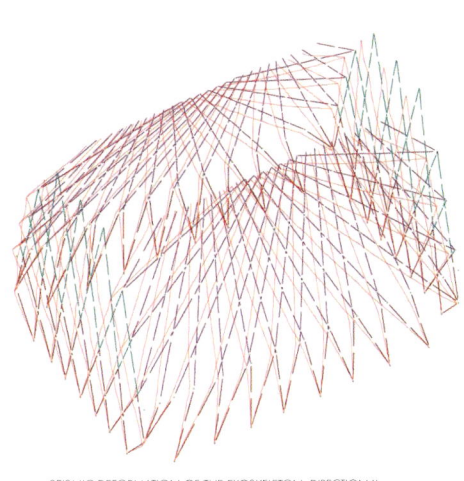

SEISMIC DEFORMATION OF THE EXOSKELETON, DIRECTION X

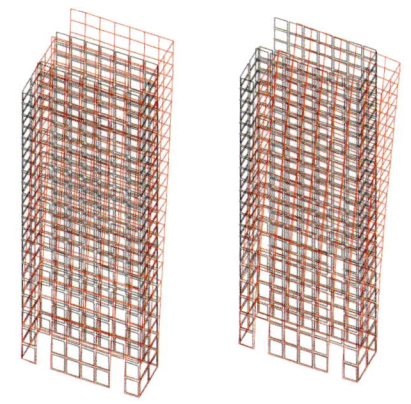

SEISMIC DEFORMATION OF THE CORE, DIRECTIONS Y AND X

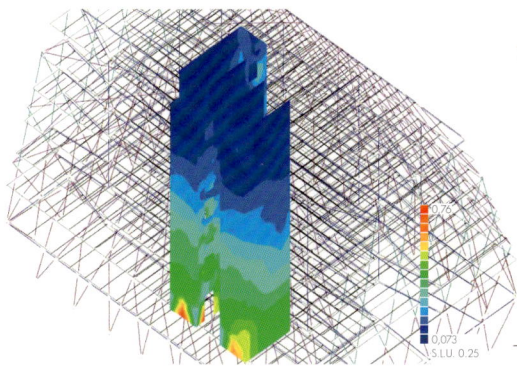

CONCRETE RENFORCED CORE, VERTICAL AND SEISMIC LOADS

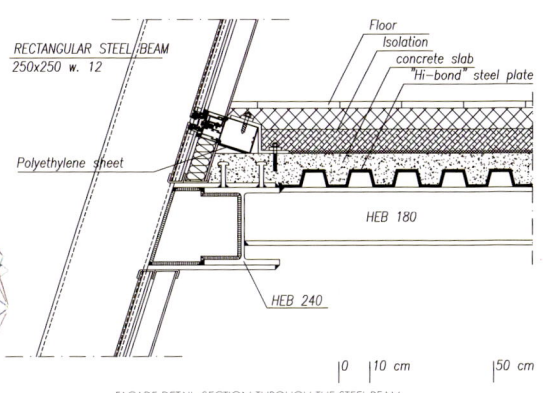

0,76

0,073
S.L.U. 0.25

RECTANGULAR STEEL BEAM
250x250 w. 12

Floor
Isolation
concrete slab
"Hi-bond" steel plate

Polyethylene sheet

HEB 180

HEB 240

|0 |10 cm |50 cm

FACADE DETAIL, SECTION THROUGH THE STEEL BEAM

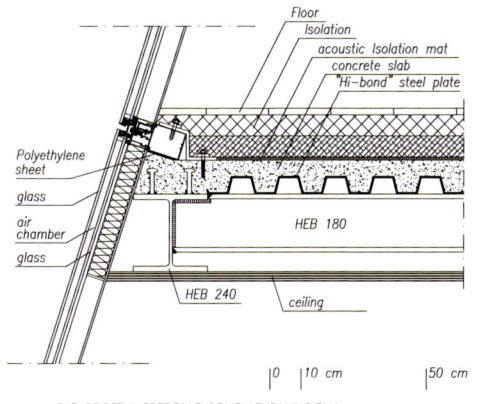

Floor
Isolation
acoustic Isolation mat
concrete slab
"Hi-bond" steel plate

Polyethylene
sheet

glass

air
chamber

glass

HEB 180

HEB 240   ceiling

|0 |10 cm |50 cm

FACADE DETAIL, SECTION THROUGH THE WINDOW

Sections and Structure

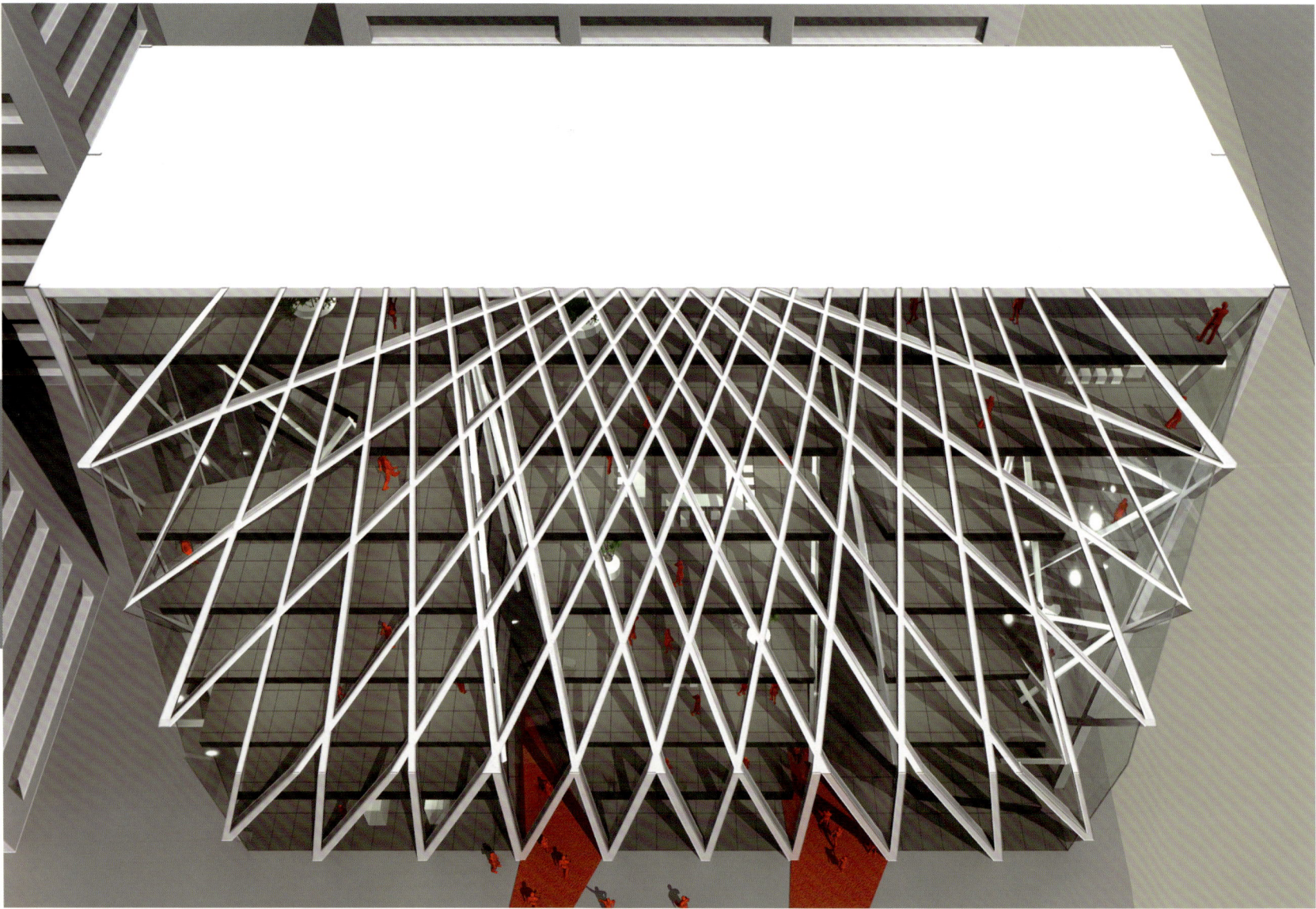

现在，在这个设计中，该事务所将其资源与能力投入到了一个建立在人类丰富的多样性上的社会。这个社会中的美丽可以同开屏孔雀的美丽相媲美。

孔雀象征着荣耀、永生和不朽，也是某些最值得人类敬仰的特质的代表，还是完整与美丽的化身。

在巴比伦和波斯，孔雀被视为是皇家的保护者。

这座由若干钢制斜梁组成的特别的"孔雀的故事"吸引着游客从各个可能的角度来探索这座建筑。虽然它的形状非常简单，但该建筑似乎有着许多秘密。拐角消失了，静态的空间感觉被一种不断延伸并包围了许多透明和半透明空间的充满动态的网格所代替。

这座建筑可以灵活的为商业活动、办公和居住提供空间。不同的楼层被电梯和楼梯紧

密地连接在一起。来这里经商、办公或居住的人们可以拥有看上去通过垂直"断面"联结在建筑平面板上的精致入口。这些"断面"提供了多层的商用空间并为办公室提供了良好的视野。

这座建筑的结构由四周的钢制斜梁交叉构成的"幕布"、设计有楼梯与电梯的混凝土核心以及6个内部立柱组成。"幕布"由立体的结构构成，可以为建筑提供巨大的横向拉力和纵向支撑力。建筑中大量不易弯曲的应力结构为建筑提供了抵御地震的能力。在钢制对角线的中心靠近建筑水平面板的时候形成的这些应力结构看起来虽然形状奇怪，但却在抗震测试中产生了巨大的水平拉力，有着很高的结构系数值。混凝土核心则保证了建筑抗拒风力和地震的稳定性。

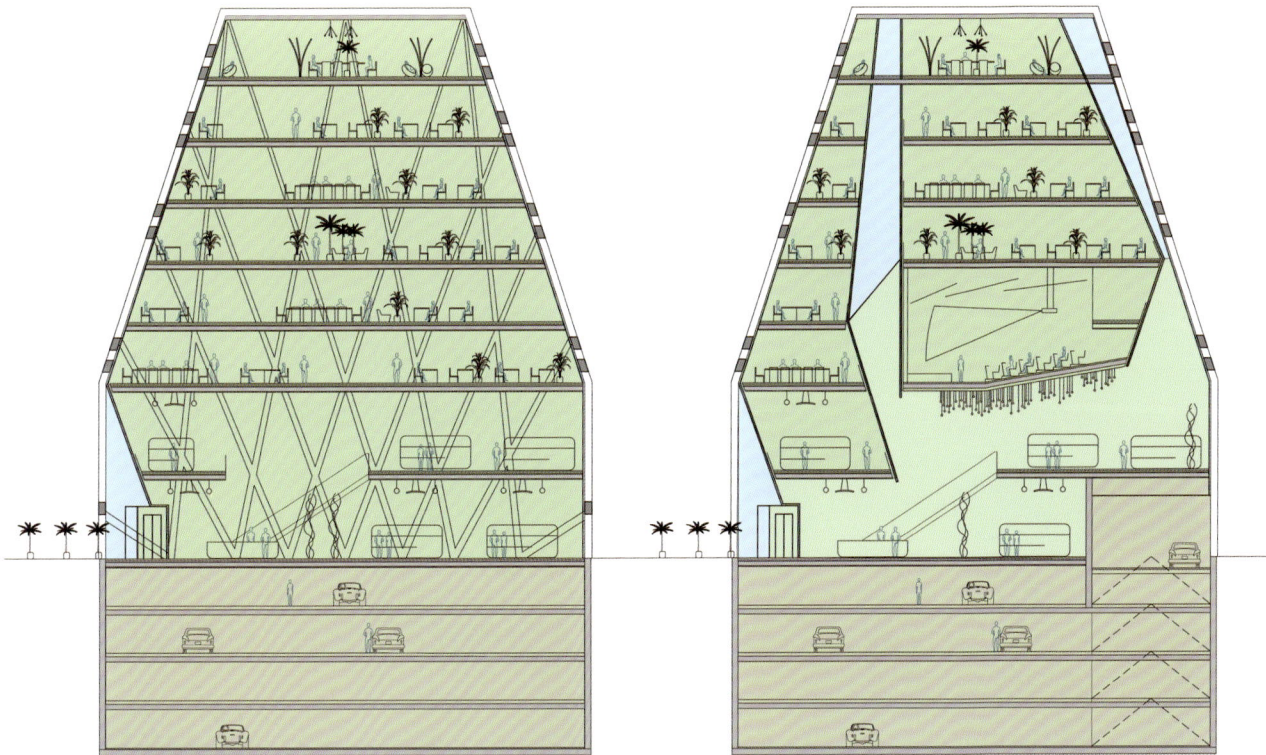

Sections

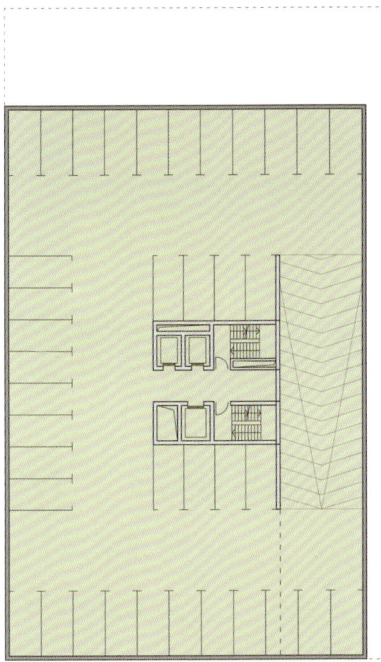

BASEMENT FLOORS - PARKING AND STORAGES

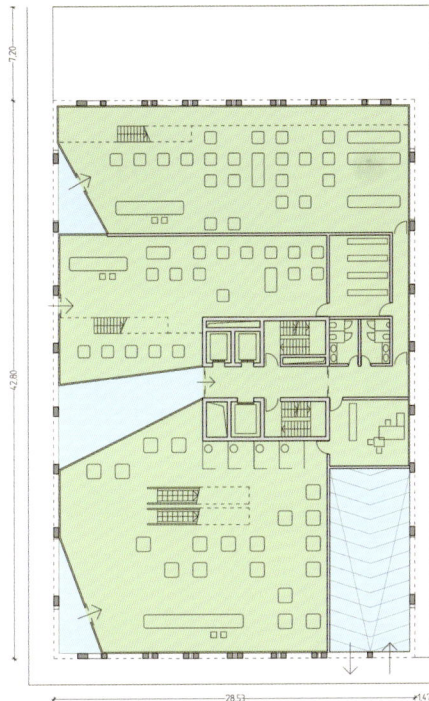

GROUND FLOOR - SHOPS

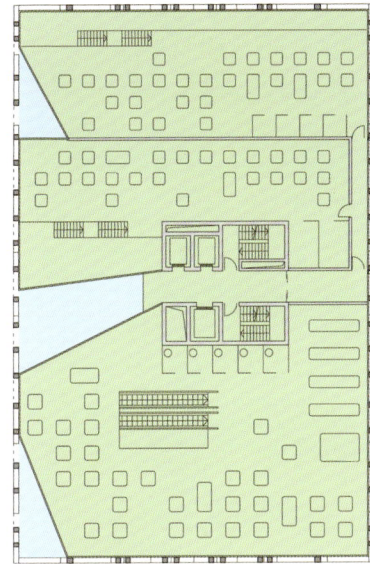

1ST FLOOR - SHOPS

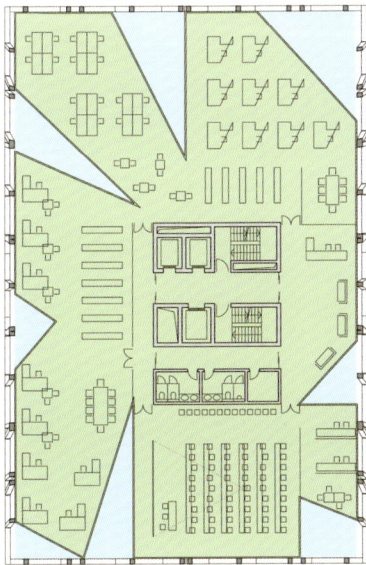

2ND FLOOR - OFFICES

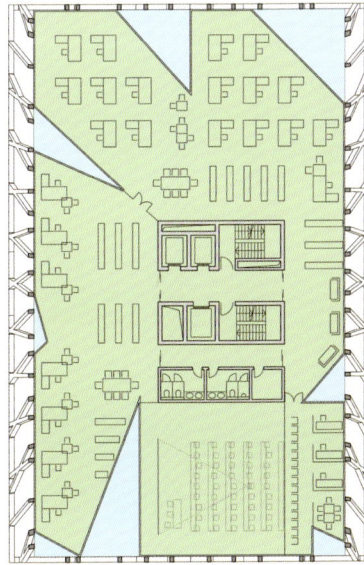

3RD FLOOR - OFFICES

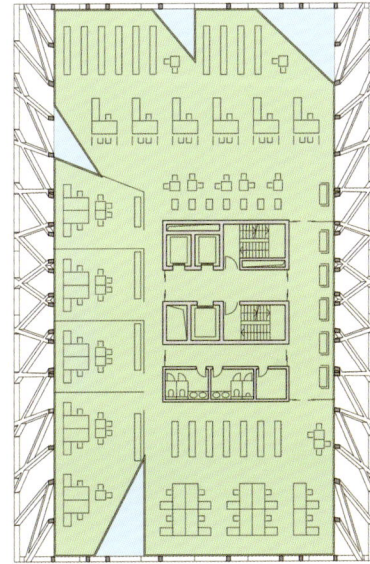

4TH FLOOR - OFFICES

5TH FLOOR - OFFICES

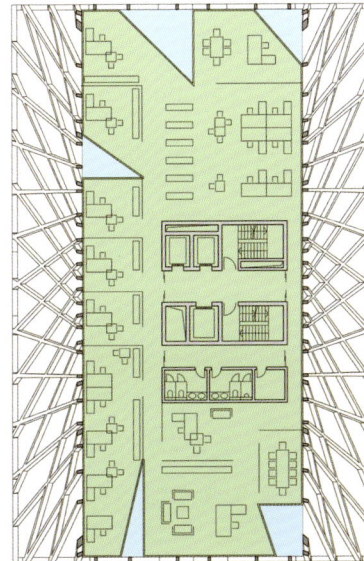

6TH FLOOR - OFFICES

7TH FLOOR - LUXURY APPARTMENTS

Car Experience

汽车体验

Architect: Francesco Gatti
Firm: 3Gatti
Location: Jiangning Area, High-Tech Zone, Nanjing, China
Area: 15,000m²
Project Manager: Summer Nie
Collaborators: Nicole Ni, Muavii Sun, Chen qiuju, Jimmy Chu, Luca Spreafico, Damiano Fossati, Kelly Han
Materials: Steel Structure, Resin Coating, Glass Partitions

"Car Experience" is a project for a building to be dedicated to the automobile: the car as an object of desire, a world to explore, a technology to study, an article to display and a means to travel around the building.

On an overall scale the area tectonically resembles a road, with a structure similar to that of an elevated motorway or a car park, but on a more human scale, the structure is as complex, ergonomic and sophisticated as the interior of a car.

The principal structure of the building is a spiral ramp with a glass partition dividing the exterior from the interior. In the internal part, reserved for pedestrians, the incline is more gradual, whereas the exterior and steeper side is for the transit of cars.

The building's typology develops sequentially, its structure similar to that of a film where the undisputed protagonist is the automobile. In fact the visitor, as the spectator of a film, is obliged, frame by frame, to follow the physical and psychological route as dictated by the museum's architect.

The visitor enters the museum with his own car and initiates the exhibition's journey. In this way visiting the museum is divided into two types of experience: the first is the experience of going up in one's own car, the second is the experience of going down on foot.

Along the surface of the ramp there are occasional glass blocks or "prisma" which protrude from the flooring and ceiling. Their size, depth and type depend on their function as each one is intended for something different, for example if the area is intended for open space functions, or if it is intended for functions that require greater privacy such as offices, meeting rooms, conference rooms, laboratories, bathrooms or kitchens.

The outer facade of the building is completely permeable and reveals on sight the interplay of the different levels and the fluidity of the internal and external spirals. The building could seem to appear as an urban car showroom, with its corners and angles filled with tempting shining automobiles.

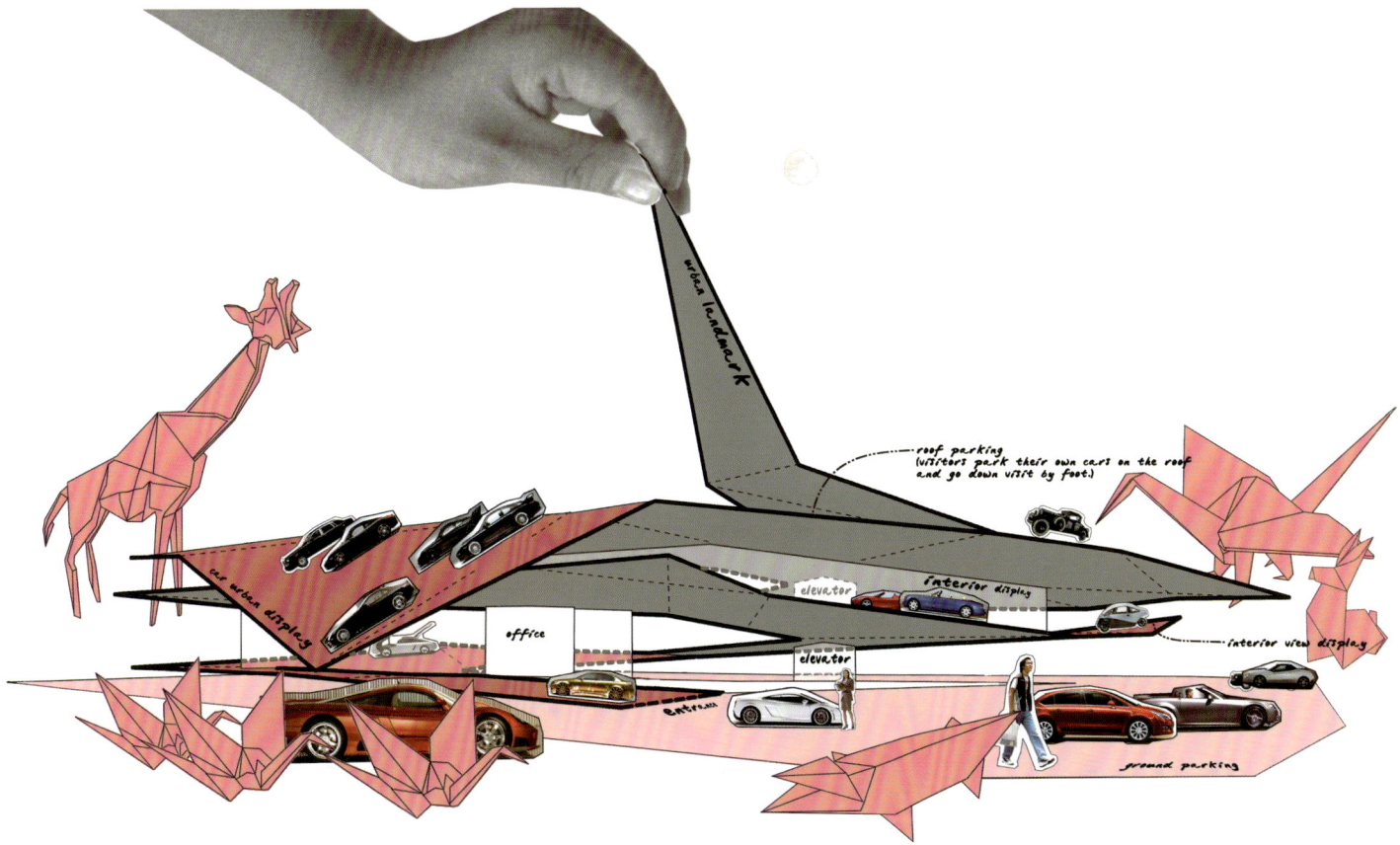

roof parking
(visitors park their own cars on the roof
and go down visit by foot.)

car urban display

office

elevator

elevator

interior display

interior view display

entrance

ground parking

urban landmark

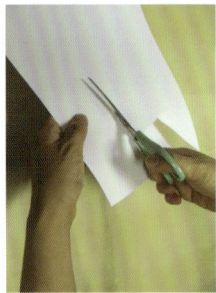

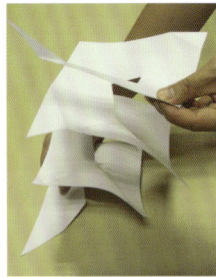

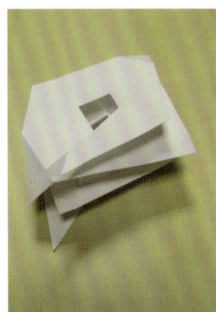

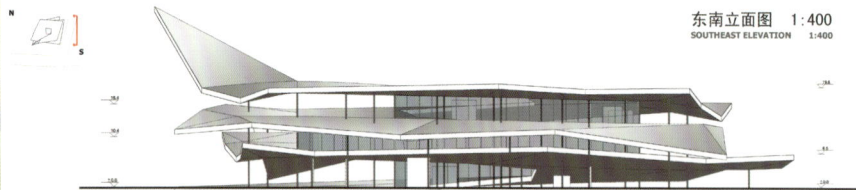

东南立面图 1:400
SOUTHEAST ELEVATION 1:400

西北立面图 1:400
NORTHWEST ELEVATION 1:400

东北立面图 1:400
NORTHEAST ELEVATION 1:400

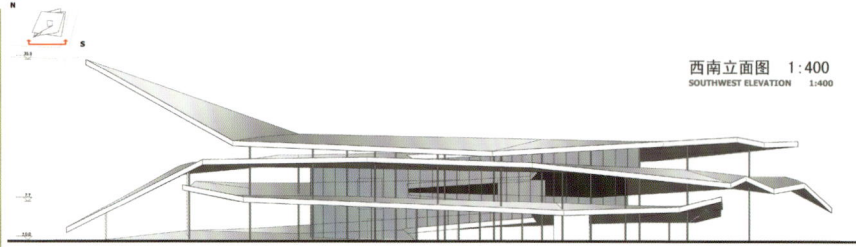

西南立面图 1:400
SOUTHWEST ELEVATION 1:400

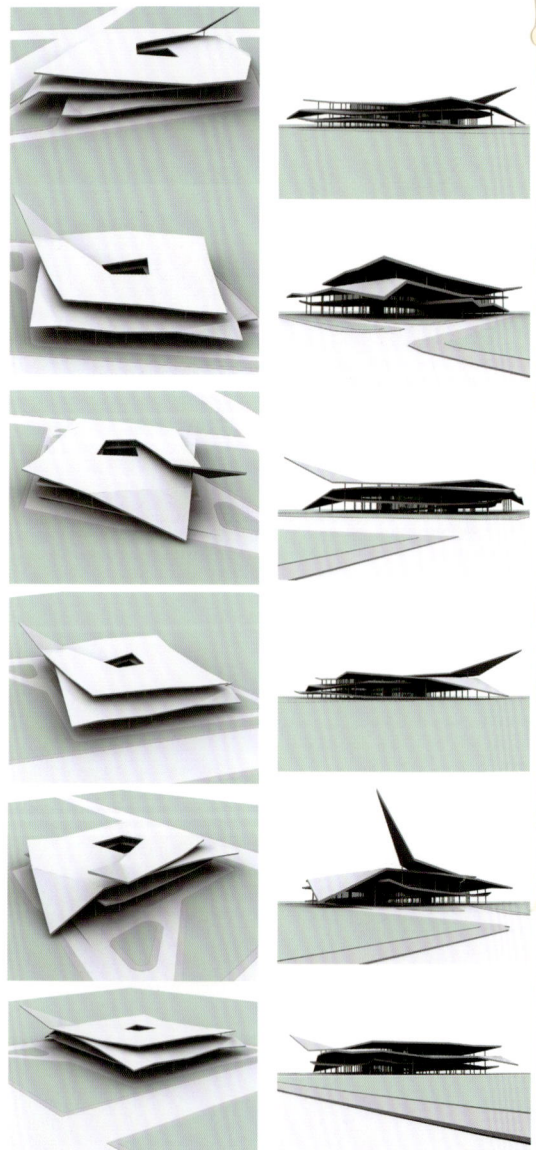

B-B剖面图　1:500
SECTION B-B　　1:500

D-D剖面图　1:500
SECTION D-D　　1:500

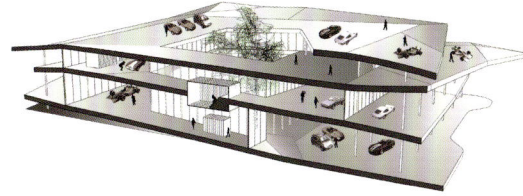

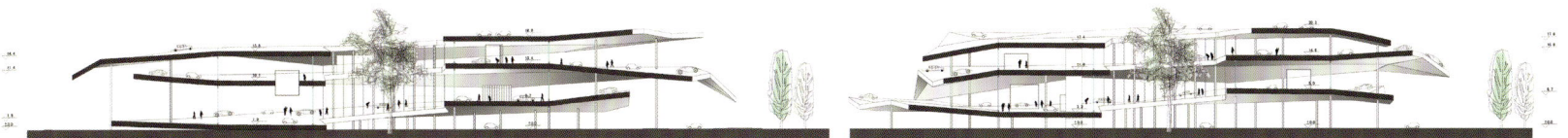

C-C剖面图　1:500
SECTION C-C　　1:500

E-E剖面图　1:500
SECTION E-E　　1:500

車博物館
CAR MUSEUM
カラー口絵

The point of this origami is the spiral cutting and the foldings on the corners.

螺旋状のはさみと角の折部分はこの折り紙のポイントです。

紙 Paper Dimension
30cm × 30cm

1
Cut from the middle of the paper
紙の中心からはさむ。

2

3
Cut according to the spuare shape.
回旋にはさむ。

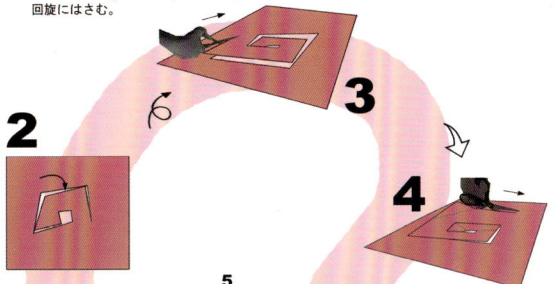

4

5
Cut the second square shape.
もうひとつの回旋をはさむ。

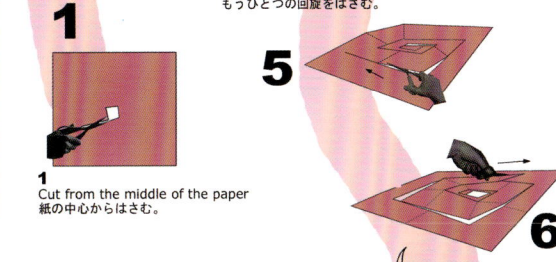

6

7
Pull it vertically.
縦に引く。

8
Fold on every corner repetitively.
角を一つ一つ折り続く。

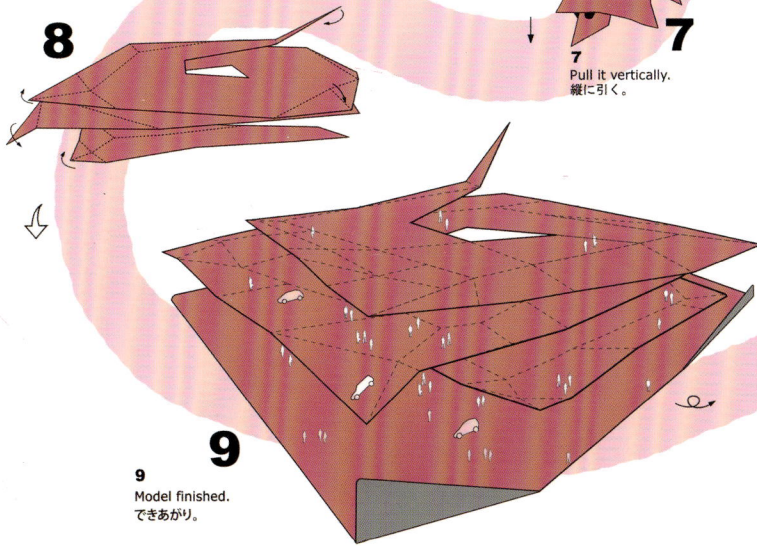

9
Model finished.
できあがり。

かざりかた

HOW TO DISPLAY

External ramp exhibition sight scheme
室外展示場の起伏和室内より激しい。

Urban view to see displayer
外の路面から見る。

Interior view to see displayer
室内から外の展示品を見る。

Conference room view　to see displayer
会議室から展示場が見える。

Description

This is the project winner of the international competition for the construction of the Automobile Museum in Nanjing.

Is the design for a building to be dedicated to the automobile, where the automobile is also the vehicle to visit the space.
You visit the first external ramp of the museum with your own private car, like a SAFARI, you park your car on the roof and visit by foot the internal ramp going down.
The building could seem to appear as a urban car showroom, with its corners and angles filled with tempting shining exposed automobiles.

12
Section plan
断面図。

11
Roof plan
屋上。

Second floor plan
二階。

First floor plan
一階。

Ground floor plan
グランド。

10
driving surface.
室外ドライブ。

walking rarea.
室内地面。

interior exhibition area.
ガラスで室内空間を分ける。

working area.
小さなガラスオックスに事務室などが配置される。

transportation core.
交通センター。

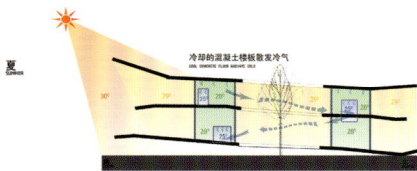

空调分析
AIR CONDITIONER

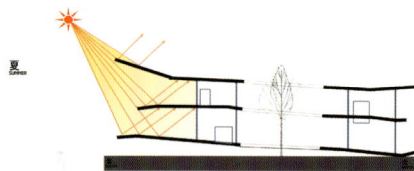

日光分析
sunlight

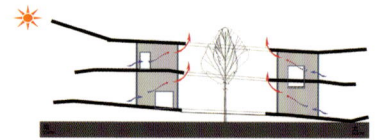

空气流通系统
ventilation

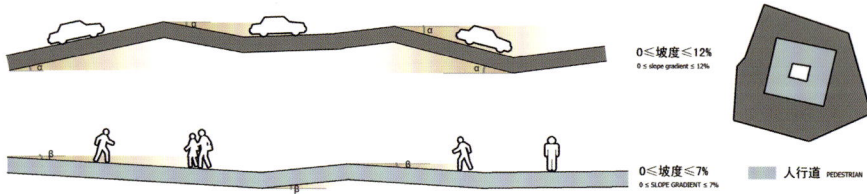

0≤坡度≤12%
0 ≤ slope gradient ≤ 12%

0≤坡度≤7%
0 ≤ SLOPE GRADIENT ≤ 7%

人行道 PEDESTRIAN
车行道 CAR

坡度分析
SLOPE GRADIENT ANALYSIS

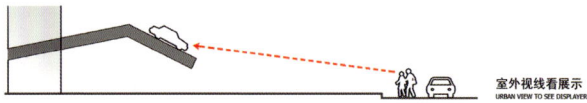

外部坡道展览观光图解
EXTERNAL RAMP EXHIBITION SIGHT SCHEME

室外视线看展示
URBAN VIEW TO SEE DISPLAYER

室内视线看展示
INTERIOR VIEW TO SEE DISPLAYER

会议室视线看展示
CONFERENCE ROOM VIEW TO SEE DISPLAYER

视线分析
VIEW SIGHT ANALYSIS

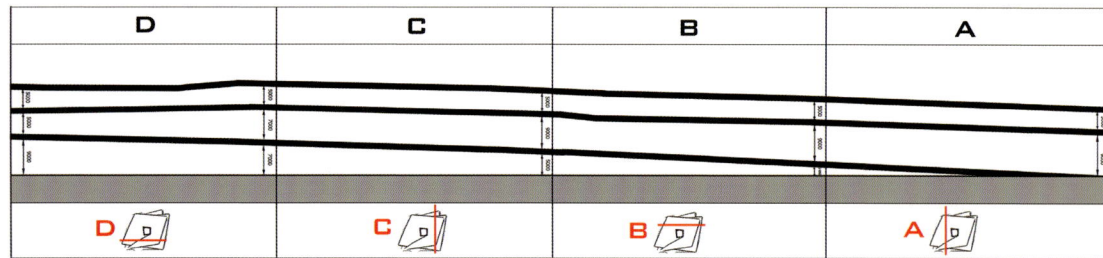

19世纪
19 centry

20世纪
20 centry

21世纪
21 centry

展示顺序
EXHIBITION SEQUENCE

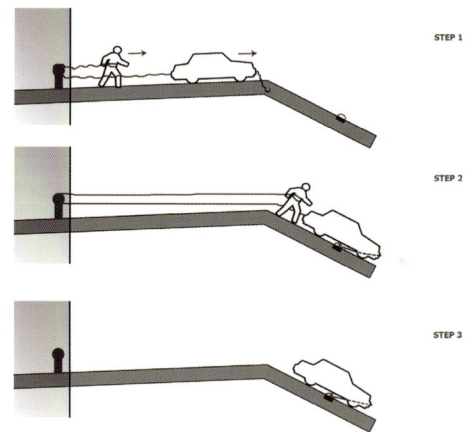

D	C	B	A
D	C	B	A

剖面展开示意图
SECTION SCHEME

STEP 1

STEP 2

STEP 3

室外展车安装固定示意图
CAR INSTALLATION SCHEME

A-A剖面图 1:200
SECTION A-A 1:200

顶层区域
ROOF LEVEL SECTOR

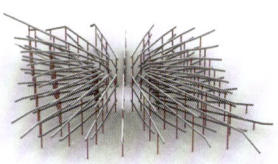

二层区域
2 LEVEL SECTOR

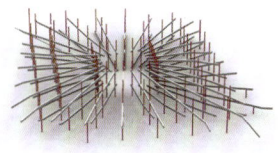

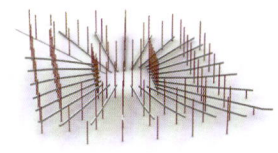

一层区域
1 LEVEL SECTOR

解构组成图
STRUCTURE COMPONENTS SCHEME

梁展开图
PLAIN DEVELOPED BEAMS SCHEME

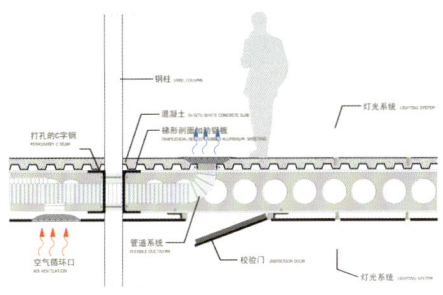

节点详图
DETAIL

隐喻
METAPHOR

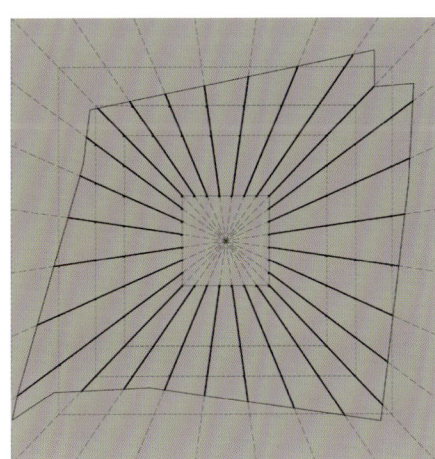

结构示意图
STRUCTURE SCHEMA

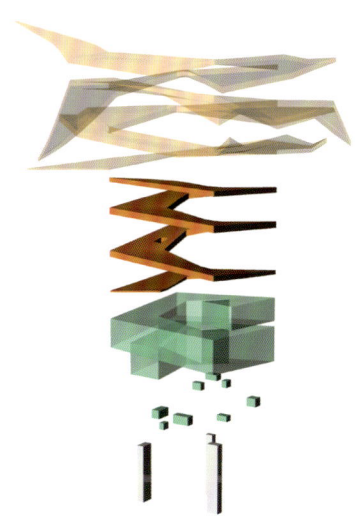

车行坡道
EXTERNAL CAR RAMP

人行坡道
INTERNAL PEDESTRIAN RAMP

主体玻璃结构
MAIN GLASS VOLUME

玻璃房间
GLASS ROOMS BOXES

核心筒
ELEVATORS AND STAIRS

建筑构成元素分析
EXPLODE MODEL SCHEME

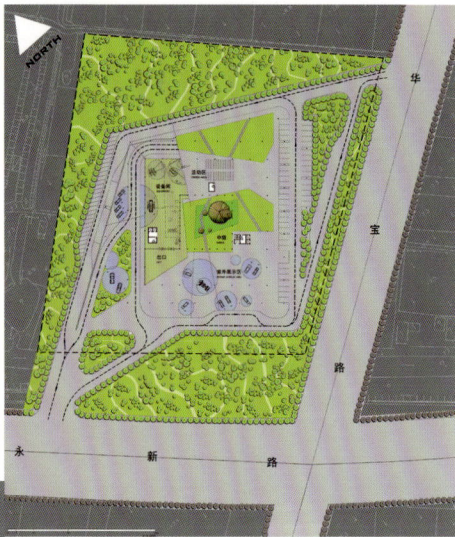

底层平面图　1:1000
GROUND FLOOR PLAN　1:1000

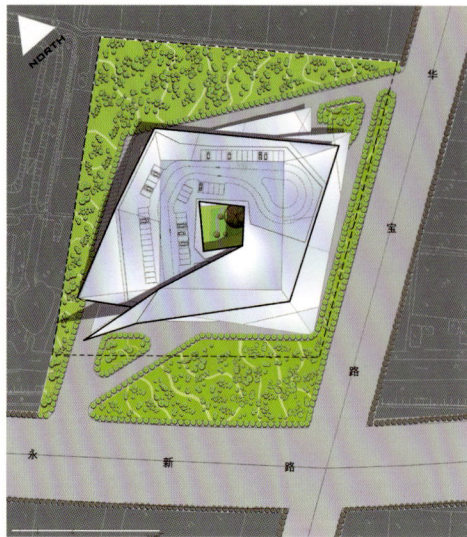

总平面图　1:1000
MASTER PLAN　1:1000

流线分析图
CIRCULATION ANALYSIS

立体流线
3 DIMENSIONS

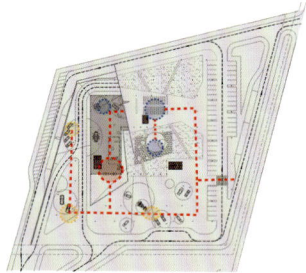

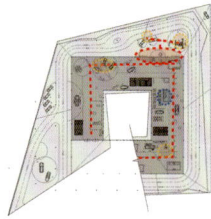

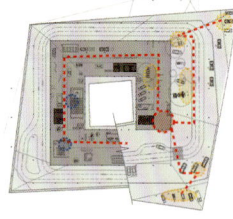

主要交通节点
次要交通节点—展览
次要交通节点—公众活动
交通线

交通节点分析图
TRAFFIC NODE ANALYSIS

总用地面积：26656平方米
建筑用地：15056平方米
室内面积：8300平方米
平室外面积：13512平方米／2＝6756平方米
建筑容积率：0.58
建筑高度：25.5米
建筑密度：51.8%
绿地率：30%
行政办公用地率：5%

底层流线
GROUND FLOOR CIRCULATION

一层流线
1ST FLOOR CIRCULATION

二层流线
2ND FLOOR CIRCULATION

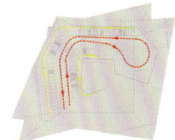

屋顶层流线
ROOF CIRCULATION

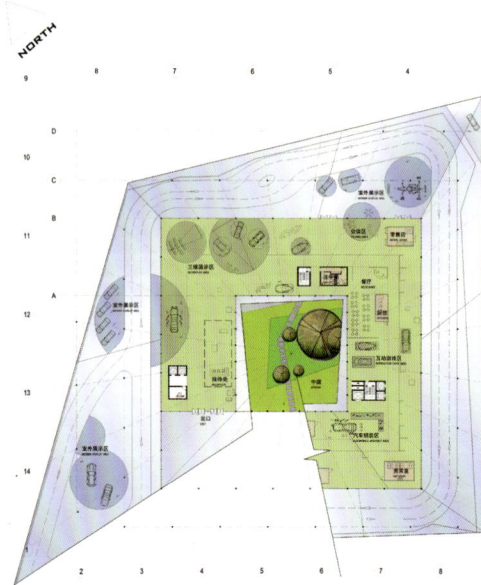

一层平面图　1:500
FIRST FLOOR PLAN　1:500

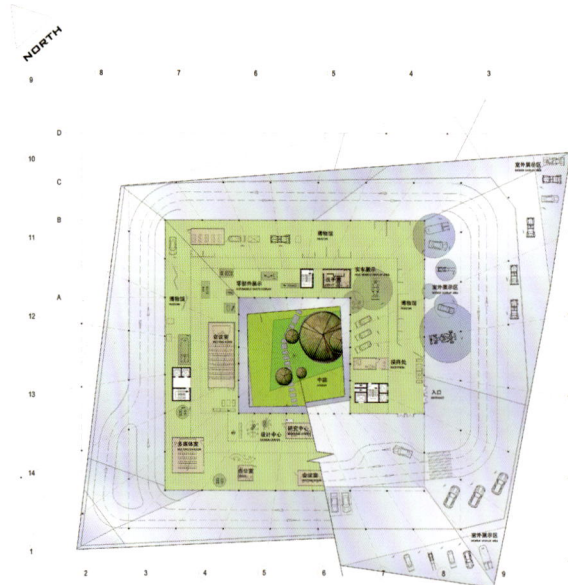

二层平面图　1:500
SECOND FLOOR PLAN　1:500

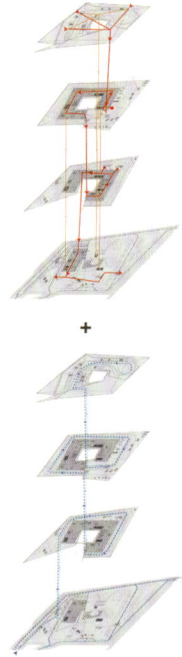

底层平面
GROUND FLOOR PLAN

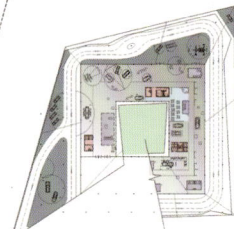

一层平面
FIRST FLOOR PLAN

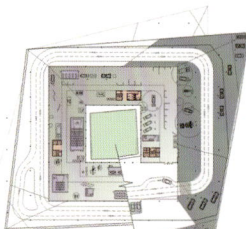

二层平面
SECOND FLOOR PLAN

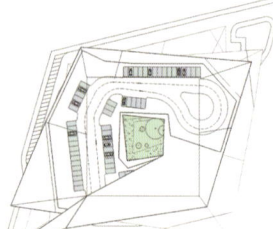

屋顶平面
ROOF PLAN

室内展示区　indoor exhibition area
室外展示区　outdoor exhibition area
垂直交通　vertical transportation
附属功能　additional function
绿地　green
停车场　parking

主要进入人流
主要离开人流
进入车流
离开车流

功能分析图
FUNCTION ANALYSIS

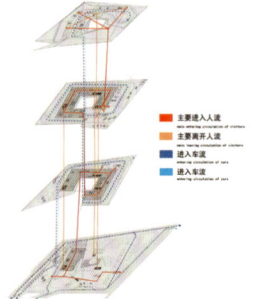

综合流线分析图
COMBINED FUNCTION ANALYSIS

048

　　"汽车体验"是一个为汽车而建设的设计项目:汽车是人们的理想,它代表着未探索的世界,代表着待研究的技术,它是一个展示出来的艺术品,也是在这座建筑旅行的工具。

　　从宏观上看,整个区域就像一条马路。这座建筑采用了一种类似升高的汽车道或者停车场的结构,但从更人性化的层面上,这种结构如同汽车的内部构造一样的复杂、精细并令人舒适。

　　建筑的主结构采用了螺旋形的走道,走道上的玻璃幕墙将建筑的内部与外部相分离。在预留了步行道的内部,地面逐渐倾斜,而外部则用作汽车行驶,故而斜度较大。

　　这座建筑在造型设计上展现出了时间上的连续性。它的结构比较类似于一部以汽车作为唯一主角的电影。事实上,参观者就如同电影的观众一样,需要一帧一帧地跟随博物馆设计者所主导的物质的与心理上的路线。

　　参观者与他们的汽车一同进入博物馆开始了展览所设定的旅程。用这种方式,参观这座博物馆被分为了两种体验:第一种是乘坐自己的车向上行驶的体验,另一种则是向下步行的感受。

　　沿着步道的表面,不时的会从地面或天花板上突出玻璃障碍物或称为"棱镜"。因为每一个棱镜都是用作不同的用途,所以它们的规格、长度和类型也因此不同,比如,某一个区域是用作开放空间,则这里的大小会不同于被用作办公室、会客室、会议室、实验室、卫生间或厨房,因后者需要更多的隐私性。

　　博物馆的外立面完全透明。这种设计表现出了不同层次间的互动以及内外两个螺旋结构的流动性。这座建筑让人们从建筑的各处都能看到迷人的汽车,而这使得博物馆成为了一个城市汽车展览厅。

The Helix Hotel

螺旋酒店

Architect: Thomas Leeser
Firm: Leeser Architecture
Location: Zayed Bay, Abu Dhabi, UAE
Area: 23,086m²
Formal Competition Name: The Pre-Concept Design of Iconic Hotel –
Zayed Bay Development
Competition Date: July 10th, 2008
Competition Scale: International
Place in the Competition: First

Elevations

With 208 guest rooms and suites arranged around a helical floor, the hotel immediately dispenses with the idea that visitors must engage in the stale paradigms of rigid hallways and atria that characterize a typical hotel stay. The floor constantly shifts in width and pitch as it rises to the top floor, keeping public spaces always in flux. No two rooms positioned across from each other have exact views to the other side, already pulling the visitor out of the pedestrian and into the hotels uniquely urban world.

Conceptually, the Helix Hotel participates in a critical dialogue between opulence and urbanness, between the variety of services offered by a small city and the demands of a five-star hotel guest. The floor suggests the curves a winding street would take through a bustling town, and many programmatic elements are open to views from across the central void. Though the void seems to offer unmitigated visibility, there are enclaves for private meetings and guest privacy. It is designed so that one activity feeds into the next rather than affecting sharp separations between each activity.

On the luxury side of vacation culture, there are playful elements that make the hotel a designer destination in an iconic setting. From the outset, it is as much a showplace for the abundance of opulent life as it is a fully incorporated urban experience. At the entry, valets drive clients cars into the car park, which, rather than being predictably aboveground or underneath the hotel, is situated instead under the bay. At the top of the Helix, the rooftop pool deck features a full sized swimming pool with a glass bottom, with the water and swimmers visible from eight floors below at ground level. In the restaurant below the lobby, the bay waves are so near to the floor plate that they lap up onto the edge of the restaurant inside of the glass curtain wall. The wall retracts, revealing a sweeping breeze.

While focusing on unique design, Leeser Architecture is also committed to sound sustainability practices and worked with consultant Atelier Ten to determine the best possible conditions and materials for heat and energy conservation. The indoor waterfall allows for the accumulation of heat inside the hotel to be minimal by filtering cool water back up into the system as it falls through the void. In the sub-lobby, a dynamic glass wall is built from the base of the second floor down into the water. The wall acts as a curtain would, opening when the weather is cool enough and closing when it is too hot for exposure to the desert air. Portions of the outside surface are clad in panels made of a new material called GROW, which has both photovoltaic and wind harnessing capabilities.

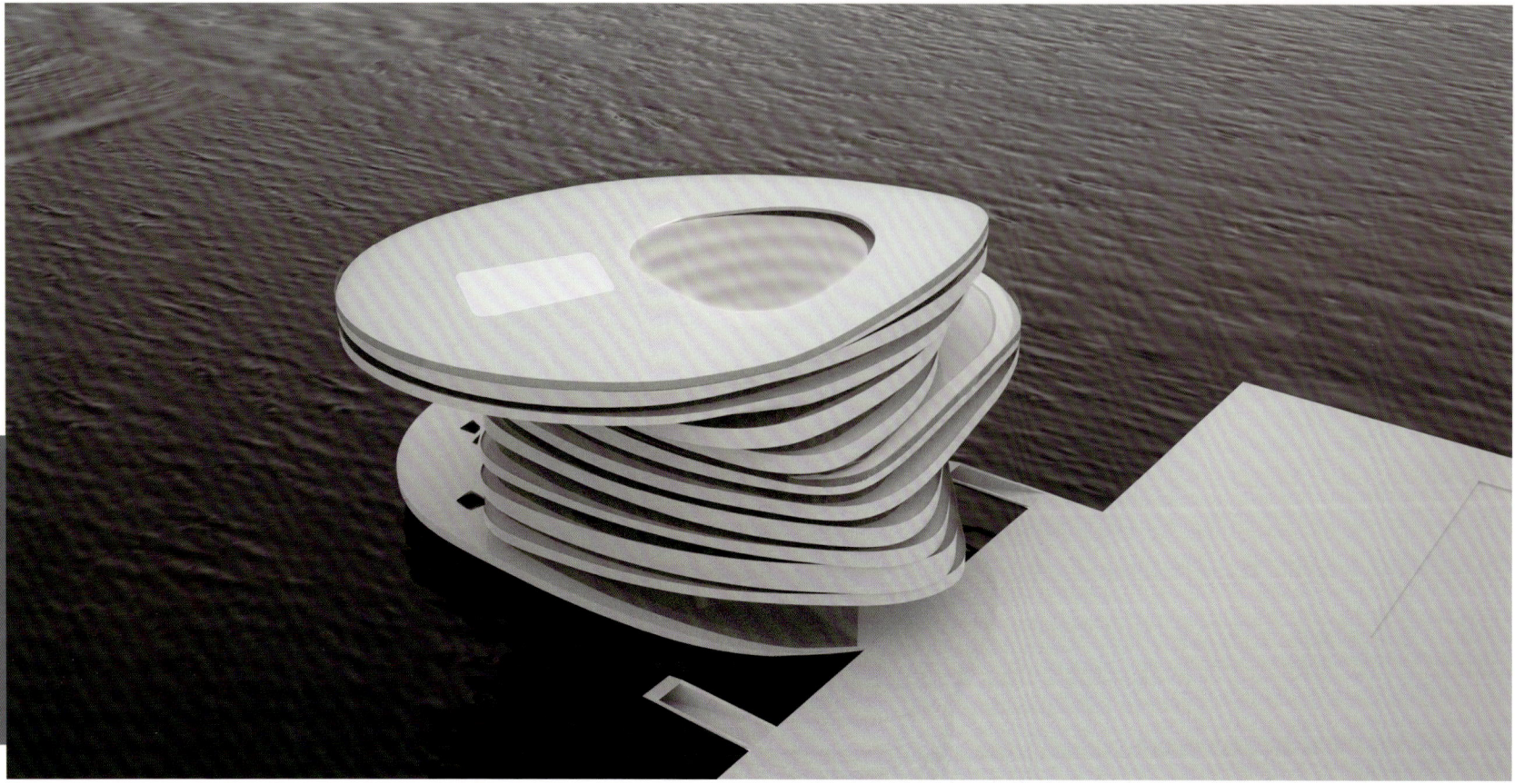

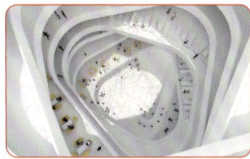

ATRIUM

SPA

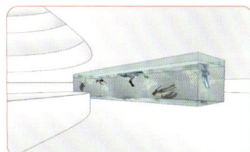

BELOW LAP POOL

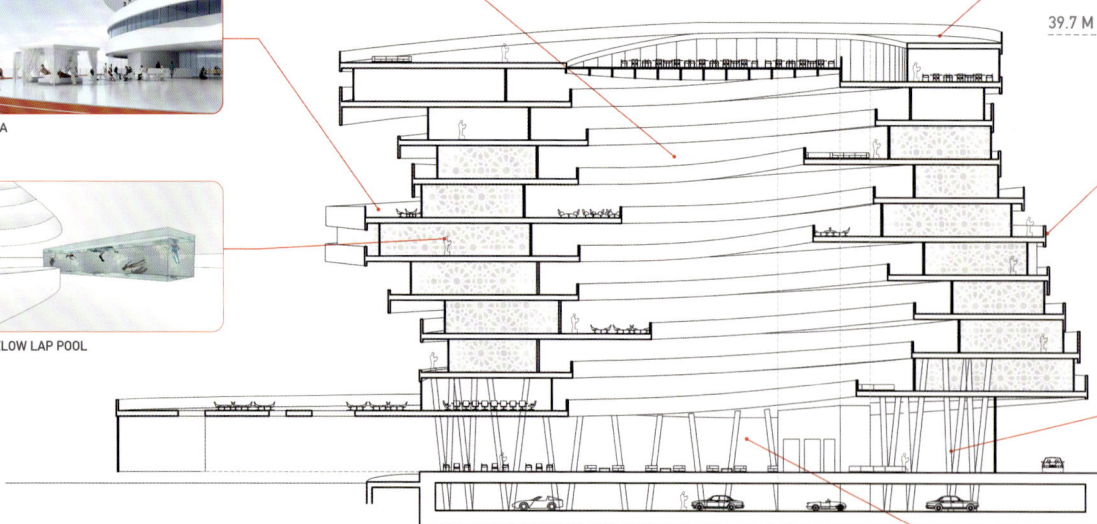

ROOFTOP

39.7 M

GROW: ENERGY HARNESING SKIN

VIEW FROM BOAT DOCK

MAIN LOBBY

Section

ROOFTOP

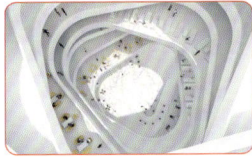

ATRIUM

GROW: ENERGY HARNESING SKIN

SPA

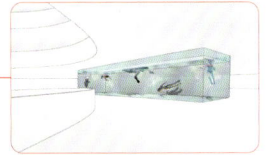

BELOW LAP POOL

VIEW FROM BOAT DOCK

MAIN LOBBY

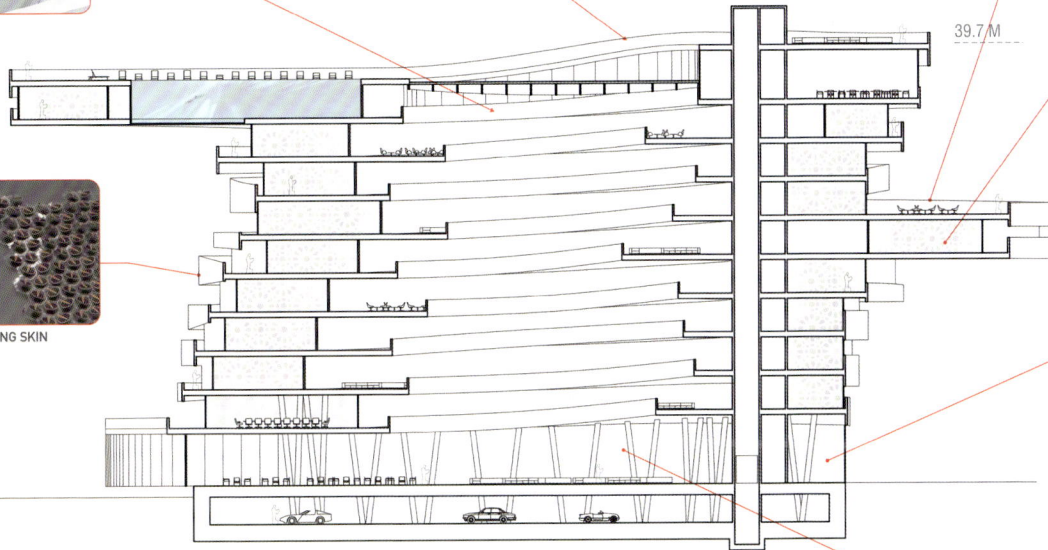

39.7 M

Section

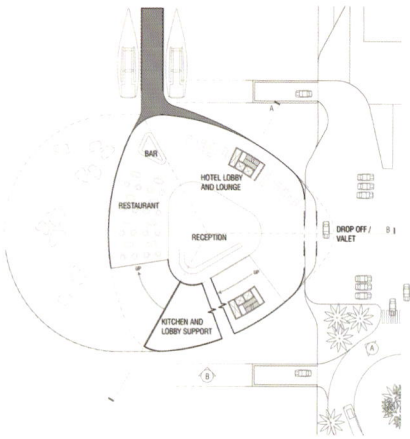

level 0

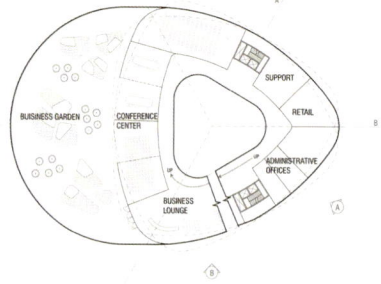

level +1

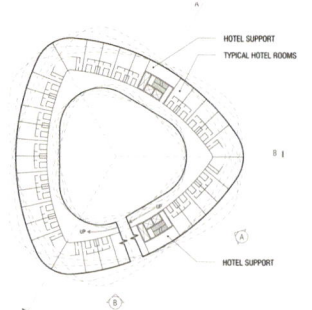

level +2

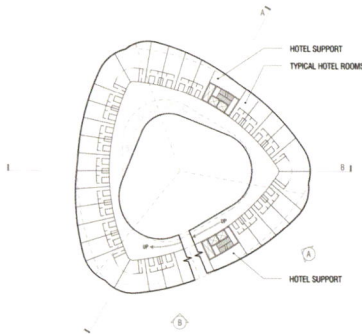

level +3

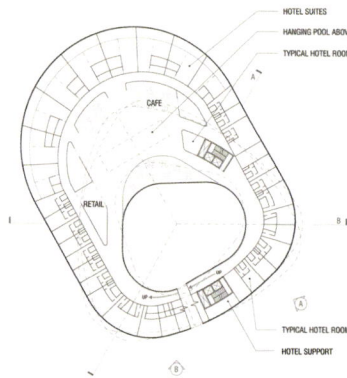

level +4

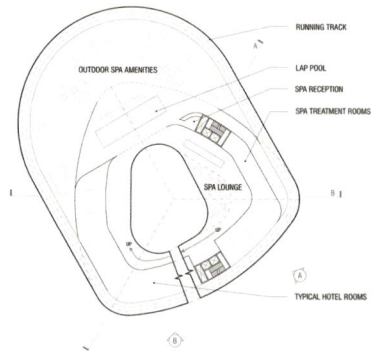

level +5

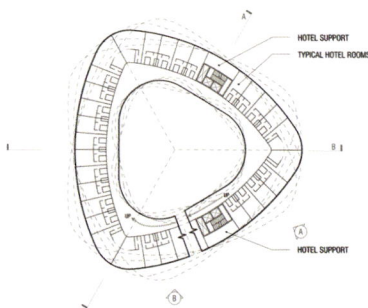

level +6

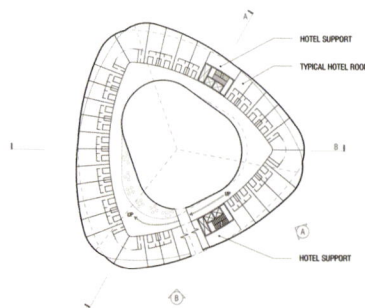

level +7

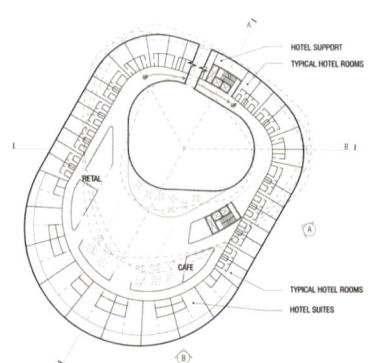

level +8

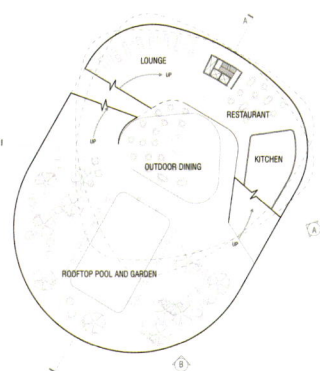

level +9   roof

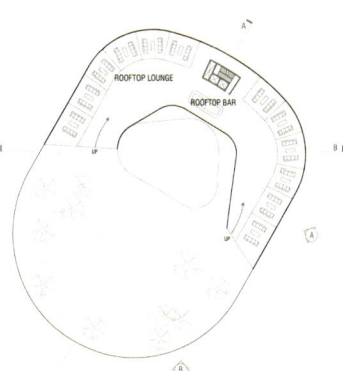

通过将酒店的208间客房和套房沿着螺旋形的楼层安排，设计师让参观者看到的不再是那种代表了酒店典型特色的一成不变的门廊和中庭。酒店的地面宽度不断的变化并在上升到顶层的过程中不断的摇摆。这种设计使得公共空间始终在不断变化。任何两个相对的房间都有着不同的视野，而这种设计让参观者不再是作为行人从外部观察，而使得他们被带入了酒店独特的城市世界。

从概念上，螺旋酒店是一种豪华性与城市性的对话，一种一座小城能够提供的丰富多彩的服务与一座五星级酒店的客人可能提出的各种需求之间的对话。酒店的地面代表了一条在喧闹的小镇里蜿蜒延伸的街道，而在中心的天井里则设计了许多元素。虽然天井提供给了客人无穷的视野，酒店仍为私人会议或客人的隐私保留着空间。整座酒店的设计初衷就是让不同的活动相互融合在一起而非将每一种活动都划上醒目的分割线。

在豪华的假日文化方面，酒店提供了大量充满乐趣的元素来使得酒店成为了满是标志性设计的目的地。首先，这座酒店既是一个豪华生活的展览厅，也同样融合

了城市生活的体验。在入口，服务生代客人将车辆停放在停车场。而这里的停车场既不是建在地上或地下，而是设计在海湾的下面。在酒店的顶层，屋顶的水池成为了一座拥有玻璃底部的全尺寸标准泳池。从酒店底层就可以看到八层之上的池水和泳者。在大厅下面的餐厅中，海湾的海浪距离餐厅的地板是如此之近以至于海浪已经可以拍打到在玻璃幕墙里面的餐厅的边缘。收缩的墙面则表现出了轻柔的微风。

在关注独特的设计的同时，Leeser Architecture也专注于完善的可持续实践，并与顾问公司Atelier Ten共同为热量与能量储存选择了最好的可行方案和材料。室内的瀑布在从上到下穿过中央大堂的空间时将冷水重新带回了整个系统，从而最大限度地降低了酒店内部的热量聚集。在下层大堂中，一个建设在第二层基础上的动态玻璃幕墙一直连接入水中。这座幕墙的作用好像是一个幕布，在天气凉爽的时候会打开，而在因沙漠空气过热的时候关闭。酒店外部的一部分覆盖有一种叫做GROW的既具有光电转换效果又有控风能力的新材料制成的嵌板。

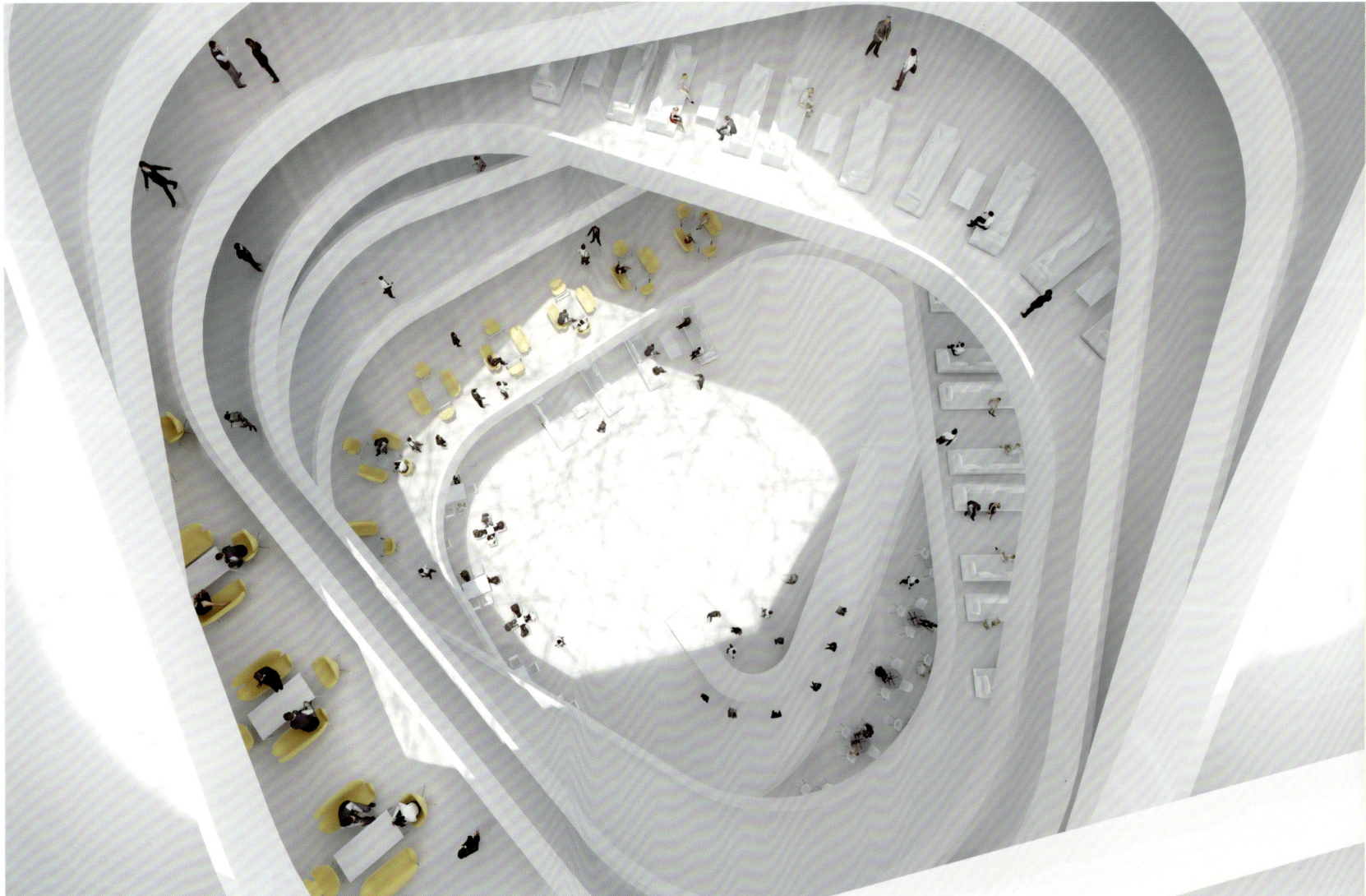

Movie Theater and Media Center of Saint-malo

圣马洛电影院和媒体中心

Firm: SERERO Architects
Location: Saint-Malo, France
Area: 4,270m²

Inspired from the particular landscape of Brittany, the project is shaped like three rocks standing on a granite esplanade. The "Grand Bé", the "Mont Saint-Michel" and large rocks, which rise up from the ocean, constitute the identity of the region by marking its history and heritage. This mediacenter and movie theaters project turns the image of rocks into an original urban system, which catalyzes the surrounding activities. From the train station to the city center, the first rock is the mediacenter, the next one the movie theaters and the last one the cultural center, which widely open themselves onto the plaza created.

These three blocks are covered with a mineral skin, as an echo to old city fortified walls city center while offering building energy efficient. Their insulation is external in order to maximize the building inertia. These rocks are not only simple objects standing on the plaza; they are lifted from the ground at specific and open up to create large entry portal. They are designed as sponges, porous to the context where they appear. These rocks are multi-faces and without particular orientation, or principal facades. The buildings have large footprint, but like a pyramid, its surface decrease with the height, limiting the shadows on the neighbor. Its higher point culminates at 19m from the ground, but 75% of the volumes surface is situated between 0 and +10m.

The volume of the rocks is generated by intersecting plans. The cantilevered effect of their volume is increased by the intersecting plans of the rocks. The architects worked on their volumes, their materiality and transparency to create complex and rich space between them. In the interstitial space between the media center and the movie theaters volume, stands the foyer, sculpted by the negative shape of this volume.

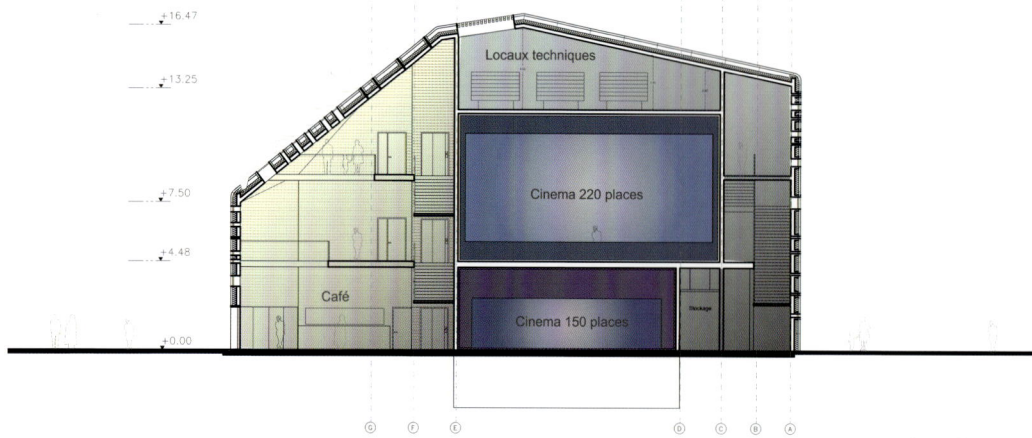

Locaux techniques

Cinema 220 places

Café

Cinema 150 places

Stockage

+16.47
+13.25
+7.50
+4.48
+0.00

Section

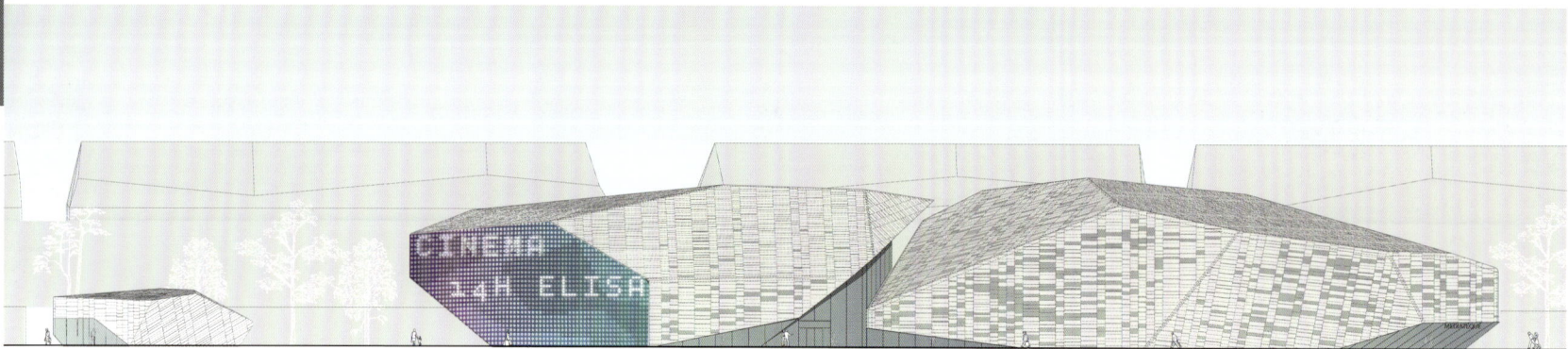

CINEMA
11H ELISA

Elevation

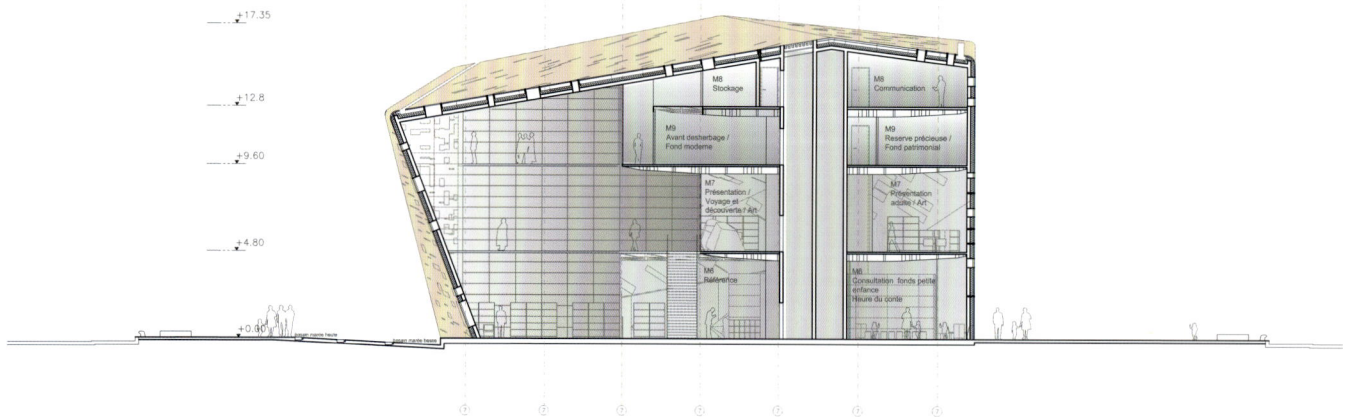

M8
Stockage

M8
Communication

M9
Avant desherbage /
Fond moderne

M9
Reserve précieuse /
Fond patrimonial

M7
Présentation /
Voyage et
découverte l'Art

M7
Présentation
adulte l'Art

M6
Référence

M8
Consultation fonds petite
enfance
Heures du conte

+17.35
+12.8
+9.60
+4.80
+0.00

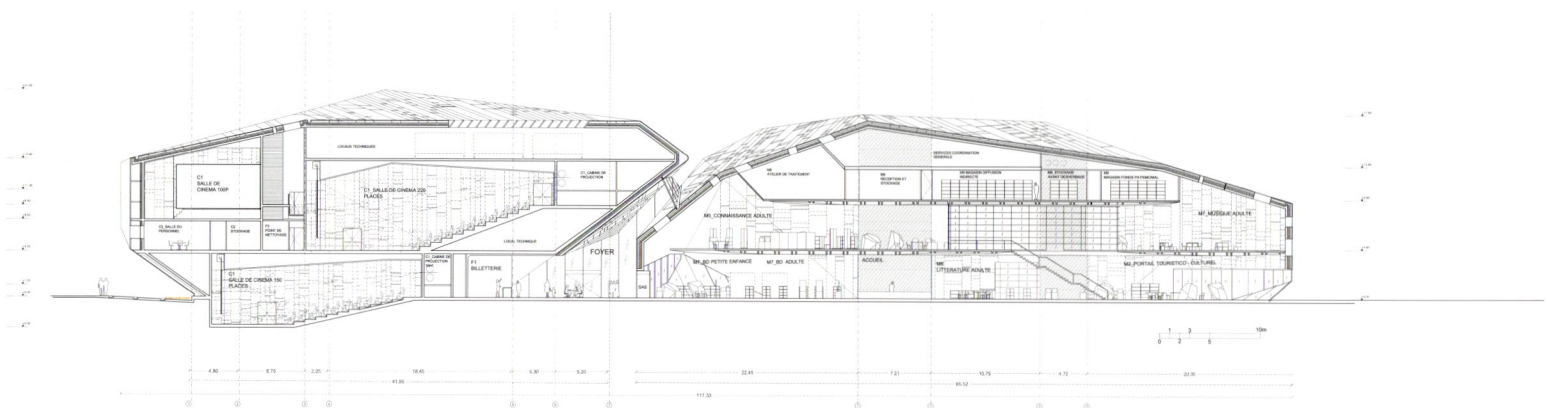

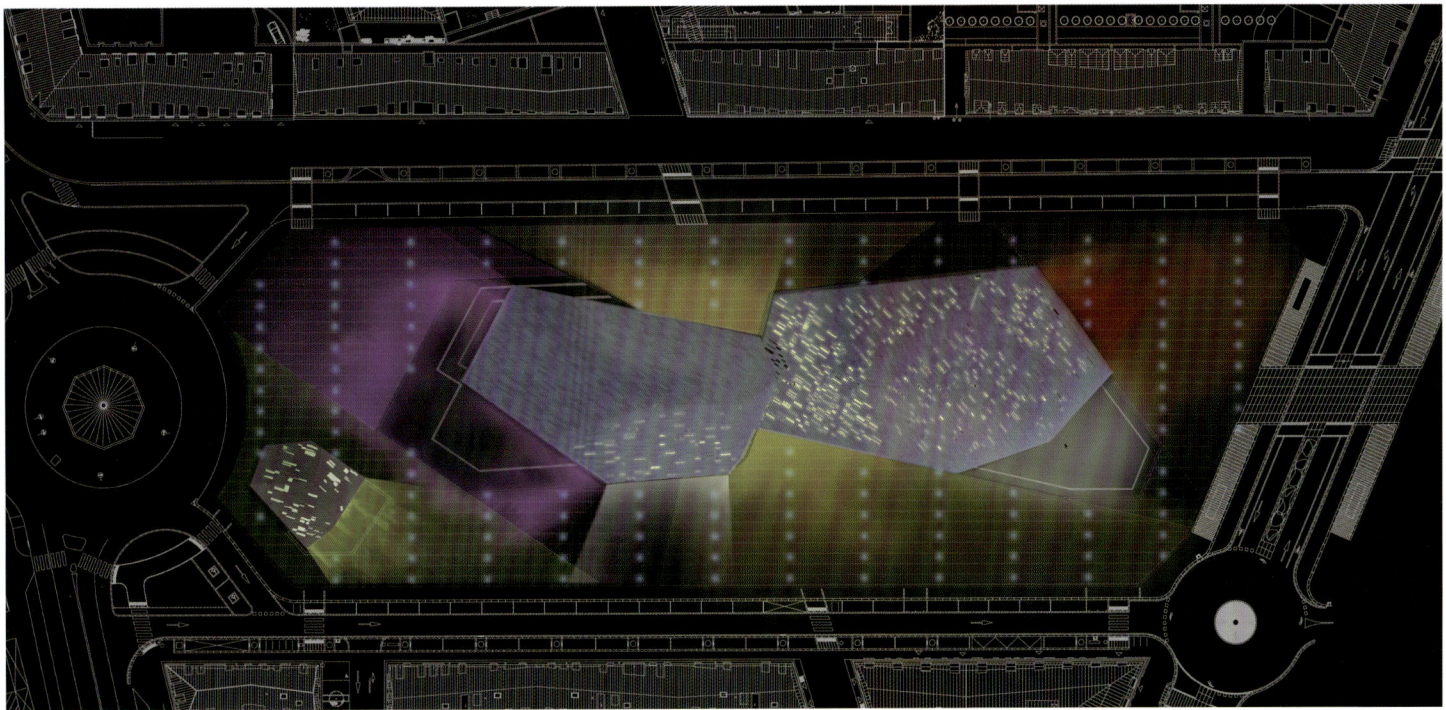

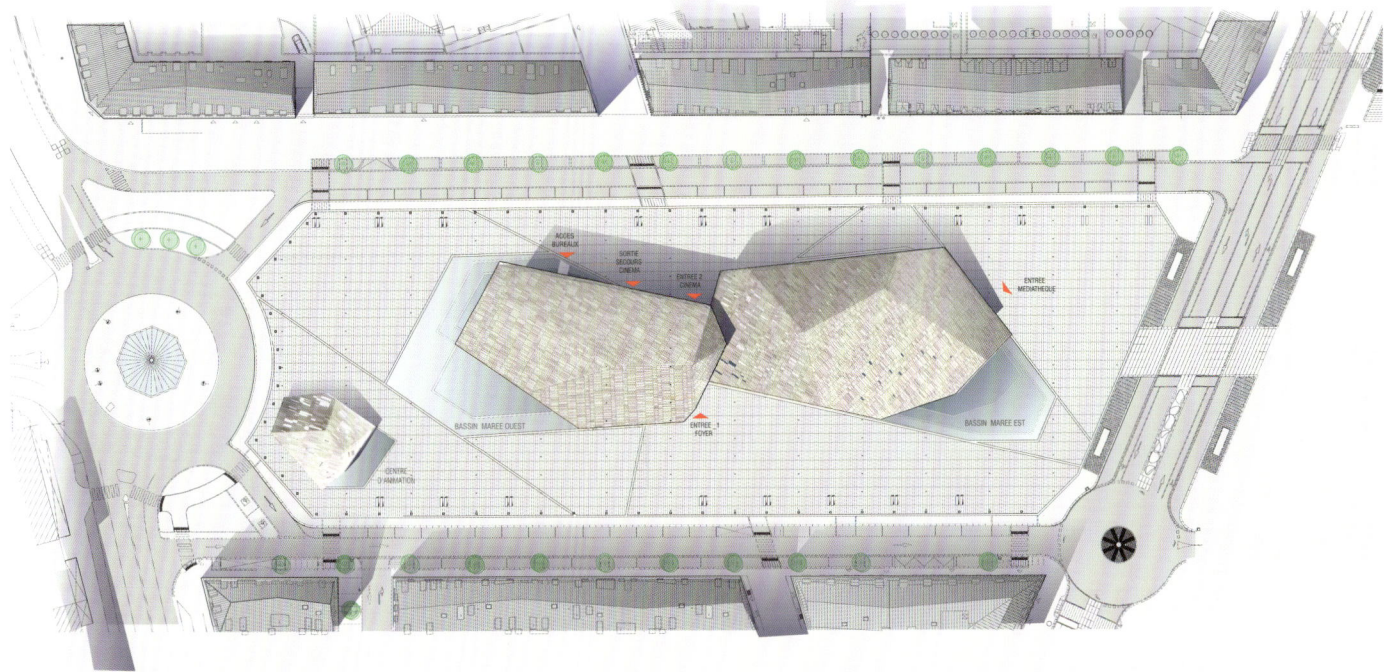

Plan

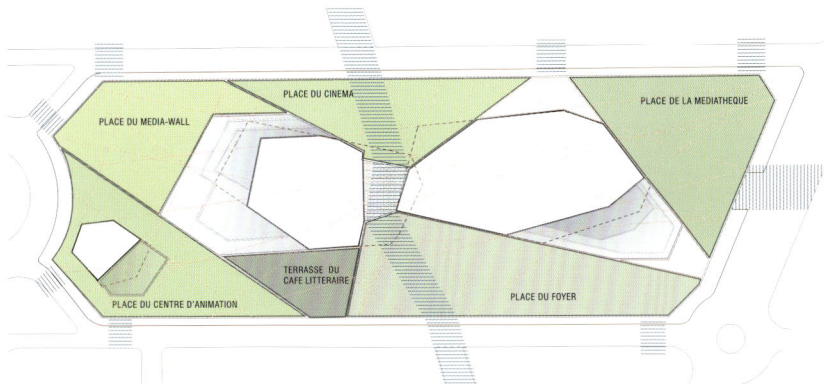

Piazza

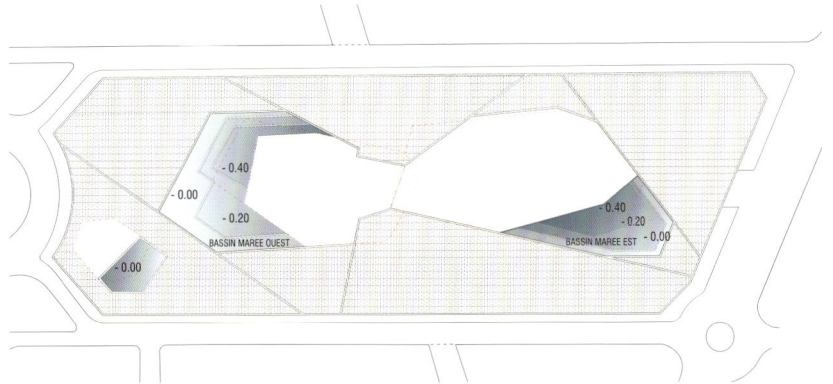

Water Pools

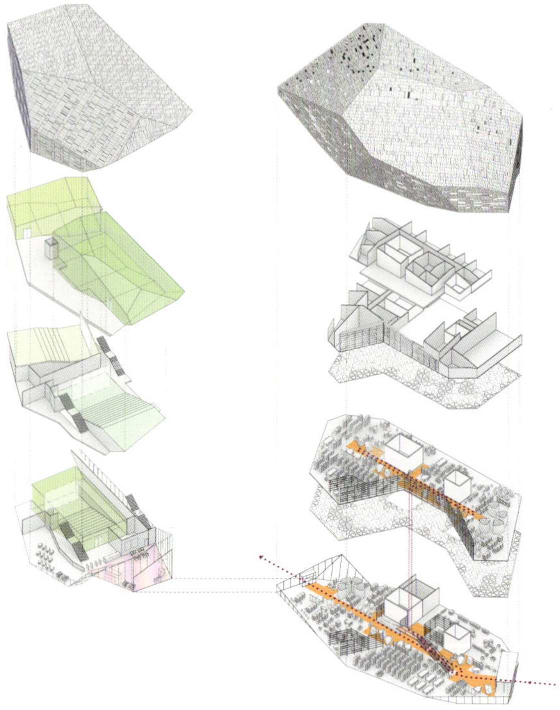

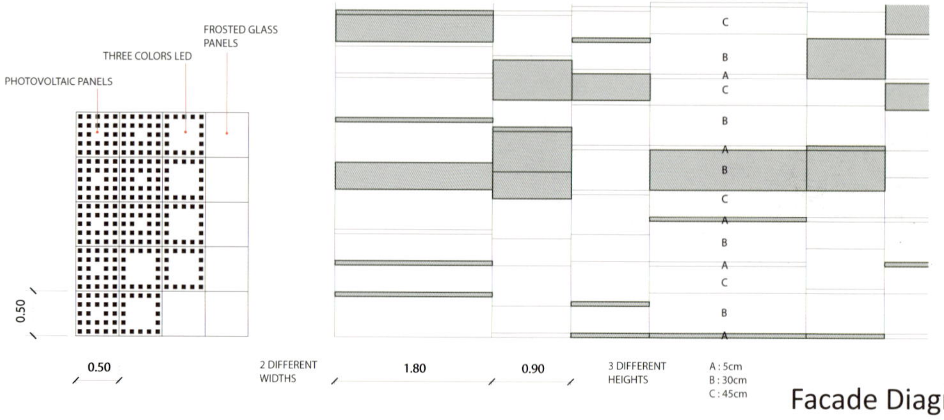
FROSTED GLASS
PANELS
THREE COLORS LED
PHOTOVOLTAIC PANELS
0.50
0.50
2 DIFFERENT
WIDTHS
1.80
0.90
3 DIFFERENT
HEIGHTS
A : 5cm
B : 30cm
C : 45cm

Facade Diagram

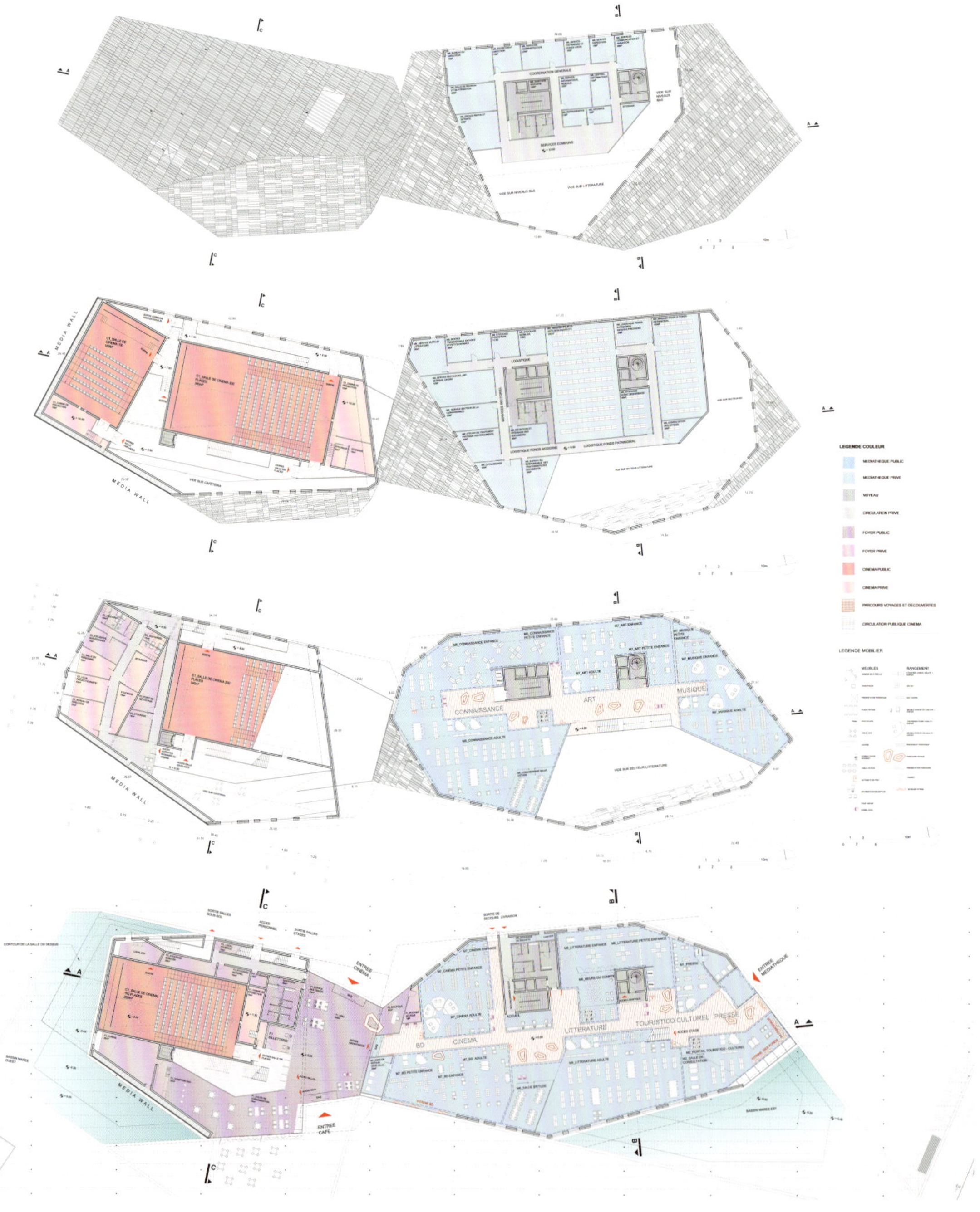
MEDIA WALL
CONNAISSANCE
ART
MUSIQUE
LITTERATURE
CINEMA
BD
TOURISTICO CULTUREL
PRESSE

Plans Final

受到布列塔尼奇特景观的启发，这个项目的造型犹如三块大石耸立在花岗岩空地之上。Grand Bé岛、圣米歇尔山和大礁石从海中升起，表明了这个地区的历史和文化遗产，构成了这个地区的身份象征。这个媒体中心与电影院项目将礁石的形象转变为一种原创的城市系统，促进了周围环境的活力。从火车站到市中心，第一块礁石便是媒体中心，然后是电影院，最后一块是文化中心，朝向形成的广场敞开。

这三座建筑由矿石外层覆盖，呼应了旧城带有强化城墙的市中心，同时也提供了建筑节能功能。建筑的隔热层被设置在外部，以使建筑惯性最大化。这些礁石不仅仅是矗立在广场上的简单物体，它们是特意从地面升起，发展创建出庞大的入口。它们被设计成海绵一样，有很多与周围环境通气的孔洞。岩石都是多面的，没有特别的方向性或者主要立面。建筑有着巨大的脚印，但是像金字塔一样，它的表面积随着高度的增高而减小了，限制了对周围的阴影的影响。建筑的高点是距离地面19m的高度，但是75%的体量表面都在地面到10m高这段高度之间。

岩石的体量是由交叉的平面所引发。体量的悬臂式效果由交叉的平面和岩石所逐渐加强。设计师们在建筑体量上进行了研究，建筑的实体性和透明性创建了这个建筑综合体以及这其间大量的空间。在媒体中心和电影院体量之间的间隙空间，容纳了门厅，由这个体量的非主要部分所塑造出来。

PJCC **Development**

PJCC开发区

Architect: Luca F. Nicoletti, Serina Hijjas
Firm: Studio Nicoletti Associati, Hijjas Kasturi sdn
Location: Kuala Lumpur, Malaysia
Area: 280,000m²

In the area of Petaling Jaya, west of Kuala Lumpur, a great urban development is under way for the establishment of a new urban centre. The mixed use Petaling Jaya Commercial City show an innovative approach to the overall strategy of the development. It has strong design qualities, sensitivity to the surrounding environment and urban community and show environmental responsibility and future sustainability of the project. The entire site is master planned to be developed and treated as an accessible park for the public with landscaped plazas, water features and seating areas, providing valuable spaces for both office tenants and general public. The main design objective is to build an iconic green mixed use development that is contemporary in design and usage of materials, timeless in form and adaptable to system technologies changes. The concept of the fragmented circle is the recurring theme throughout the development. A circle when fragmented, generates a series of different shapes that uniform the general image of the whole master plan. The sail shaped buildings generate together a coherent strong architecture that overcomes the classic concept of commercial city and balance between every day business activities and leisurely pursuits. A holistic environmentally responsive design solution maximise the concept of energy efficiency and sustainability. Low energy systems and controls renewable resources are integrated with the building forms and site massing in order to reduce the reliance on delivered energy. The appropriate design of interiors will allow occupant control over their comfort and immediate environment with a high level of flexibility in use. Sustainability general design strategies of structure facade and building services will reduce the demand of energy whilst improving the users' health comfort and environment.

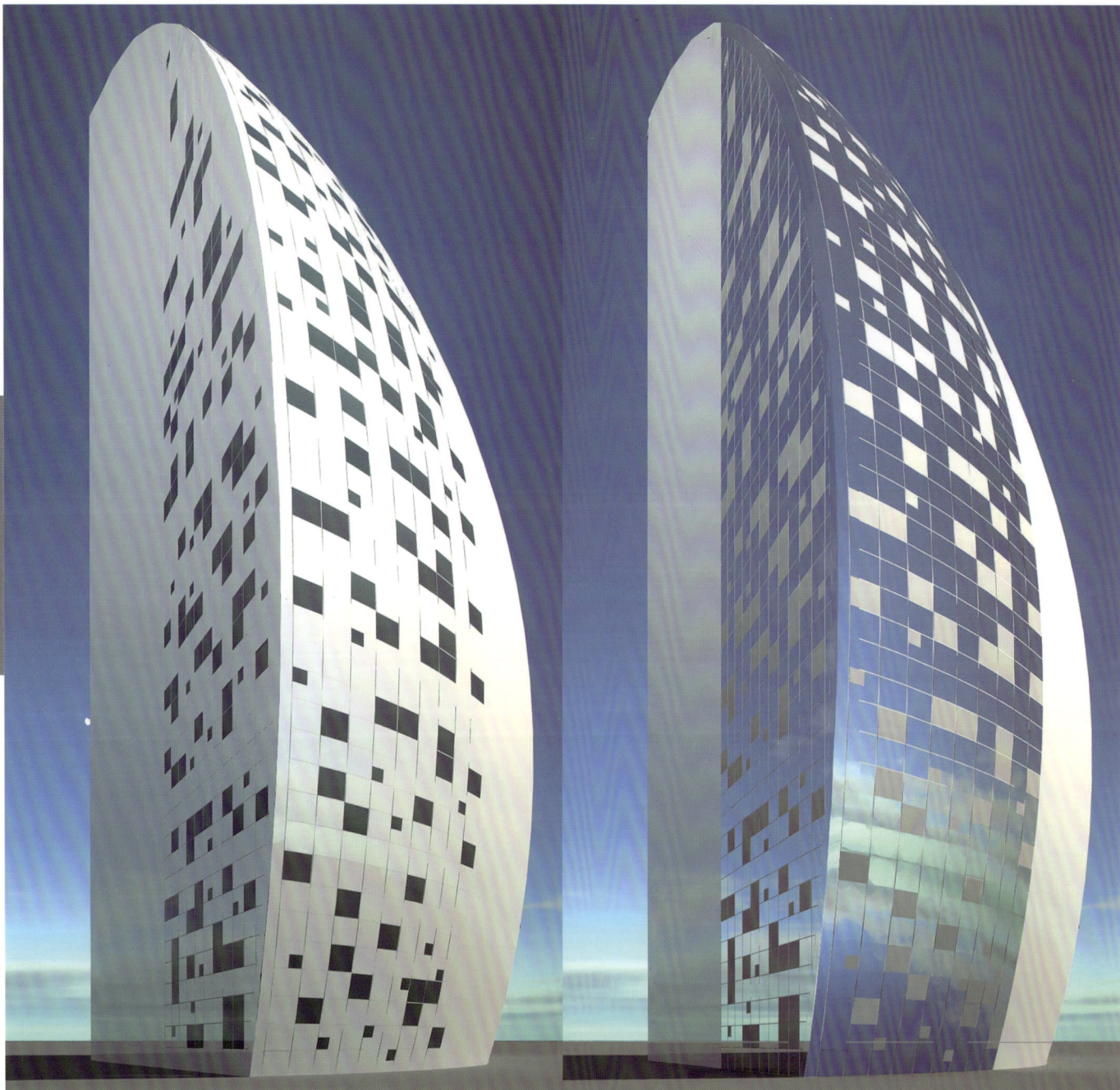

吉隆坡以西的必打灵查亚（Petaling Jaya）地区正在为建设新的市中心而兴建一个巨大的城市开发区。多功能的必打灵查亚商业城展现了这一开发项目的总体战略。该设计具有优秀的质量，对周围环境和城市社区也有敏感的呼应，同时体现了对环境的责任及项目未来的可持续性。整个区域是被作为向公众开放的一处充满景观广场、水体造型和休息区的公园来进行规划的，被设计成既能满足写字间使用者的需要，也能满足普通公众的需要。主要的设计目标是建造一座混合用途的绿色开发项目，并要运用当代特色的设计和材料且对系统技术的改变要具有适应能力，而建筑的外形也要求永远不会过时。零散的圆形的概念是整个开发项目设计中不断重复的主题。一个零散的圆周形式生成了一系列与整体设计的大致轮廓相同的不同形状。这些像船帆一样的建筑体合在一起产生了一个清晰而强大的建筑作品。它超越了经典的商业城的定义并在日常商业活动与休闲要求之间达成了平衡。一个整体的环境解决设计方案最大限度地拓展了节能与可持续性的概念。低能量系统和可再生资源控制被整合到了建筑的外形和所在区域中以达到减少对外部能量的依赖。内部空间的设计则可以让使用者按照他们最舒适的方式和周围环境的情况进行最灵活的配置。对于结构立面和建筑服务的可持续性总体设计方案将在提高使用者舒适度和保护环境的同时减少对能量的需求。

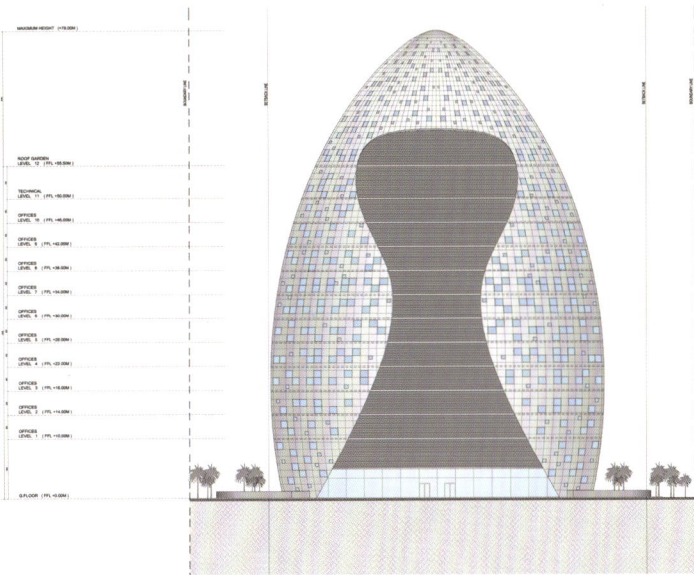

Elevation A

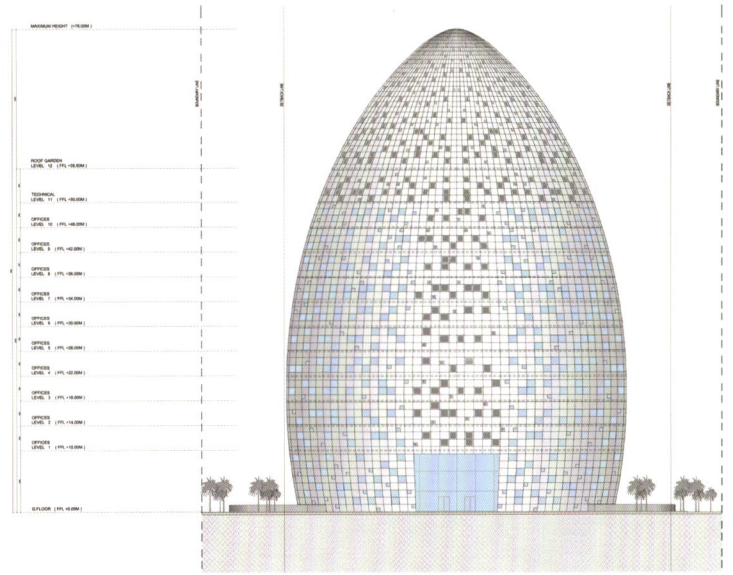

Elevation B

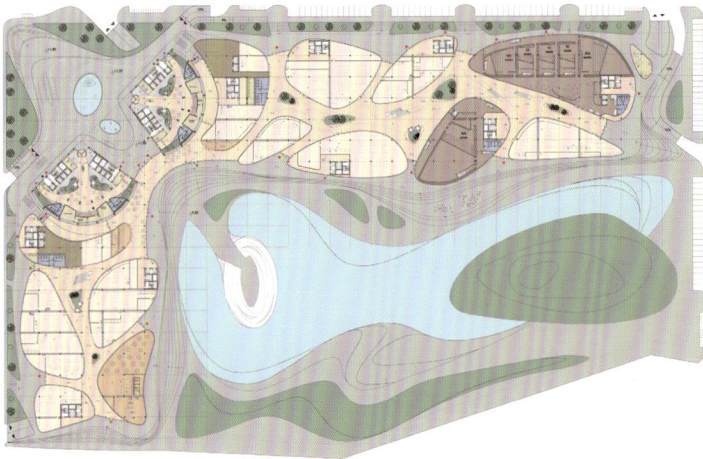

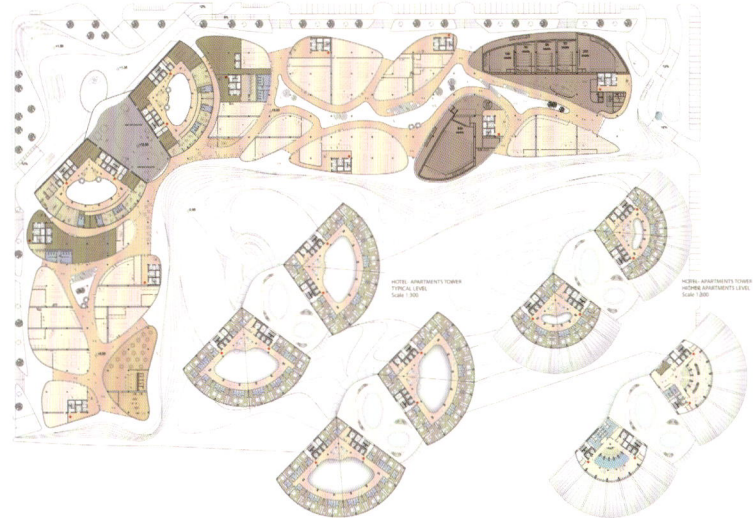

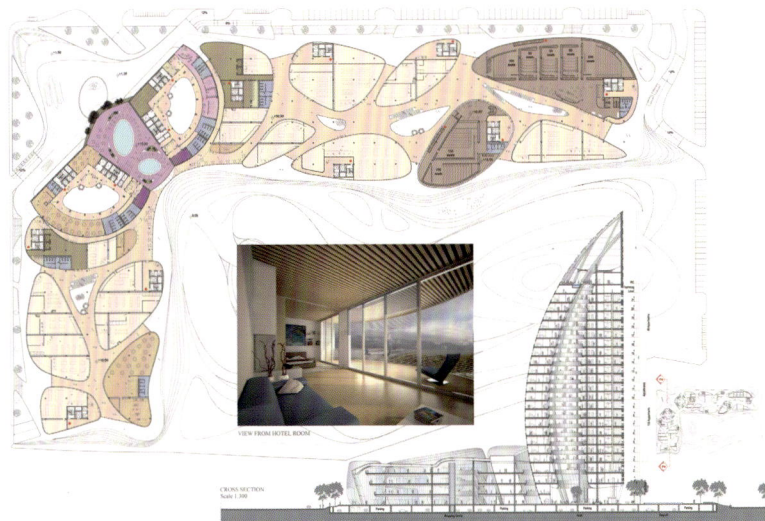

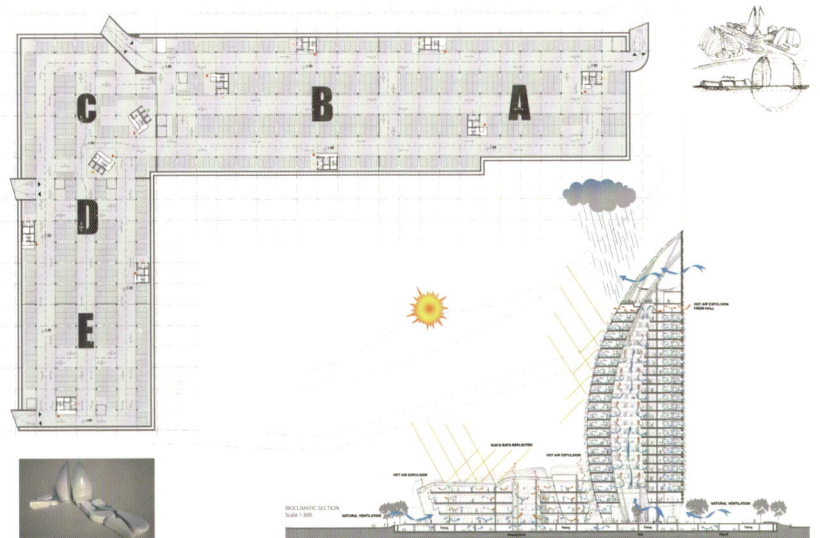

Taipei Performing Arts Center/Studio Nicoletti Associati

台北表演艺术中心

Architect: Luca F. Nicoletti
Firm: Studio Nicoletti Associati
Location: Taipei, Taiwan,China
Area: 40,000m²
Structural Design: RFR Stuttgart

The Kite of Sound was entered into the competition for the Taipei Performing Arts Center but was not shortlisted. A glass & steel light coverage that alludes to the socio-functional unity of the complex is a metaphor of a big sound vibration, a Kite of Sound, dragged by the wind. This dynamic and unifying shell is necessary to give identity to varied unity of functions, each one having different and specific formal characteristics, immerged in a urban frame that is also strongly fragmented, diversified and subject to turbulent traffic.

Without the "Kite of Sound Roof" the Performing Arts Center would be absorbed in the anonymity of the urban texture, while with it, it finds its unitary and unmistakable identity. That identity is neither autonomous nor extraneous in comparison from the surrounding Taipei cityscape because it constitutes its iconic, qualifying and characteristic heart, and none the less a calm island, away from the never stopping traffic where it is immersed and given back to the life and sociality of the Taipei inhabitants.

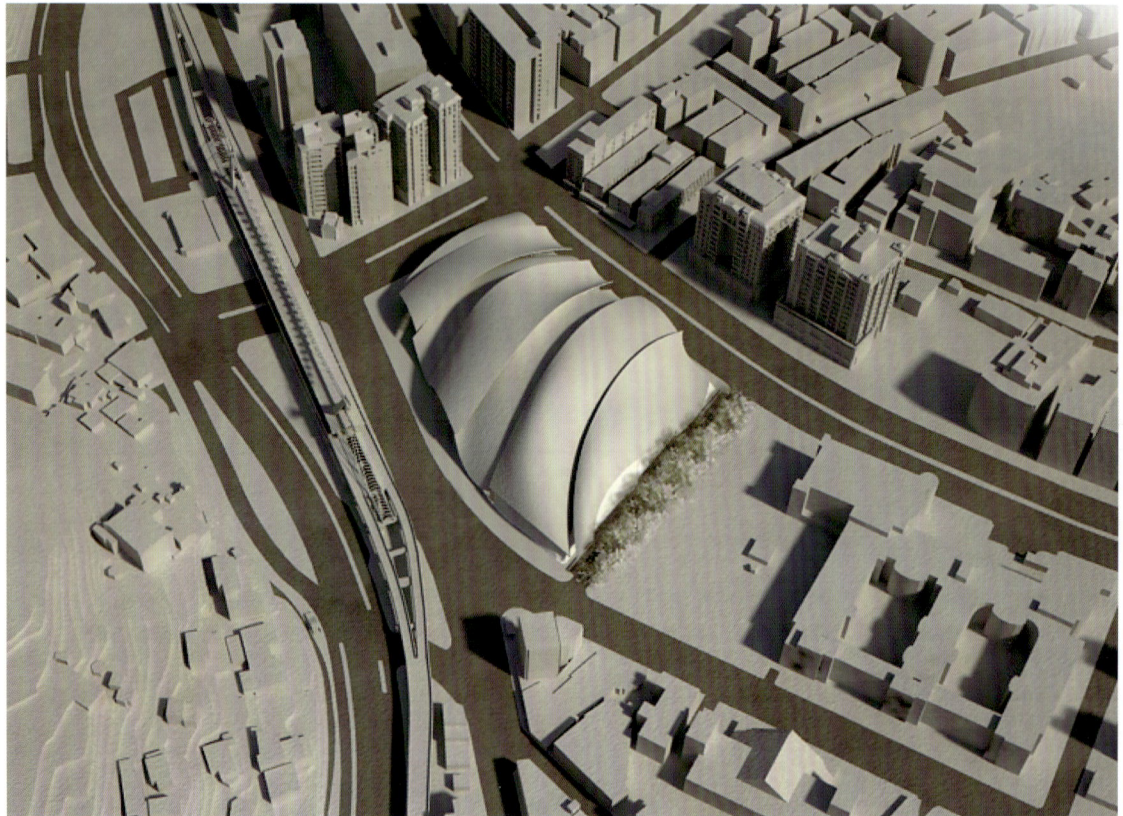

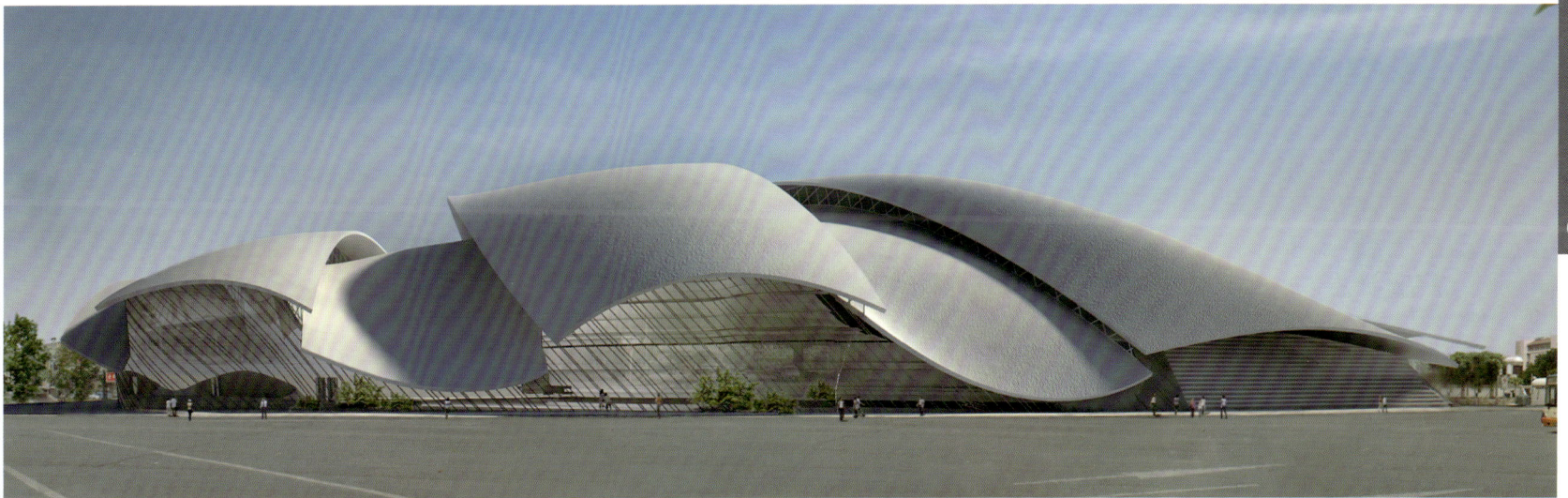

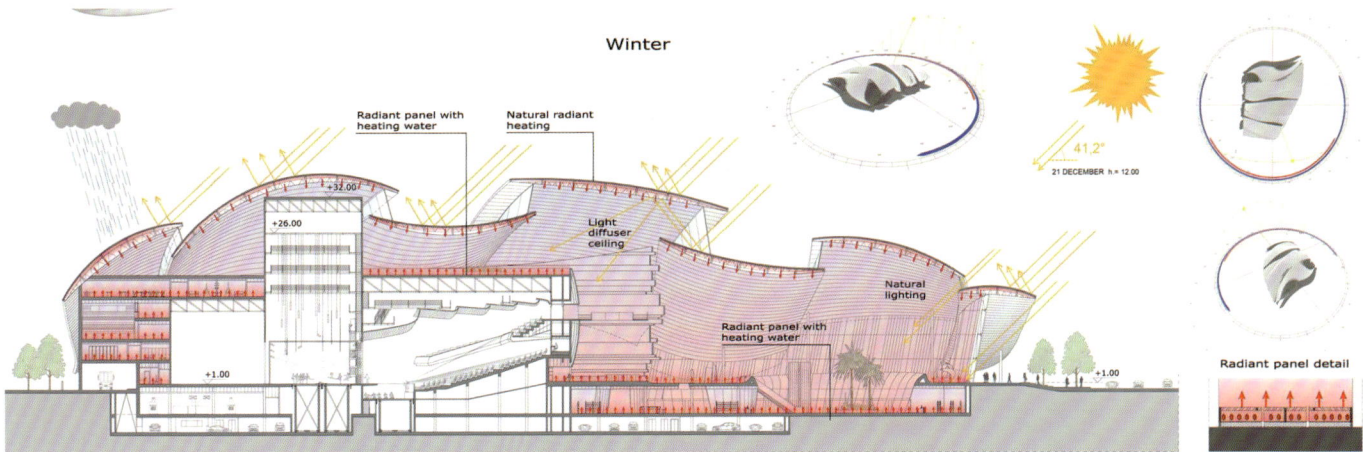

Winter

Radiant panel with heating water

Natural radiant heating

Light diffuser ceiling

Natural lighting

Radiant panel with heating water

+32.00

+26.00

+1.00

+1.00

41.2°
21 DECEMBER h = 12.00

Radiant panel detail

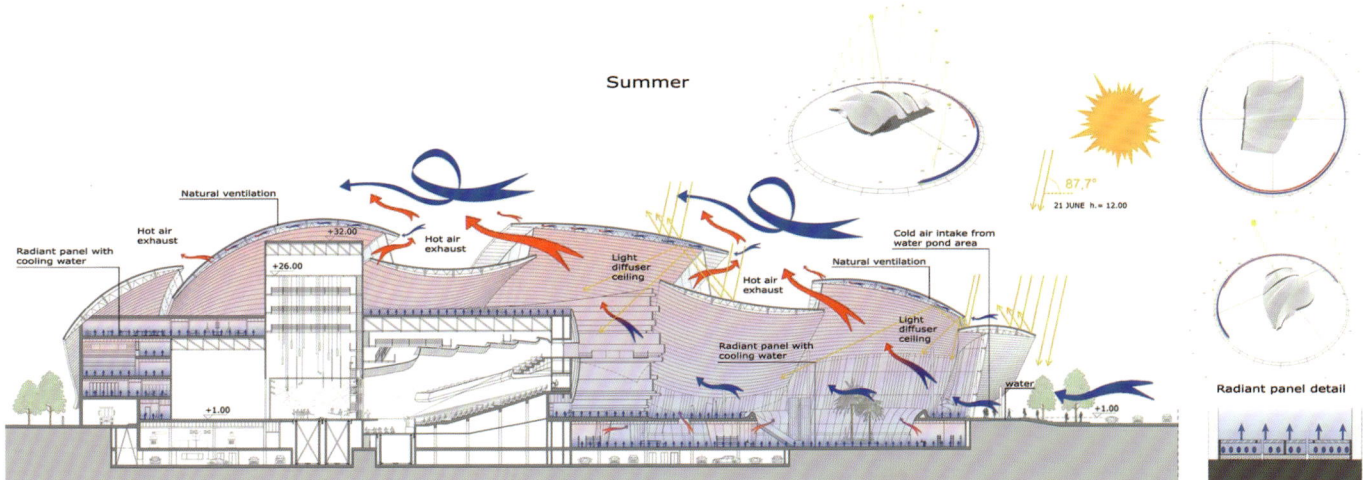

Summer

Natural ventilation

Hot air exhaust

Radiant panel with cooling water

Hot air exhaust

Light diffuser ceiling

Hot air exhaust

Cold air intake from water pond area

Natural ventilation

Light diffuser ceiling

Radiant panel with cooling water

water

+32.00

+26.00

+1.00

+1.00

87.7°
21 JUNE h = 12.00

Radiant panel detail

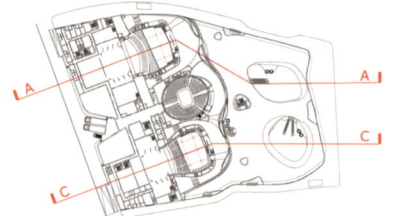

0 6 30

+36.00
+32.00
+26.00
+19.00
+15.50 Administration
+12.00 Rehearsal Studios
+8.50
+5.00 Dressing room and storages
+1.00
0.00
-4.00
-7.00

Section A -A

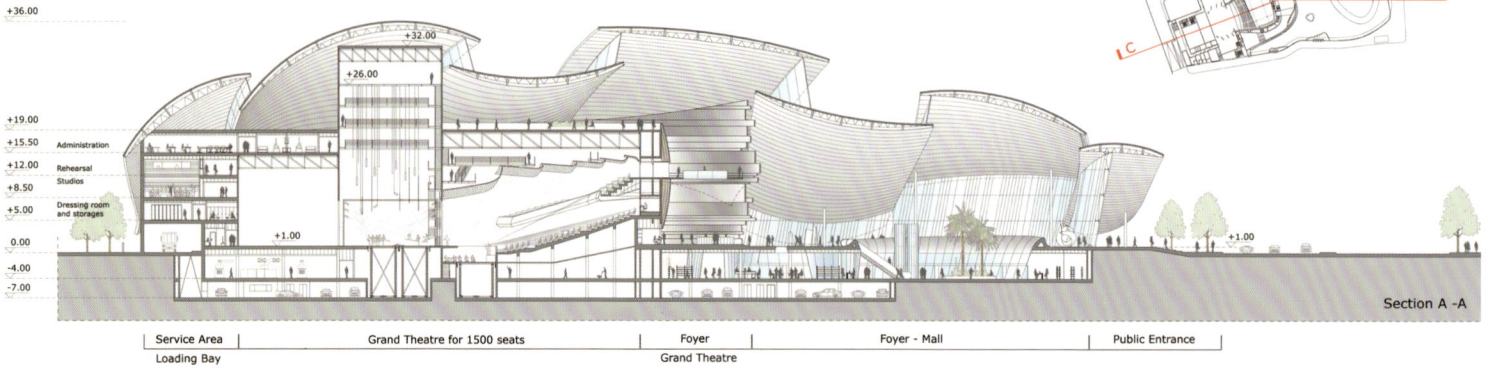

| Service Area | Grand Theatre for 1500 seats | Foyer | Foyer - Mall | Public Entrance |
| Loading Bay | Grand Theatre | | | |

070

+36.00
+24.50
+19.00
+18.50
+15.50 Library and Conference Rooms
+12.00 Rehearsal Studios
+8.50
+5.00 Dressing rooms
+1.00
0.00
-4.00
-7.00

Section C -C

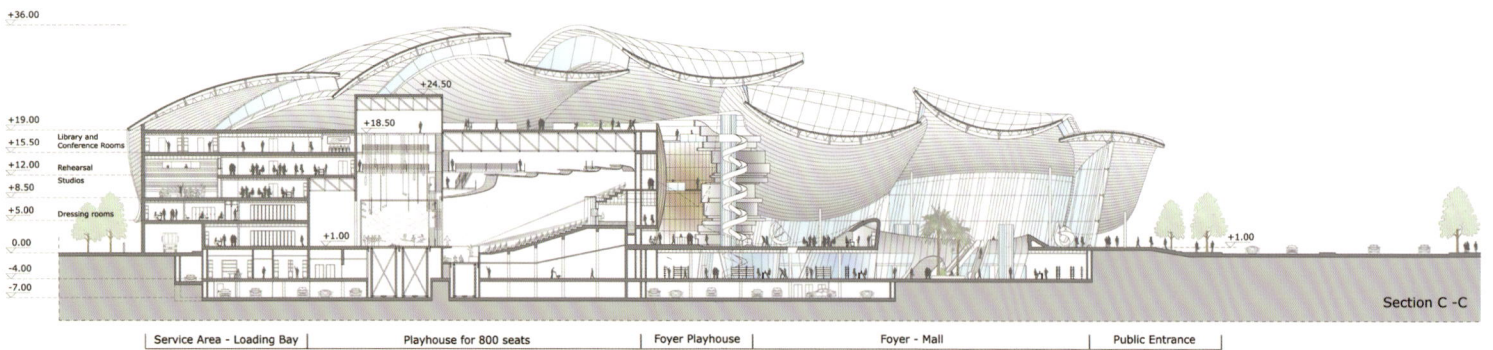

| Service Area - Loading Bay | Playhouse for 800 seats | Foyer Playhouse | Foyer - Mall | Public Entrance |

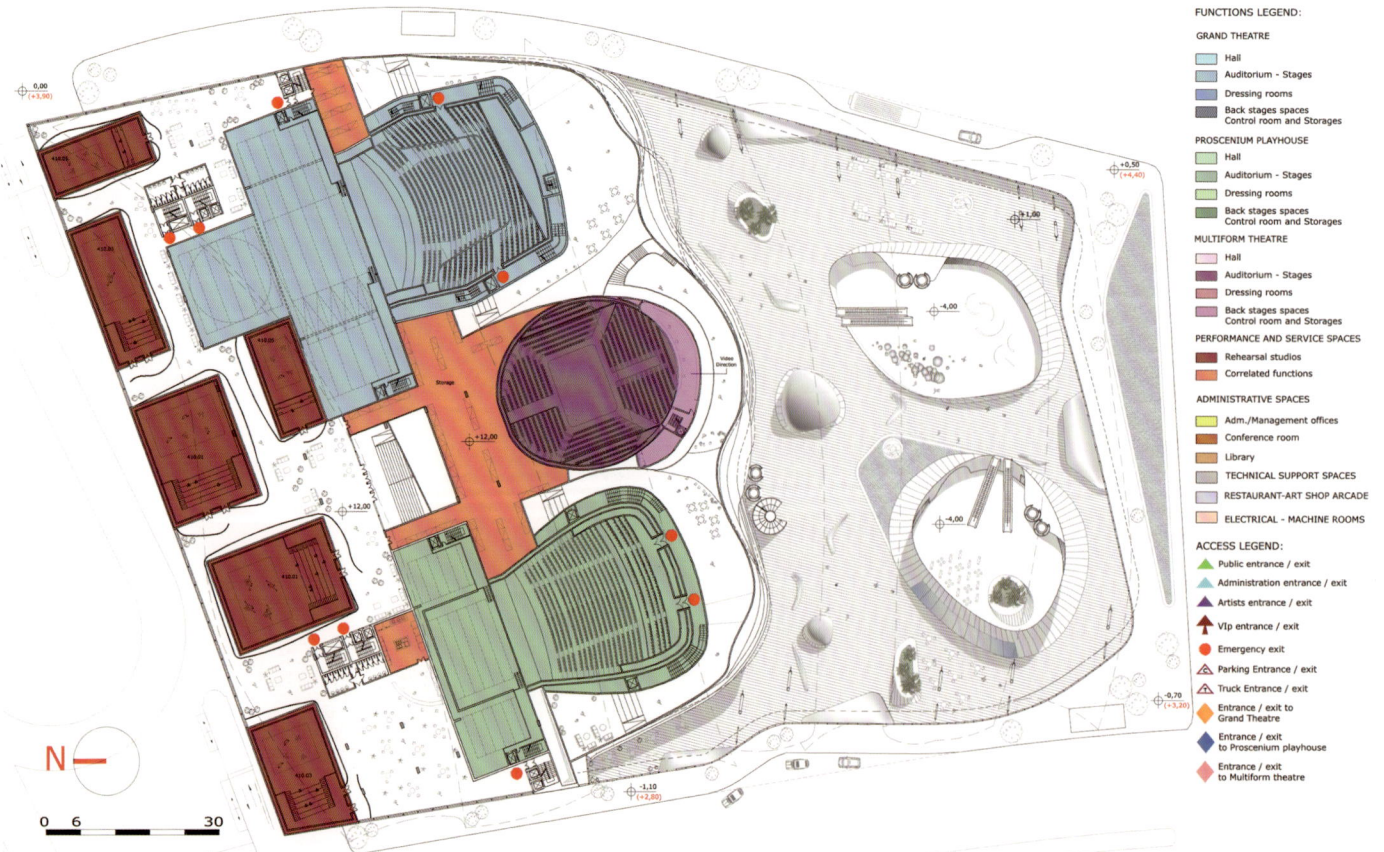

N

0 6 30

FUNCTIONS LEGEND:

GRAND THEATRE
- Hall
- Auditorium - Stages
- Dressing rooms
- Back stages spaces Control room and Storages

PROSCENIUM PLAYHOUSE
- Hall
- Auditorium - Stages
- Dressing rooms
- Back stages spaces Control room and Storages

MULTIFORM THEATRE
- Hall
- Auditorium - Stages
- Dressing rooms
- Back stages spaces Control room and Storages

PERFORMANCE AND SERVICE SPACES
- Rehearsal studios
- Correlated functions

ADMINISTRATIVE SPACES
- Adm./Management offices
- Conference room
- Library

TECHNICAL SUPPORT SPACES

RESTAURANT-ART SHOP ARCADE

ELECTRICAL - MACHINE ROOMS

ACCESS LEGEND:
- Public entrance / exit
- Administration entrance / exit
- Artists entrance / exit
- VIp entrance / exit
- Emergency exit
- Parking Entrance / exit
- Truck Entrance / exit
- Entrance / exit to Grand Theatre
- Entrance / exit to Proscenium playhouse
- Entrance / exit to Multiform theatre

本项目"声音的风筝"参加了台北表演艺术中心的设计竞赛，但最终没有入围。暗示了建筑群社会经济功能的统一的玻璃与钢的轻型覆盖，隐喻了一个巨大的声音振动，一个由风拖着的"声音的风筝"。这个动态的、一体化的框架对于赋予不同功能的统一起到了必要的作用，每一部分都有着不同的特殊的形式特征，没入与混乱的交通一样混乱的多元化的城市框架中。

如果没有"声音的风筝"这个设计，表演艺术中心将隐匿于城市肌理中；而有了这个设计，它就找到了它单一的、确定无疑的特征。这种特征与台北周围城市环境相比既不是完全独立的也不是与其他无关的，因为它构成了其标志性的、具有资质和特性的中心，并且它也仍然是一个平静的岛屿，远离那从未停止的车流，在那里它沉浸在并且回馈于台北居民的生活和社会。

Centers for Disease Control Complex

疾病控制中心大楼

Architect: Luca F. Nicoletti
Firm: Studio Nicoletti Associati
Location: Taipei, Taiwan, China
Area: 42,800m²
Sustainability: HOARE LEA, London, Thomas Lefevre
Plants: Parsons Brinckerhoff International, Inc. Taiwan,China

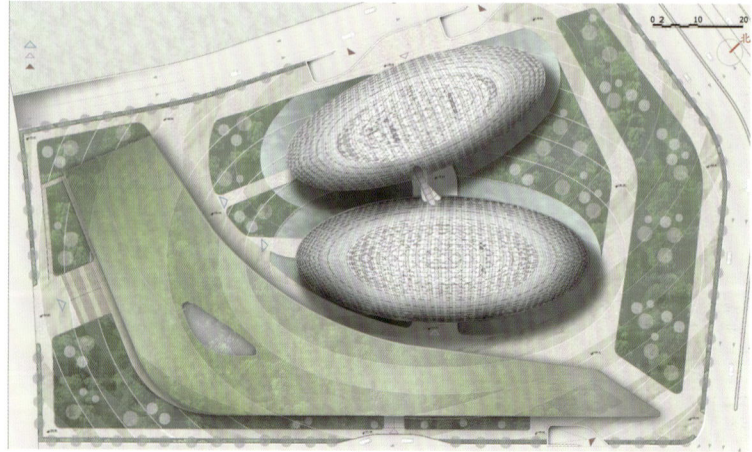

Swietokrzyska Tower

Swietokrzyska塔楼

Architect: Lukasz Wawrzenczyk
Firm: OneByNine studio
Location: Warsaw, Poland
Area: 56,930m²

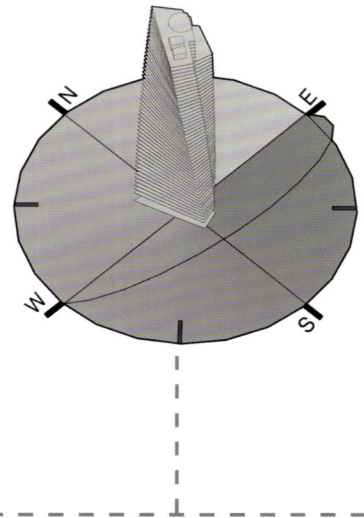

Scale

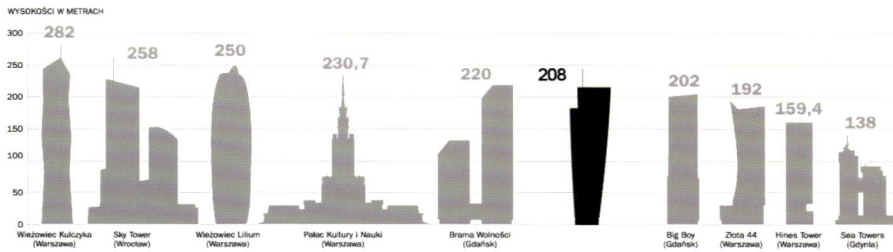

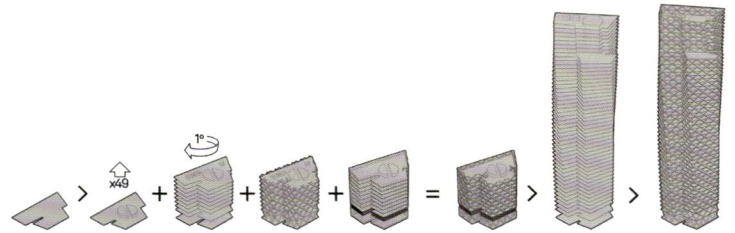

The Swietokrzyska Tower is located in the heart of Warsaw, Poland, next to the main artery of the city center, on the corner of Marszalkowska and Swietokrzyska street. A 49-storey building of 50,000 square meters, 208 meters high is dedicated to office space and is a distinct contemporary accent over the substantially lower development. The Tower is situated on the exact axis of Marszalkowska street. The idea was create a point of reference, visible from a large distance.

The eye catching shape of the building was generated through rotating each floor by one degree. The construction is adjusted to the direction of sun movement and consists of two simple blocks revolving on one's axes. The structure is based on bracing on the facade allowing avoiding structural columns inside of the building in order to get more office space and flexibility. The architects tried to minimize the southern facade which is most exposed to the sun. The western facade is also protected from the sun by outside blind system.

The building opens up to the street and is directly connected with 'Swietokrzyska' subway station by an independent entrance.

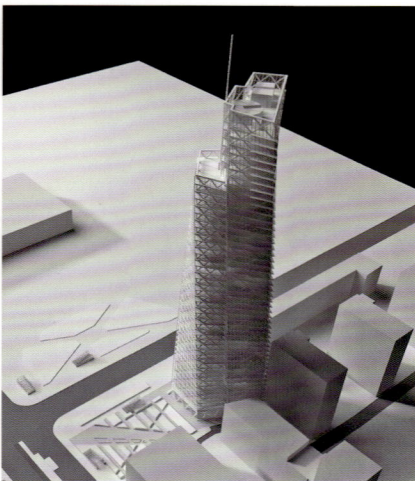

　　位于波兰华沙市中心的Swietokrzyska塔楼毗邻市中心的主要公路，坐落在Marszalkowska和Swietokrzyska大街的转角处。这座写字楼总面积为50000平方米，共有49层楼，总高度为208米。

　　这座建筑在周围大量低矮普通的建筑中成为了一栋突出的当代建筑的代表。这座塔楼建在了Marszalkowska大街的正中心线上。这一设计的用意是让它成为一座能够在很远距离就可以看到的地标建筑。

　　这座建筑最吸引人之处就是每一层楼都围绕中轴线旋转了一度。整个建筑由两个可以围绕各自的中轴旋转的楼体，且其朝向与太阳的运动相协调。塔楼的结构基于立面上的支撑设计。这种设计使得整个建筑内部避免了结构承重立柱从而获得了更多的办公空间和灵活性。设计师尽力在建筑的南侧减少建筑外立面的面积，因为这一方向受到太阳直射最多。建筑的西侧也通过设置外部遮阳系统来避免阳光直射。

　　这座建筑向街道开放并通过一个独立入口直接与Swietokrzyska地铁站连接。

Ground Floor

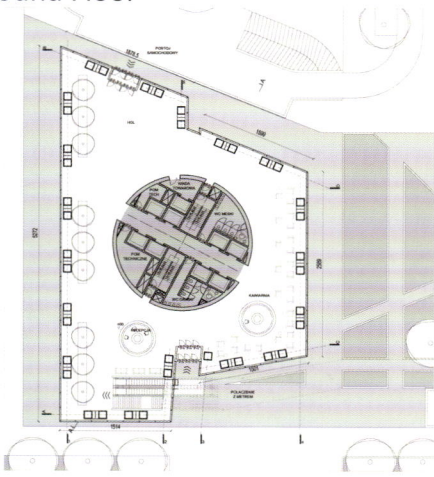

Level 10

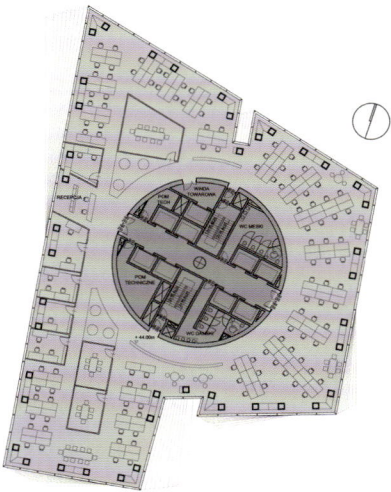

Level 20

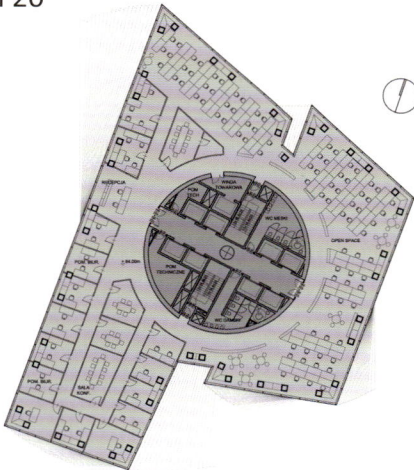

Roof

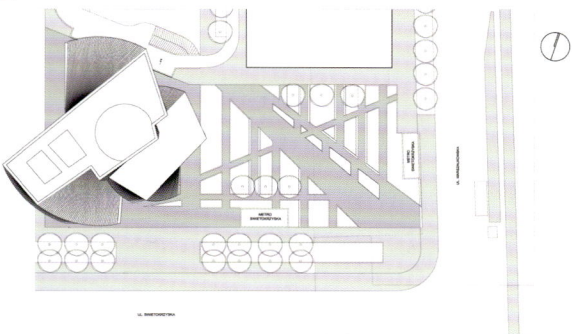

TEN: Campus Tenova

十：特诺华园区

Architect: Leonardo Consolazione, Gaia Maria Lombardo, Giorgio Pasqualini, Laura Perri, Maria Adele Savioli
Firm: Labics
Location: Castellanza, Varese, Italy
Area: 30,000m² (Gross Floor Area)
       27,180m² (Net Floor Area)

The project for the Tenova campus is the project of a two-fold transformation: on the one hand it is supposed to ensure the improvement of the working conditions for the Pomini employees, on the other hand, it represents the headquarters of the Tenova group, to which the Pomini company belongs. Therefore the project aims at meeting material needs, linked to the wellbeing of all those working in the campus, and immaterial needs, linked to the transformation of the area from a mere place of production to a place to represent production, that is the place to represent the identity of a group that passes over the physical space of the place itself. This doubleness of intentions, making up the real richness of the intervention, has been absorbed and summarized by the project without hierarchies nor distinctions; the project, with a unique radical gesture organizes the whole area as the symbol and manifest of a way to conceive work, the relationship with Nature and the quality of the built space, the physical expression of the mission of the Tenova group around the world.

Three strategies have been used to express the values, meaning and identity of Tenova:

- The issue of environmental sustainability, seen as the relationship between the quality of the natural environment and the psychophysical wellbeing of people, being convinced that a more natural environment may substantially contribute to the individual and collective quality of life.
- The identity: a one-building with one stairway, at a glance, an elevated and floating volume, a singular and dynamic object.
- Innovation: a practical visionary that guided all the choices; a radical and experimental vision while transforming the landscape; a glance projected into the future, a vision that is not happy to transform places, it rather tries to define new places.

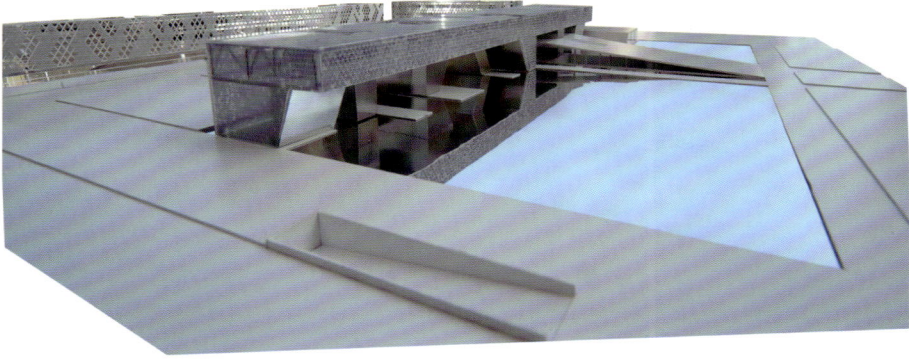
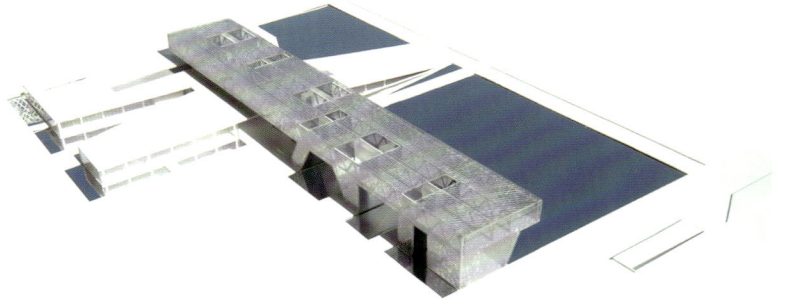

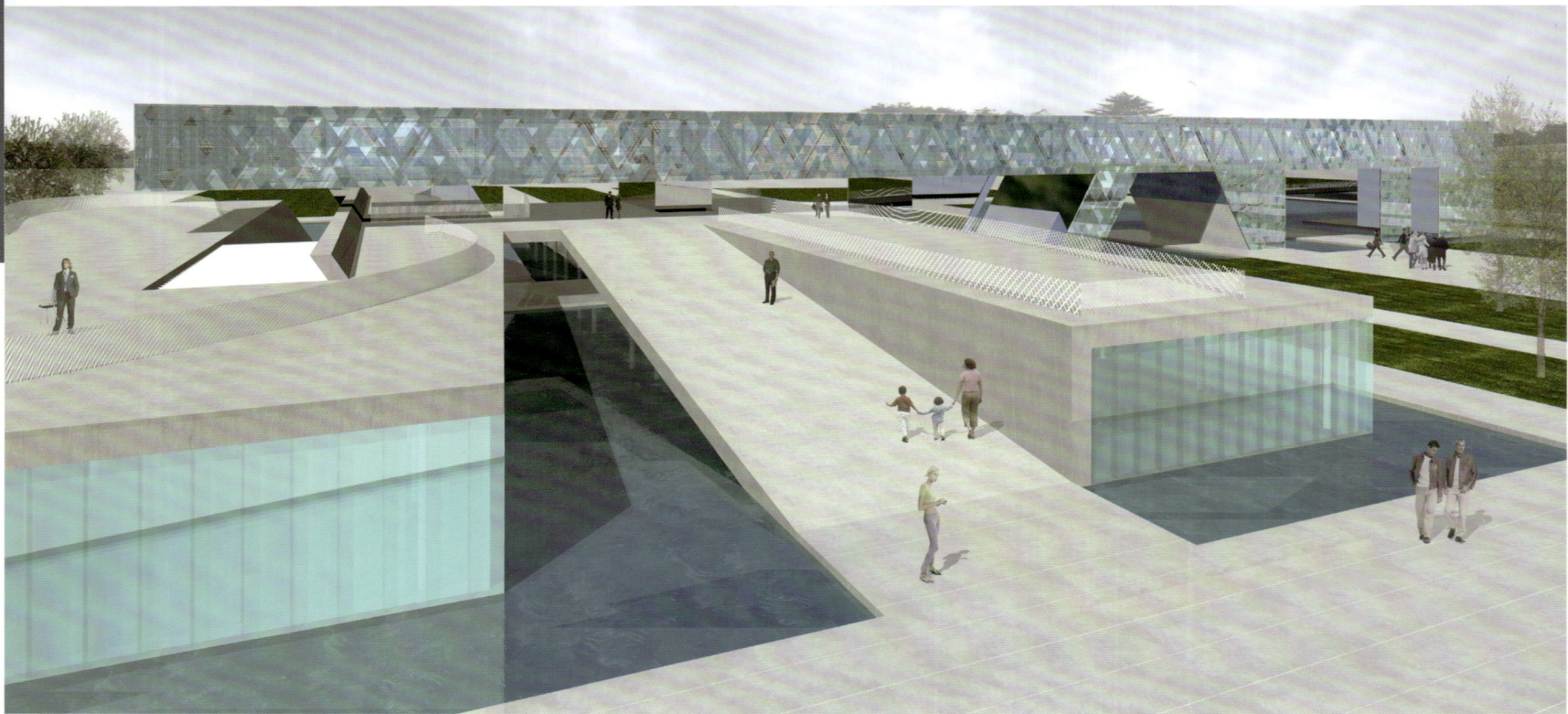

这个为特诺华园区设计的项目是一个双重转型：一方面它应该保证改善波米尼雇工的工作环境；另一方面，它代表了波米尼公司所属的特诺华集团的总部形象。因此，这个项目目的在于满足物质需求，使所有在园区中工作的人都享受到这一福利；满足非物质的需求，使这里从一个单纯的生产环境转换为一个代表了生产的环境，也是一个超越了这里本身物理空间而代表了集团身份的环境。这个双重性的目的，构成了真正的丰富的互动，由这个项目吸收并且总结出来，而没有区分层次和差别。这个带有激进的形态的特别项目将整个区域进行规划，标志和体现了设想工作的方法、与自然的关系和内置空间的品质，更是全球特诺华集团使命的客观表达。

为了表达特诺华集团的价值观、思想和特征，设计采取了三种策略：

——环境可持续发展问题，这被视为自然环境质量和人类心理健康状态之间的关系，说明一个更加自然的环境将对个人和集体的生活品质有着持久的贡献。

——特征：一座建筑配备一座楼梯，一眼望去，一个是高架的浮动的体量，一个是奇特的充满活力的物体。

——创新：一个指导着所有选择的有远见的事实；一个在改造景观设计时不同凡响的，实验性的远见卓识；一个投射到未来的眼光；一个不仅仅是改变环境，而是试着重新定义这个环境的愿景。

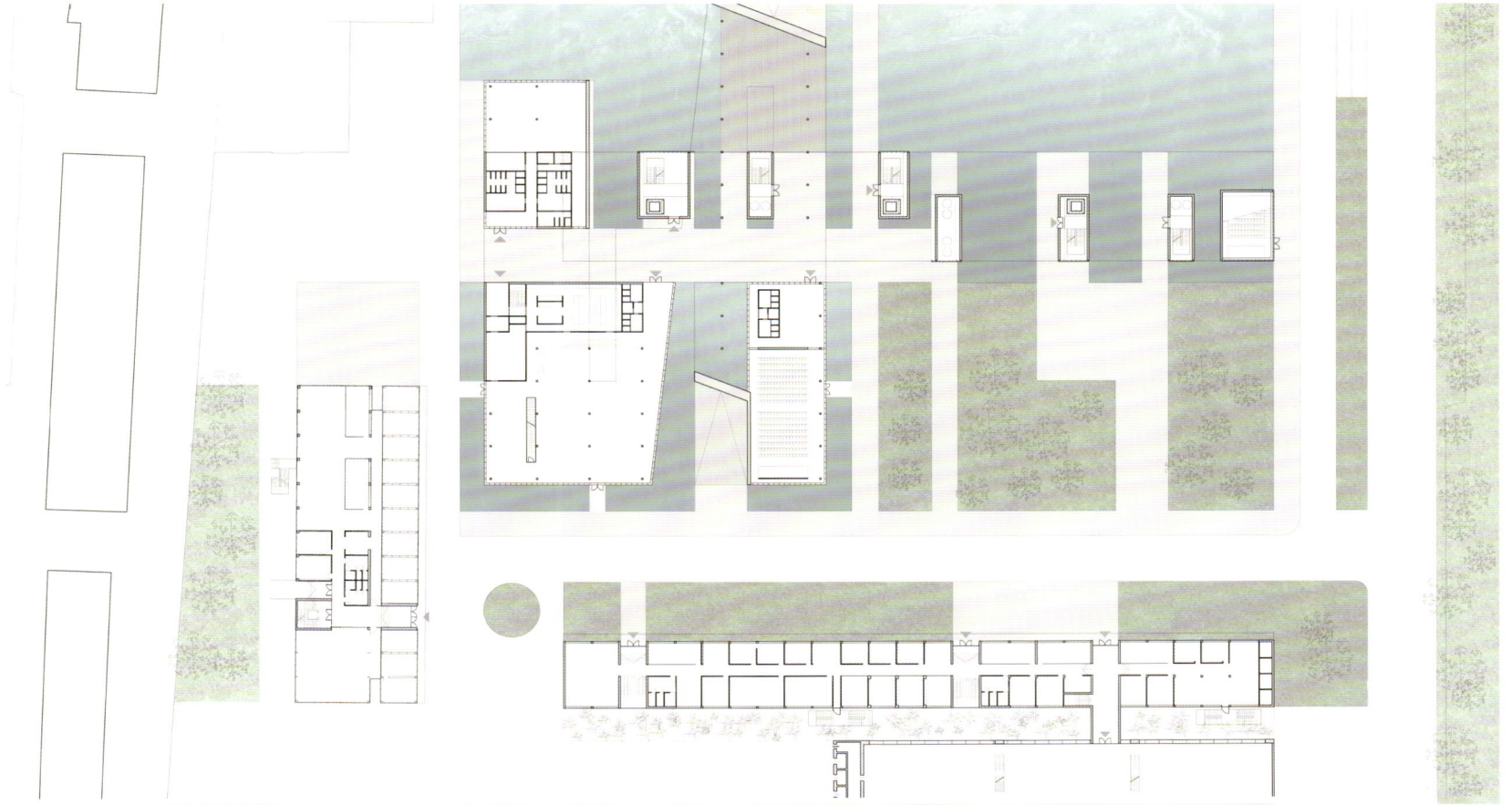

BNT: Designing in Teheran

BNT: 在德黑兰的设计

Architect: Maria Claudia Clemente, Francesco Isidori
Firm: Labics
Location: Teheran, Iran
Area: 23,850m²
Project Team: Paola Bettinsoli, Leonardo Consolazione, Annalucia Scarascia

This project aims to uncover a design strategy that could be used to exemplify the innovative, socially-active, and provocative aspects of Benetton's identity while providing a basis for regionally-modifiable variations on these global themes, transformable for multiple sites, climates, cultures, and audiences.

The fundamental design premise begins with an idea of the city as a multicultural urban landscape, and in this particular case of Teheran's Grand Bazaar and its surrounding district, which is perhaps the most vibrant expression of Teheran's social and commercial life. As with a ribbon of fabric, the spatial facets of the bazaar are folded, hemmed, pleated, and shaped into a new multi-storied, three-dimensional object that then carries within its figures pieces of the city where a multiplicity of cultural activities are recorded and can be enacted.

The patterning of the object is determined by aspects of the site's geometry, its solar and wind orientations, and its principal public access routes. As it is culturally and climatically plausible, multiple entry points can be provided, as well as an embrasure of street-level public spaces within the site. The interior spaces are configured so as to keep the air circulation at an optimal level throughout most seasons.

In the case of Site A, the ribbon of urban fabric is folded outward, forming an interior 'chasm' of public space, and pleated across the irregular geometry, with articulated programmatic elements shaping its exterior. Circulation makes frequent appearances around the external surfaces.

In the case of Site B, the ribbon of urban fabric is folded inward, forming a smooth exterior surface on two sides, with a cascading series of articulated spaces open to each other on the interior, through which most of the circulation passes and across which multiple views are enacted.

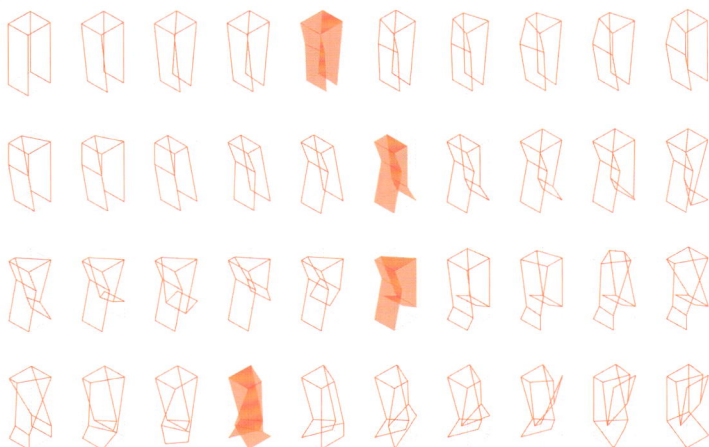

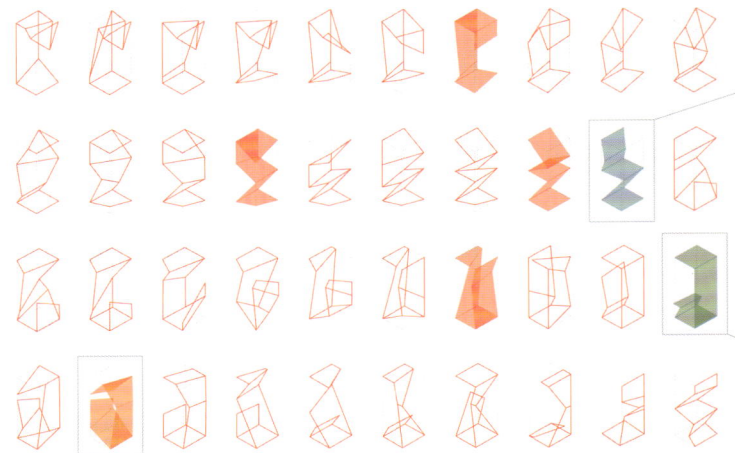

这个项目目的就是揭示一种可以展现Bentton特征中创新、活跃与激情等方面的设计方案表达给公众，同时为这些全球性的主题提供了一个地区性多变元素的基础，而这种基础可以为各种各样的场合、气候、文化和观众而变化。

设计的核心发源自将这座城市视为多元文化城市景观的理念，并融合在了此次对作为德黑兰社会商业生活最集中表现的"德黑兰大集市"及周围环境区域的设计案例中。如同采用了纺织物制成的丝带一般，大集市的空间立面被整体折叠、弯曲边缘并增加了褶皱，最终形成了一个新的立体的多层建筑体。这栋建筑因此承载了这座记录并发生着多种多样的文化活动的城市的片段。

建筑所在地点的几何形状、阳光与风的方向以及主要人群往来的通道决定了这栋建筑的造型。由于在文化和气候上的可能性，设计师提供了多样化的入口以及连接地块中的街道高度的公众空间的楔形开口。内部空间被设计成可以在多数季节中保证最佳的空气流通效果。

在A区，城市结构所构成的丝带向外折叠，从而在公众空间中形成了一个内部的"凹陷"，并且越过不规则的几何地形。在其外部则是由规则的铰链结构构成。围绕外部表面可以感受到流动的气流。

在B区，城市结构所构成的丝带向内折叠，在两边形成了一个光滑的外部表面，大量在内部连通的交缠的空间体从上方向下延伸。大部分的气流都穿过这些结构，且产生了多样化的视角。

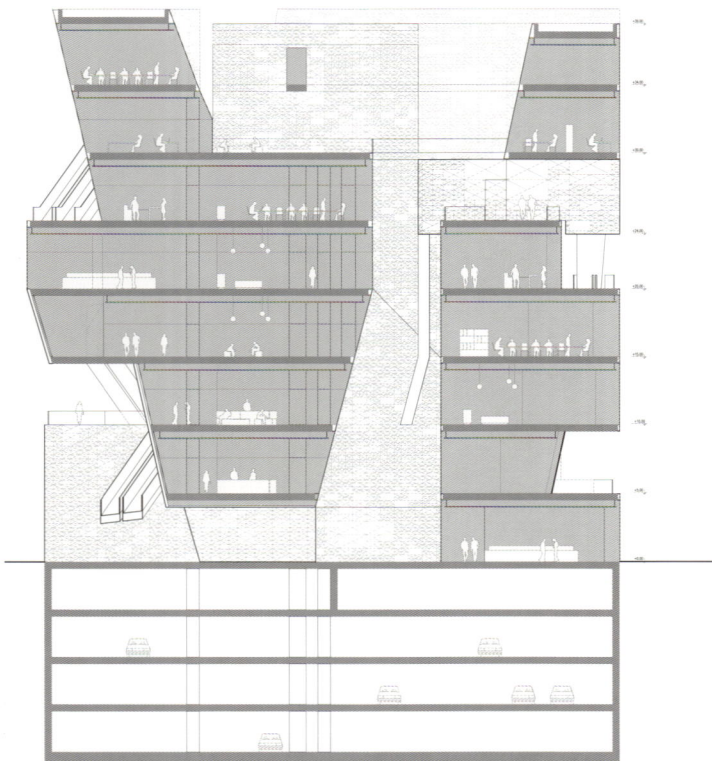

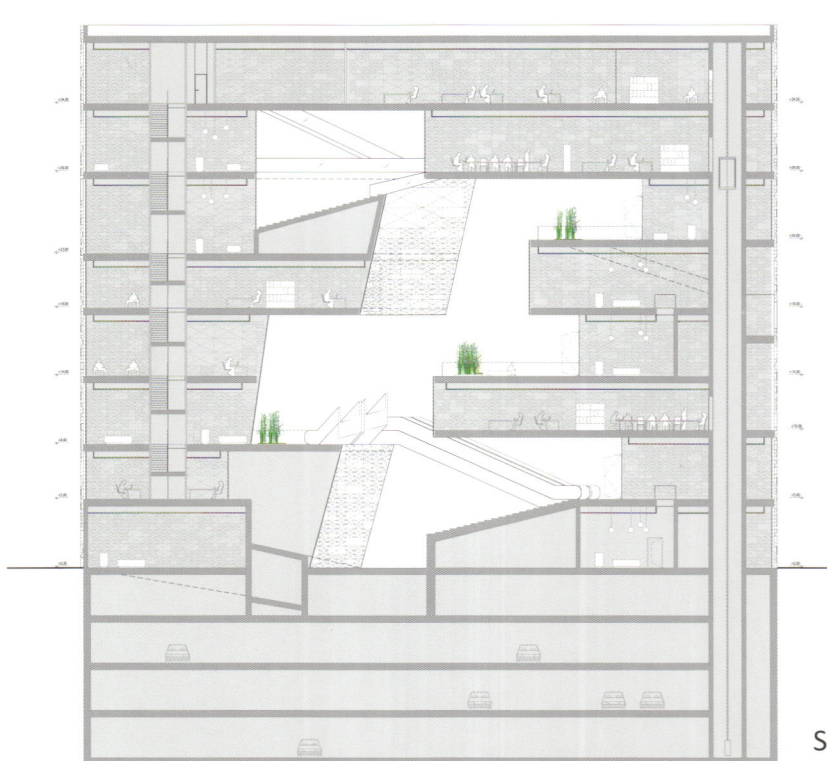

Sections

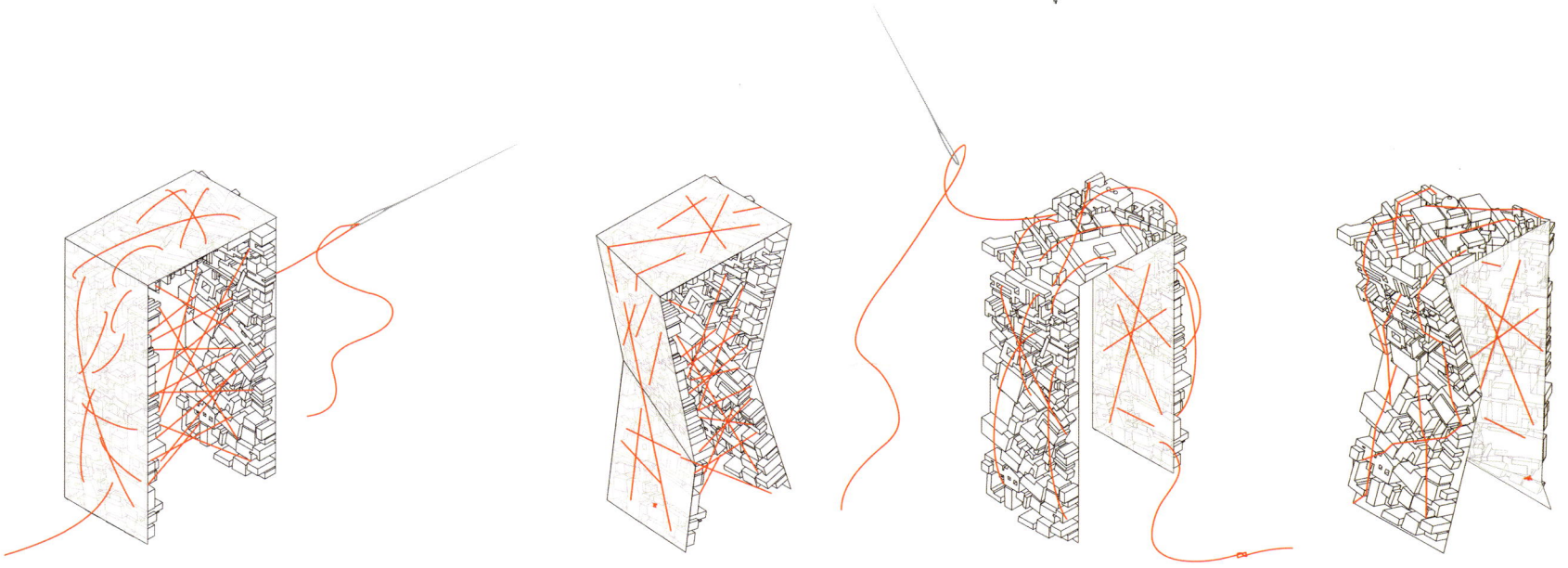

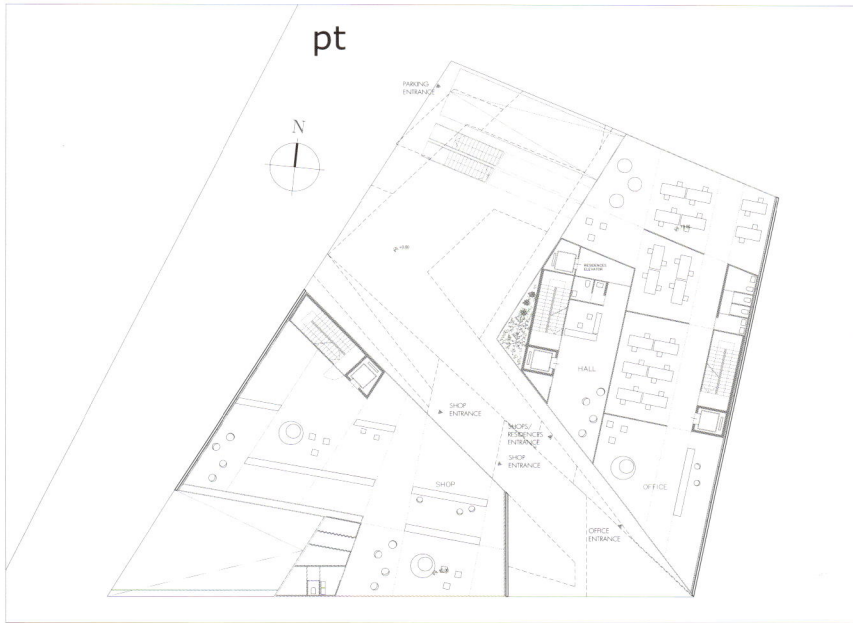

Ground Floor A

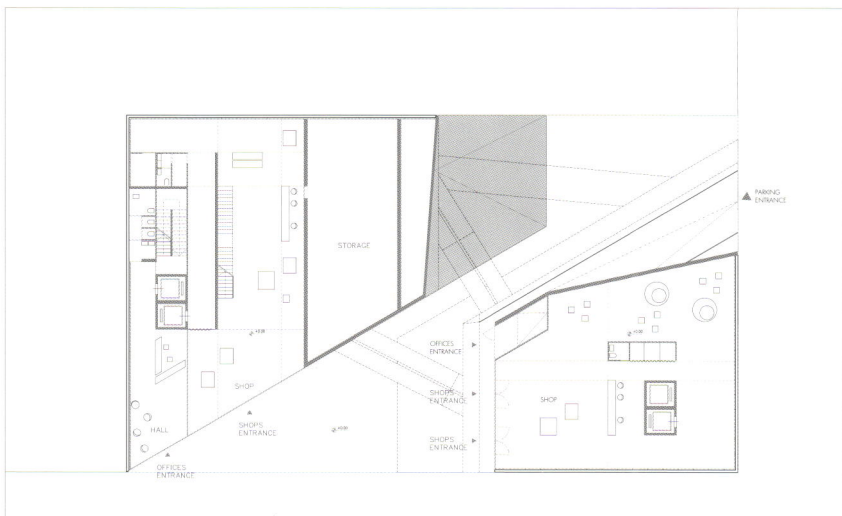

Ground Floor B

LCF Farnesina High School

法尔内西纳LCF高中

Architect: Maria Claudia Clemente, Francesco Isidori
Firm: Labics
Location: Rome, Italy
Area: 2,430m²
Project Team: Fabio Balducci, Leonardo Consolazione, Manuela Gentile, Domenico Santoro
Domenico Santoro
Consultants: 3TI Progetti Italia, Eliana Cangelli

The project's aim was to develop an innovative relationship between the existing school's architecture and the proposed intervention. The design starts by considering the new classrooms not solely as simple pavilions or smaller parts of the whole, but as specific moments in the architecture, each with individual tectonic identities, resulting in unexpected relationships between the existing school structure, the new classrooms and the surrounding site.

The construction methodology of the existing building defined a conceptual basis that informed the design of the school extension. Constructed from a cascading series of classroom-dimensioned slabs, the existing school blends into the surrounding landscape in an unobtrusive manner, creating a soft background with respect to the natural topography of the site. The new school design follows this strategy by becoming a neutral object, the horizontality and geometric regularity of the new volume follows the contours of the ground plane, creating a stronger relationship with the landscape while liberating a wide external space. While, the building morphology appears rigorous, the classrooms have been transformed and rotated to optimize views and the comfortable feeling of openness to the surrounding landscape.

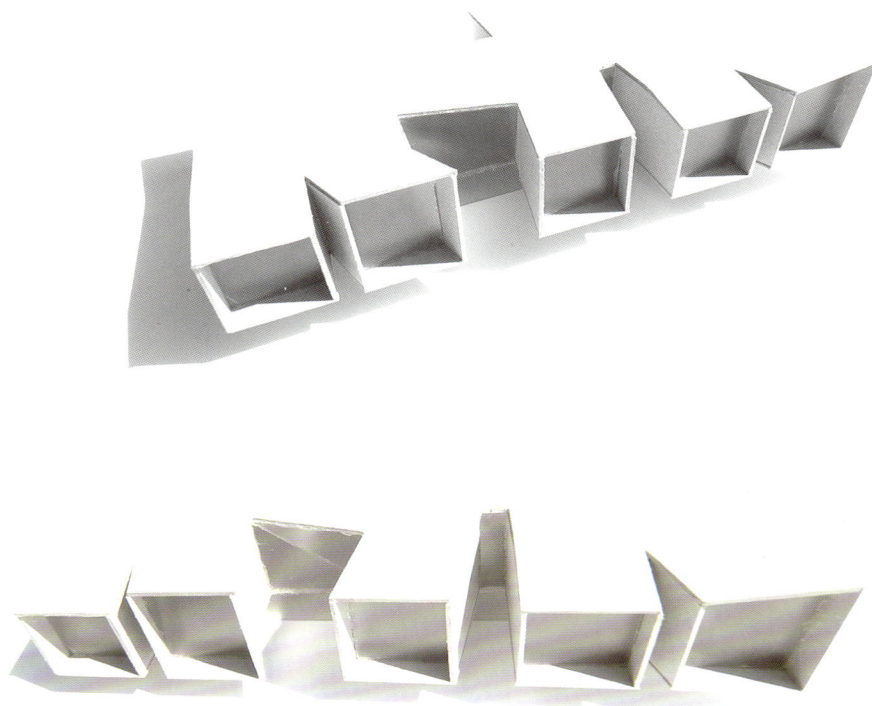

　　该设计项目的目的是发展现存的学校建筑方式与设计师提出的干预方案之间的创新性关系。设计开始于不再把新的教室独立作为简单的大厅或整体结构中的一部分，而将教室作为建筑中的特定时刻，每一间教室都有各自的构造特点。这种设计在已有的学校结构、新的教室与周围环境之间创造出了一种出乎意料的关系。

　　现有建筑的建设方法定义了一种影响了学校扩建工程设计的概念基础。从建设级联层叠的教室尺寸大小的厚板开始，现有的学校建筑用一种不显张扬的方式成为了周围景观的一部分，并成为了与该幅土地自然地形相适应的柔软的背景。新学校的设计将学校变成中立客体，延续了这种方式。新外型采用水平与规则的几何设计遵循着地平线的轮廓，在建筑与景观之间建立了更牢固关系的同时释放了外部自由的空间。同时，建筑形态十分严谨，教室也被变形与旋转以便赋予人们开放的最佳的视野与舒适的感受来观察、体会周围的景色。

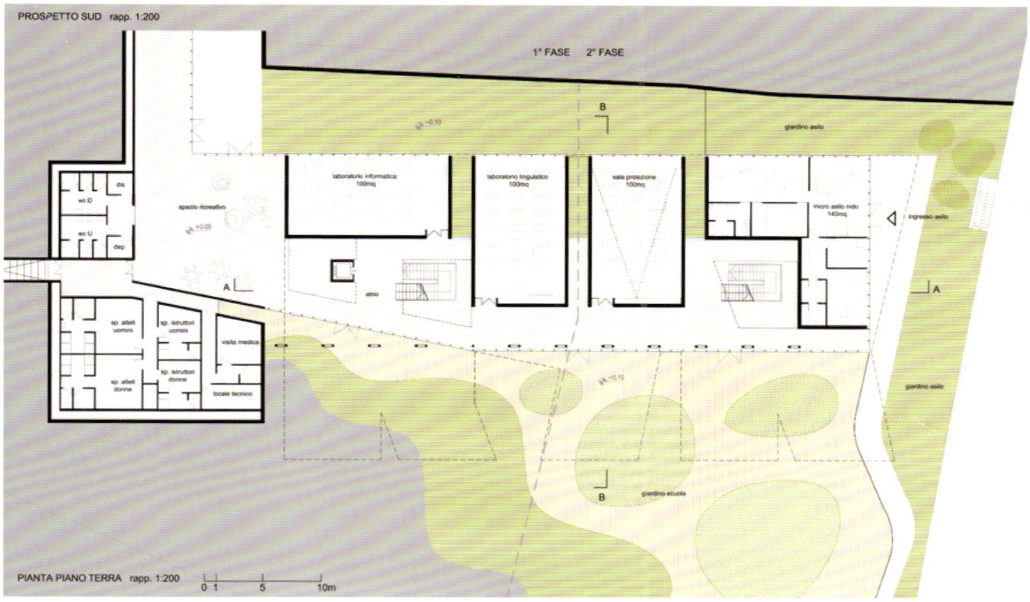
PROSPETTO SUD rapp. 1:200
1° FASE 2° FASE
giardino asilo
laboratorio informatica
100mq
laboratorio linguistico
100mq
sala proiezione
100mq
micro asilo nido
140mq
ingresso asilo
spazio ricreativo
+2.00
-0.10
dis
wc D
wc U
dep
atrio
sp. atleti uomini
sp. istruttori uomini
visita medica
sp. atleti donne
sp. istruttori donne
locale tecnico
giardino scuola
A
B
PIANTA PIANO TERRA rapp. 1:200
0 1 5 10m

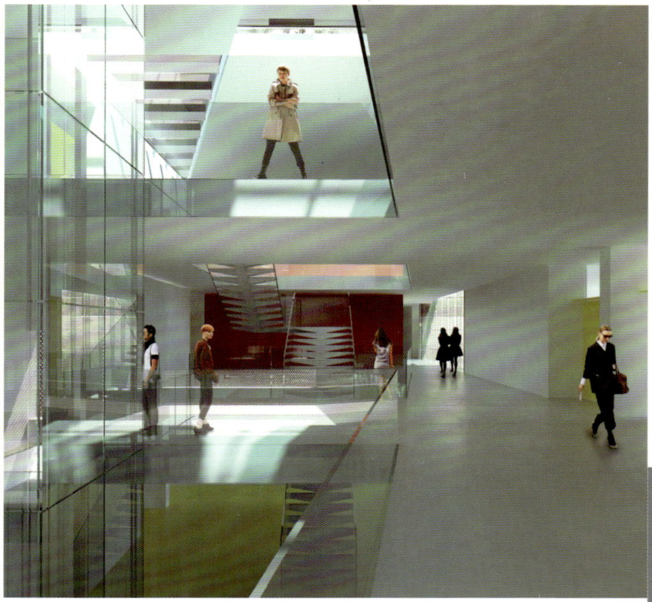

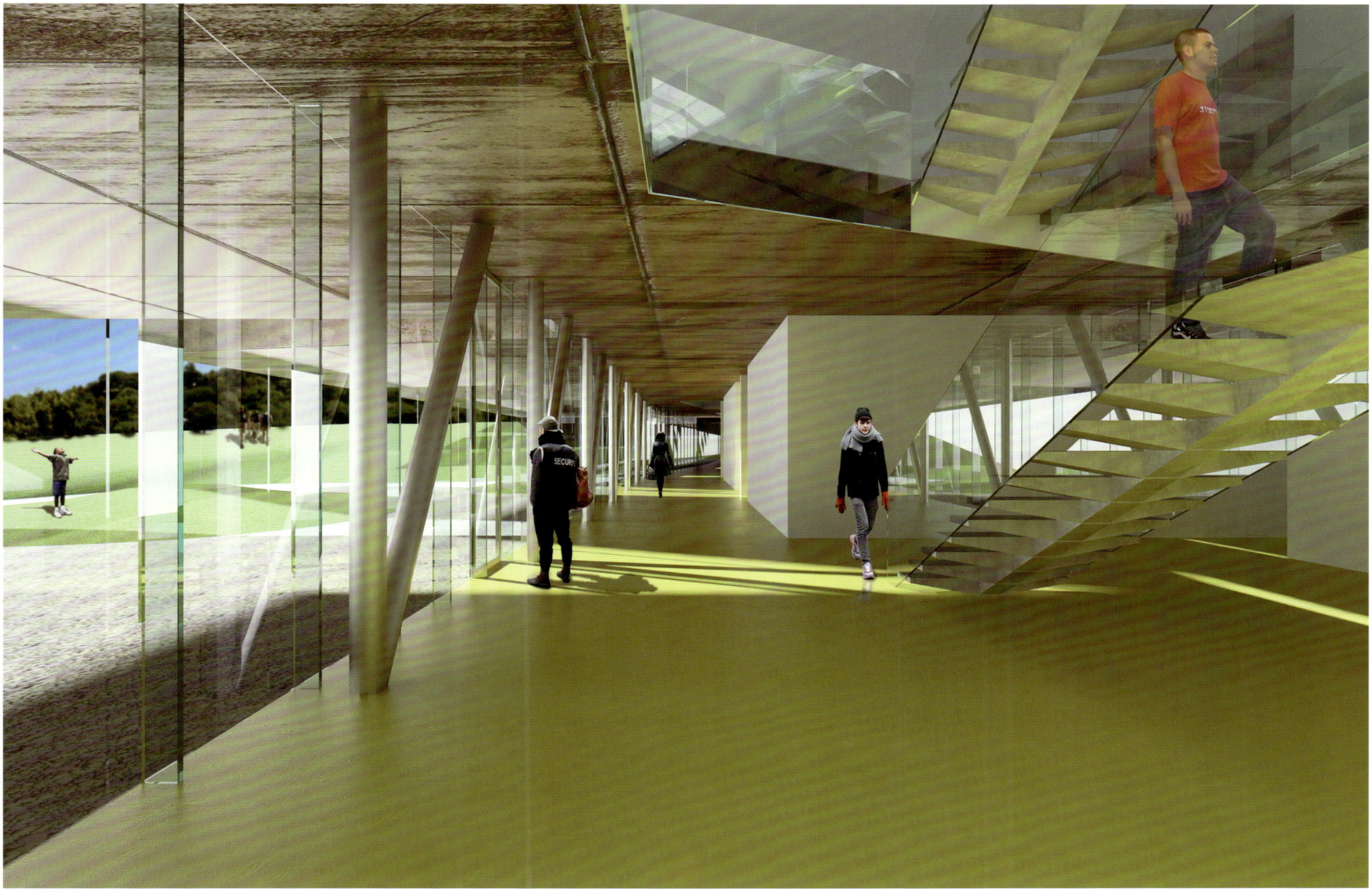

Putrajaya Waterfront Residential Development

普特拉贾亚滨水区住宅开发

Architect: Luca F. Nicoletti, Serina Hijjas
Firm: Studio Nicoletti Associati, Hijjas Kasturi sdn
Location: Kuala Lumpur, Malaysia
Area: 280,000m²

The key of the overall planning for the waterfront is the relationship to the waterfront or lakefront. Whilst the buildings on the boulevard predominantly reinforce the alignment of the boulevard, the waterfront planning should tie the boulevard back to the waterfront. The current masterplan creates large building block footprints that remain heavy a reminiscent of old American models. The conceptual approach is relative to context and tropical climate. Three factors influence the sketch planning approach: allow permeable building blocks, smaller block sizes and radiate the buildings to enhance the visual corridors and links between the boulevard and the lakefront; varying the building height of ten-storey restrictions heightening at nodal points for massing hierarchy; unify the buildings by introducing a canopy roof uniquely tropical.

The new urban plan places integration with landscaping and view as priorities. It encourages creating "fingers" of developments towards the water body and infused with park, creating a series of green forecourts.

The urban plan suggests the building orientation and massing provide maximum view toward the water and development across the lake. The massing is broken into smaller components with different heights. This approach will ensure the integration of landscape and building massing. The landscape and public areas will be larger and located in between the building fingers, which will create an interesting "journey" towards the waterfront.

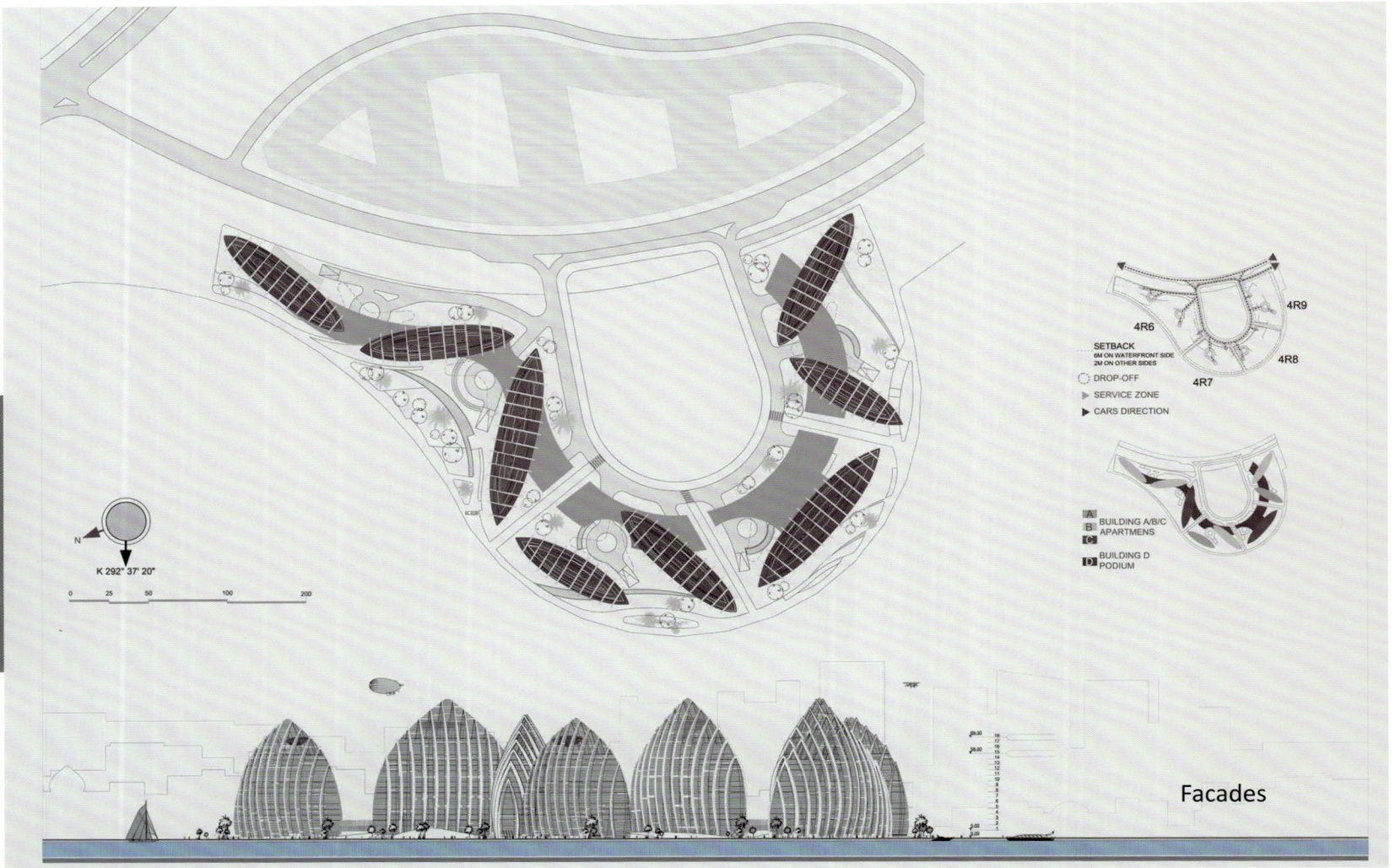

SETBACK
6M ON WATERFRONT SIDE
2M ON OTHER SIDES

○ DROP-OFF
▶ SERVICE ZONE
▶ CARS DIRECTION

4R6 4R9
4R7 4R8

A
B
C
BUILDING A/B/C
APARTMENS

D BUILDING D
PODIUM

0 25 50 100 200

N

Facades

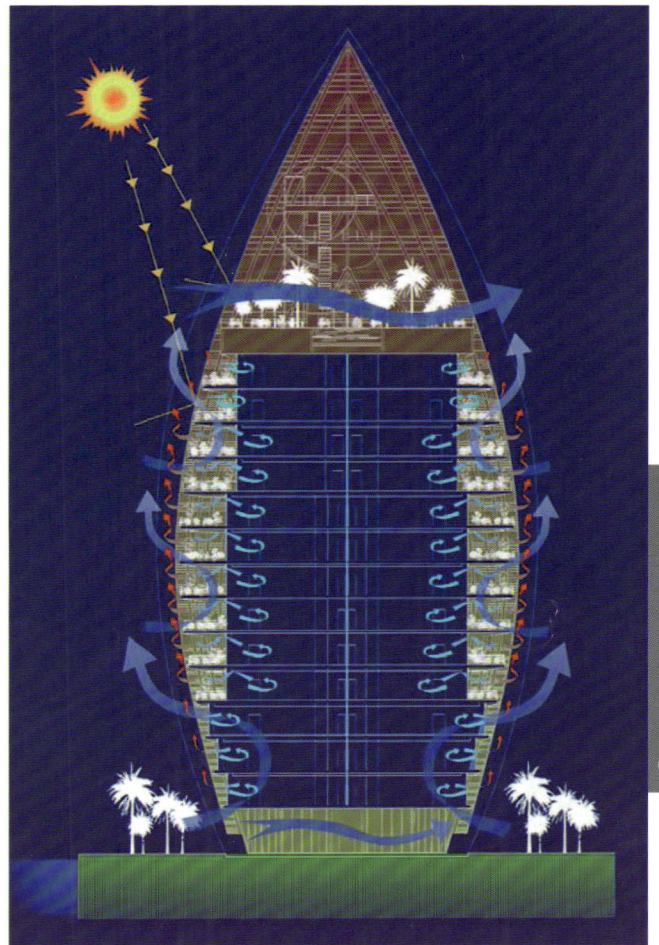

Sustainability

整个滨水区规划设计的关键
于与滨水区或者说是湖畔区的关
。尽管大道两侧的建筑加强了大
整齐的感觉，但是滨水区规划应
将大道联结回水岸边。目前的总
规划创造了大型的建筑组件的占
面积，保持了浓厚的传统美式建
模式回忆。设计的概念方法与周
环境和热带气候相关。三个主要
素影响了规划草图：允许建筑构
的穿插，减小建筑的面积并且将
筑发散式安排以加强视觉走廊和
道与滨水区的联系；改变建筑十
高的限制，提高节点的集结层
；利用热带独有的屋顶天篷统一
些建筑。

这个新的城市规划将景观与
野作为优先整合的对象。它鼓励
向水体开发建造，并与公园相融
，创造出建筑前一系列的绿色前
。

城市规划建议了建筑的朝
，并且提供了朝向水景的最大化
视野和横跨湖面的开发建造。整
规划部分被分割成不同高度的更
的组成部分。这种方法将保证景
与建筑层面的融合。景观与公共
域将会变得更大，而且在建筑之
，这将会创造出朝向湖水出发的
趣的"旅途"。

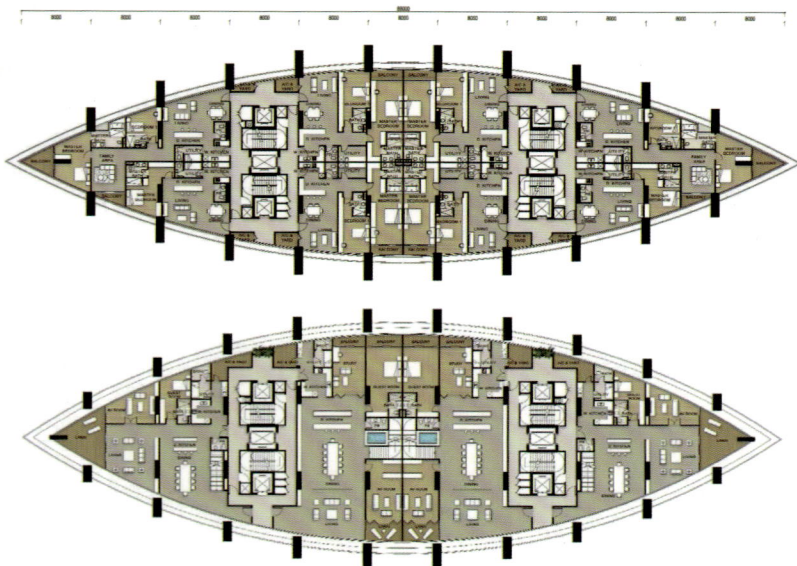

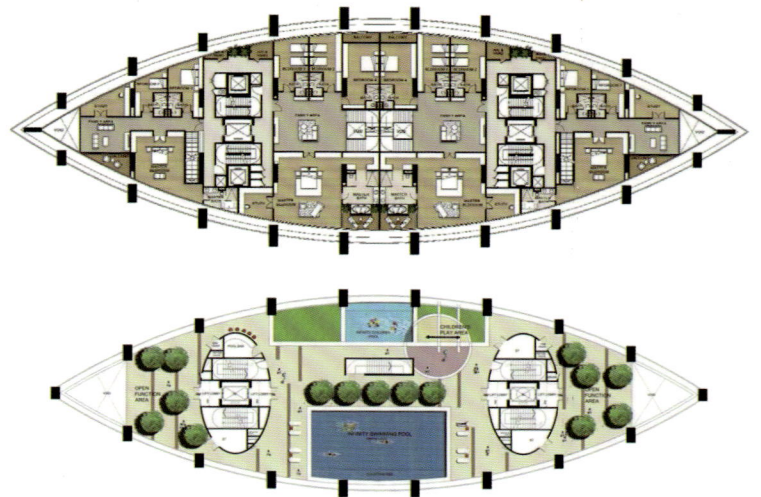

Edil Tomarchio Commercial Park

Edil Tomarchio商业公园

Firm: modostudio | cibinel laurenti martocchia architetti associati
Location: Aci Sant'Antonio, Italy
Area: 14,950m²

It's difficult to deal with an area so rich in history, traditions, and an area extremely fascinating, between Etna and the sea. An area in which recent years has had a chaotic development.

The intervention will propose a recovery of the surrounding agricultural tissue and will create a pedestrian and cycling connecting the center of the village with the new intervention which will host a big underground parking. The parking, in addition to serving customers company, will be used as parking public by improving the livability of the age which will be partially pedonalized.

The circular volumes want to propose the architecture very recognizable, but linked to the contemporary context. If in fact deliberately circular forms identify a representative image of the intervention, the coating materials are stone, local lava for the floor and the facade, a warehouse and two other buildings used sandstone. All the materials rooted in traditional local building.

The intervention is based on a clear division of functions with the proposed volumes. The bigger and north part located circular building hosts a large parking lot in its basement, while the ground floor hosts the main store; th commercial building hosts the administration and sale area.

The southern building, close to the roundabout and connected with a canop to the commercial building hosts the cafeteria, a conference hall, the librar the gym and the guest rooms. Such clear division of functions allows th possibility of using the buildings individually, saving energy consumption an avoiding overlapping functions without public and private interference.

A great importance is given to the proposition of solutions which ensur natural construction materials, low power consumption and minimum energ supplies.

The local stone and its particular implementation through perforated pane will provide a shading natural casting for the for sale and services building which will use low emission glass.

The covers will host an installation of solar thermal and photovoltaic system and will facilitate the recovery of rainwater that will be used for sanita purposes and irrigation.

Percorsi ciclabili e pedonali
Viabilità autostrada
Viabilità principale
Viabilità secondaria
Viabilità di servizio

Recupero del tessuto agricolo

Tessuto urbano

Visuali

PIAZZA PUBLICA

AREA GIOCO BAMBINI

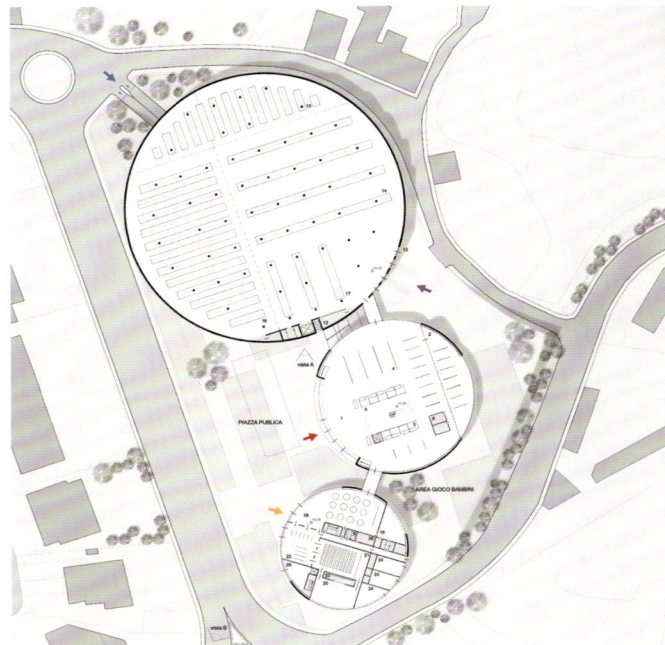

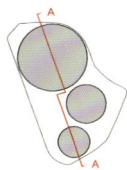

SEZIONE LONGITUDINALE A-A 1:250

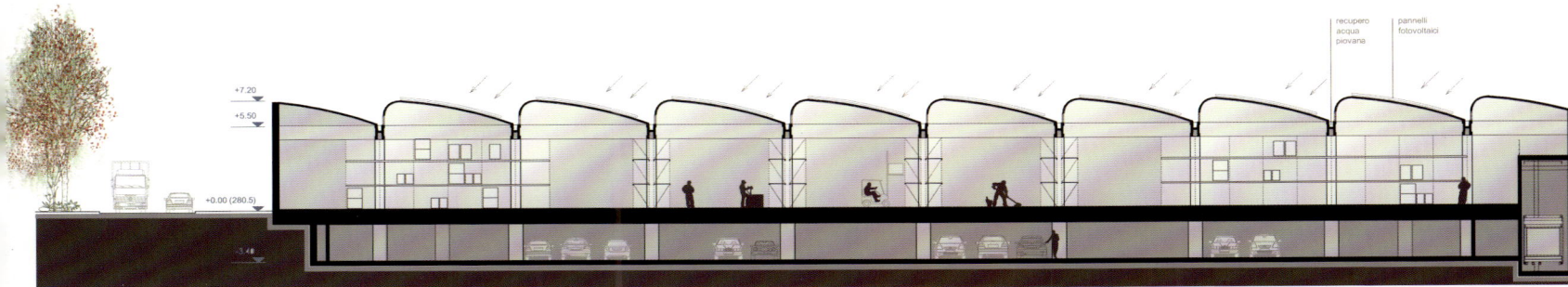

recupero acqua piovana   pannelli fotovoltaici

+7.20
+5.50
+0.00 (280.5)
-3.4

EDIFICIO DEPOSITO MERCI / PARCHEGGIO

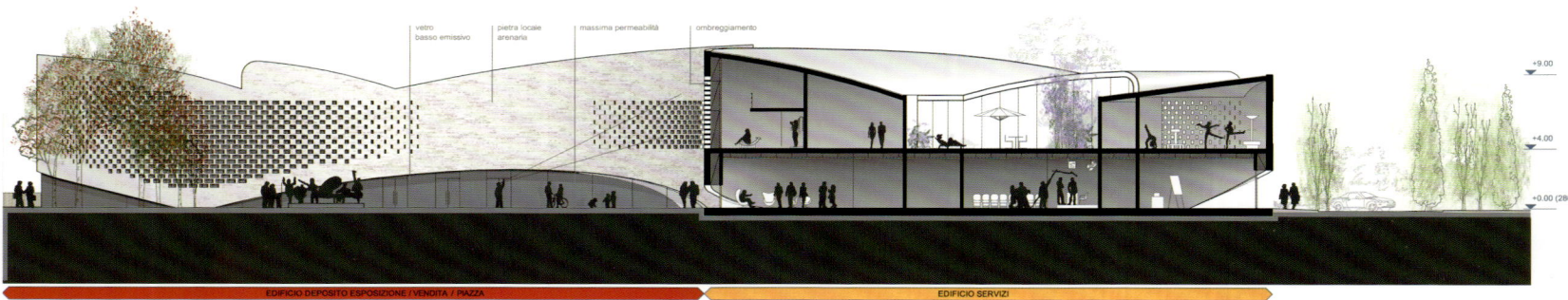

vetro
basso emissivo   pietra locale
arenaria   massima permeabilità   ombreggiamento

+9.00

+4.00

+0.00 (280.5)

EDIFICIO DEPOSITO ESPOSIZIONE / VENDITA / PIAZZA   EDIFICIO SERVIZI

ACCESSI PEDONALI
■ Percorsi pedonali

ACCESSI CARRABILI
■ Percorso auto
■ Percorso merci

FASI D'INTERVENTO
○ Demolizioni
○ Nuove costruzioni

FASE 1: PARCHEGGIO E DEPOSITO

FASE 2: AREA ESPOSITIVA E VENDITA

FASE 3: SERVIZI E SPAZI ALL'APERTO

处理一个有着悠久历史和丰富传统的地区是很困难
，而且这里景色还非常的迷人，位于埃特纳和大海之
。这个地区在最近几年的发展有些混乱。

新项目将带来周围农业组织的复原，并会创建出行
区和自行车区，将即将修建大型地下停车场的新项目与
庄中心连接起来。除了为客户的公司服务之外，停车场
将作为公共停车场使用，将会改善这里未来的适居性。

环形的体量希望能使这个建筑具有高度的可辨识
，但是仍能与周围环境相融合。如果实际上这个环形结
确定了新项目的代表形象，建筑的外层使用的是石材，
仓库的铺地和外立面使用当地火山岩，另外两座建筑使
的是砂岩，都是传统当地建筑中常见的材料。

新的项目建立在一个功能区分明确的拟建的体量
。位于地基上稍大一些的北向的圆形建筑内是大型的停
场，一层是主要商业店铺，商业建筑内是行政中心和销
区域。

靠近环岛南部的建筑与商业建筑外的伞篷相连接，
商业建筑内包括有一个餐厅、一个会议大厅、图书馆、
身馆和客房。这样清晰的功能区分使建筑的每一个部分都可以单独使用，能够节约能源消耗并避免公共和
人区域之间功能上的重复。

设计非常重视设计的解决方案，以保证使用天然的建筑材料、低电力消耗和最小的能源供应。当地的石
，特别是作为穿孔板上的补充材料，为使用低辐射玻璃的销售和服务建筑提供了天然的阴凉。

建筑的外立面将安装太阳能热能和光伏系统，还将安装雨水收集装置，收集到的雨水可用作卫生用途和
溉。

DETTAGLIO SEZIONE VERTICALE
EDIFICIO MOSTRA / VENDITE   1:50

1_ Carter in lamiera metallica.
2_ Pannello prefabbricato in cls con isolamento termico.
3_ Pannello composto da blocchi in pietra arenaria locale.
4_ Controsoffitto in cartongesso.
6_ Pavimento galleggiante con finitura in gomma naturale.
7_ Vano per posizionamento impianti e luci.
8_ Pilastro con finitura naturale.
9_ Serramento a taglio termico con doppio vetro.
10_ Facciata con profilo a taglio termico e vetro basso emissivo.
11_ Pavimento in cls con finitura in resina.
12_ Pannello di copertura sandwich con doppia coibenza
    acustica e termica

0   1.00m

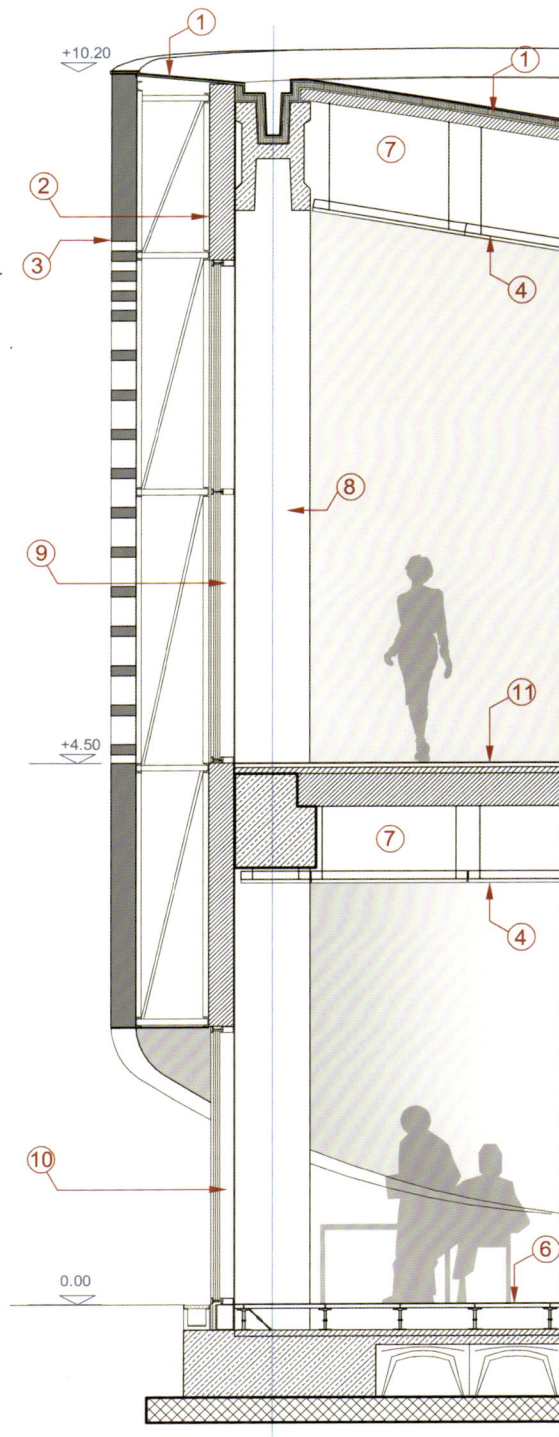

Ex Fonderie Riunite

Ex Fonderie Riunite新规划

Architect: MODOSTUDIO - Cibinel Laurenti Martocchia architetti associati + CCDP + Studio Cattinari
Location: Modena, Italy
Area: Faculty of Design-DAST and Services 23,000m²,
        Commercial and Offices 2,000m²,
        Residential 8,000m²

The competition brief was the regeneration of an urban area called Ex Fonderie Riunite, a strategic area for the city of Modena with important historical value. The basic idea was the recovery of a former industrial building and a new detailed urban plan in the other areas of the competition. The proposal for this existing industrial complex, characterized by historically protected office buildings, was to keep the identity as a symbol for the history of the Italian workers movement. The regeneration includes new uses and functions and the new spaces of the DAST faculty (Design, Art, Science and Technology).

It proposed to create buildings of different heights with plans that follow the structure of the industrial building and preserve the existing fascinating spaces of the old plant. These new volumes will be detached from the ground and they will be covered, on the north and south facades with a golden perforated metal sheet. This particular material is the same that will characterize the exterior skin of the roof throughout the rest of the building.

The golden color of this new roof, the harshness of the metal material and at the same time, the softness of the perforated pattern will be able to evoke the past; a metal smelting complex building full of history, and at the same time to communicate unambiguously the birth of the new feature.

The cladding will be seen from the railroad and from an adjacent bridge that connects the two parts of the city and will become a symbol for the city of Modena. The cultural destination of DAST, such as the School of Design, will give a new feature to the building as a vital place pushing research and innovation. The complex will also contain other features such as commercial areas and services in support of the DAST, creating an important centre for the city, available throughout the day.

The new urban design provides maximum accessibility for new functions with the creation of a square outside the former building, which will become a commercial, residential and public park.

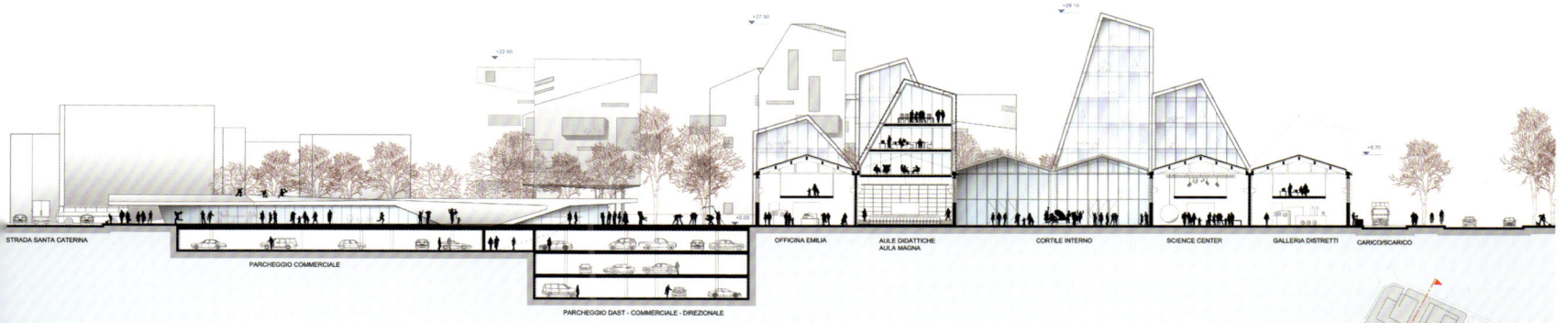

STRADA SANTA CATERINA   OFFICINA EMILIA   AULE DIDATTICHE   CORTILE INTERNO   SCIENCE CENTER   GALLERIA DISTRETTI   CARICO/SCARICO

PARCHEGGIO COMMERCIALE   AULA MAGNA

PARCHEGGIO DAST - COMMERCIALE - DIREZIONALE

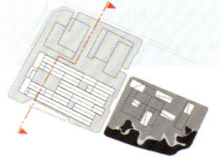

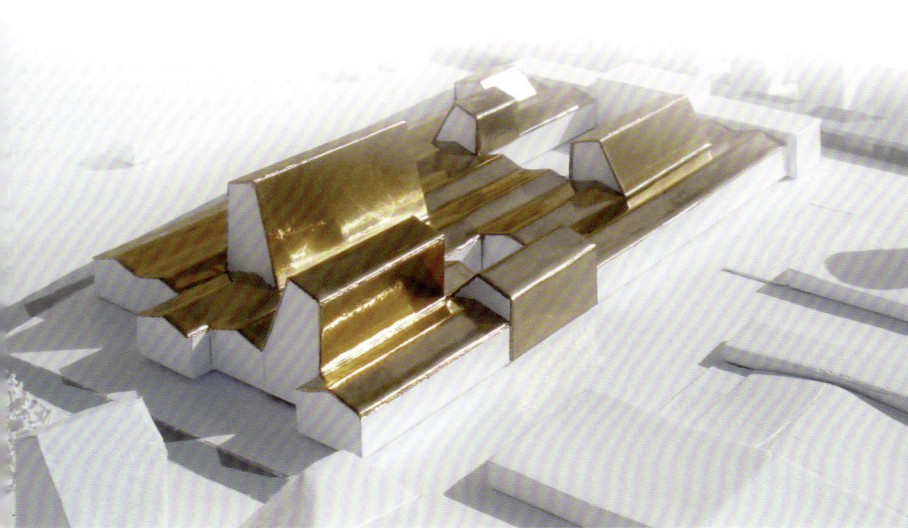

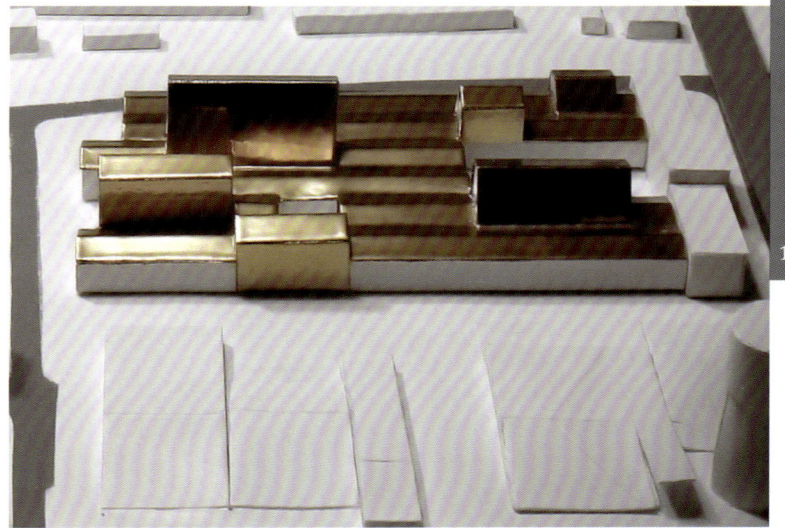

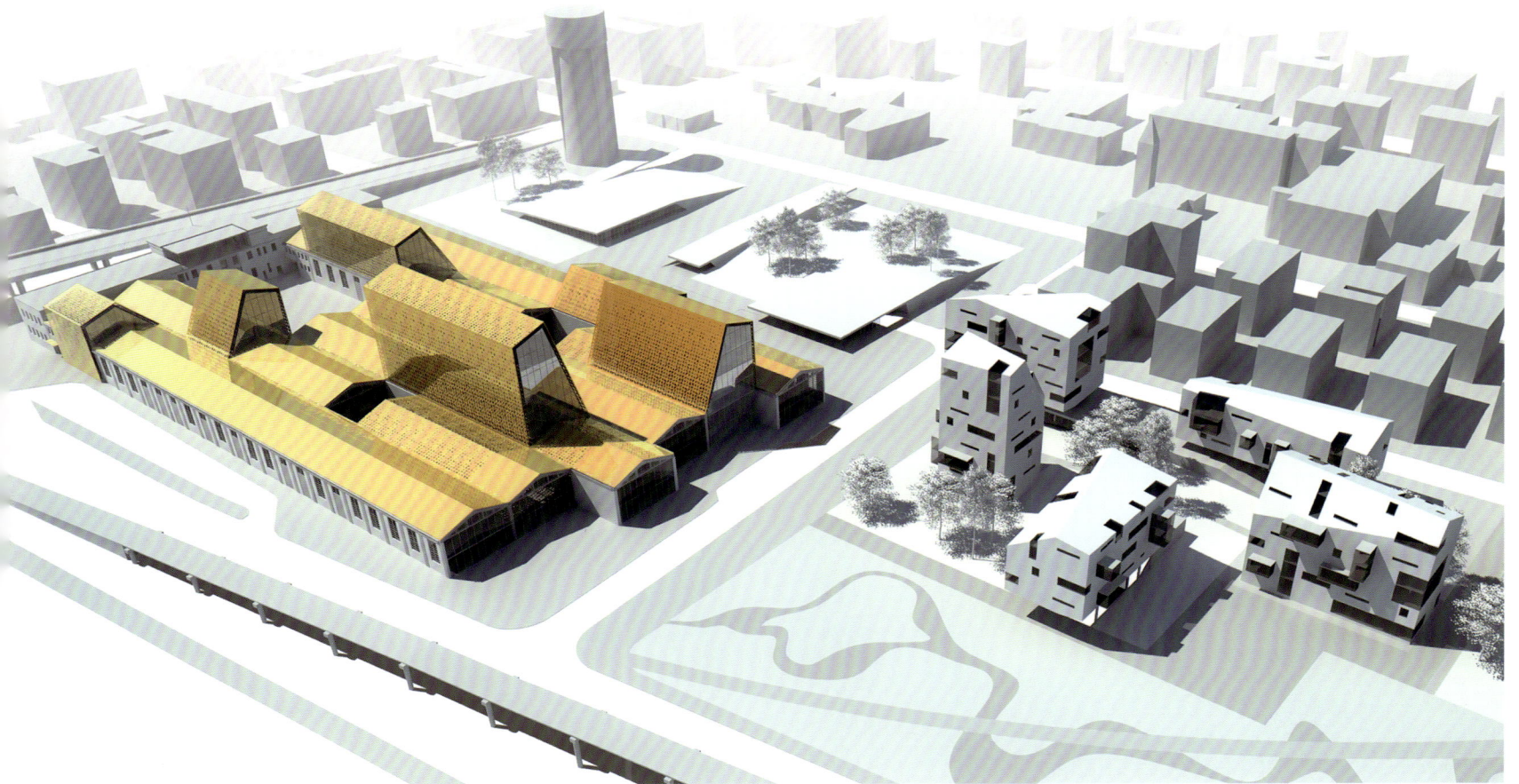

schema delle funzioni

RESIDENZIALE:

A   6 piani + p.t.
B   6 piani + p.t.
C   3 piani + p.t.
D   4 piani + p.t.
E   5 piani + p.t.
F   8 piani + p.t.

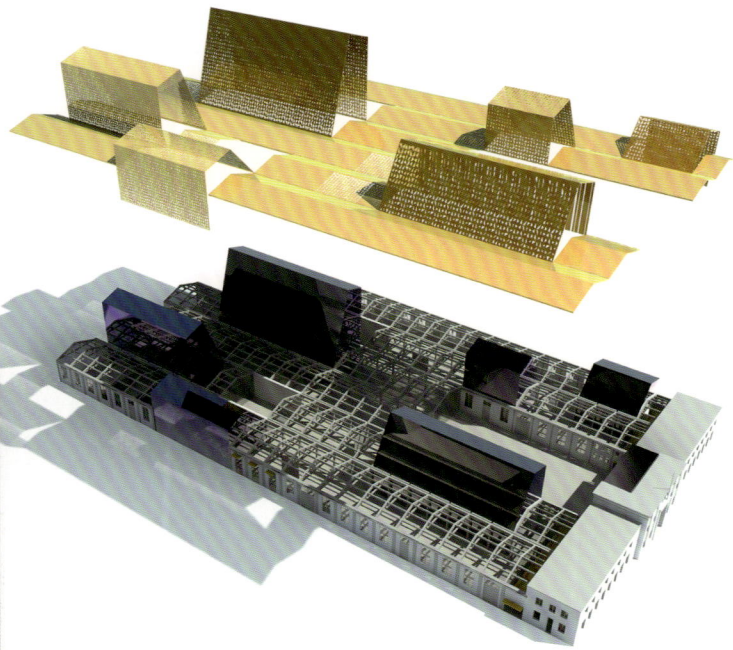

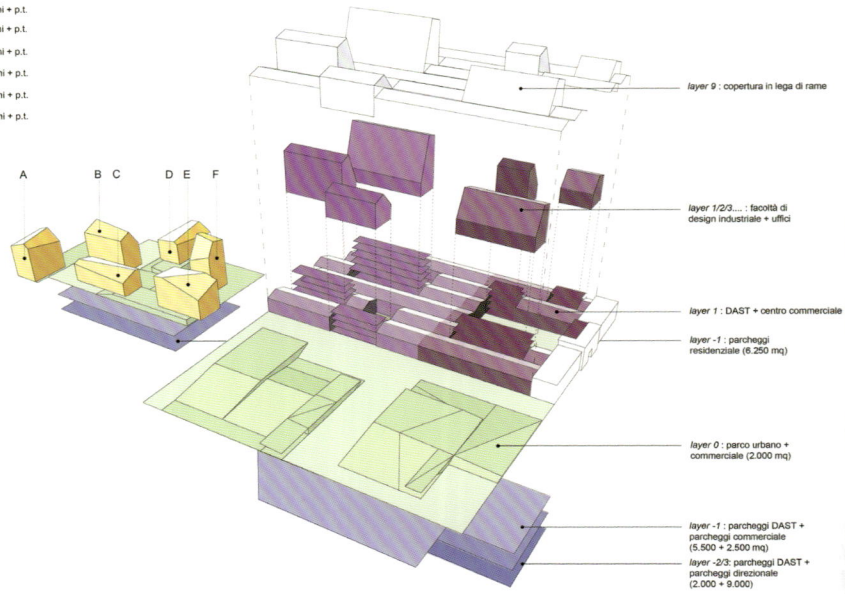

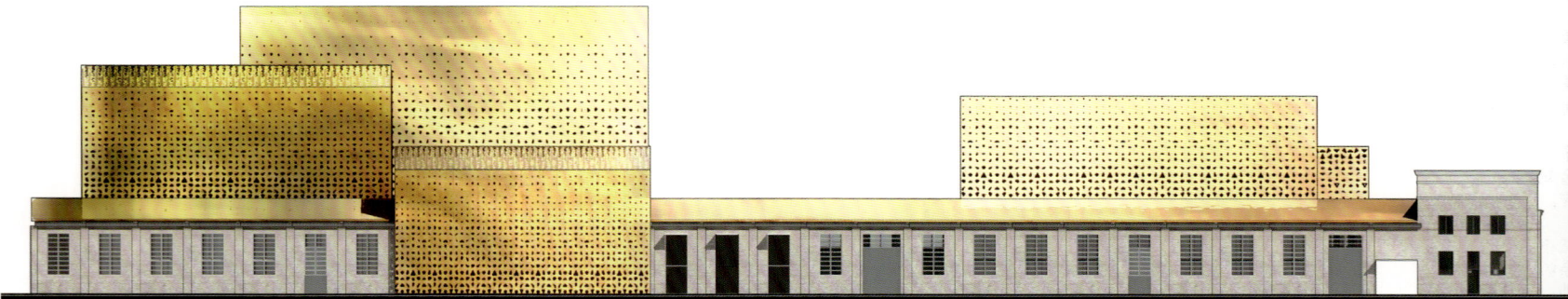

layer 9 : copertura in lega di rame

layer 1/2/3.... : facoltà di design industriale + uffici

layer 1 : DAST + centro commerciale

layer -1 : parcheggi residenziale (6.250 mq)

layer 0 : parco urbano + commerciale (2.000 mq)

layer -1 : parcheggi DAST + parcheggi commerciale (5.500 + 2.500 mq)

layer -2/3 : parcheggi DAST + parcheggi direzionale (2.000 + 9.000)

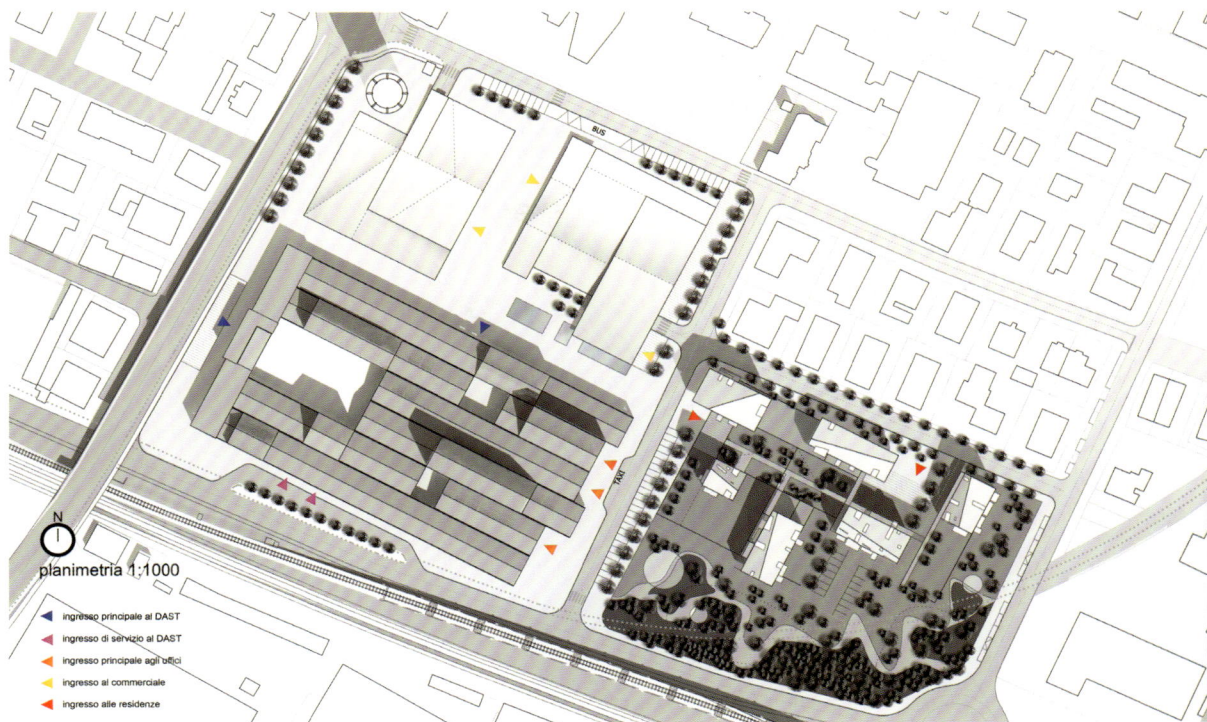

planimetria 1:1000

▶ ingresso principale al DAST
▶ ingresso di servizio al DAST
▶ ingresso principale agli uffici
▶ ingresso al commerciale
▶ ingresso alle residenze

比赛的目的是将一个名为Ex Fonderie Riunite 的市区重新规划，这里是摩德纳市带有重要历史 价值的战略要地。设计的基本构思是将一个工业 建筑重新规划，并在竞赛中的其他地区做一个 详细的城市规划设计。以历史性的办公建筑为特 征，设计的提案保持了现存工业建筑，作为意大 利工人运动这段历史的标志性。建筑的再生包括 了新的使用和功能以及新的设计、艺术、科学技 术空间。

建筑师们计划依据工业建筑的构造建造不同 高度的建筑，保持这一地块原有的迷人空间。这 些新的体量将脱离地面，其北向和南向的外立面 将会由金色的穿孔金属片覆盖。这种特别的材料 将会使建筑的屋顶像其他建筑一样富有特点。

新屋顶的金色、金属材料的冷酷感和穿孔 金属片的柔和感将会唤起过去，这座充满历史感的 金属冶炼大楼，同时与新的未来作出了明确的呼 应。

建筑的覆层能够从铁路上及一架相邻的连接 城市两个部分的桥梁上看到，将会成为摩德纳市 新的标志。这个设计、艺术、科学技术文化目的 地，比如设计学校等，将会给这座建筑带来一个 新的未来，作为推动试验和创新的重要场地。这 栋大楼还会涵盖商业区，设计、艺术、科学技术 的服务区，为城市创建了一个新的重要中心，全 天提供服务。

新的城市规划最大限度地提供了新功能的可 接近性，包括在原有建筑外建造了一个广场，将 会规划为商业、住宅和公共公园用地。

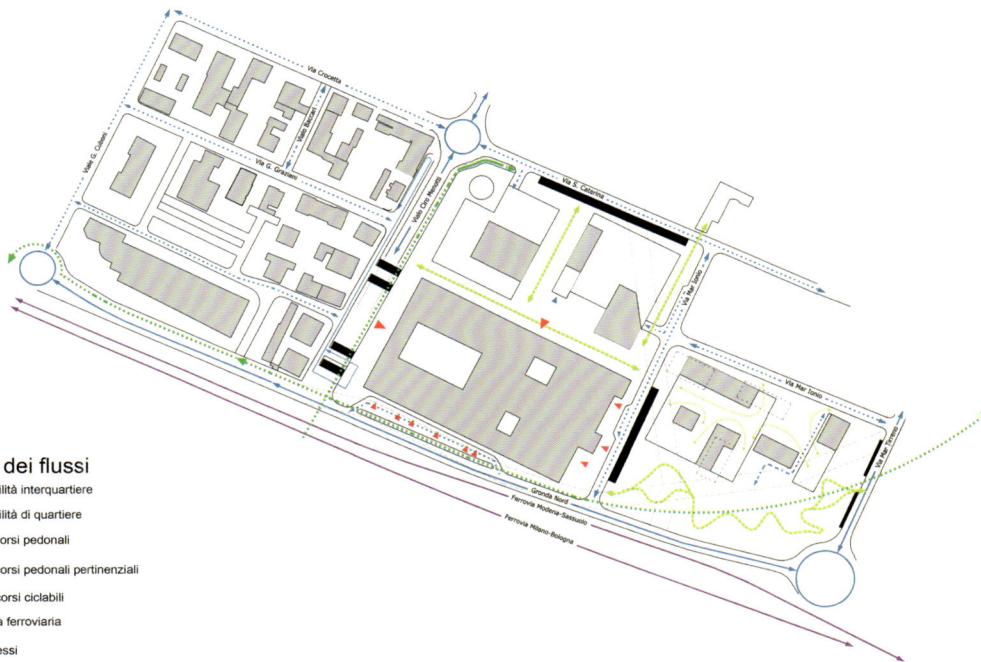

schema dei flussi
— viabilità interquartiere
---- viabilità di quartiere
••••• percorsi pedonali
••••• percorsi pedonali pertinenziali
••••• percorsi ciclabili
— linea ferroviaria
◀ accessi
■ parcheggi a raso

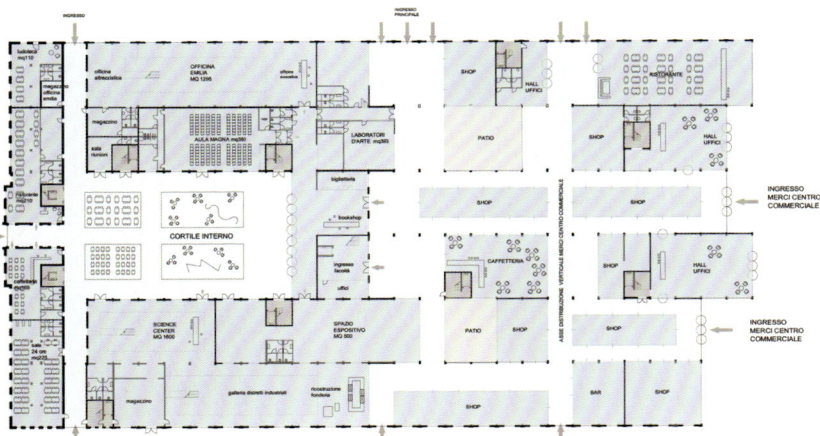

pianta piano terra 1:500

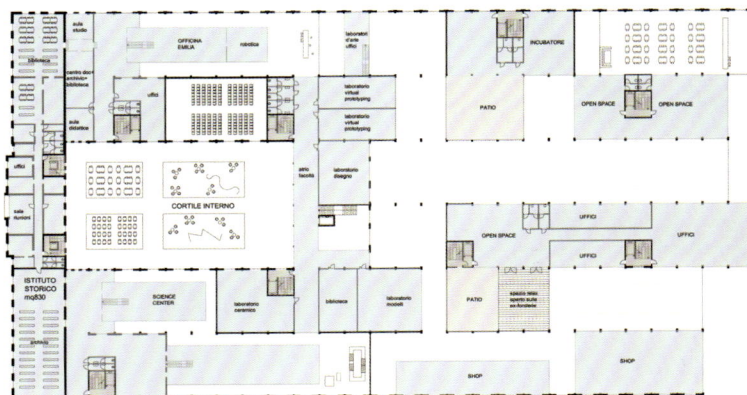

pianta piano primo 1:500

Seoul Performing Arts Center - Nodeul Island

首尔表演艺术中心

Firm: Zerafa Architecture Studio
Design Team: Jason M Zerafa, Scott Springer AIA, Hughy Dharmayoga, Luis Carmona, Fritz Johnson AIA
Location: Nodeul Island, Seoul, Korea

The design for the Seoul Performing Arts Center was generated through careful consideration of several important factors: its prominent island site in Seoul and the island's history of flooding, its role in Korean and world cultures, its complex program, and its potential to act as an iconic symbol of Korea's growing global influence.

Nodeul Island, located in the Hangang River, is suspended between the cultural realities of modern Seoul to the south and the old city to the north. The island can be considered the geographical center of Seoul, the place where traditional Korean culture meets the modern aspirations of contemporary Korean culture. Seoul is bisected by the flow of the Hangang River and encircled by an inner ring and an outer ring of granite mountains. These mountains provide a continuous backdrop for views from within the city. The island's history of flooding was viewed as an important indication of nature's great power, one that should be celebrated in the project as evidence of man's eternal and sublime quest to live in harmony with the natural environment.

Early in the design process, the team considered the nature of the twenty-first century opera house. Although the facility will be used for a variety of international music events, the architects determined that it was critic to get a better understanding of some the indigenous music likely to performed within. This led them to study the traditional Korean Panso opera, of which only five of the original twelve survive. As the etymolo of the word Pansori is based on the terms pan, which means place performance, and sori, which means sound, they felt that this was a appropriate point of departure for a large complex dedicated to the art musical performance.

Unlike Western opera, Pansori is performed not by a large ensemble musicians, but by a solo singer accompanied by a drummer. In addition providing drum beats, the drummer also provides chuimsae, or verbal soun of encouragement. The direct interaction between the singer (sorikkun) a drummer (gosu) suggested itself to be an analogy appropriate to a proje with two major programmatic spaces, an Opera House and a Concert Ha Given the importance of the drum in Pansori, they used the drum form as basis for the initial formal investigations and allowed the complex program manipulate these forms into a project that they believe can serve as a pote symbol of Korea's traditions and global aspirations.

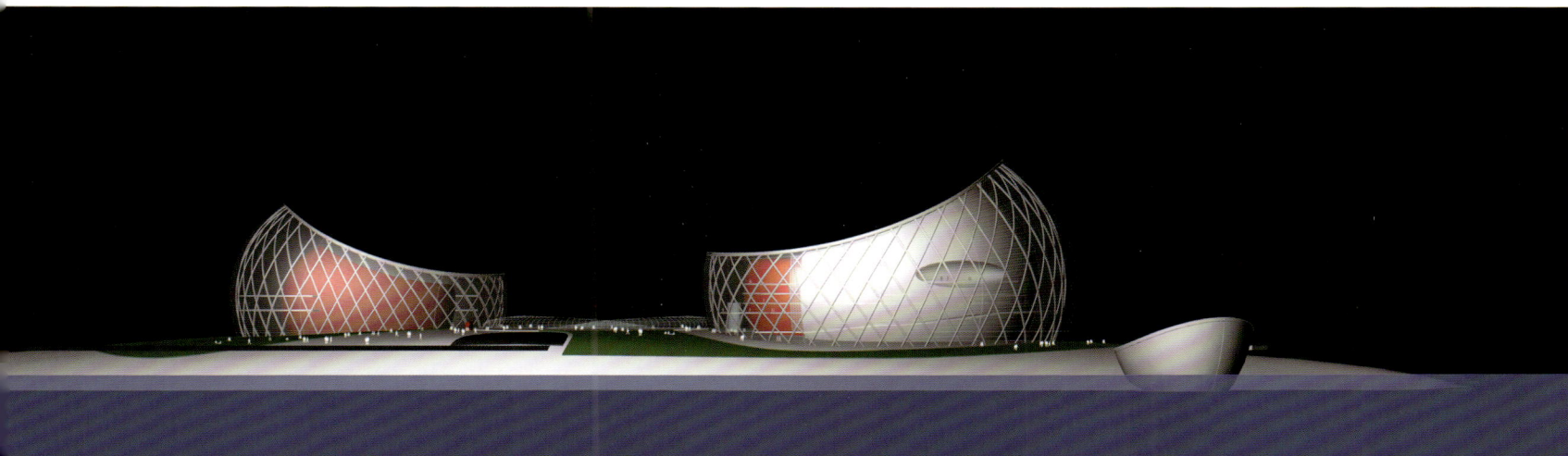

SOUTH ELEVATION

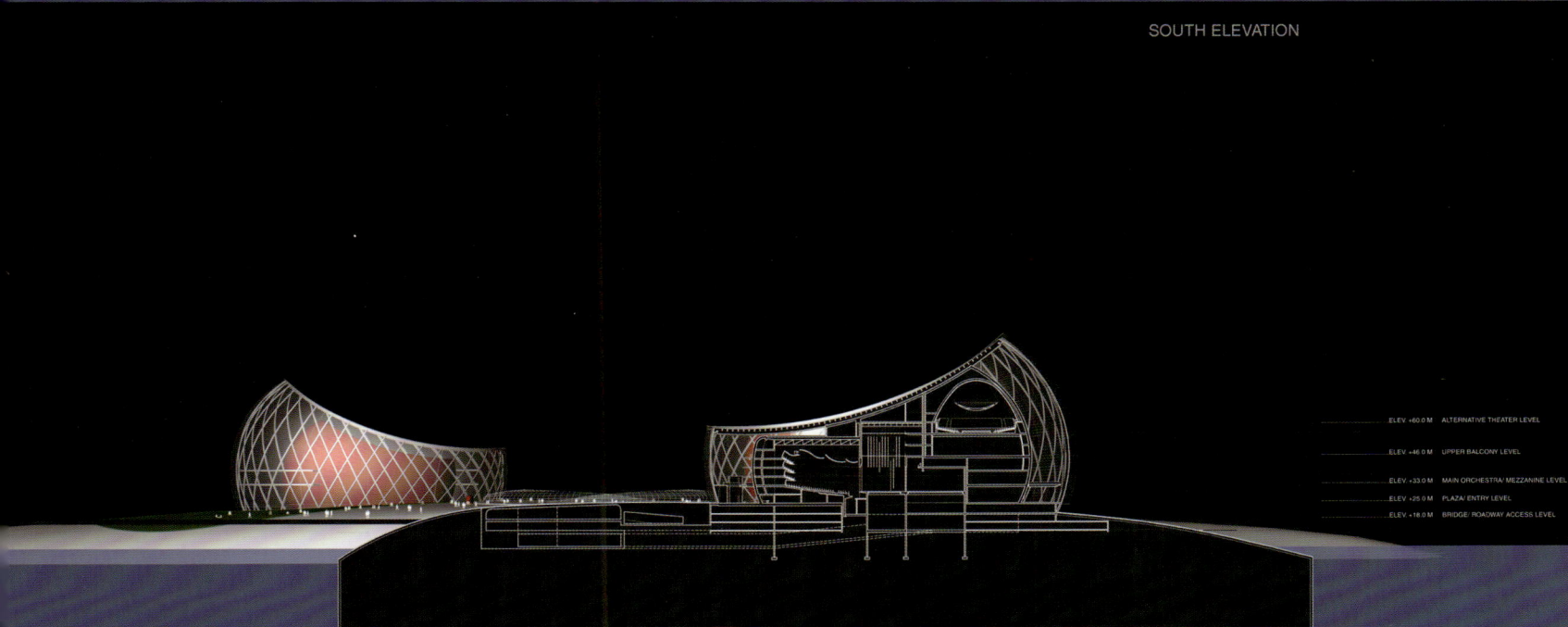

ELEV +60.0 M   ALTERNATIVE THEATER LEVEL

ELEV +46.0 M   UPPER BALCONY LEVEL

ELEV +33.0 M   MAIN ORCHESTRA/ MEZZANINE LEVEL
ELEV +25.0 M   PLAZA/ ENTRY LEVEL
ELEV +18.0 M   BRIDGE/ ROADWAY ACCESS LEVEL

OPERA HOUSE SECTION LOOKING NORTH

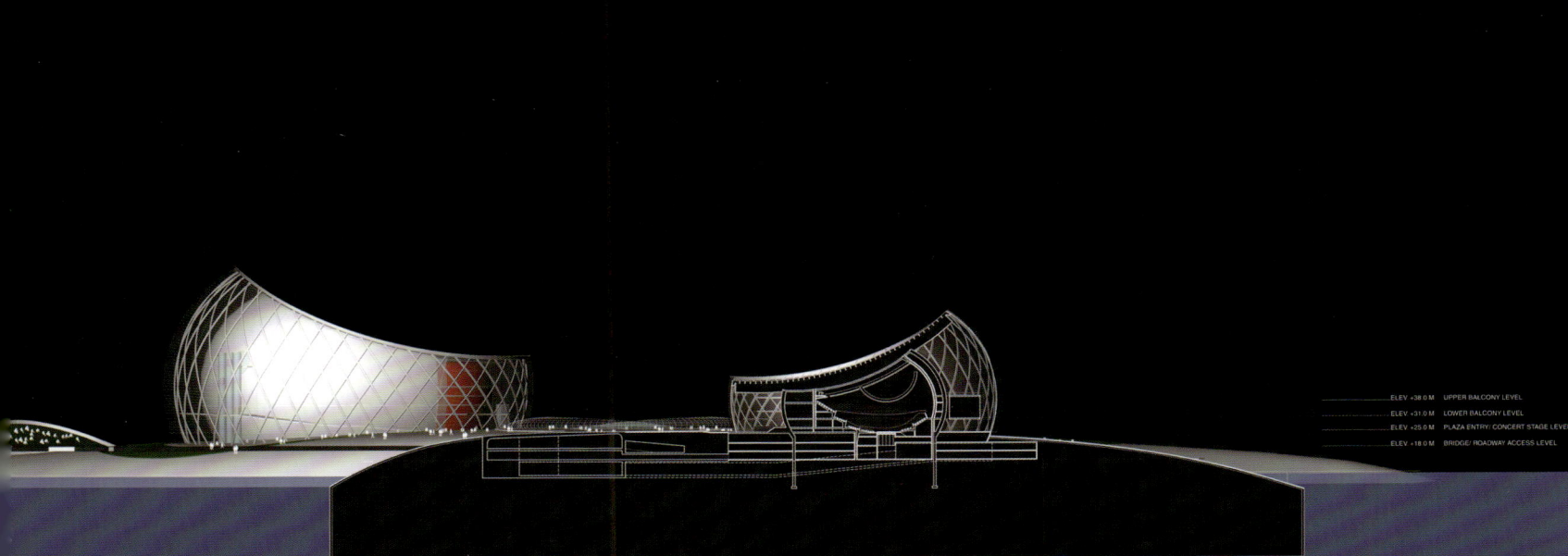

ELEV +38.0 M   UPPER BALCONY LEVEL
ELEV +31.0 M   LOWER BALCONY LEVEL
ELEV +25.0 M   PLAZA ENTRY/ CONCERT STAGE LEVEL
ELEV +18.0 M   BRIDGE/ ROADWAY ACCESS LEVEL

CONCERT HALL SECTION LOOKING SOUTH

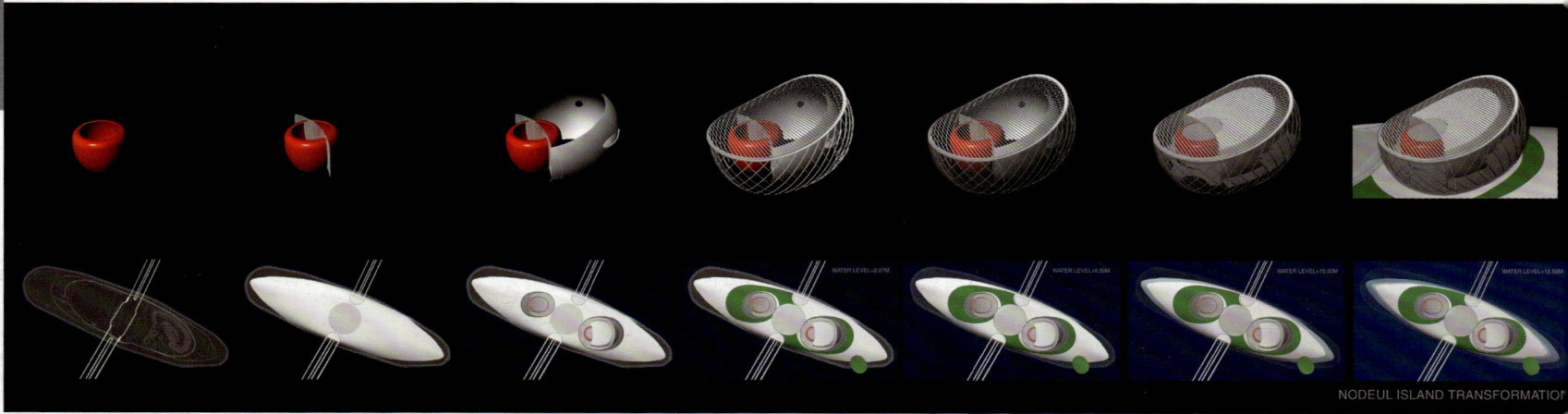

NODEUL ISLAND TRANSFORMATION

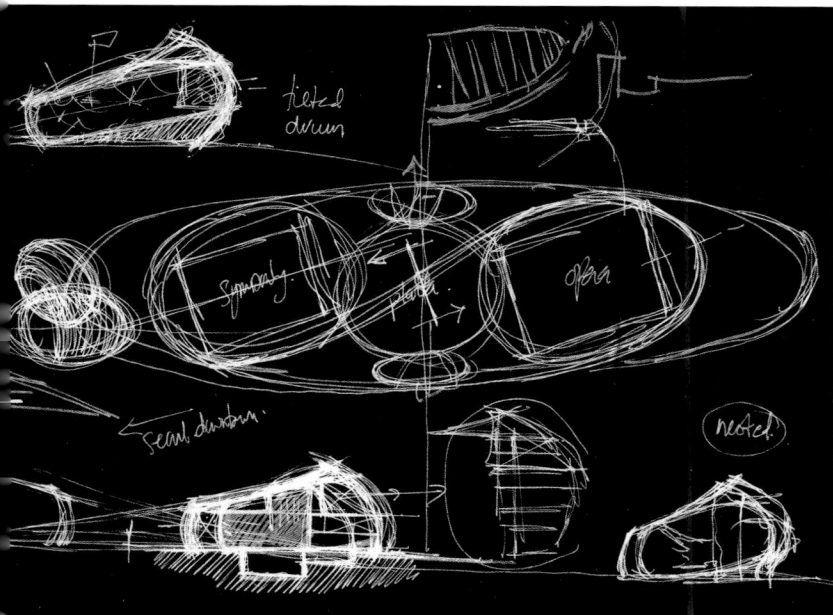

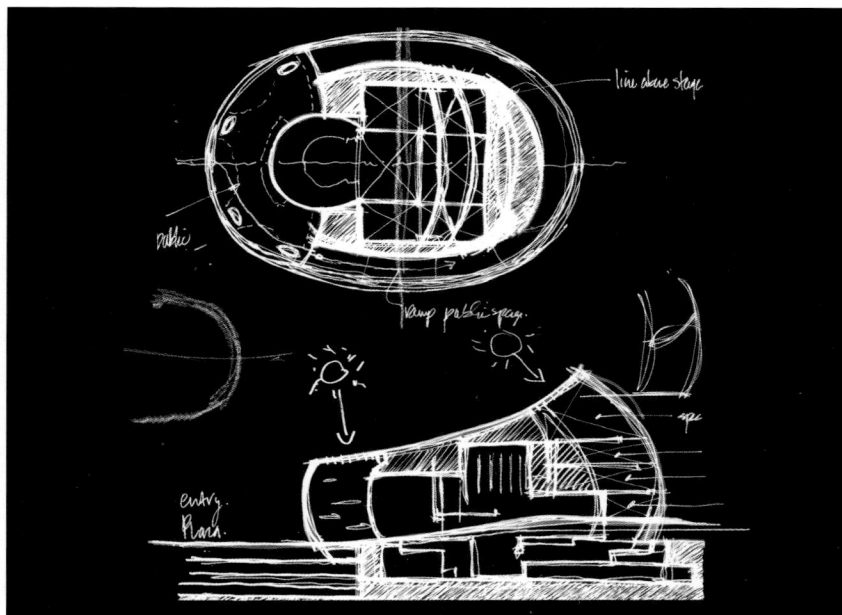

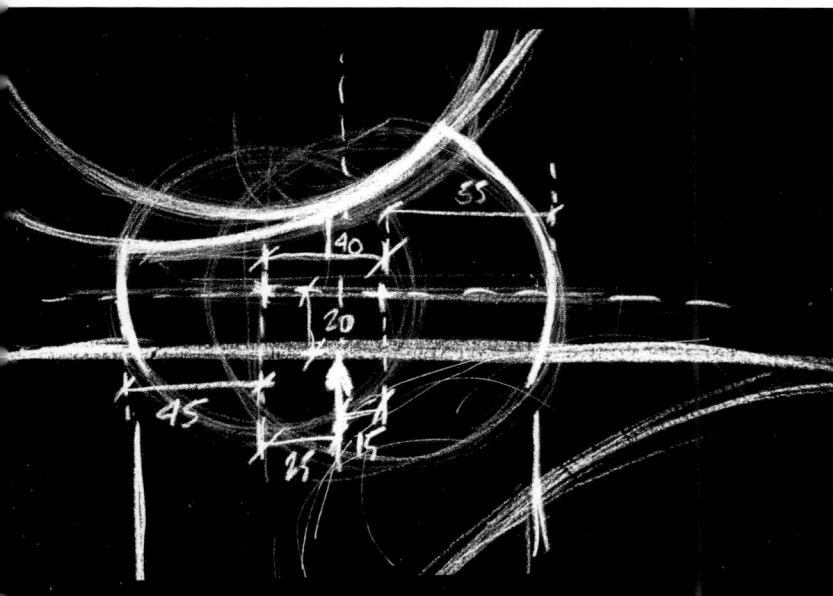

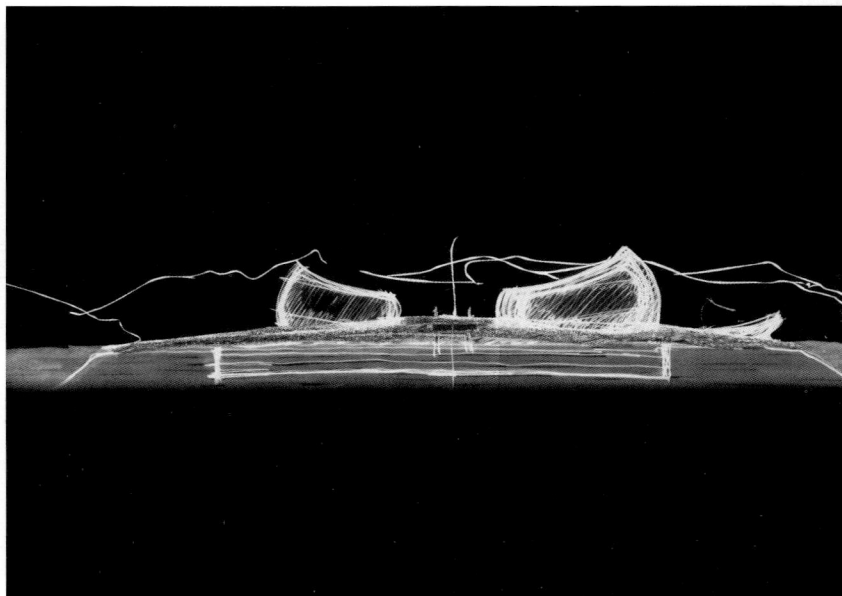

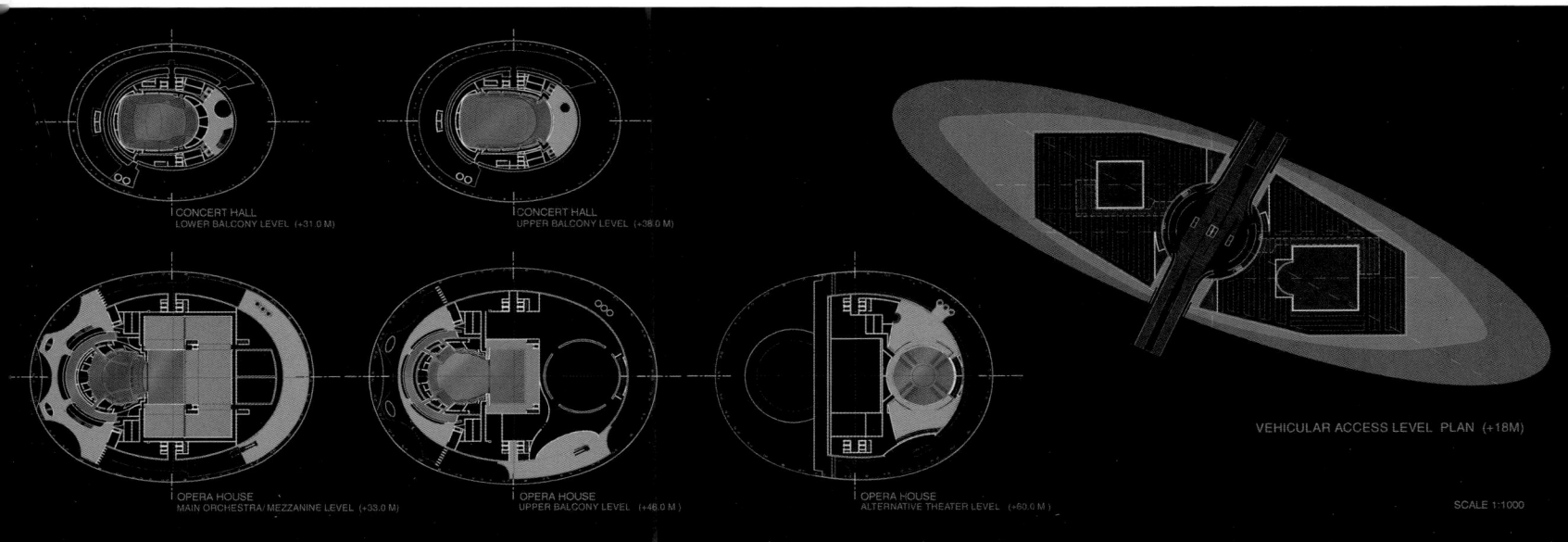

CONCERT HALL
LOWER BALCONY LEVEL (+31.0 M)

CONCERT HALL
UPPER BALCONY LEVEL (+38.0 M)

OPERA HOUSE
MAIN ORCHESTRA/ MEZZANINE LEVEL (+33.0 M)

OPERA HOUSE
UPPER BALCONY LEVEL (+46.0 M)

OPERA HOUSE
ALTERNATIVE THEATER LEVEL (+50.0 M)

VEHICULAR ACCESS LEVEL PLAN (+18M)

SCALE 1:1000

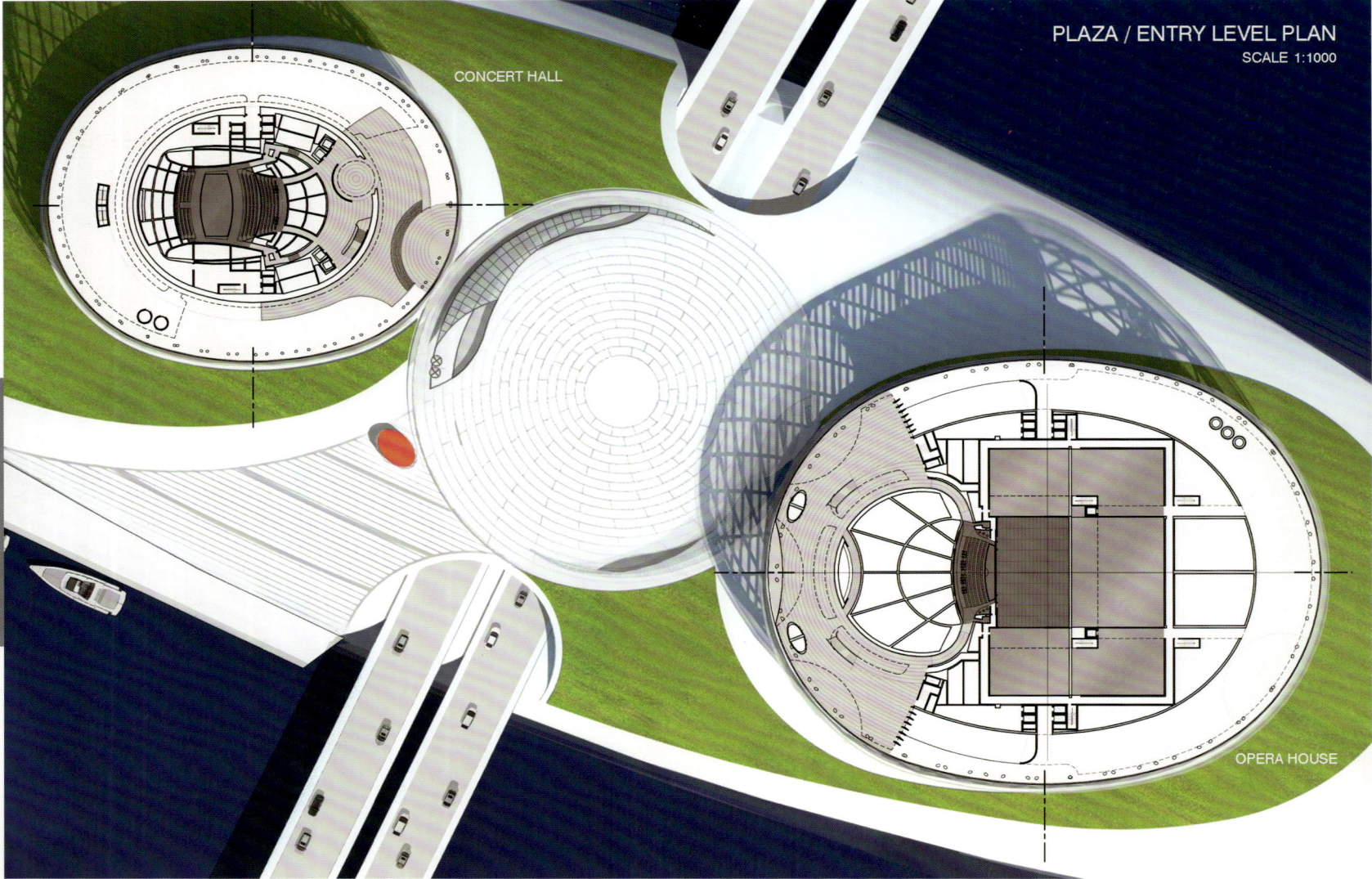

CONCERT HALL

OPERA HOUSE

　　首尔表演艺术中心的设计考虑到了以下几个重要因素：位于首尔的突出的岛屿选址和岛屿曾发生过洪水的历史，它对于韩国及世界文化的作用，它复杂的规划和它在韩国逐渐增长的世界影响性中有潜力成为代表性建筑。

　　鹭德岛位于汉江之中，是韩国南部的现代城市和北部的老城之间的文化连接。鹭德岛可以说是韩国的地理中心，是传统韩国文化与收到现代启发的当代韩国文化之间交融的地方。首尔被汉江一分为二，被花岗岩山脉的内环和外环环绕着。这些山脉为城市提供了延绵不绝的背景景色。岛屿曾经发生过的洪水灾害被看作是大自然伟大力量的重要象征，应该在这个项目中得到表现，以作为人类不断追求与自然环境和谐共存的证明。

　　在设计的初期，设计团队曾考虑过21世纪音乐中心的基本特征。尽管各种设施将会在各类国际音乐活动中使用，设计师们认为更好的理解很有可能在这里举行演出的本土音乐是头等重要的。这使设计师们开始研究传统的韩国清唱剧，原本有12出的歌剧，现在只留存下了5出。清唱剧（Pansori）一词的词源是单词pan，意思是表演的场地，而sori意思是声音，设计师们认为这是为音乐表演艺术设计大型场地的一个适当的出发点。与西方歌剧不同，清唱剧不是由大型剧团来表演，而是由单独的一名歌手表演，用鼓来伴奏。除了以鼓点伴奏以外，鼓师还会以口头发出的鼓励声（chuimsae）来伴奏。演唱者（sorikkun）和鼓师（gosu）之间的直接交流就好比是一个项目空间中的两个主要部分，一个歌剧厅与一个音乐厅。鉴于清唱剧中鼓的重要作用，设计师们使用鼓的形式作为最初的正式设计，使这个复杂的规划利用这些形式共同构成一个项目，设计师们认为这些形式有潜力成为韩国传统与全球化渴望的象征。

Stadium Kajzerica

Kajzerica体育场

Firm: 3LHD
Location: Zagreb, Croatia
Project Team: Sasa Begovic, Marko Dabrovic, Tatjana Grozdanic Begovic, Silvije Novak, Tin Kavuric, Irena Mazer, Krunoslav Szorsen, Dragana Simic, Matija Crnogorac, UEFA, FIFA standards advisor
Project Team Collaborators: Ljiljana Dordevic, Vibor Granic, Jelena Mance, Ana Safar
Landscape Design: Ines Hrdalo
Lighting Design: Zlatko Galić, Novalux d.o.o.
Urban Planning: Dražen Pejković, IGH PC Split
Electrical Engineering: Branko Čorko ing. el., IPZ Elektroinženjering-22 d.o.o.
3D: Boris Goreta

The city stadium Kajzerica in Novi Zagreb represents a new landmark in the town.

The stadium, which seats more than 50,000 spectators, is located on an elevated plateau next to the sports and recreational zone of the racetracks and it dominates the surrounding area, the green zone of Sava and Bundek. The stadium is visible from all surrounding areas on both sides of the Sava River.

The coverage area is divided into three zones, consistent with the project's program, and into three phases, consistent with the financial framework of construction.

The northern zone, the first building zone, enables the undisturbed building of the stadium, independent of the surrounding settlements, parcels and Zagrebački Velesajam. The middle zone, the second building phase is crucial for the development of the stadium because this is where different services will be situated, as well as the infrastructural elements used by all three

zones (public garages, bus parking and similar). The southern zone, i.e. the third building phase, encompasses the whole area of Velesajam and all of its buildings. It is for the construction of new businesses, commercial and residential buildings as well as a new commercial center. Within the new formed cassettes – "blocks" there will be hotels, a congress center, business centers and obligatory residential areas. Integration of these elements ensures livelihood of the new center and brings a long-awaited downtown feeling and an opportunity for socialization to Novi Zagreb.

The inspiration for the circular arena came from several suitable forms; the so-called Magnus' effect, one of the most difficult and attractive moves in soccer which occurs when a ball changes its course, the name of the neighborhood "kajzerica", which is a name for a certain kind of a bun whose shape is reflected in the stadium. In the context of a contemporary sports arena, this design evokes the saying from ancient times about "bread and circuses".

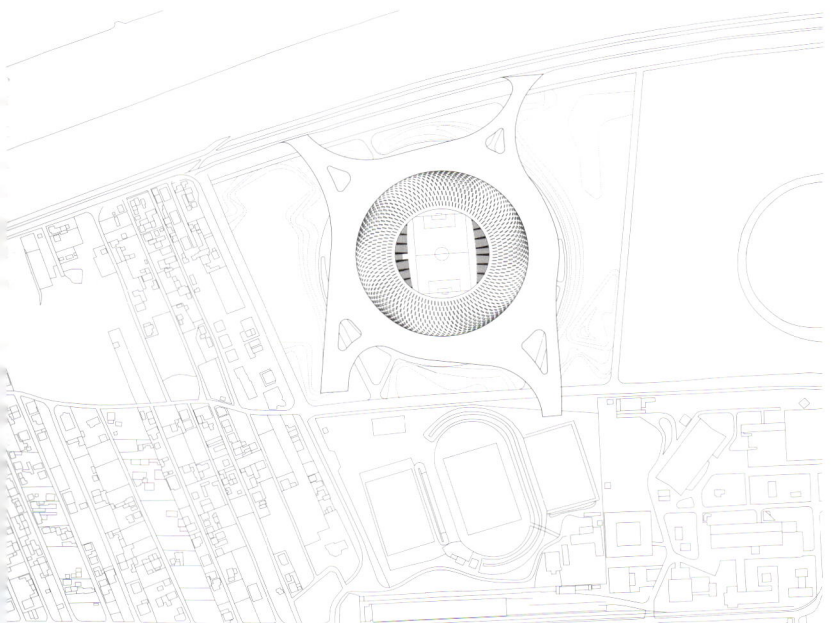

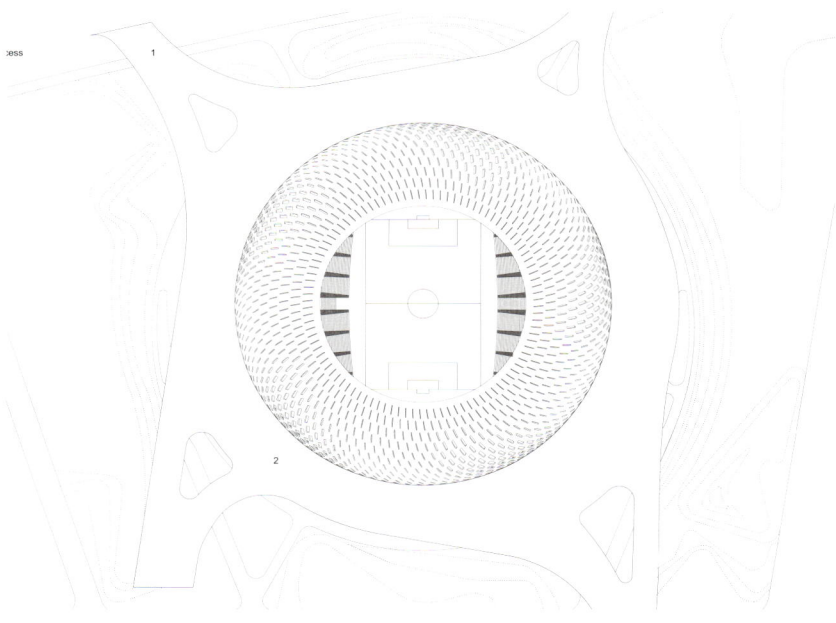

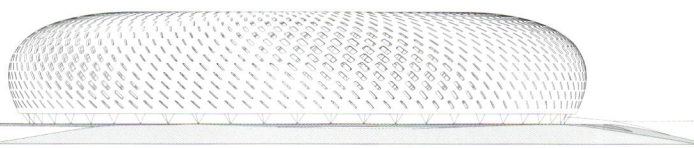

Elevation East

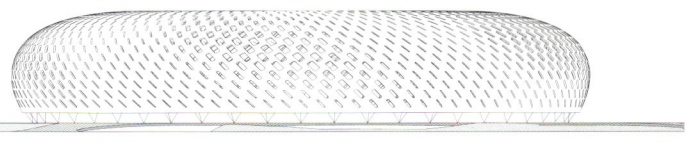

Elevation North

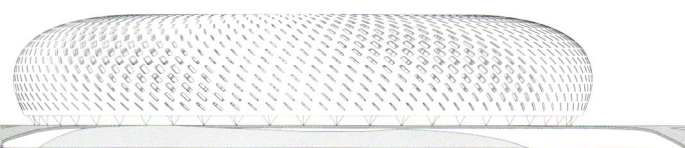

Elevation South

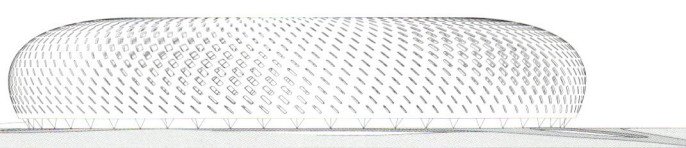

Elevation West

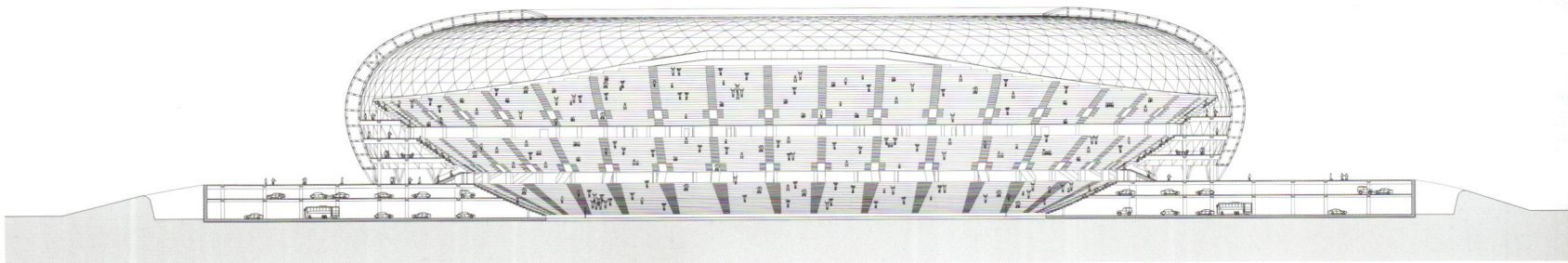

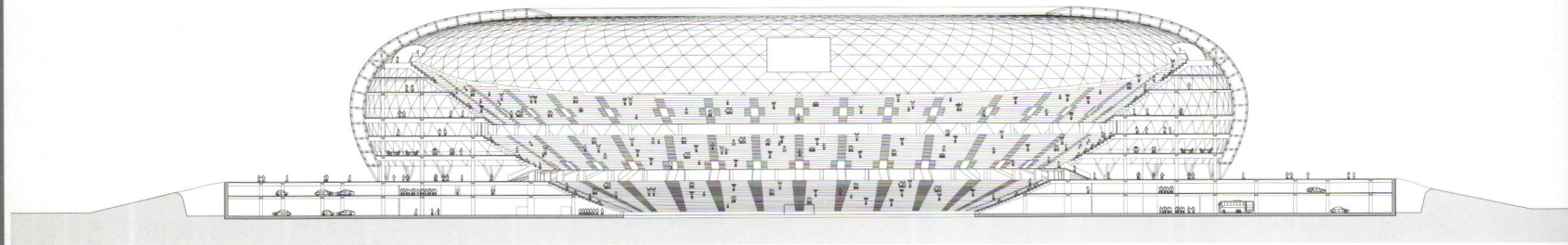

Sections

位于诺维萨格勒布的Kajzerica城市体育场代表了这座城镇的新地标。

这座超过五万个座位的体育场坐落于一处逐渐升高的高地，紧邻着赛车娱乐场地和体█区。这座体育场在周围的沙瓦河与Bundek湖的绿化区中占据着极其重要的位置。从沙瓦█两岸的任何周边地带都可以看到这座体育场。

整个建筑区域分为与整个项目计划一致的三个分区。同时为了与建设的融资方案相一█，体育场也被分为了三个阶段进行建设。

北区是首先开始建设的位置。这一区域保证了体育场与周边的居住区、地块以及萨格█布世界贸易中心相分离，使得整个建筑有了一个相对独立的环境。中区是第二阶段开始█设的部分。这一部分对于体育场的发展十分重要，因为这一部分将为全部三个区域提供█种不同服务以及所有的基础设施（如公共停车场、公交车站与其他类似服务）。南区是

第三阶段建设的部分。为了建设新的商业、贸易和居住建筑以及一个新的商业中心，这一部分将穿过整个世界贸易中心及其所属的建筑。在这些被称为"立方体"的方盒形建筑中将会有酒店、商业中心、保障性居住区以及一座会议中心。这些元素的整合赋予了新的中心以生活气息，也带来了久违的市中心商业区的感觉，还令诺维萨格勒布市有了更多的社会生活感。

设计这座圆形竞技场形的建筑的灵感来自于多种合适的形象，比如马格努斯效应——在足球运动里指一种最有难度与吸引力的球的运行方式，其发生时球会改变运动的轨迹；还有体育场所在地区的"kajzerica"的名字，这个名称本是指一种小圆面包，而体育场正是带有它的形象。作为一个当代的体育竞技场，它的设计唤起了人们对于那个充满"面包和马戏"的古老年代的回忆。

Entrance - VIP
Entrance - VIP i administration
Entance - media
Recepcion
Garage entrance
Parking
Conference room
Utility spaces
Entrance
Lobby
Entrance - teams
Locker room
Gym
Warmup space
Wellness
Referee's room
Coach
Dopping test
Physical
Flash interview
Delegates
Technical area

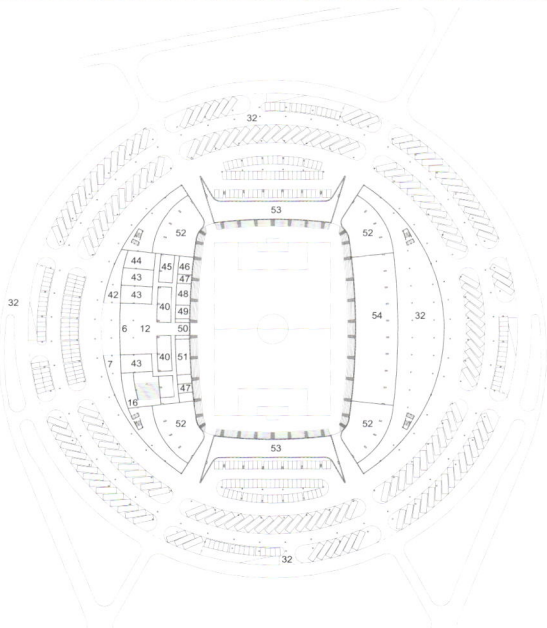

18	Conference room
20	WC
31	Garage entrance
32	Parking
33	Lobby
34	Press - welcome zone
35	VIP - welcome zone
36	Media rooms
37	Web editing
38	Photografs
39	Press
40	Conference room
41	Utelity room

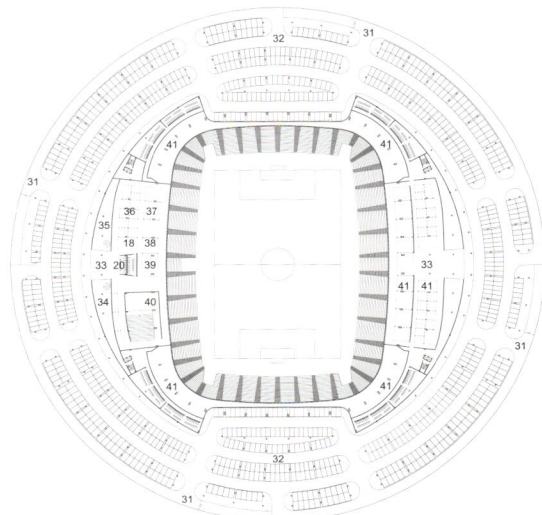

1   Pedestrian access
2   Pedestrians
3   Entrance upper stand
4   Entrance - VVIP
5   Entrance - staff
6   Entrance - VIP
7   Entrance administration
8   Media
9   Entrance
10 Lower stand entrance
11 Utility room
12 Entrance - media
13 Caffe
14 Space for disabled visitors

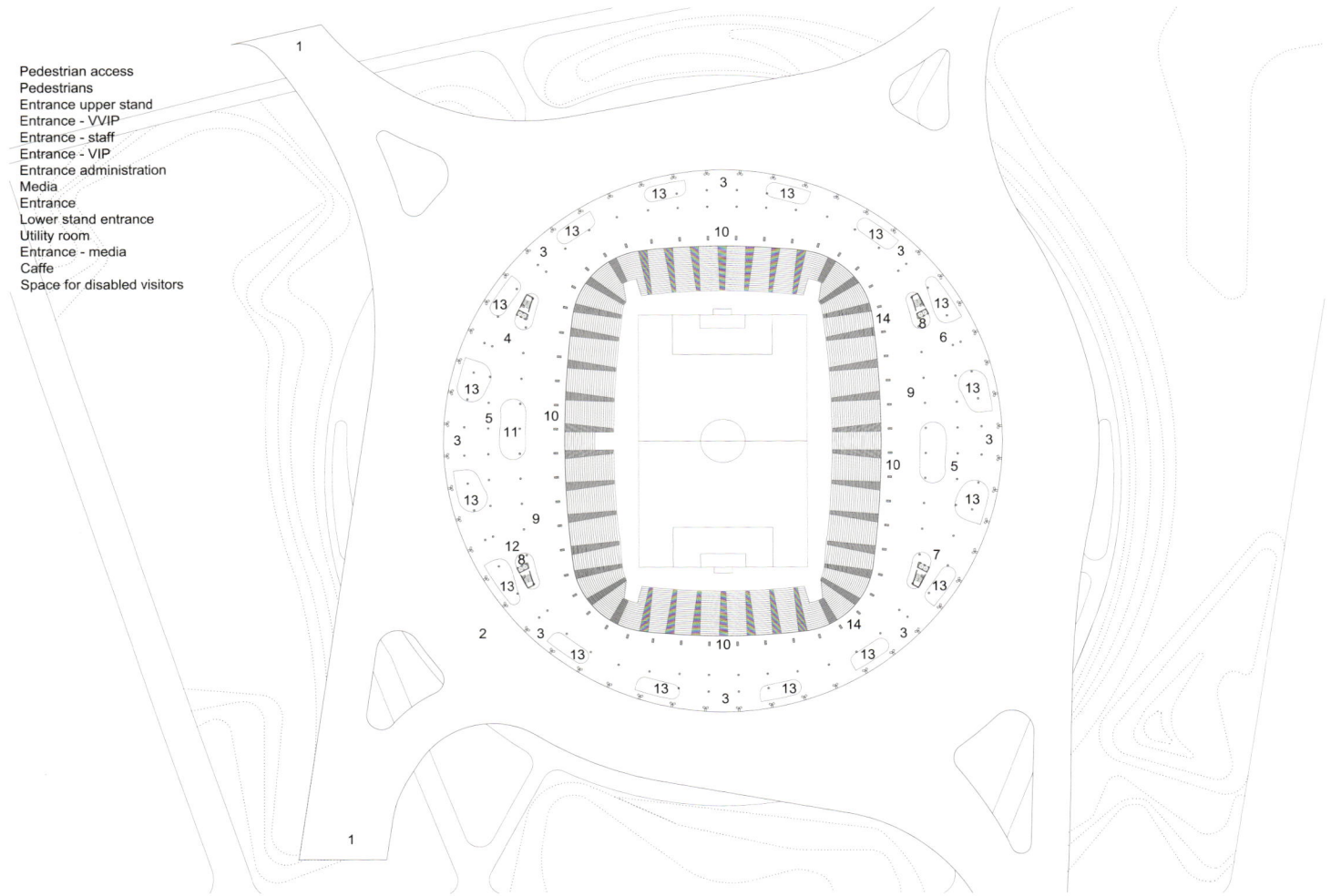

Entrance - VIP
Entrance - administration
Caffe
Lounge
Recepction
VIP wardrobe
Conference room
Bussiness lounge
WC
Staff wardrobe
Office
Restaurant

6	Entrance - VIP
14	Space for disabled visitors
24	VVIP bar
25	VIP restaurant
26	VIP
27	Concert mixing stage
28	Sky studio

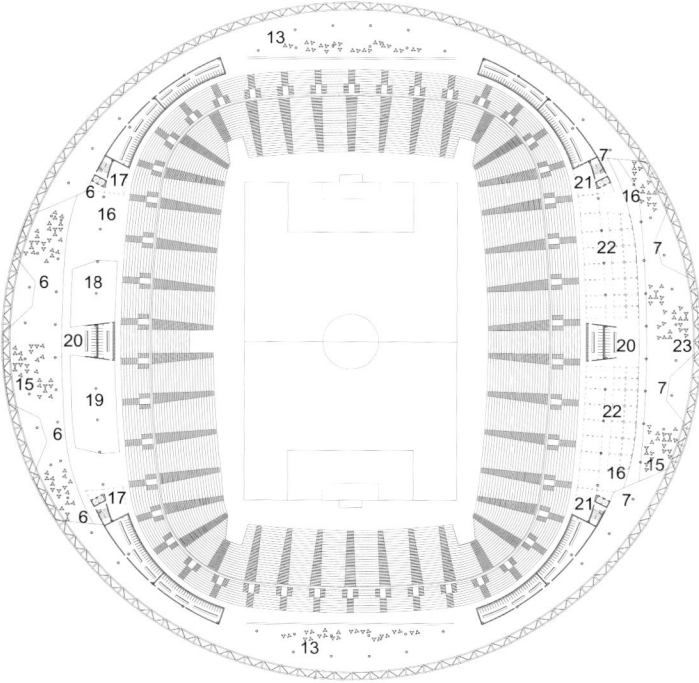

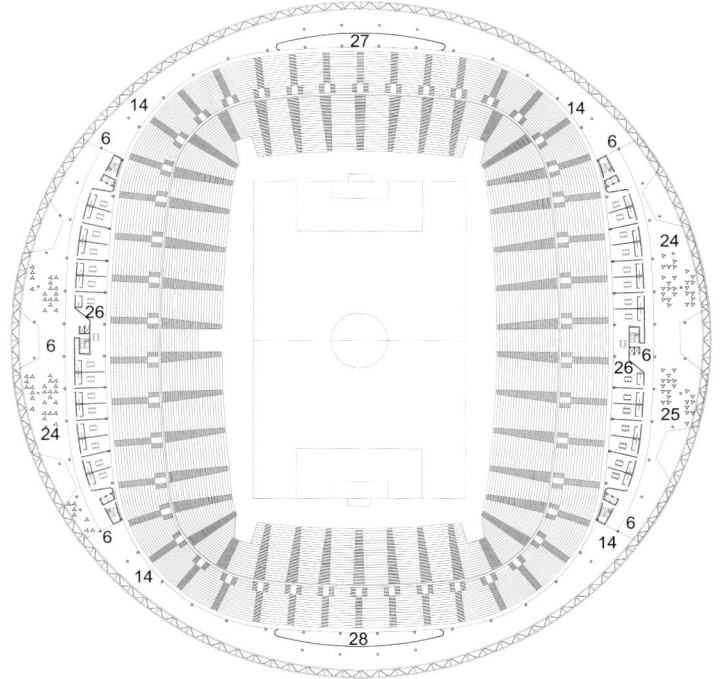

| | Caffe |
| | WC |

| 29 | Commentators |
| 30 | Police |

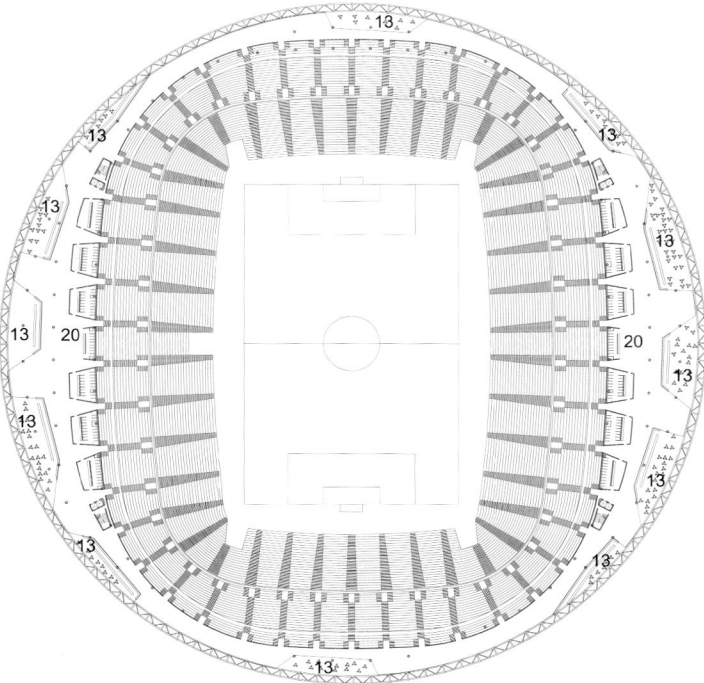

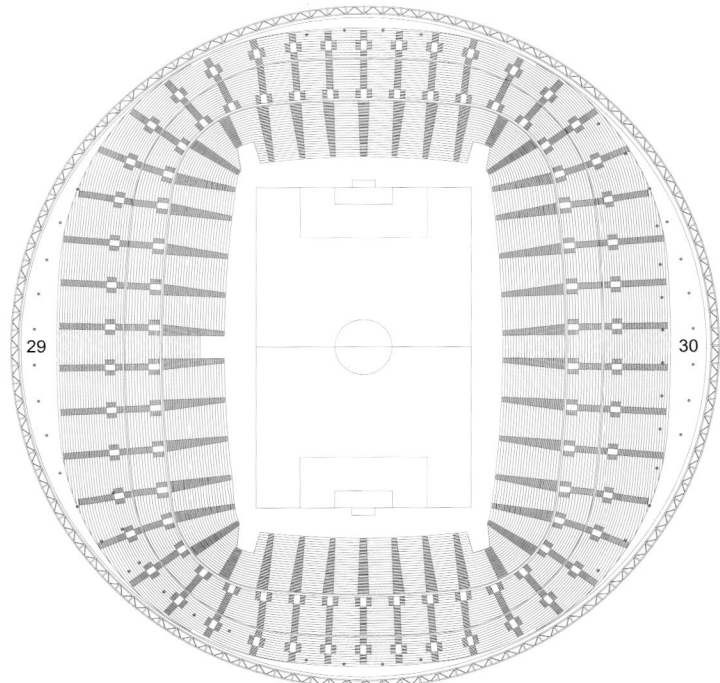

Maribor Art Gallery

马里博尔美术馆

Architect: Vlado Valkof, Anne Valkof, Rumen Yotov
Firm: design initiatives
Location: Maribor, Slovenia
Area: 15,180m²
Designer: Anne Valkof
Rendering: Rumen Yotov
Date: February, 2010

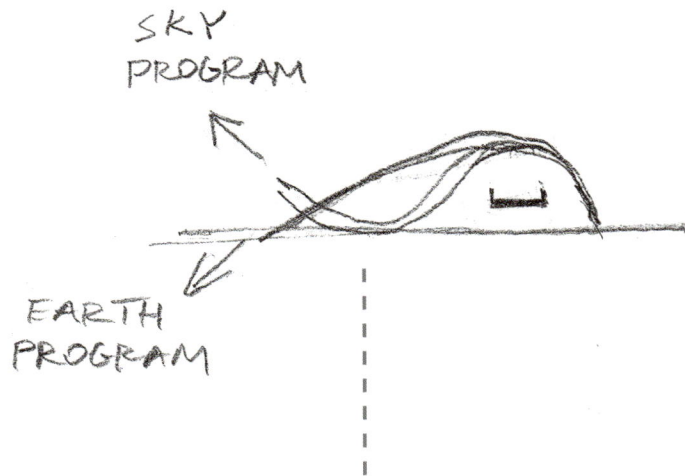

SKY PROGRAM

EARTH PROGRAM

This is an entry in international competition for 17,800m² new building of the Maribor Art Gallery in Maribor, Slovenia. The complex and heterogeneous program includes galleries for permanent collection of modern and contemporary visual art and periodical exhibitions (5,200m²), children's center (750m²), design center (450m²), live-work creative center (1,700m²), lecture room, library, catering (700m²), offices, depository, and underground parking garage.

In addition to animate forms the architects manipulate movement in order to induce the production of new urban life.The urban scheme of the museum building is inspired by the typical Maribor city block of perimeter buildings surrounding enclosed plaza/courtyard. Gradual spaces of covered plaza/lapidarium and open outdoor plaza connect the different parts of the building and are partially open towards the embankment and the pier. One wing contains the "sky" program with the primary function of periodical and permanent exhibition galleries and the other wing – the "earth" program with supporting and secondary functions of children's center, design center,

library, etc. Each one of the wings has its own vertical circulation core. The one of exhibition galleries and management offices is enclosed and with controlled access. The other one has a set of escalators and is open from the plaza to all other functions. Two-storey plaza with open vertical core concept allows lecture room, catering, children's center/ playground, library, design center, and creative center to operate independently after museum hours and to serve not only the museum visitors but also an outside crowd. At the entry wedge of the plaza there is a removable screen or stage for outdoor performances at summertime.

In the Program there is a clause of preserving the existing east-west street which divides the site in two. The proposed building bridges over and tunnels under the existing street and physically connects the two parts of the site. Three dimensional loop interconnects periodical exhibitions galleries which have different illumination, heights, and proportions, suitable for variety of exhibitions and artists/curators requirements.

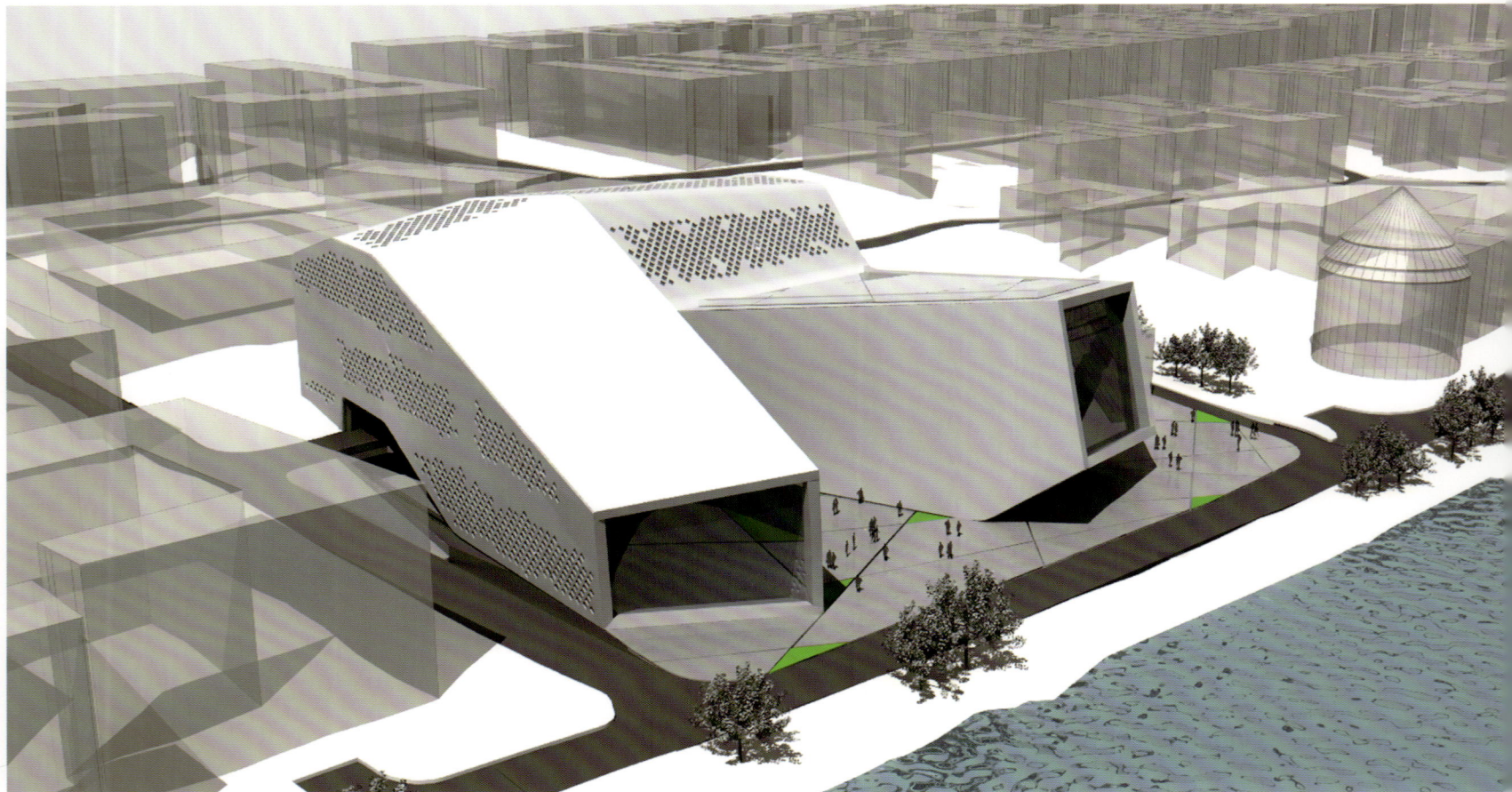

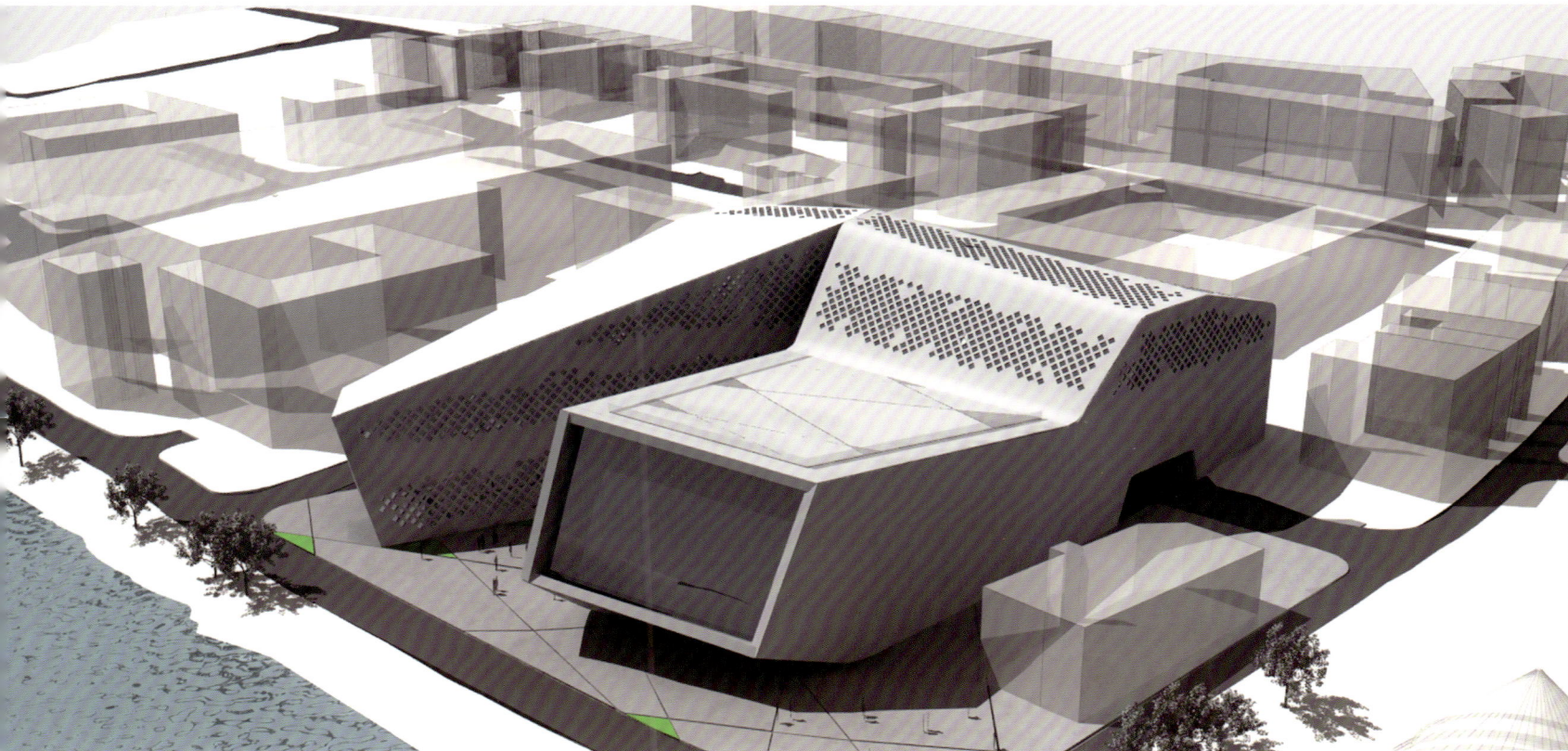

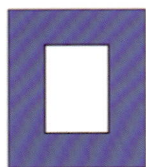

typical Maribor block   open passage   open to the river   The New Maribor Art Gallery

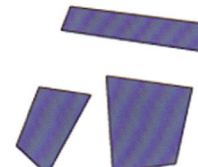

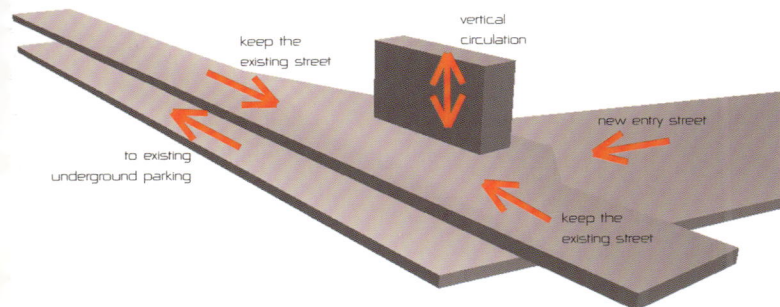

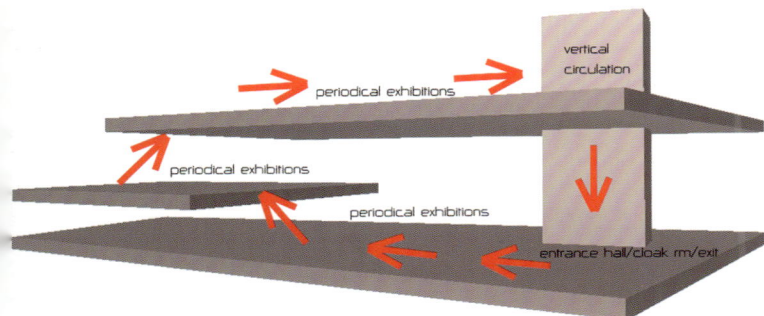

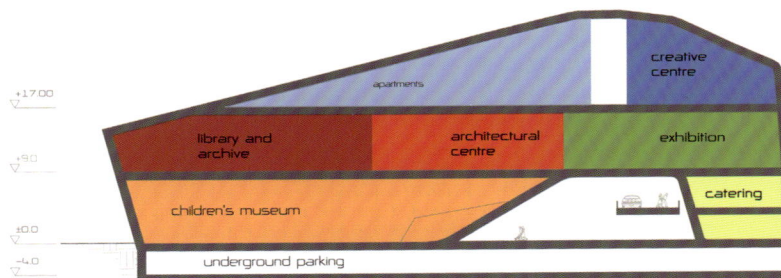

keep the existing street

vertical circulation

to existing underground parking

new entry street

keep the existing street

+17.00

apartments

creative centre

+9.0

library and archive

architectural centre

exhibition

±0.0

children's museum

catering

-4.0

underground parking

section 1-1   0   10m

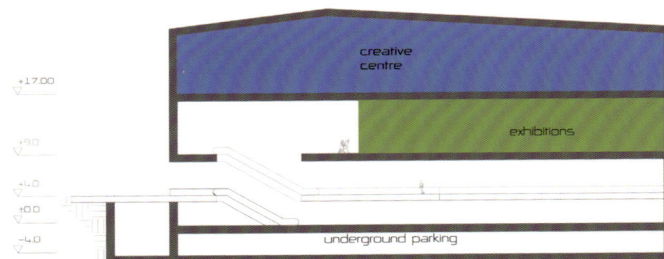

vertical circulation

periodical exhibitions

periodical exhibitions

periodical exhibitions

entrance hall/cloak rm/exit

+17.00

creative centre

+9.0

exhibitions

±0.0

-4.0

underground parking

section 3-3   0   10m

这个设计是斯洛文尼亚马里博尔市17800平方米的美术馆竞赛项目的参赛品。建筑和多样化的项目包括了为现代和当代视觉艺术永久性展览及定期展览艺术馆（5200m²），儿童中心（750m²），设计中心（450m²），现场创作中心（1700m²），讲座中心、图书馆和餐厅（700m²），办公室、仓库和地下停车场。

除了有活力的结构设计以外，设计师们还控制了动线以促使新城市生活的发生。博物馆建筑的城市规划是受到典型的马里博尔市广场周边建筑物的启发。漫步被覆盖的广场空间和开敞的露天广场将建筑的不同部分连接起来，并且局部向路堤和码头方向开放。其中一座侧楼被称作"天空"部分，主要是定期和永久性展览厅；另一座侧楼"大地"部分，包含有支持性和次要功能区，比如，儿童中心、设计中心和图书馆等。这两座侧楼都有其各自的垂直循环核心。带有展览和管理办公室的一座大楼是封闭的，并且设有门禁。另一座大楼配有自动扶梯从广场部分开始一直到其他各个功能空间都是开放的。两层高的广场使用开放的垂直核心理念进行设计，使讲座室、餐厅、儿童中心/游乐场、图书馆、设计中心和创意中心在美术馆闭馆后可各自独立使用，而且也使这些功能不仅为美术馆的参观者服务，还可以为其他的外来人群服务。广场的入口处有一个可移动的幕，或者也可以叫做舞台，是为了夏天的户外表演设置的。

在整个比赛项目中，有一条规定要维持已有的将整个场地划分为二的东西向街道。这个设计所提议的是在原有的街道上架设桥梁，街道下方设置地下通道，这样从根本上将两个部分连接起来。三维立体式的环线使具有不同照明、不同度和不同规模的定期展览厅可以适应各种不同的展览及艺术家和策展人的要求连接起来。

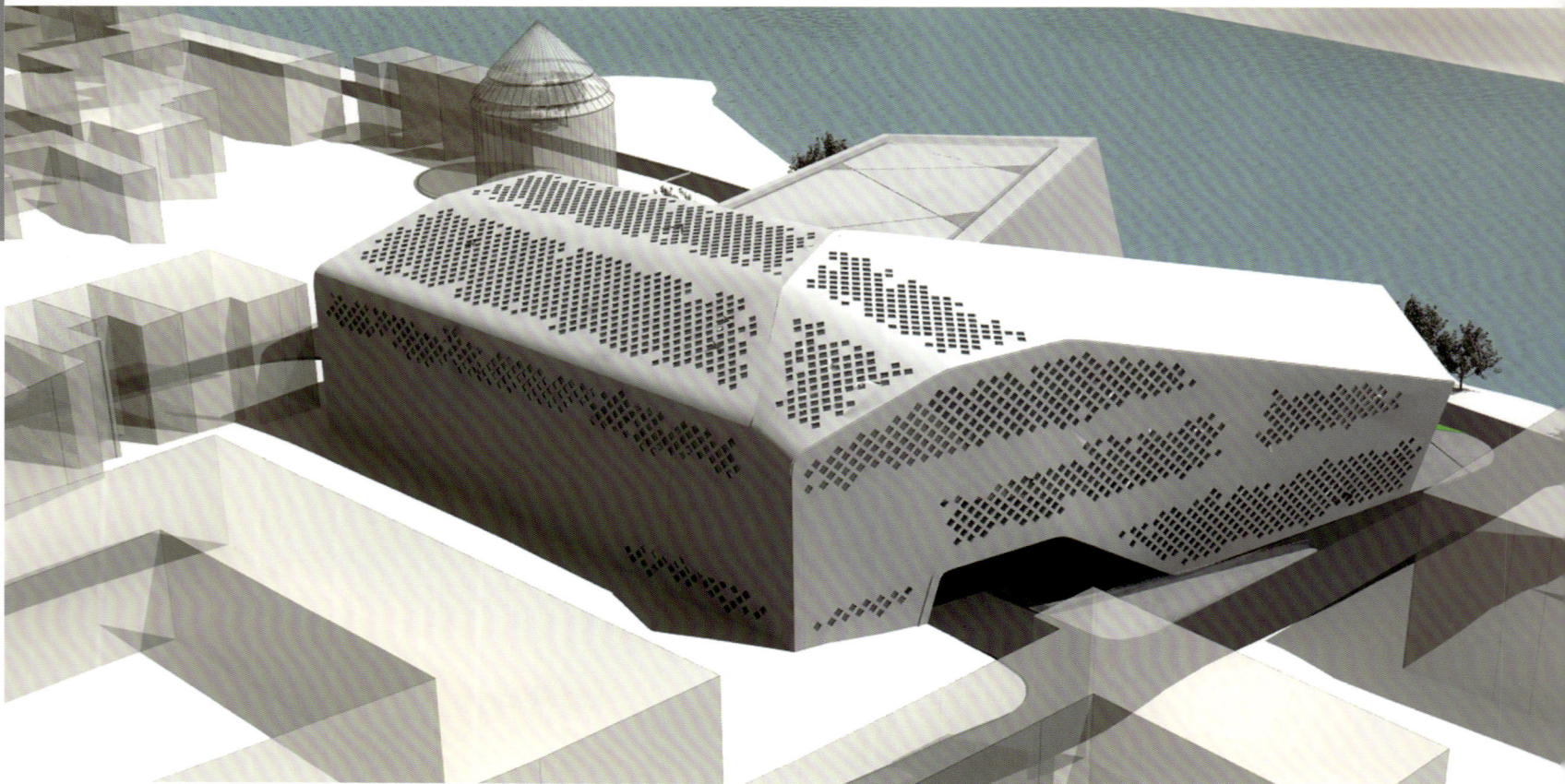

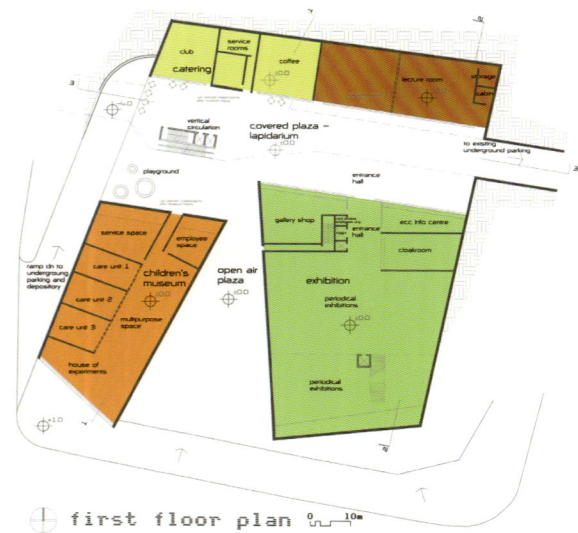

first floor plan   0———10m

second floor plan   0———10m

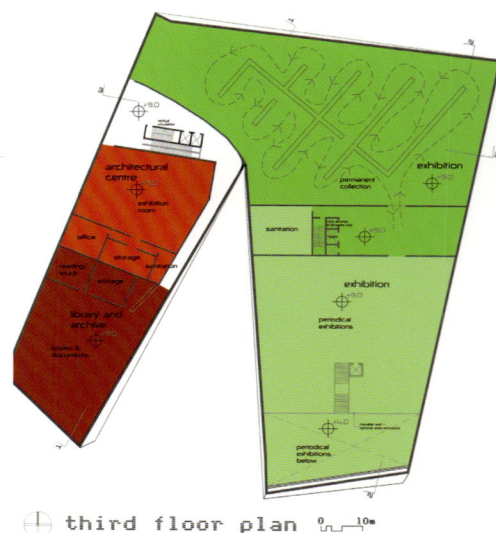

third floor plan   0———10m

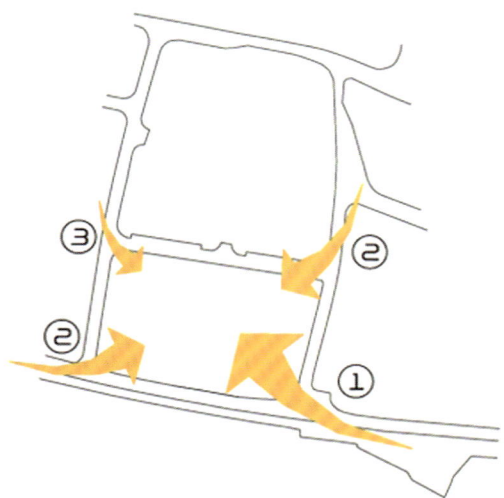

fourth floor plan ⌖ 0   10m

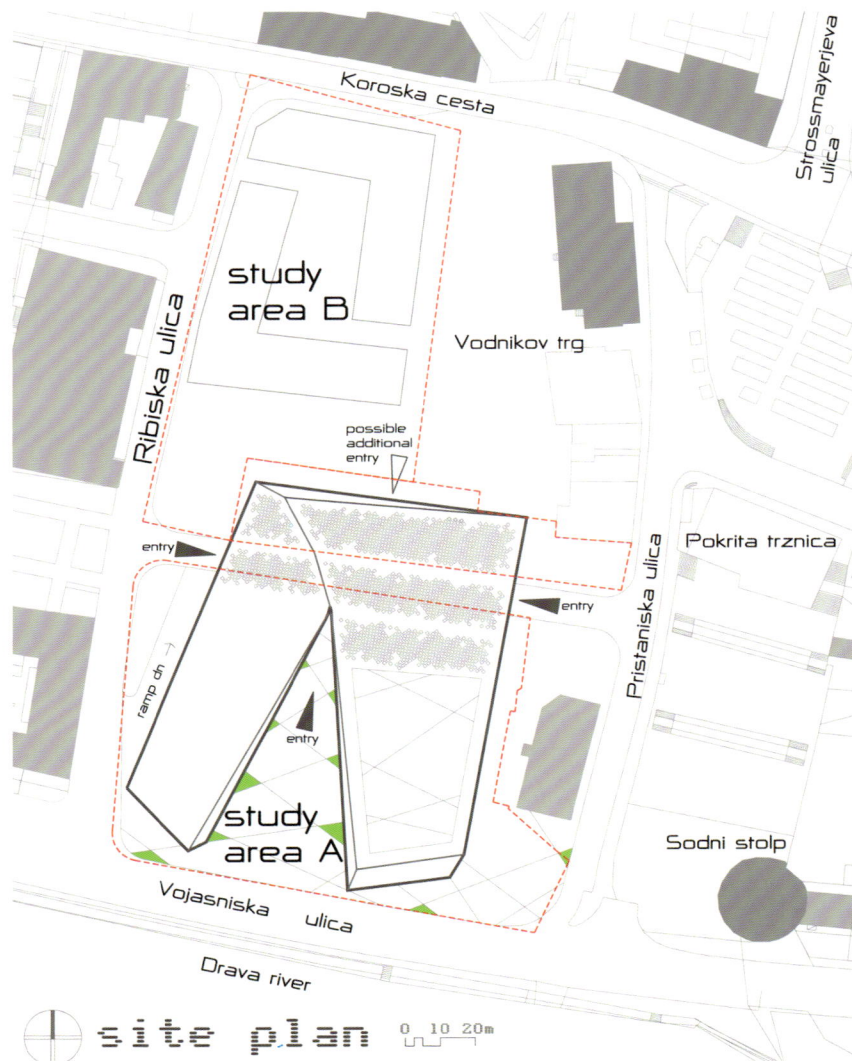

study
area B

Koroska cesta

Strossmayerjeva ulica

Vodnikov trg

Ribiska ulica

possible
additional
entry

entry

entry

Pristaniska ulica

Pokrita trznica

ramp dn

entry

study
area A

Sodni stolp

Vojasniska   ulica

Drava river

⊕ site plan   0 10 20m

The Crowd in the Cloud

云中的人群

Architect: Vlado Valkof, Anne Valkof, Rumen Yotov
Firm: design initiatives
Location: Greater Noida, Delhi, India
Area: 62,750m²
Designers: Anne Valkof, Rumen Yotov
Rendering: Martin Nikolov
Type: mixed use
Date: January, 2011

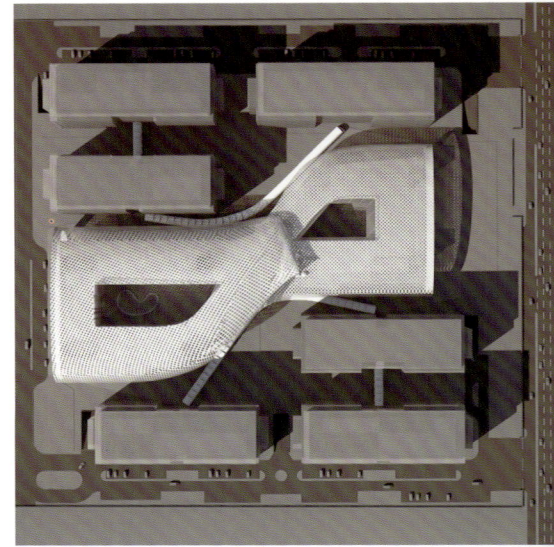

124

This project responds to the context which is planned to be build by the time the project is completed, and to the dynamically twisted site. The attempt is to increase density and pedestrian traffic. The architects strongly believe that high-rise typology is an oxymoron of sustainable development in its base and is in contradiction with promoter's goal for a green complex. From the program they identify five major functions: Research Institute and Knowledge Centre, Offices and Incubation Centre, Hotel, Residential, and Retail-Food-Beverage. they clearly differentiate four of them in four wings grouped around two inner courts in two separate buildings (Research Institute and Offices in one building and Hotel and Residential in the other). At the area of conjunction of the two buildings they design outdoor distributing hub/plaza/promenade which connects to all functions and the underground parking garage. It contains the fifth function of shopping and dining and combines the similar function of Research Institute food court with the F+B and retail outlets into one shopping center like hub. Because of the specificity of Noida climate the plaza is not enclosed, just covered by the buildings above. The plaza is open to serve not only the new building's population but also people from the IT Park and from outside. With the creation of the outdoor plaza they break the 220m long "barrier" of the site for better circulation.

The inner court scheme is inspired by the Indian tradition. One of the courts is developed around a water pool and serves both Hotel and Apartments.

The Hotel and Apartments also share Banquet Facility/Service Area and common drop off area which is an additional access point. The other cou is surrounded by the thematically grouped Auditorium, Lecture Hall, Galler and Library and could be easily transformed into Open Air Theatre. Function like Auditorium, Lecture Hall, Gallery, Library, Bank, and Lobbies are locate at the ground level and are directly accessible to the public and buildir occupants. Each one of the four wings with restricted access has an individu entry accessible through the inner court and vertical core in the corner. Bot buildings connect at fifth shopping level and sixth level where the tall building leans above the lower building and covers a natural shaded space a food and beverage terrace on top of the lower building.

The pattern on the sustainable double / secondary skin of the facades inspired by traditional carved stone lattice screens of Hoysala Temple Belur.

The architects also propose lightweight tent structures against sun and ra to connect the new building with the office buildings from the complex. The could be executed in next stages.

For landscaping, instead of a regular grass lawn they propose areas wit agricultural crops typical for the fields around Noida — Indian tea, rice, cor and sugar cane.

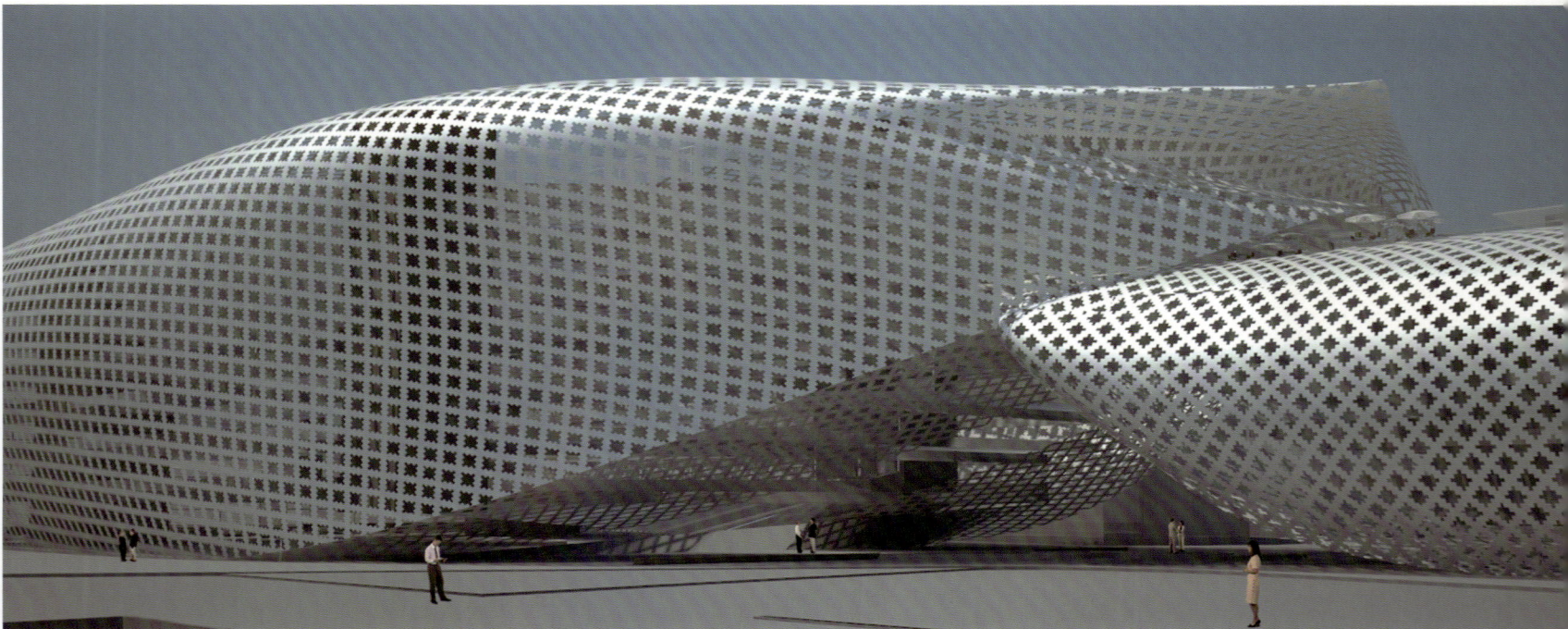

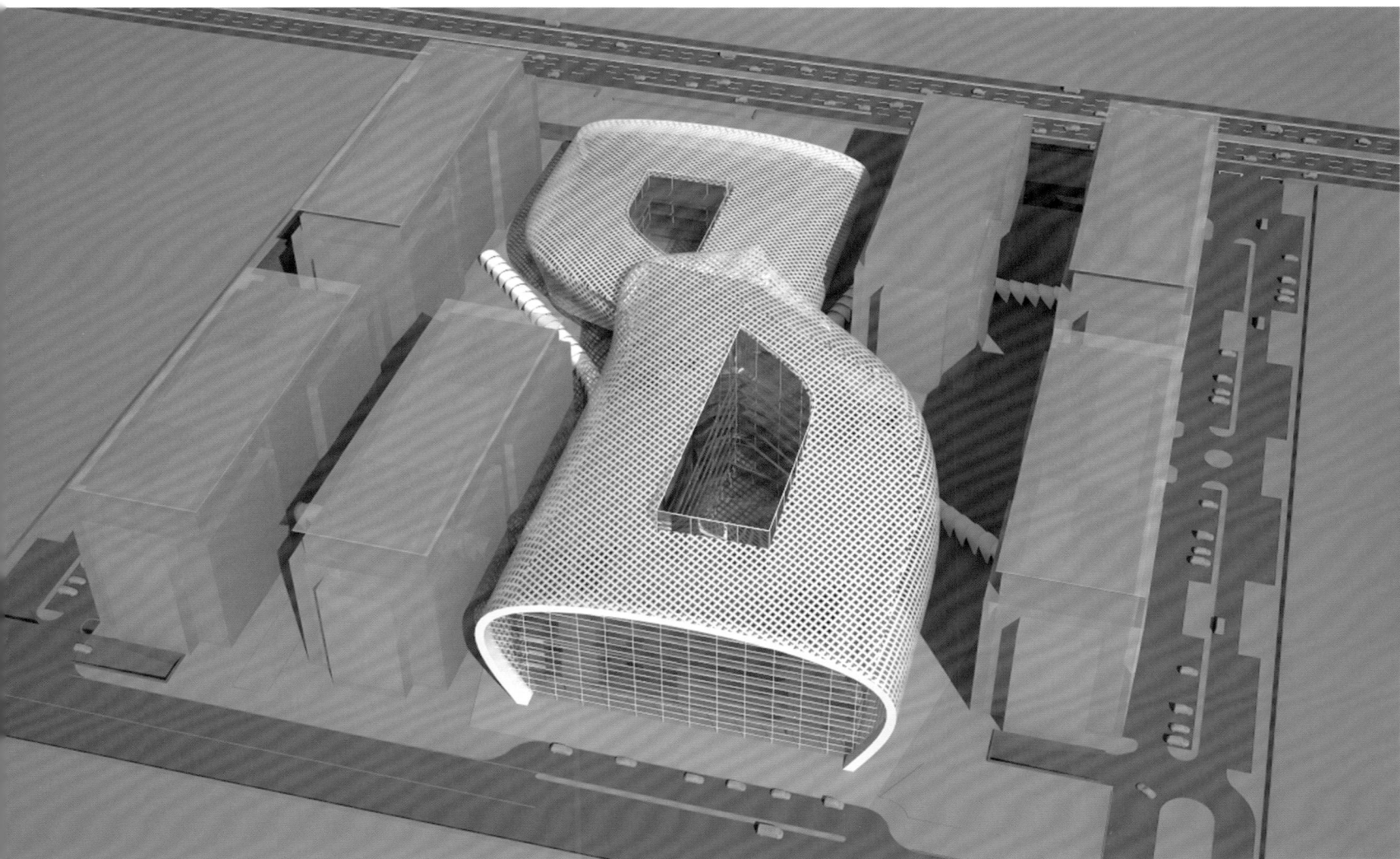

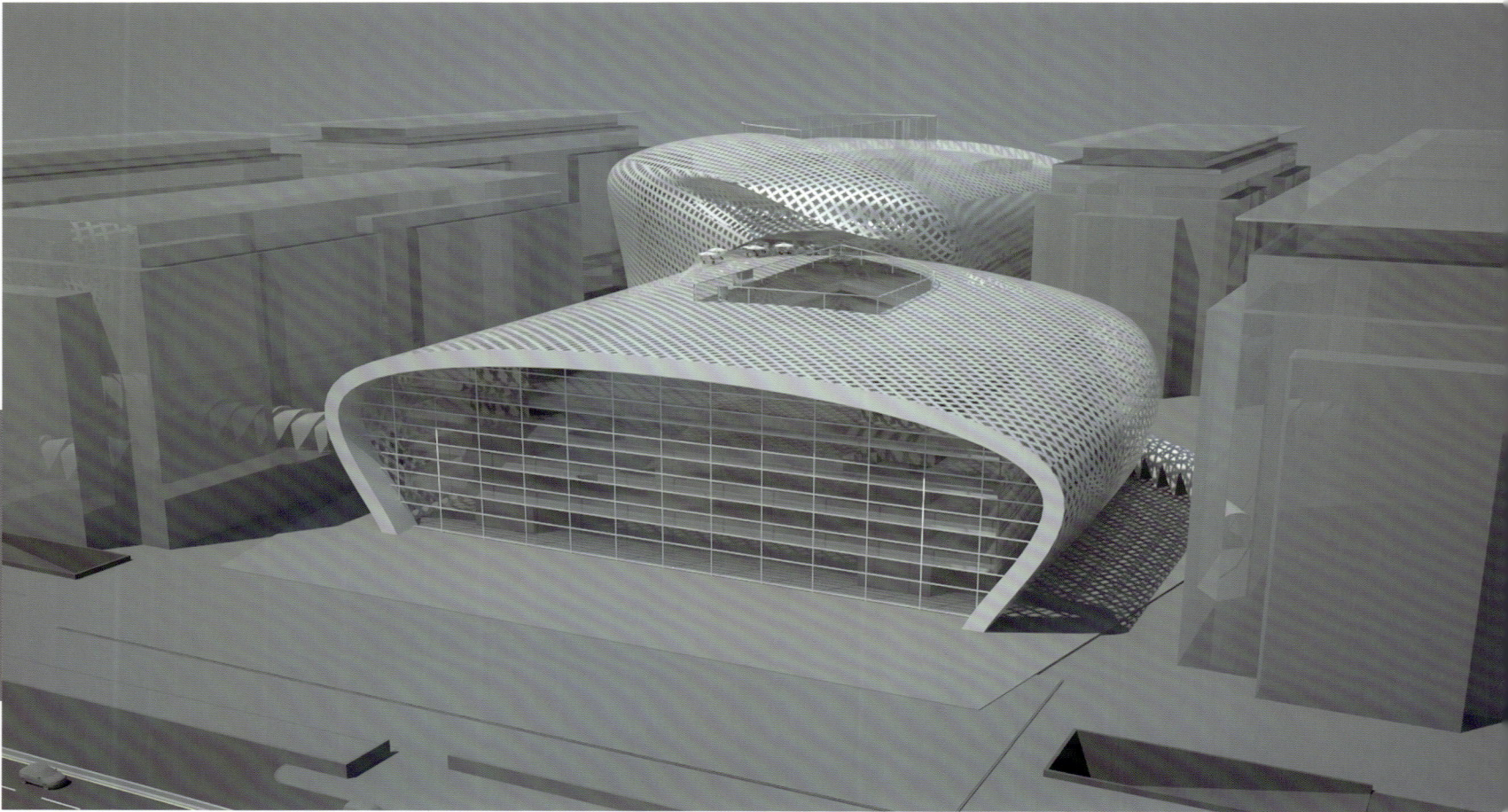

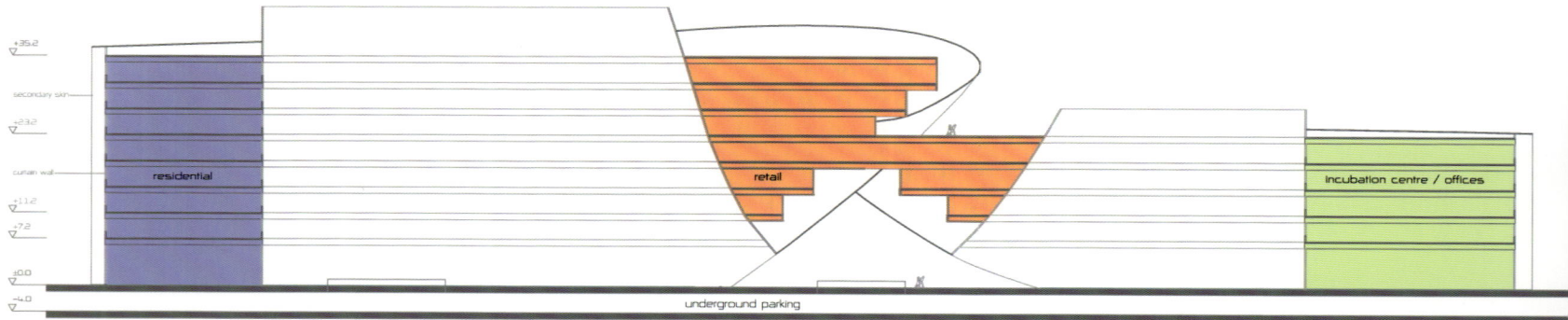

+35.2

secondary skin

+23.2

curtain wall

residential

retail

incubation centre / offices

+11.2

+7.2

±0.0

-4.0

underground parking

Longitudinal Section A

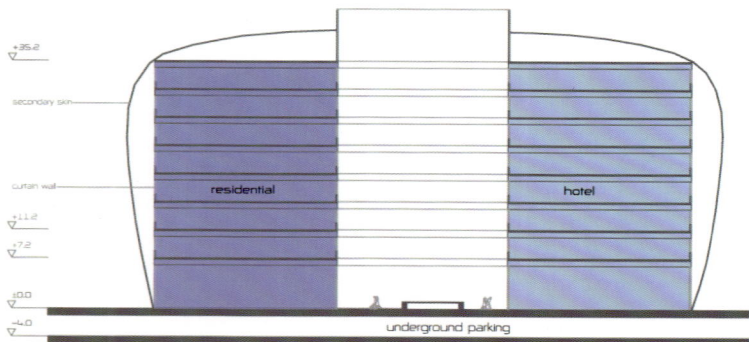

+35.2

secondary skin

curtain wall

residential

hotel

+11.2

+7.2

±0.0

-4.0

underground parking

Cross Section B

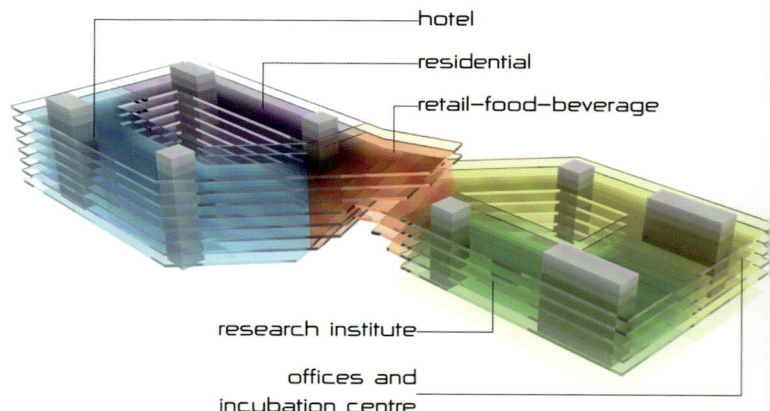

hotel

residential

retail-food-beverage

research institute

offices and
incubation centre

Program Distribution Diagram

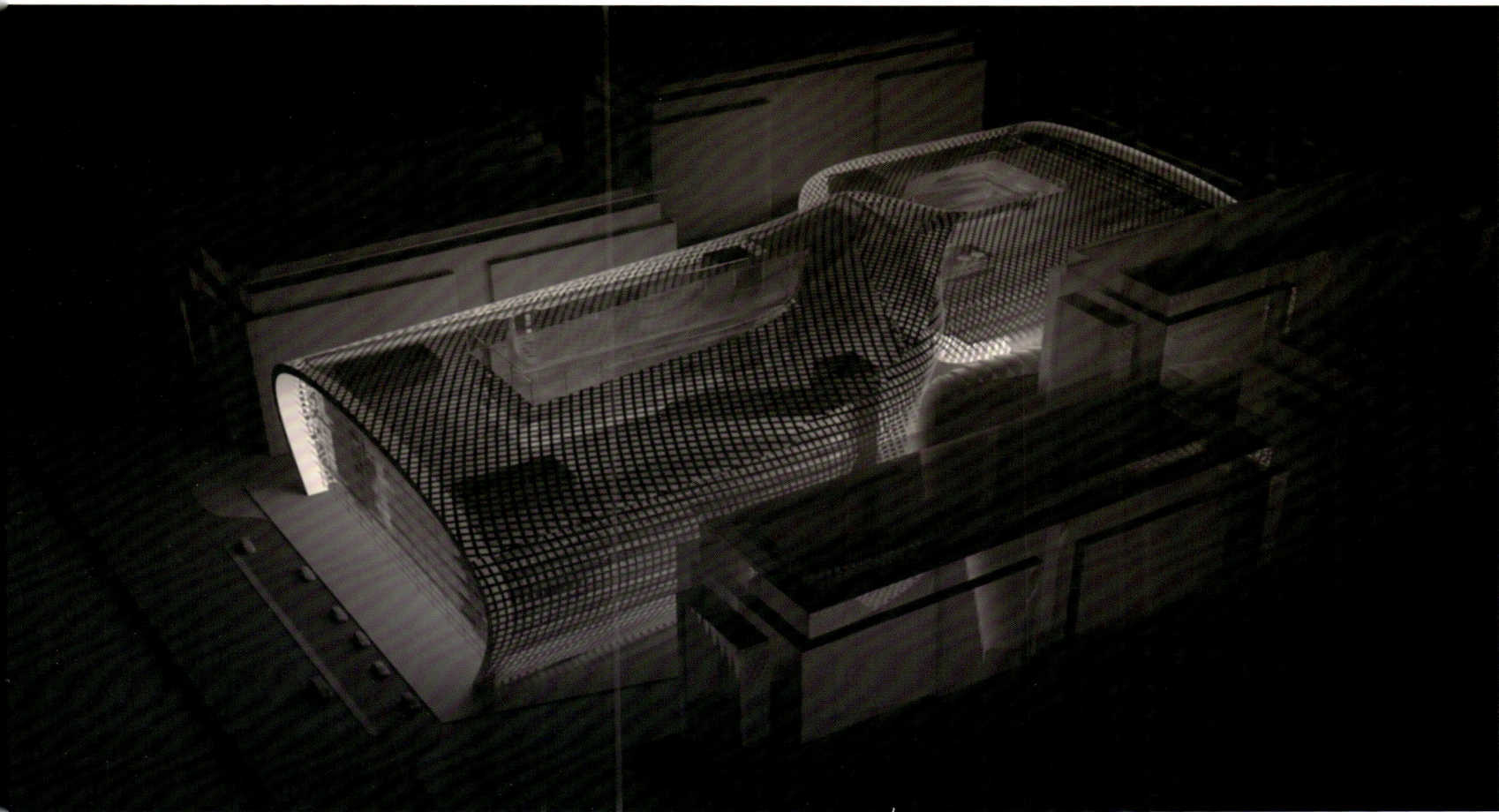

1st floor/site plan

5th floor plan

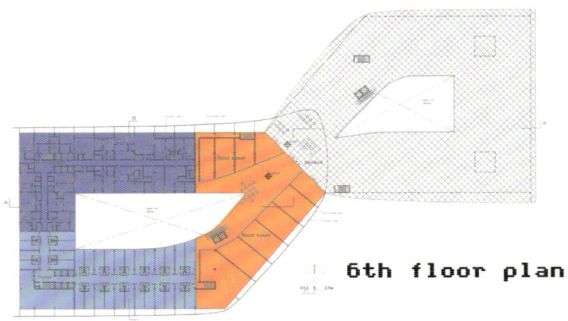

6th floor plan

与周围环境相呼应的这座建筑计划是在整个项目完成时建成，并与其动态弯曲的项目选址相搭配。设计师们试图增加人口密度和行人交通。他们强烈地相信，高层类型学在它的基础和在与发起者为建造一栋绿色建筑的目标相矛盾方面是可持续发展的矛盾修辞法。从项目开始，设计师们就确立了五种基本功能：试验所与知识中心、办公室与孵化中心、酒店、住宅和商业—餐饮。他们清楚地将四种功能分配在围绕着有两个内庭的两座独立建筑的四栋侧翼建筑中（试验所和办公室在同一座建筑中，酒店和住宅在另一座建筑中）。在两座建筑的连接处，建筑师们设计了一个户外的分布中心/广场/散步区，这个区域与所有的功能区以及地下停车场相连。这里包含了第五种功能——商业和餐饮，而且将试验所美食广场与餐饮和零售店结合成一个购物中心。由于诺伊达的特殊气候原因，这个广场并不是封闭的，而是由上面的建筑所遮盖。这个新建的广场并不仅仅为新建筑中的人员服务，也为周围的IT园区和外来人员服务。这个户外广场的建立打破了这个场地220米长的"障碍"，为这里创造了更好的环境。

内庭的规划受到了印度传统的启发。其中一个大厅沿着一个水池分布，作为酒店与公寓区。酒店与公寓区也同时共享宴会设施及服务区和一个额外的接入点的落客区。另一个大厅由专题分组礼堂、演讲厅、展览廊和图书馆所环绕，并且可以轻松变化为露天剧场。诸如视听、演讲、展览、图书馆、银行和大厅等功能都被安排在一层，公众和大楼用户都可以直接访问。四座限制访问的侧翼建筑中的每一座都与内庭和竖向中心有着单独的通道。两座建筑在购物中心的第五层和第六层相连，高一点的建筑向另一座建筑倾斜，在矮一点的建筑顶层形成一个天然的遮蔽区域，作为餐饮天台。

建筑的环保双层及第二层立面上的图案受到了贝鲁尔葛萨拉神庙中格子屏风上传统的雕刻石头的启发。

设计师们还使用了轻型帐篷结构来抵御阳光和雨水，将新的建筑与原建筑群的办公大楼连接起来。它们可以在下一个阶段再进行设计。

在景观方面，设计师们并没有使用常规的草坪，相反，使用了诺伊达有代表性的农作物，例如，印度茶、稻米、玉米和甘蔗。

Ordos UNVEIL

鄂尔多斯UNVEIL办公楼

Firm: ///byn
Area: 40,000m²
Principals: Nicolas SALTO DEL GIORGIO & Bittor SANCHEZ-MONASTERIO
Team: Segolene DUBERNET, GUO Dan, DUAN Jun, HUANG QiShan, GUO ZhiChuan & LI Min
Engineering & Environmental: CCDI, Shanghai
Rendering: ZhuJin for ///byn
3D printing: LTModel
Photographer: ///byn

NVEIL is an office complex, part of ORDOS 20×10, the new financial district Ordos (1,800,000m²)in the third planning growth of the Dongsheng strict, in the Northwest area of the City of Ordos.

is design proposal is inspired in the origins of the Inner Mongolian culture d its own recognition.

e weather in Inner Mongolia is severe and strict. The cold temperature d the sand storms teach people how to live in relationship with nature d its components: protection, sheltering, environmentally intelligent nstruction techniques, changeable temperature performance along the y... all these aspects will drive the architectural guidelines. In the past the il was a resource of temperature stability. Here the architects consider this ement in a new way: potential for sharing. By occupying its depth they can nnect the different buildings with a shared and sheltered public space. This ntinuous space will host the entrance lobbies from the public plazas and e program of shared public facilities. It also links to segregated building obies to ensure a secured management.

is indoor public space will be pierced with generous openings to ensure e necessary needs of day light. This weather-proof space also provides alternative indoor circulation to the outdoor pedestrian paths along the siness park.

e architects propose a rational typical floor plan that is flexible enough allocate all sizes of office spaces. As the public amenities are located at e public shared floor, the office space can be fully usable. One company n hold one entire building, one entire floor or only a part of one floor. The trance lobbies are visually connected to the public network space, giving

them the added value of a wider representative exposure that a detached small building could not afford.

The above ground area will be dedicated uniquely to office work areas. The vertical core and the vertical atrium will help to distribute the space. The cores have been located and designed accordingly to the orientation and size of the buildings. In the smaller buildings (26mx26m) the cores have been pushed to the north facade providing a bigger office area facing south. In the bigger buildings the core has a central position dividing two long strips of office bays.

The verticality of a chimney that points to the sky represents for people the meaning of pride of a better and prosperous future. From the public spaces the visitors can discover the blue sky of Ordos through SKYEYES that link the public space in the basement to the square shaped office plates. These vertical atriums are shaped according to sun orientation and prevailing winds direction to ensure natural lighting, passive heating and ventilation.

The facade is wrapped with an exterior fabric facade. The PVC based fabric skin is not parallel to the building envelope giving to the building freedom from its urban planning "cubic" requirements. As well the different deformations facilitate spatial conditions as entrance canopies or connections to the building base. The fabric exterior skin can roll up and down depending of the weather conditions. During day time it will be open allowing the building to gain heat, during night time it will roll down to keep the temperature. This movement will bring the identity of the buildings: in the morning the buildings UNVEIL to welcome the users, at night the buildings shut down as a metaphor of the end of the working day.

STAGE 1	STAGE 2	STAGE 3	STAGE 4	STAGE 5	STAGE 6

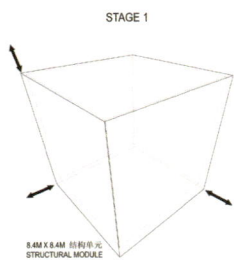 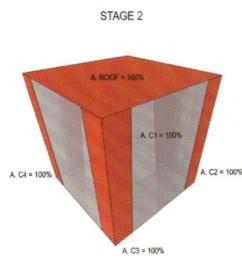 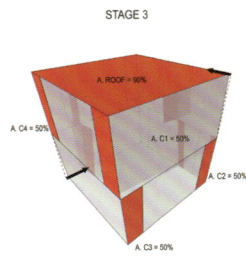 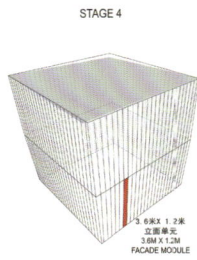 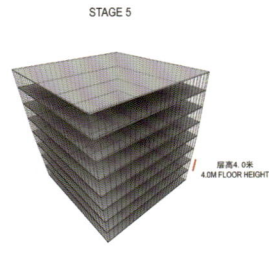 

8.4M X 8.4M 结构单元
STRUCTURAL MODULE

A. ROOF = 100%
A. C1 = 100%
A. C4 = 100%
A. C2 = 100%
A. C3 = 100%

A. ROOF = 50%
A. C4 = 50%
A. C1 = 50%
A. C2 = 50%
A. C3 = 50%

3.6M X 1.2米
立面单元
3.6M X 1.2M
FACADE MODULE

层高4.0米
4.0M FLOOR HEIGHT

织物立面在不同的方向
上均遵以1.2米为单元
FABRIC FACADE MODULE
1.2M IN DIF. DIRECTIONS

1/ 将建筑的结构单元定成1.2米
2/ 将建筑的悬臂宽度限制在3米以内

1/ ADJUST BUILDING FOOTPRINT TO 1.2M
STRUCTURAL MODULE
2/ LIMITE CANTILIEVER UNDER 3M

1/ 计算屋顶面积使其满足最小的50%
2/ 计算转角处的面积使其满足最小的50%

1/ CALCULATE ROOF AREA TO MATCH AT A
MINIMUM OF 50%
2/ CALCULATE CORNER AREA TO MATCH AT A
MINIMUM OF 50%

1/ 将建筑立面在一半高度处一分为二
2/ 在较低的地方两个侧面分别向里面退进0.6米
3/ 在较高的地方两个侧面分别向里面退进0.6米

1/ DIVIDE FACADE AT HALF HEIGHT
2/ PUSH IN 0.6M TWO SIDES IN THE LOWER HALF
3/ PUSH IN 0.6M TWO SIDES IN THE UPPER HALF

1/ 两立面分割成以1.2米宽的板材为单元的若
干等份
2/ 限制尺寸不规则的板材，使其排列在转角处

1/ DIVIDE FACADE IN PANELS OF 1.2M WIDTH
2/ LIMIT IRREGULAR PANELS TO CORNER ZONES

1/ 纵向将建筑立面以4米高为单位进行等分

1/ DIVIDE FACADE IN 4M HIGH FLOOR SPANS

1/ 将宽度小于1.5米的织物外立面覆盖在
外部的龙骨上面。

1/ ADD FABRIC FACADE WITHIN AN OUTER
BOUNDARY OF LESS THAN 1.5 M

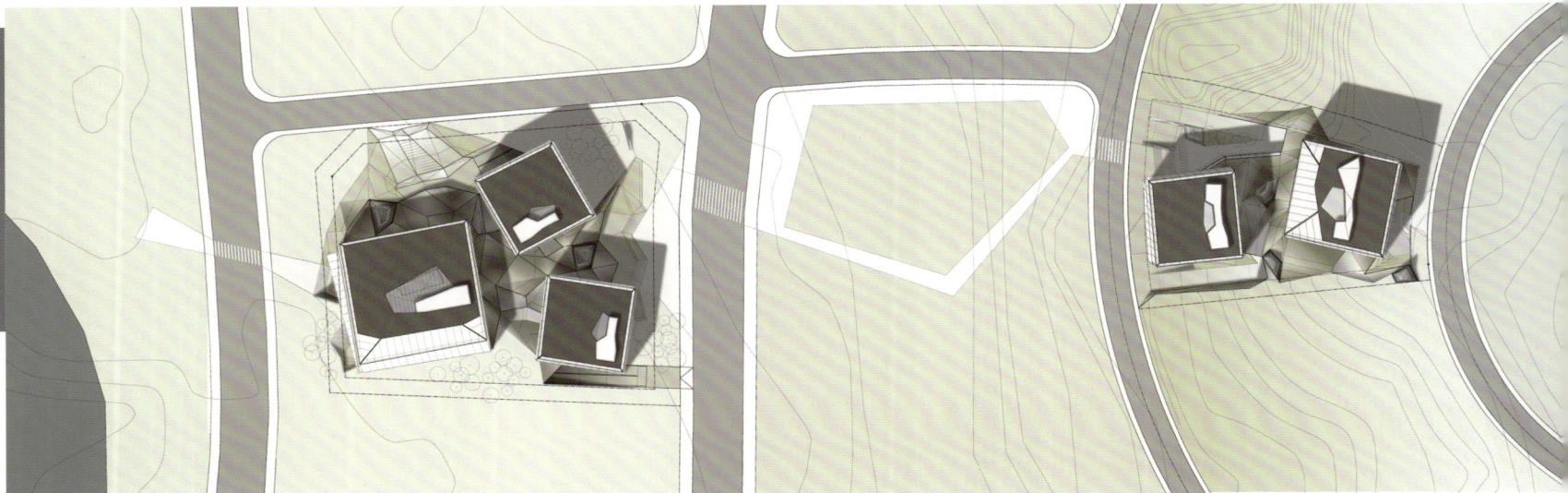

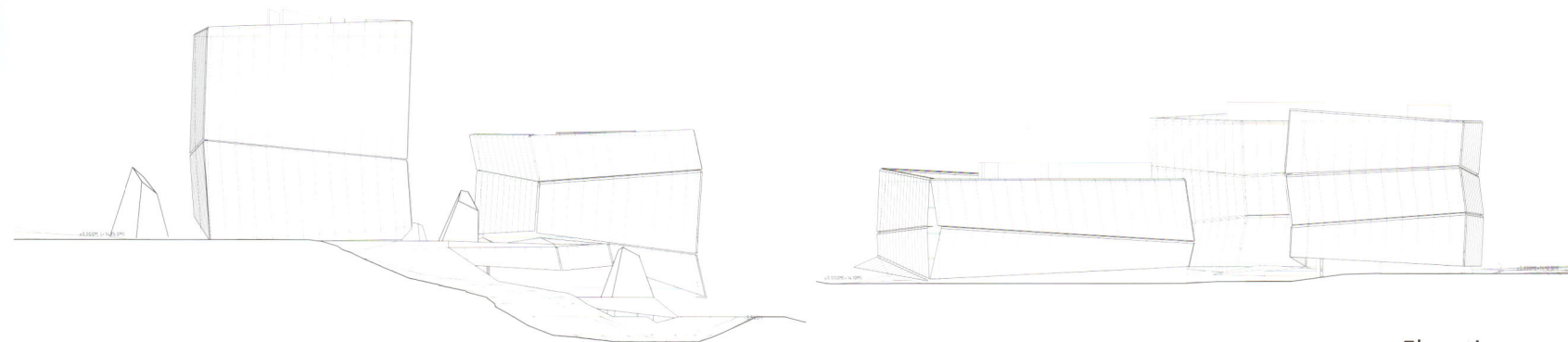

Elevations

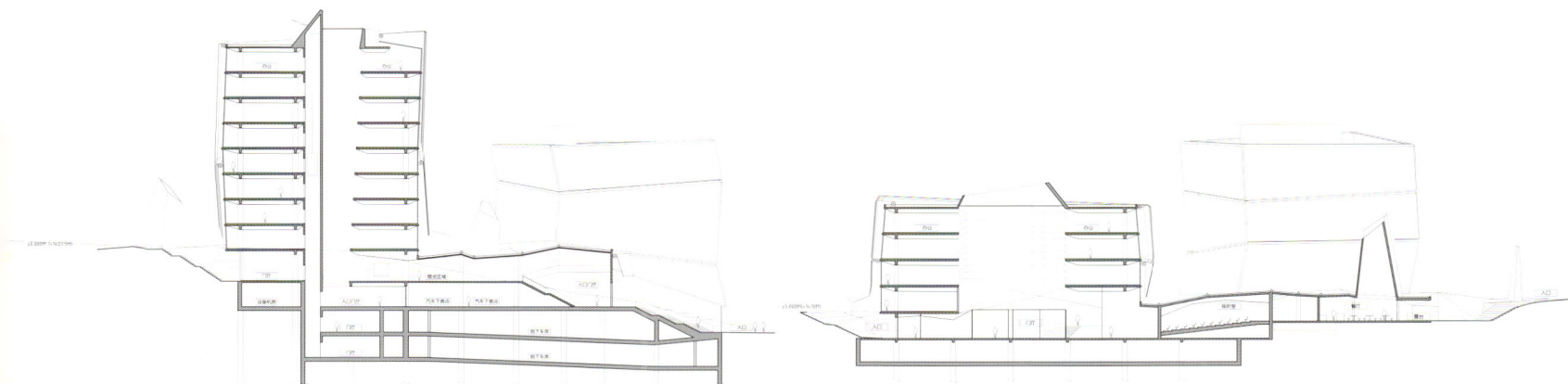

Sections

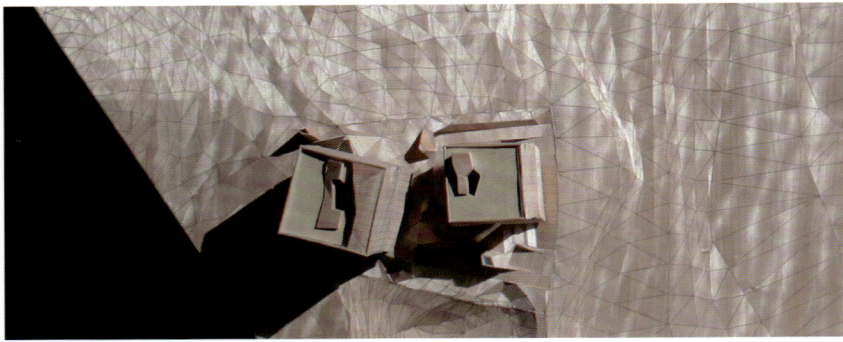

Elevation

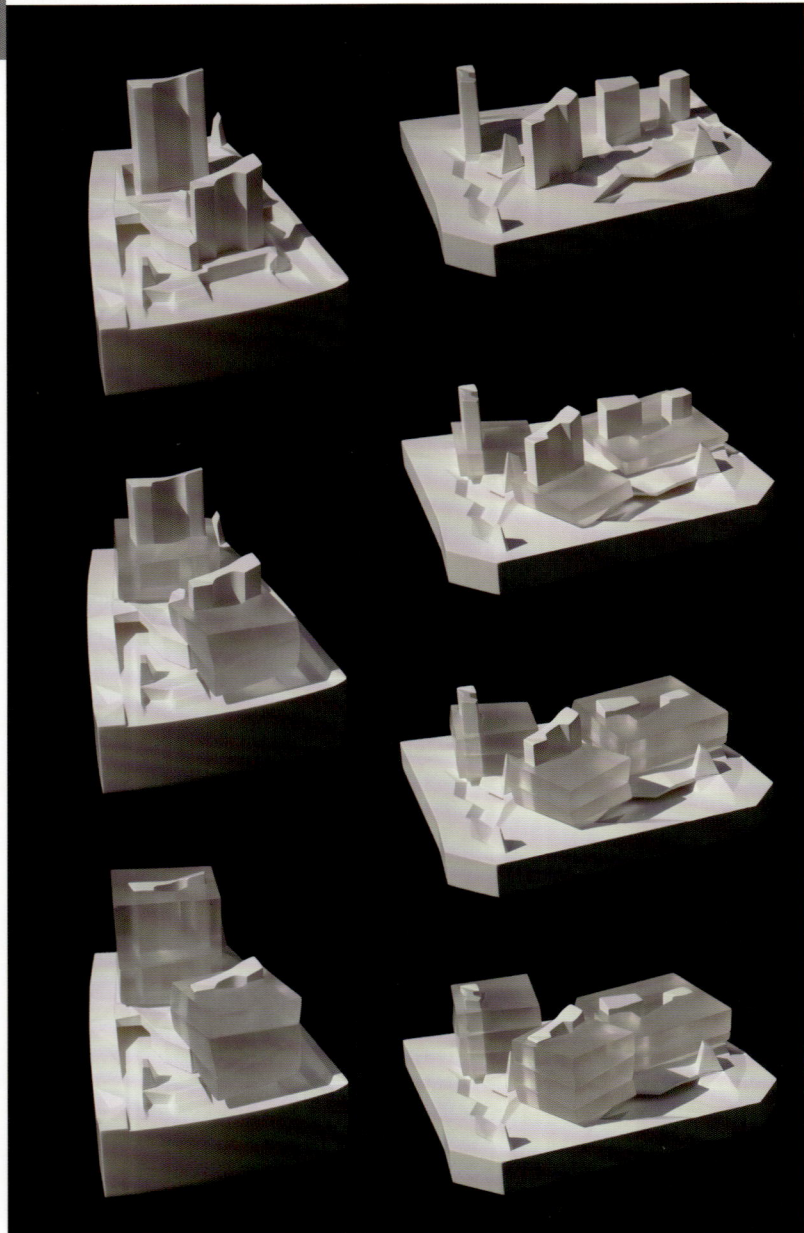

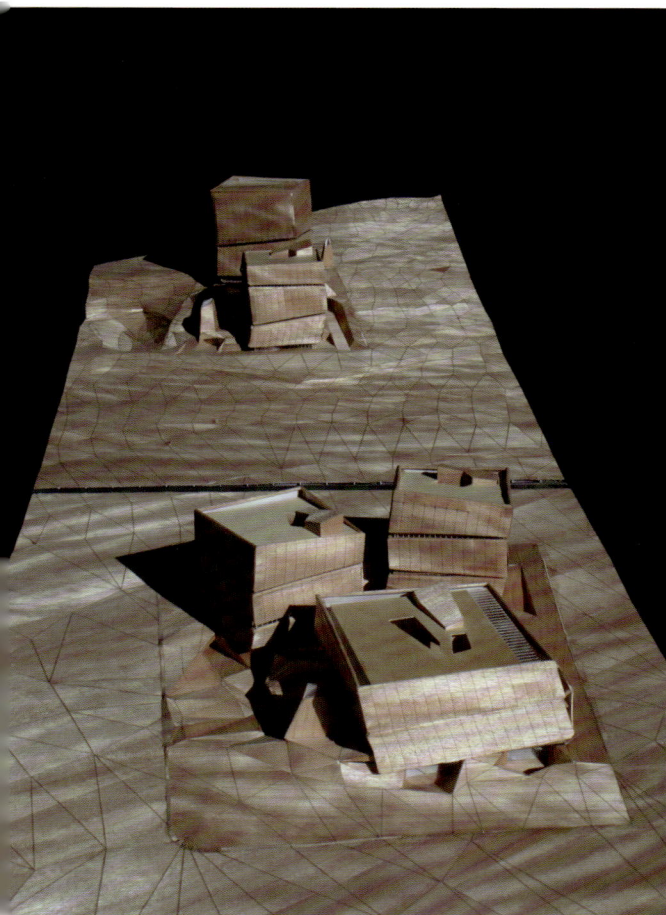

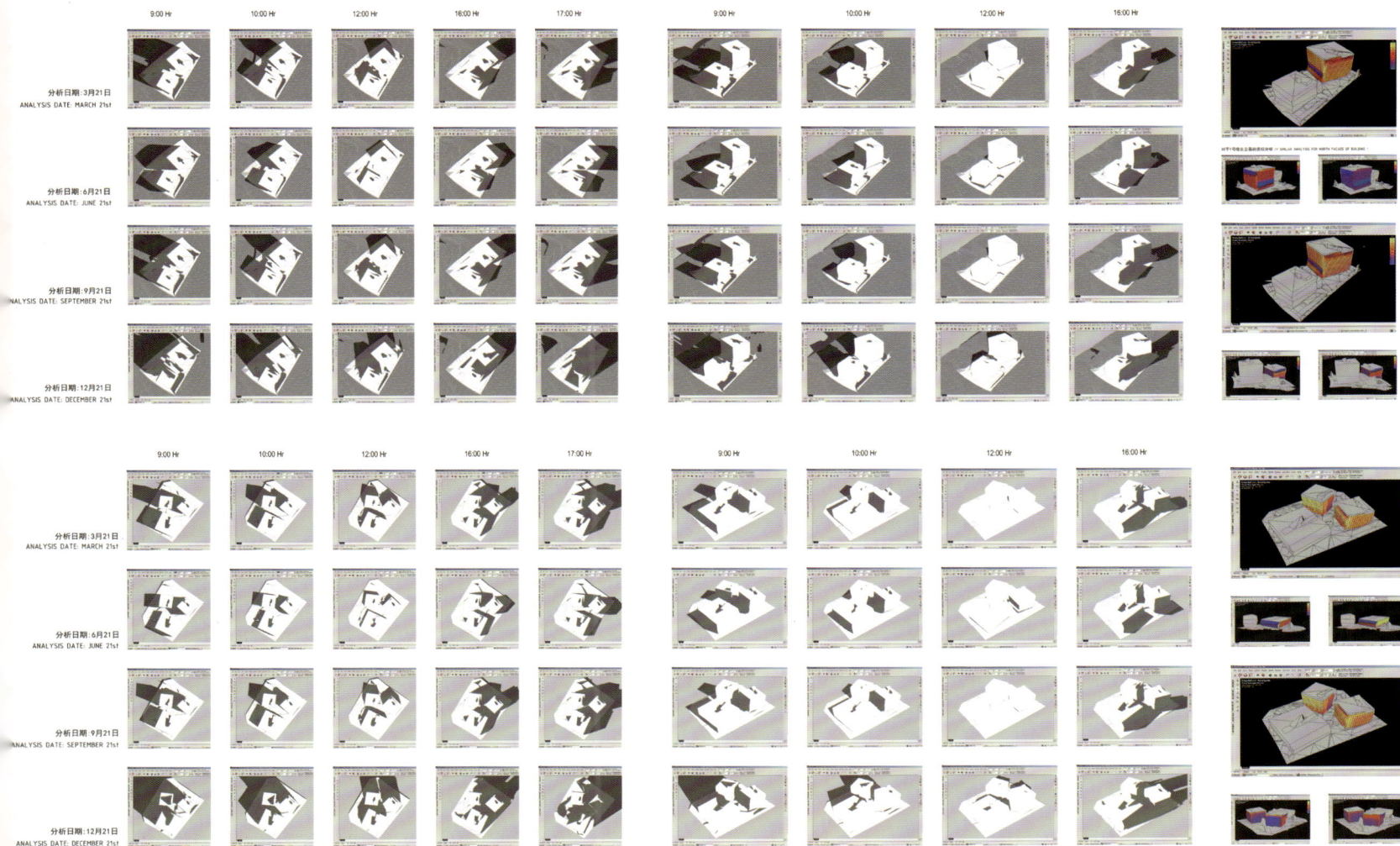

	9:00 Hr	10:00 Hr	12:00 Hr	16:00 Hr	17:00 Hr
分析日期：3月21日 ANALYSIS DATE: MARCH 21st					
分析日期：6月21日 ANALYSIS DATE: JUNE 21st					
分析日期：9月21日 ANALYSIS DATE: SEPTEMBER 21st					
分析日期：12月21日 ANALYSIS DATE: DECEMBER 21st					

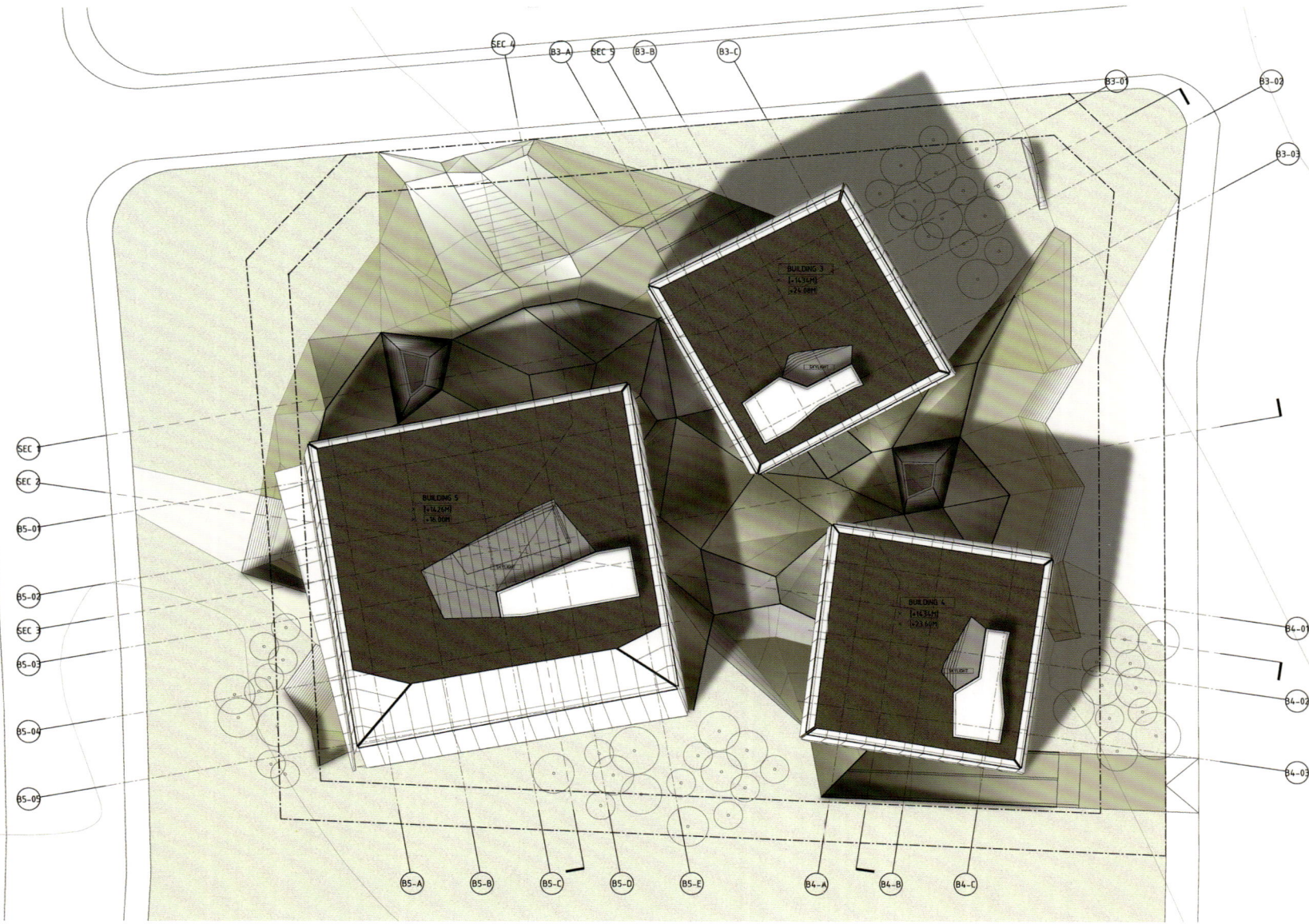

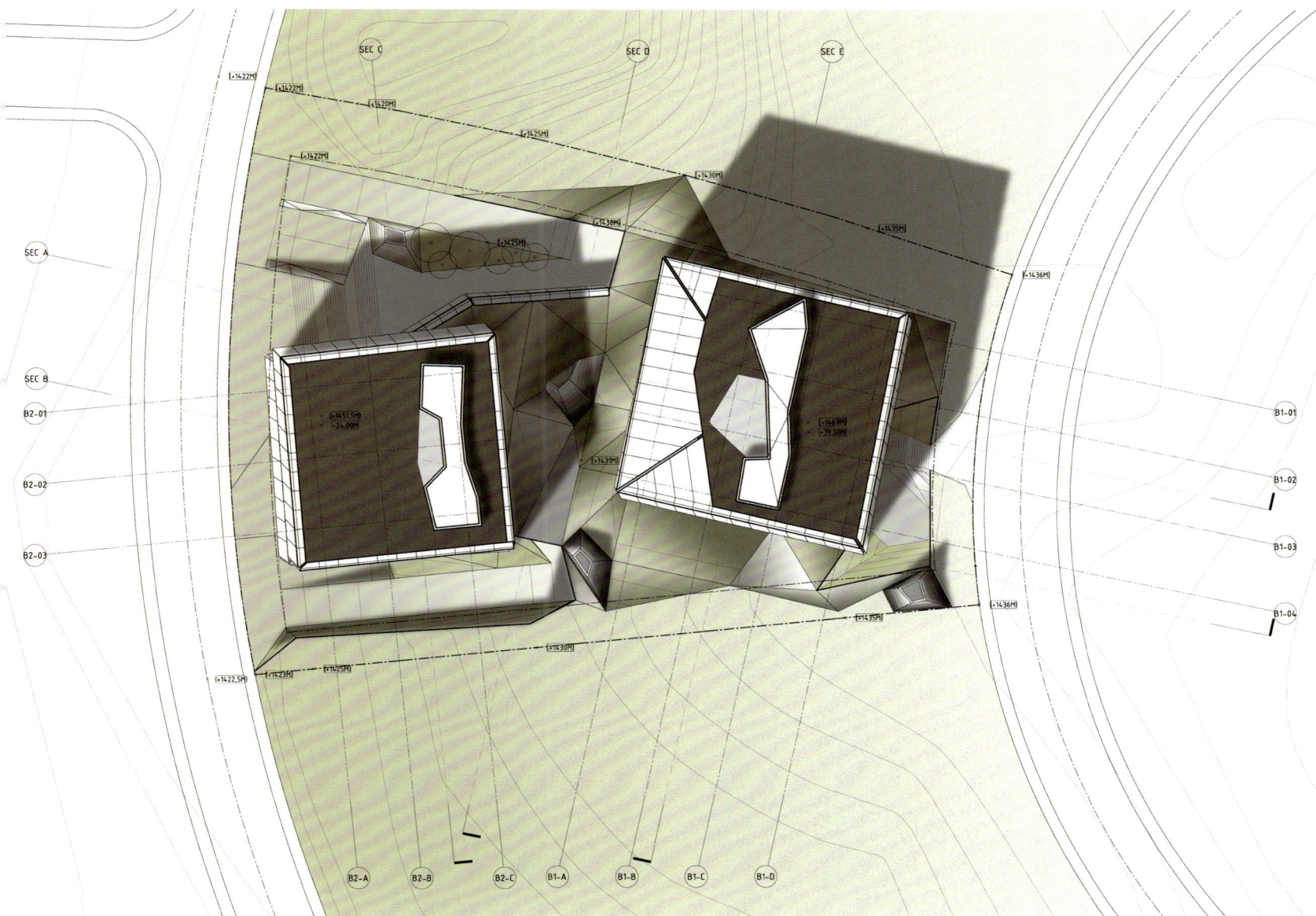

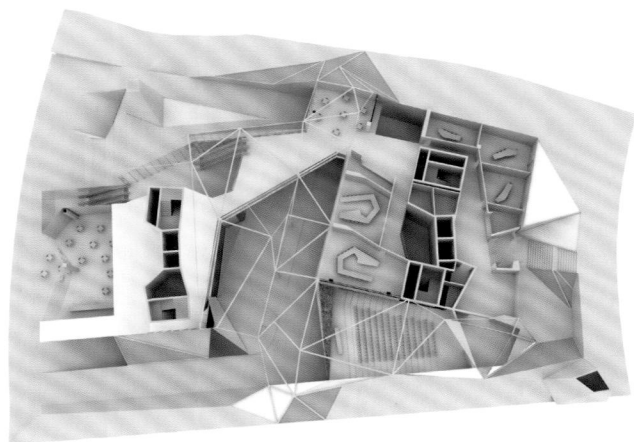

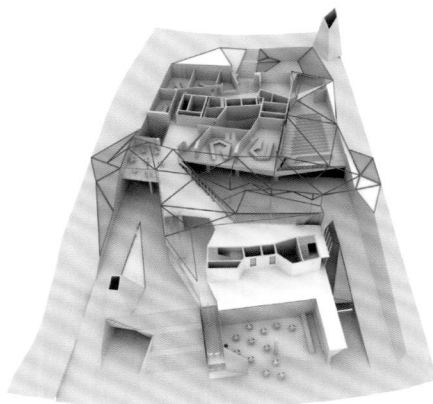

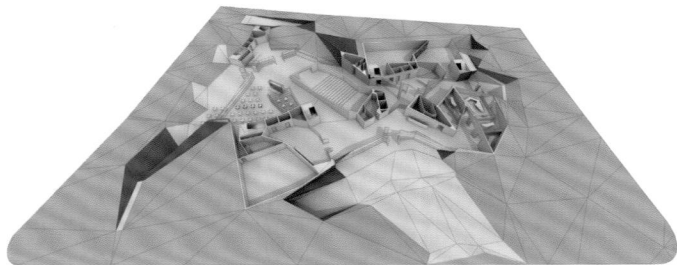

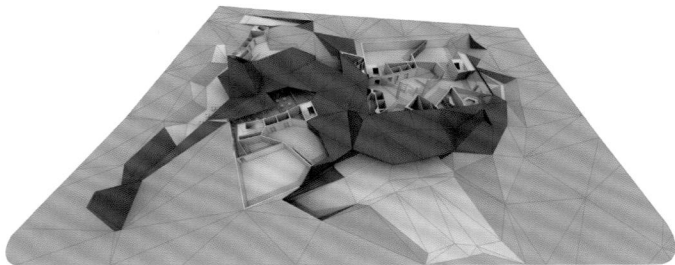

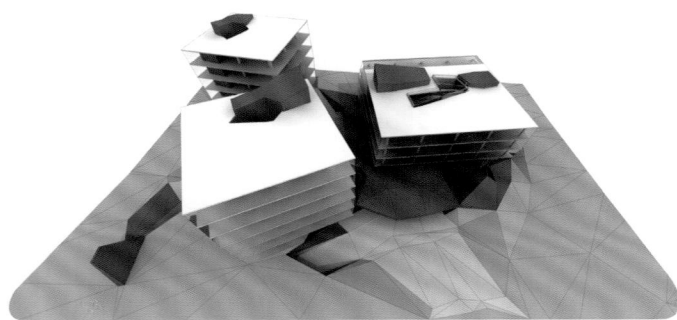

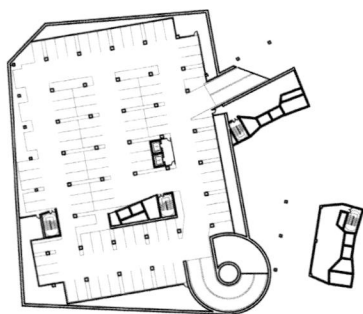

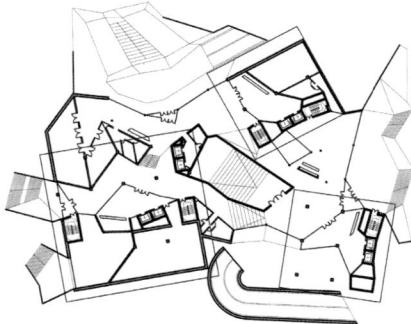

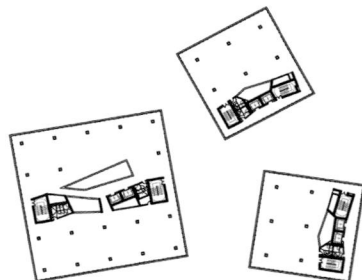

UNVEIL是一座办公建筑综合体，是鄂尔多斯20×10设计活动的参赛作品之一，它位于鄂尔多斯1800000m²的新金融区，是鄂尔多斯西北部东胜区第三个增长规划区。

本案的设计灵感来源于内蒙古的文化起源和自身的认可。

内蒙古的气候条件不是很好。寒冷的天气和沙尘暴教会人们如何与自然及它的组成部分相处：保护、遮蔽、环境智能建筑技术和每天多变的气温等等，这些方面将会主导建筑的设计。在过去，土壤是温度稳定性的一种资源。在本案中，建筑师们将此观点重新诠释，体现在分享的潜力这方面。通过开拓出深度，设计能够利用一个可共享的受到遮蔽物保护的公共广场空间将不同建筑连接起来。这个连续的空间将设有入口大厅、公共广场和共享公共设施。它还将被隔离起来的大堂连接起来，以保证一个安全的管理。

室内公共空间将会配有开阔的开口以保证室内有充足的阳光。这个不受天气影响的空间为室外沿着商业园区的行人通道提供了另一种室内的道路选择。

建筑师们提供了一种合理的典型楼层设计，可灵活适应所有大小不同的办公空间。由于公共设施都被安排在公共共享楼层，所以办公区域可以被完全利用起来。一间公司可以占据整个建筑，整个楼层或者是一层楼的一部分。各个入口大厅从视觉上连接到公共网络的空间，更广泛地增加了曝光机会，为它们增加了额外的价值，这些是小型建筑所无法提供的。

地面上的区域将会特别专用于办公区域。竖向的中心和中庭将会起到分配空间的用。中心部分都已根据定向和建筑的大小安排位置并设计。在小一点的建筑（26m×26m中，中心部分被推句北向的立面，为南侧的提供了更宽敞的办公区域。在大一点的建中，中心部分被安排在中心位置，将空间分隔为两个长条的办公空间。

垂直指向天空中的狭缝代表了一个更好、更加繁荣的未来的骄傲内涵。从公共空间始，到访者便可通过"天眼"发现鄂尔多斯的蓝色天空，这个"天眼"是连接地下室空间和方形办公空间的。这些竖向的中庭按照太阳的走向和盛行风的方向塑形，保证了然采光、被动加热和通风。

建筑的立面由一面织物幕墙包裹着。以PVC材料为主的织物幕墙并不是与建筑围护构平行的，按照城市规划"立方的"要求，给了建筑一定的自由空间。同时，不同的变也赋予了空间条件作为入口遮雨棚或者建筑基地连接。外部的织物层可以根据天气状况起或者放下。白天，它将会被卷起，使建筑物能够得到热量。而在夜晚，它将会被放下以保持建筑物内部的温度。这个变化将会为建筑带来一个特征：早上UNVEIL办公楼欢迎家的到来，晚上建筑关闭，象征着一天的工作结束。

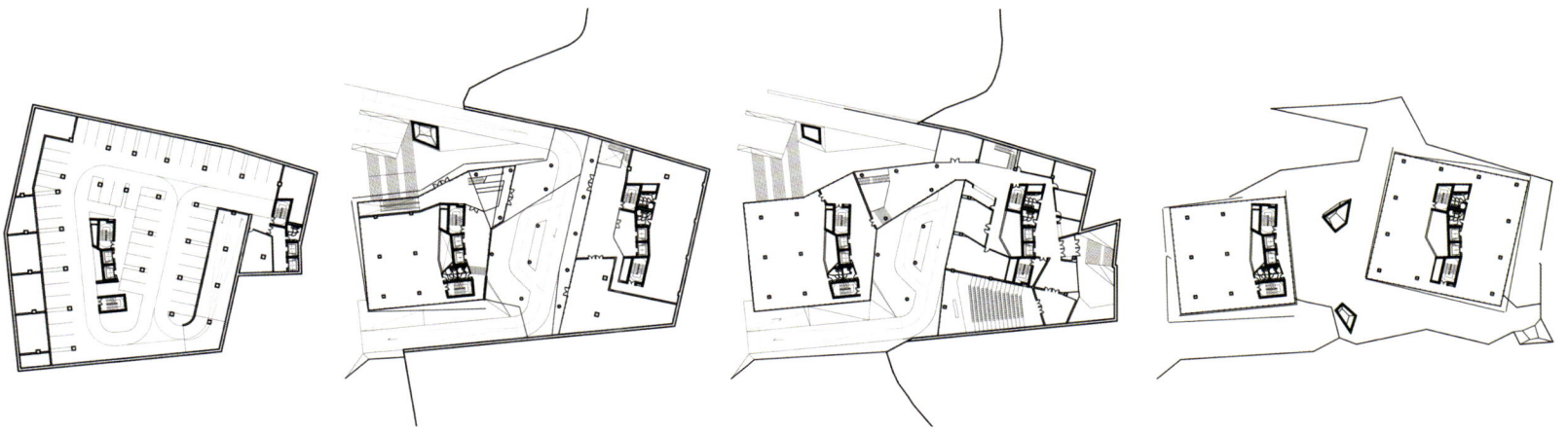

Innsbruck

因斯布鲁克

Firm: studioMDA
Location: Austria
Area: 2,787m²

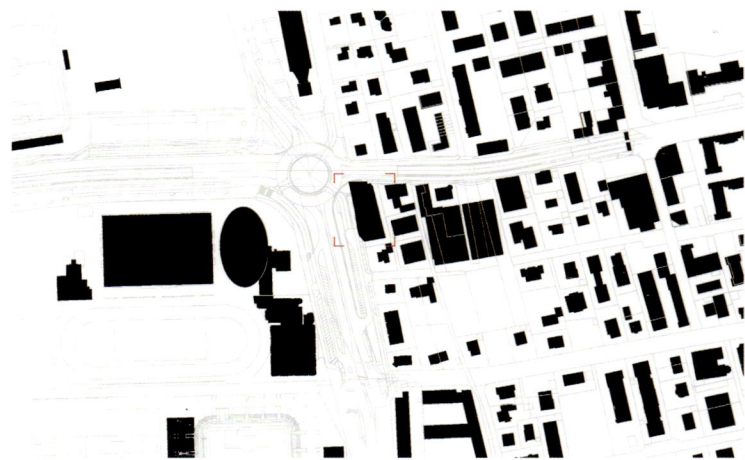

This iconic conference center in Innsbruck is a private development with a visual connection to the mountains. The design signifies the changing beauty of the surrounding nature, embodied in a reference to melting ice. This visual analog creates a still of an element reinventing its physicality; the interior freeing itself as it dissolves it surroundings. As an object signifying a past, one sees the erosion that must have taken place to form the smooth concrete envelope. It remains as the firmitas of a structure able to withstand the natural forces implied, similar to canyon walls, or glacial cirques.

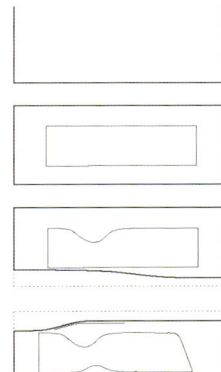

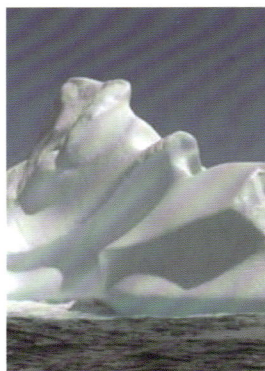
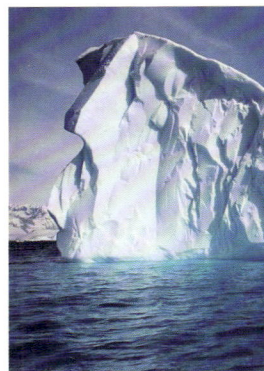
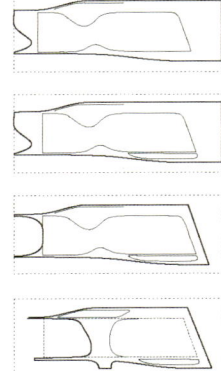

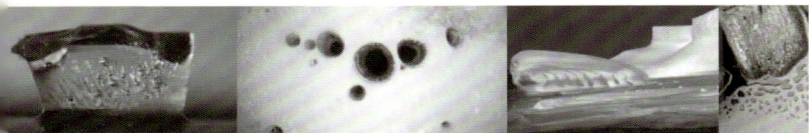

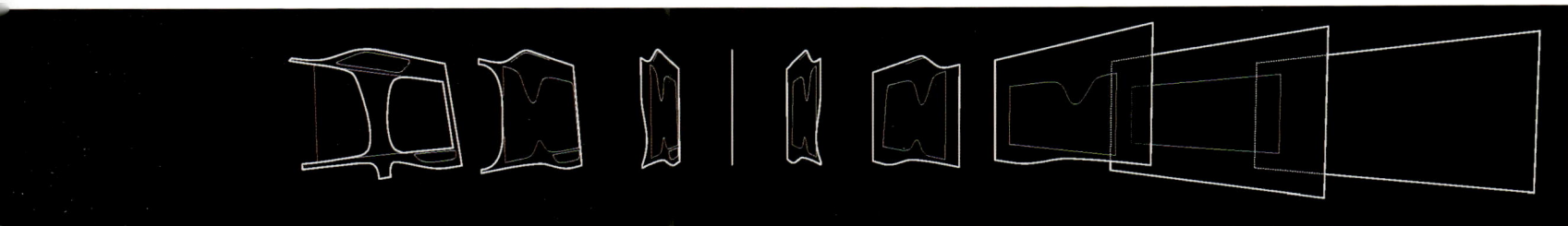

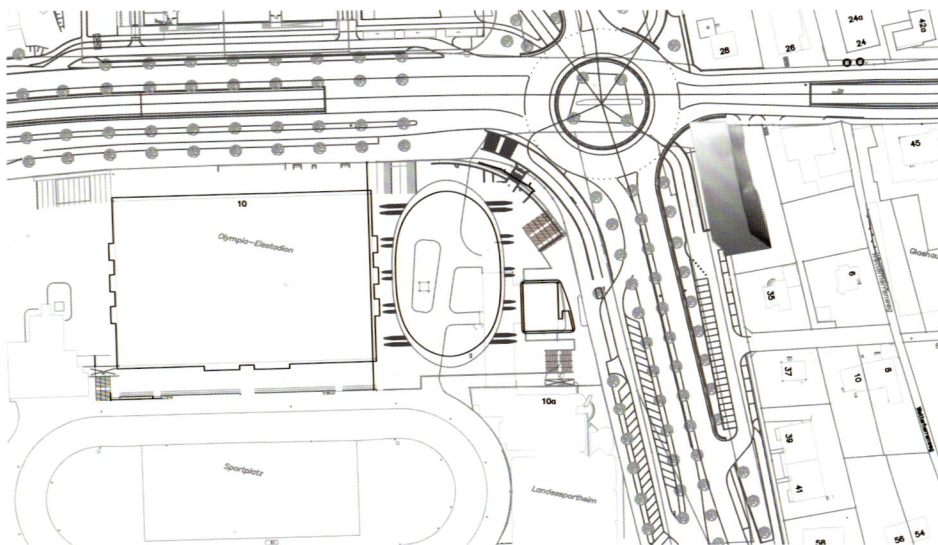

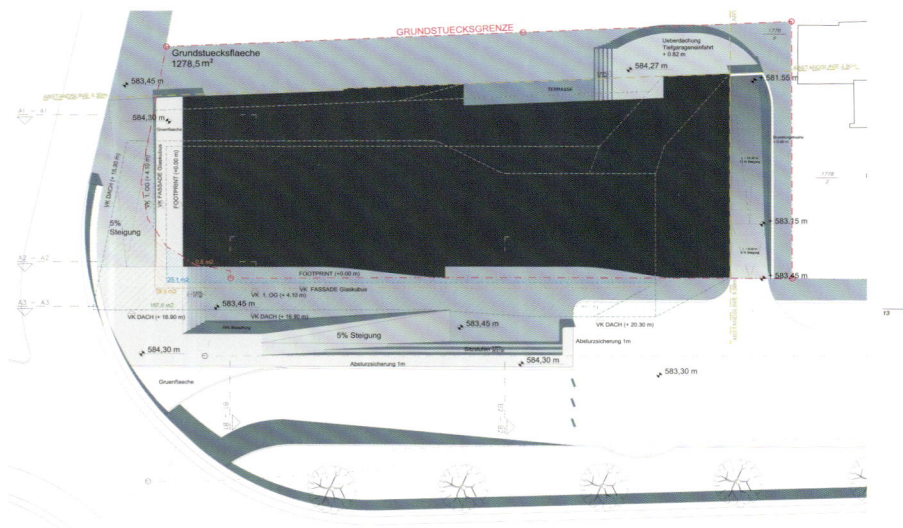

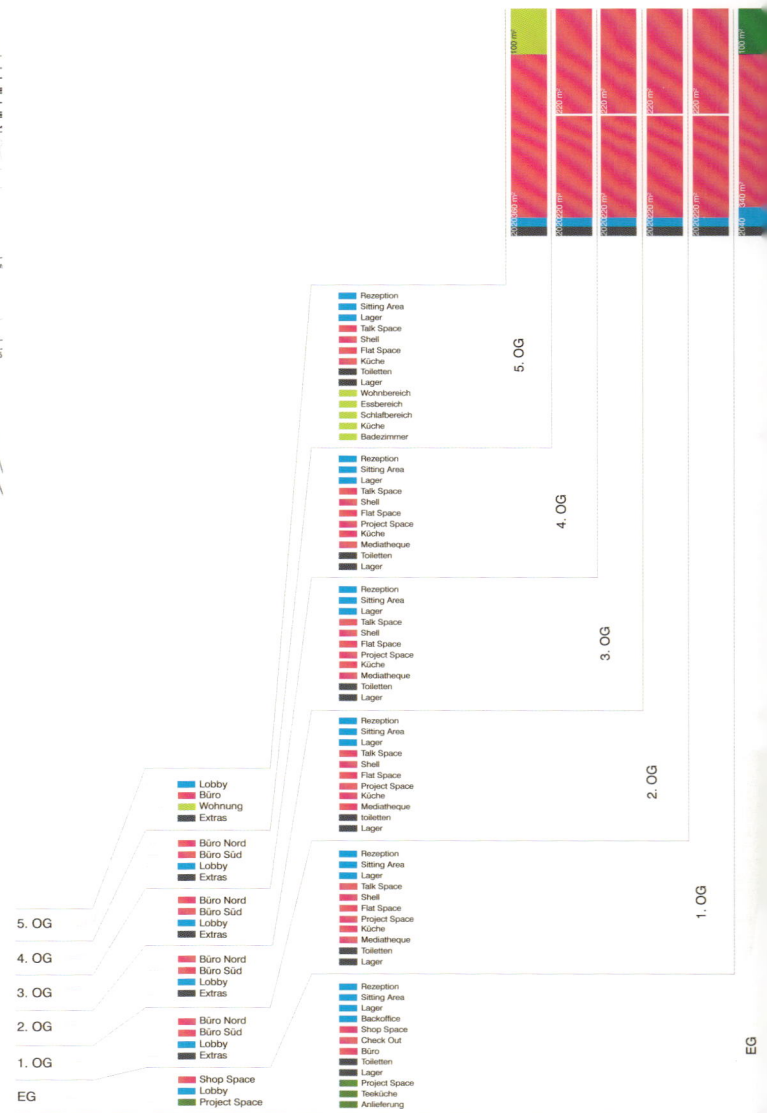

AUSSENBEREICH NORD-WEST 1:100 A1

GRUNDSTÜCKSGRENZE NORD

AUSSENBEREICH WEST o.M. A3

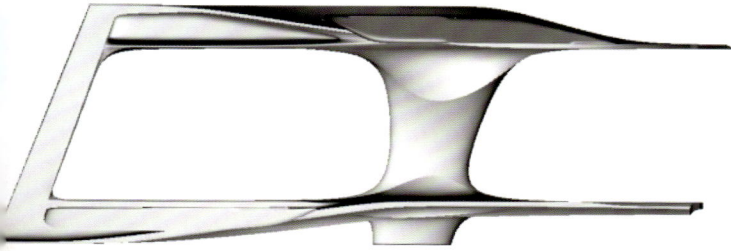

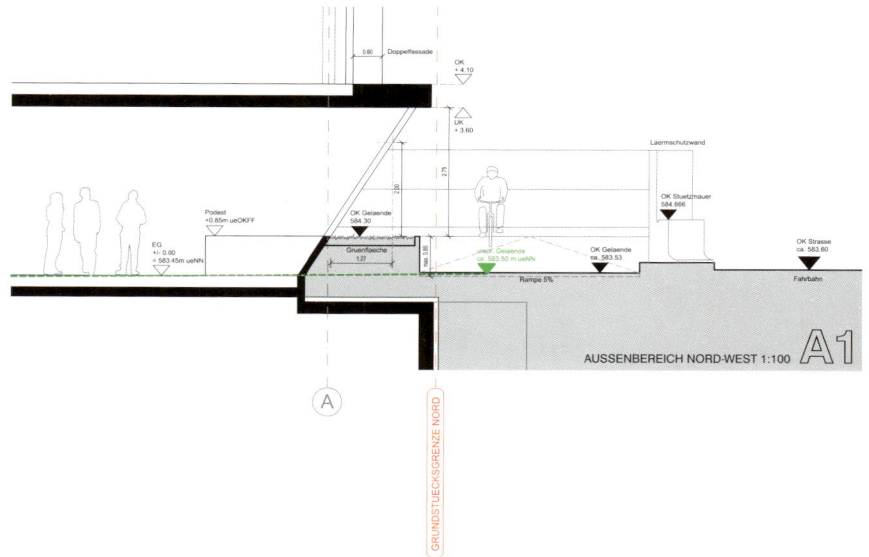

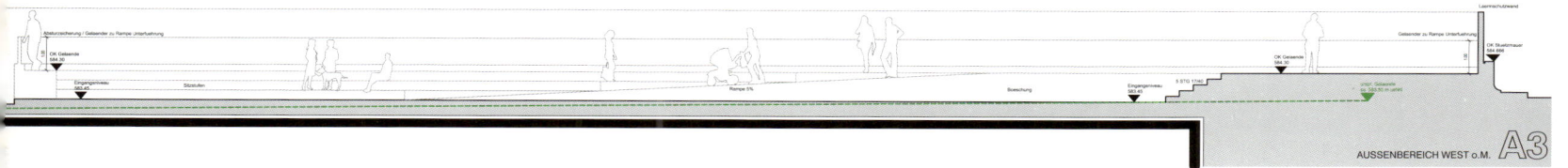

BÜRO ORGANISATION 2

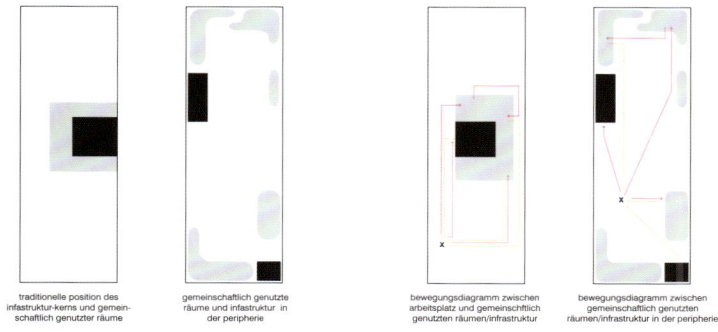

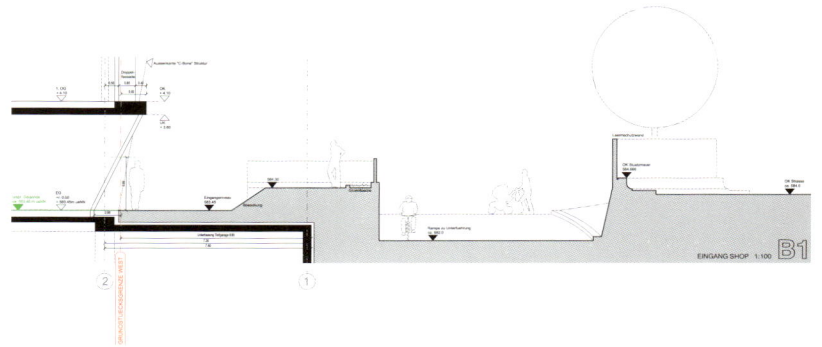

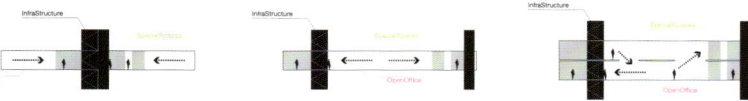

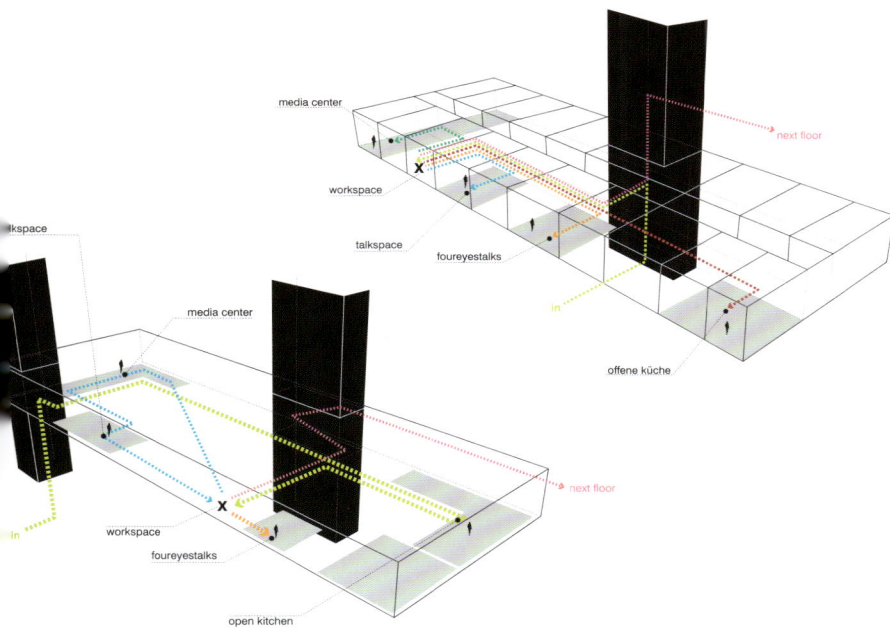

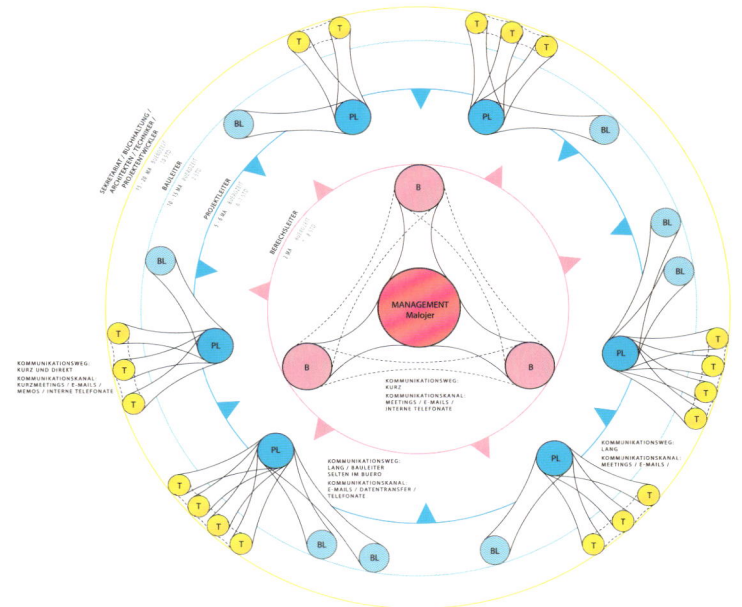

traditionelle position des infastruktur-kerns und gemeinschaftlich genutzter räume

gemeinschaftlich genutzte räume und infastruktur in der peripherie

bewegungsdiagramm zwischen arbeitsplatz und gemeinschftlich genutzten räumen/infrastruktur

bewegungsdiagramm zwischen gemeinschaftlich genutzten räumen/infrastruktur in der peripherie

EINGANG SHOP 1:100 B1

media center

workspace

talkspace

foureyestalks

next floor

open kitchen

offene küche

MANAGEMENT
Malojer

TYPOLOGIE	PROGRAMM	POSITION		SICHTBARKEIT			AKUSTIK	
				durch Separation	durch Typologie	durch Entfernung	durch Separation	durch Entfernung
BUERO	Talk Space		Tageslicht sehr wichtig, keine Fensterläden. Aussicht wichtig. Nah zur Infrastruktur. Keine diagonalen Wege. • Seminarräume • Besprechungsräume					
	The Shell		Tageslicht sehr wichtig, keine Fensterläden. Aussicht. Ruhiger Bereich, Keine Geräusche. • abgeschotteter, ruhiger Arbeitsbereich für Telefonate, o.ä.					
	Project Space		Auf Grund von Projektionen und Medien kein Tageslicht notwendig. Nah zur Infrastruktur					
	Flat Space		Offene Bürofläche					
	Küche		Nah zur Infrastruktur, Sichtbar von allen Positionen im Büro, Hoch frequentierter Bereich					
	Mediatheque		Tageslicht , Aussicht • Bibliothek • Kataloge • Materialien • Digitale Datebank					
LOBBY	Rezeption							
	Lobby							
	Lager							

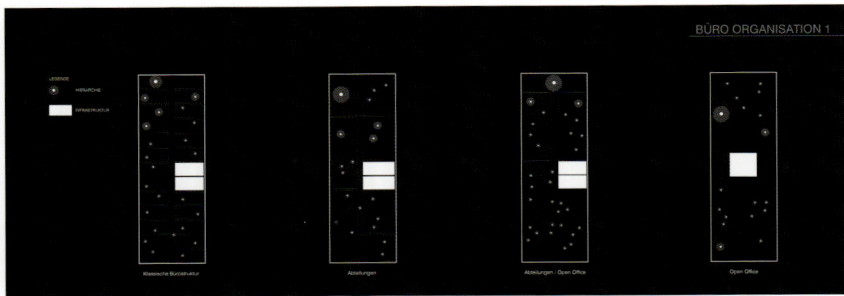

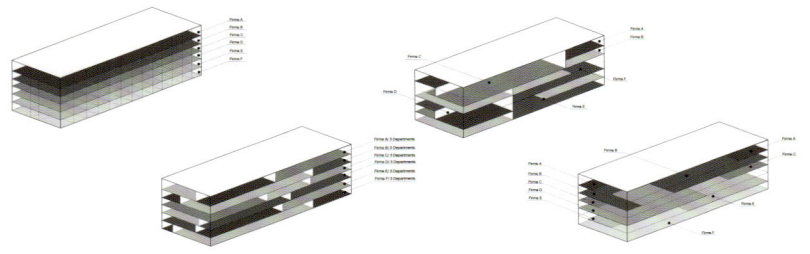

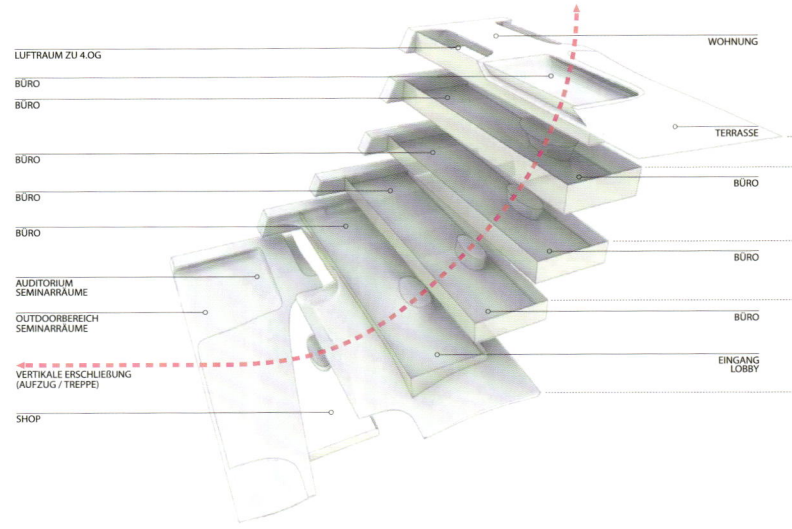

BÜRO ORGANISATION 1

FASSADE	BUFFER ZONES • EAST AND / OR WEST •			EXTERNAL SHADING • EAST / WEST • // SELF SHADING • WEST		WINTER GARDEN • SOUTH •		
	winter day	winter afternoon / night	summer day	external shading	self shading	winter day	winter afternoon / night	summer day
BESCHREIBUNG	passive solar heat to space	closed for insulation		lowers in summer, opens in winter (but then glaze control is needed inside). can also serve only as glaze entre (no winter passive glaze heating benefit).	reduces solar again, but fully glazed facade will still need external shade. partly glazed ok.			
BESCHREIBUNG	shade can be admitted with deep zone (overhang effect) at least one facade needs good insulation "ZONE" can be as thin as 10cm, or amy size larger							

KATALOG KLIMA

这个位于因斯布鲁克的标志性会议中心是一个私人发展项目，建立了一种与山峦的视觉联系。设计呼应了周围环境的自然变化之美，体现在以融化的冰作为设计参考。这种视觉模拟创造了一种重塑其形体的静止的元素，在与周围环境相融合的同时其内部本身也得到释放。作为一个象征着过去的物体，人们可以看到光滑平整的混凝土立面上那些仿佛被侵蚀了的痕迹。它就如同是这个建筑的签名般，抵御着自然力量的暗示，就好像峡谷壁或者冰川。

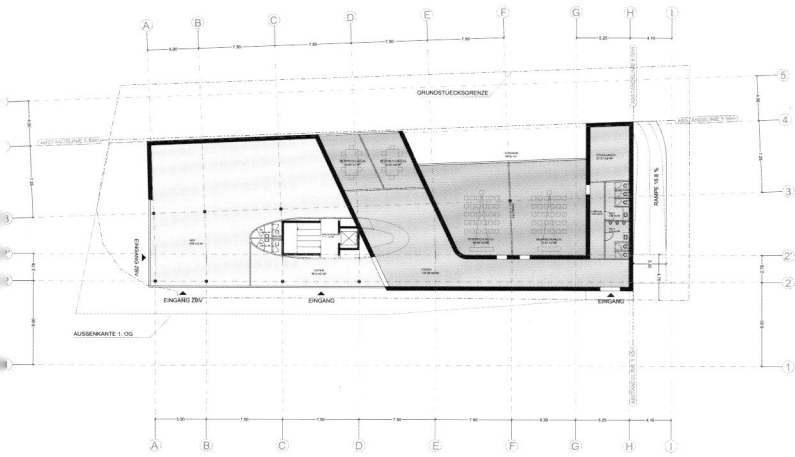

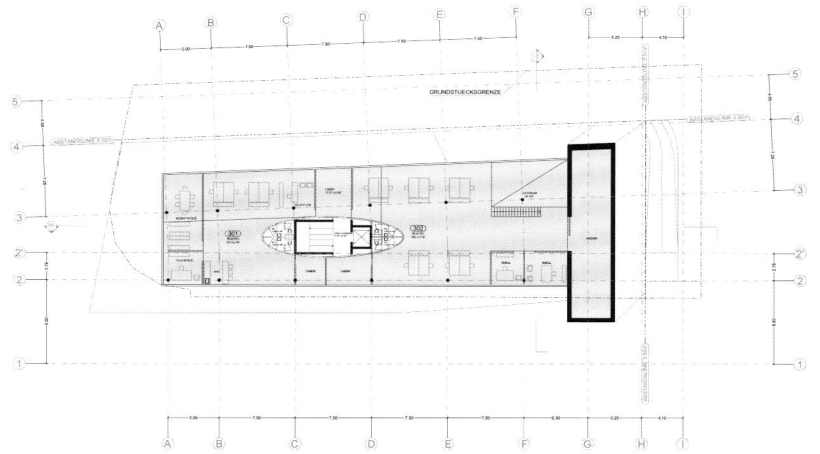

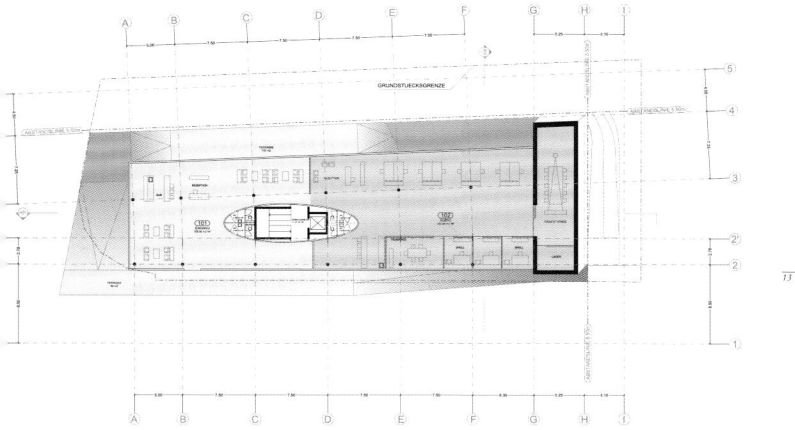

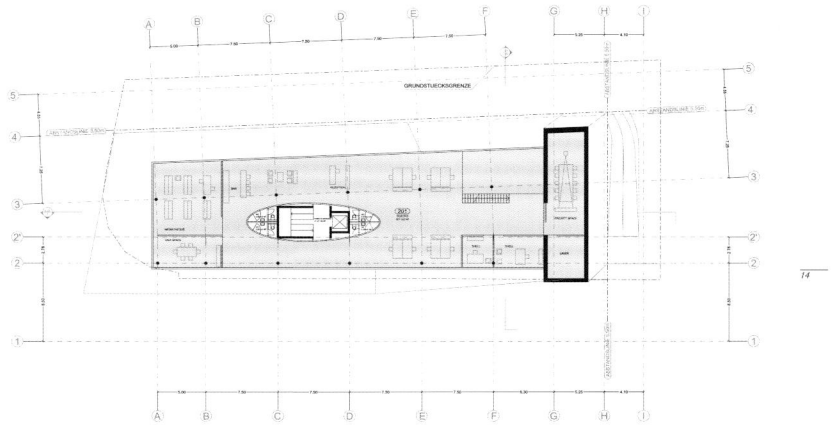

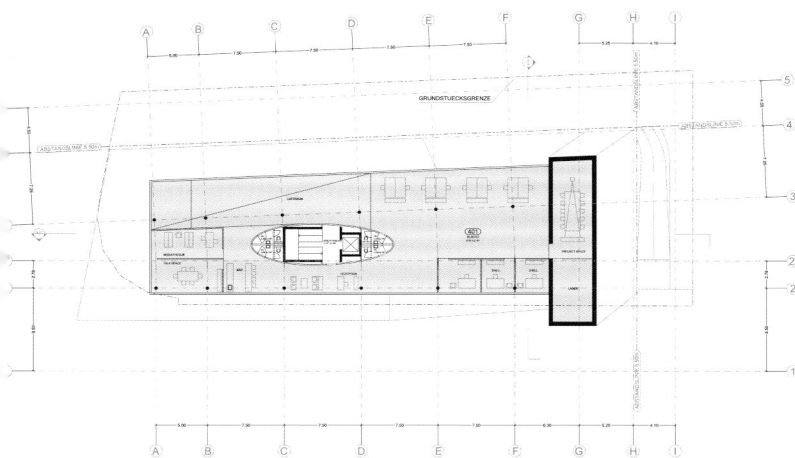

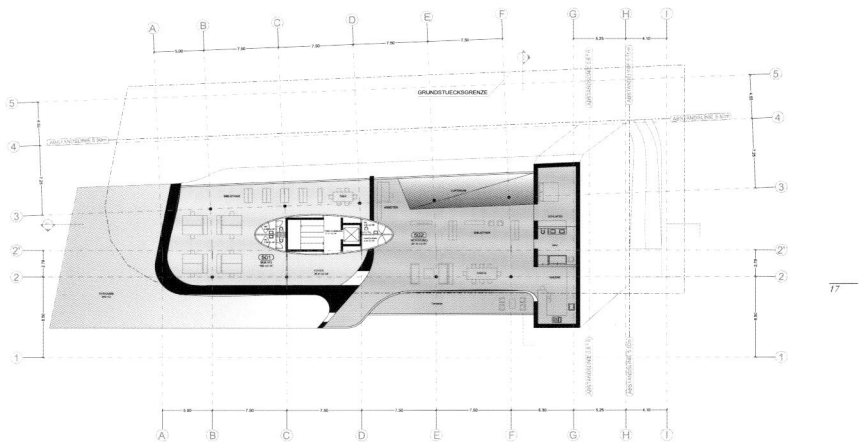

The New Earth

新的地球

Architect: Timur Bashkaev., Mikhail Krymov, Alexey Goryainov
Firm: T.Bashkaev's architectural bureau
Location: Shanghai, China
Area: 4,130m²
Cooperator: Arch Group

144

The theme of EXPO 2010 is the modern city and possibilities of developing and improving the quality of life. The proposed design reflects the theme as it represents a new urban structure segment on a small scale and serves as a symbol of a modern approach to urban development and reconstruction.

The pavilion's idea is based on an innovative concept, which is revealed through the exhibition in the atrium gallery.

This concept is a principled and universal proposal for modernizing any megalopolis. It has been created using the experience in designing numerous transportation and transfer points and rebuilding Moscow industrial zones. The necessity of creating a new approach to megalopolis development is long overdue, and the architects see it possible to use their concept for creating the ideology and image foundation for the Russia Pavilion at EXPO 2010.

The pavilion comprises three zones, each having different meaning and space:

Covered exhibition space under the arc (3,300m²) is intended for the main exhibition, forums and events. The absence of vertical supporting units and the use of light mobile stained glass partitions allow for any hall configuration and for outlay.

Open green arc cover is intended for walking on, and sightseeing of both the pavilion and the exhibition from, a different, green recreational level. Also the entrance to the capsules is located on the same level. The body of the arc contains a 400m² round hall. The exhibition dedicated to urban development ideas is built along the perimeter of the hall, which includes 4 niches.

Third level is located inside the four capsules. One capsule is built into the arc, and three capsules are flying. The exhibitions in the capsules approximately follow the topics of sports, education, arts, culture, and the main subject, family values.

The proposed design of the Russia Pavilion at EXPO 2010 is aimed at creating an attractive image of the nation and reflects the exhibition's main theme, modeling a city of the future.

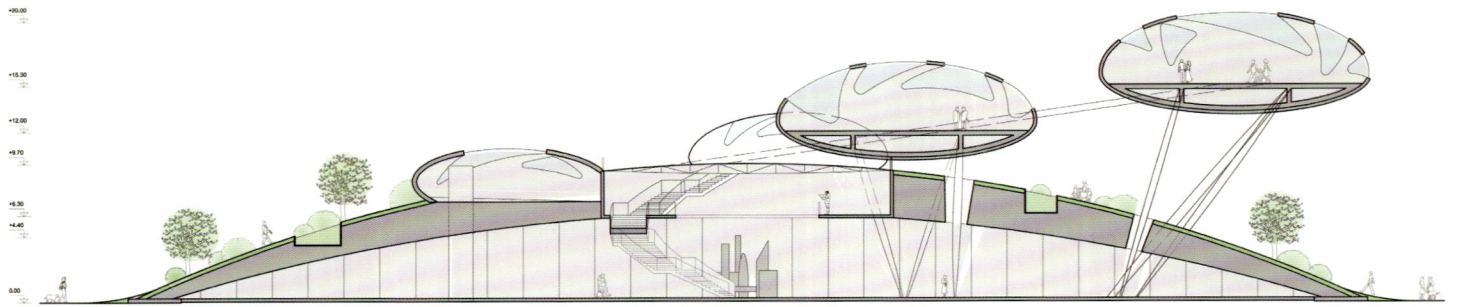

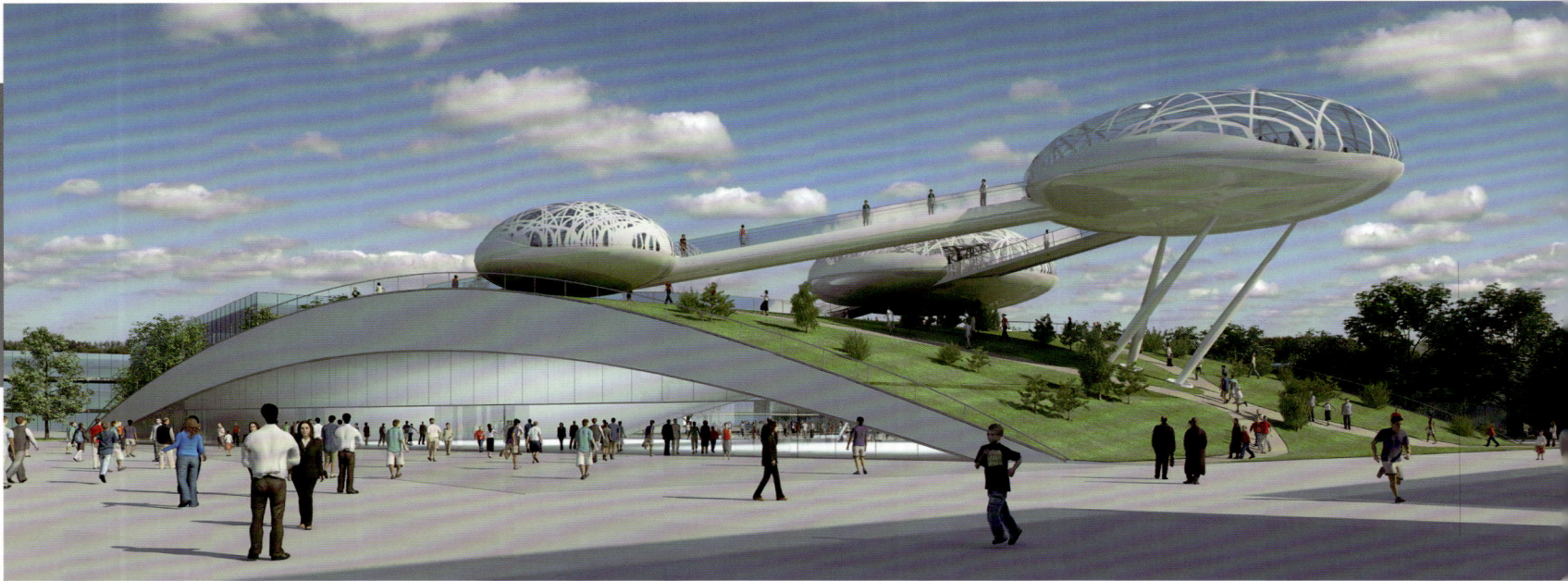

2010年世博会的主题是现代城市与发展和改善生活质量的可能。设计师所提出的方案体现了这个主题，这是一个小规模的新城市结构的代表，并作为城市发展与重建的现代方式的代表。这座展馆的主题是以一个创新的概念作为基础的，而这种概念可以通过在天井画廊的展览体现出来。

这一概念作为一种具有很强原则性并被普遍接受的提议用以加快大都市的现代化发展。这种概念来自与于对设计大量运输和交通节点以及重建莫斯科工业区的经验的应用。长久以来，大都市的发展就一直需要一种新的方式，而在设计2010年世博会俄罗斯馆的概念和形象基础的过程中，建筑师们看到了采用这种新方式的可能性。

整个展馆分为三个区。每一区都有不同的含义和空间。

在弧形顶棚下面的展览空间（3300平方米）被设计成为主要展览、论坛和其他活动提供空间。无垂直支撑结构的设计以及采用轻质可移动式玻璃隔断可以使展厅采用任何一种格局并节省了开支。

开放式绿色弧形顶棚可以供参观者在上面行走，从一个与众不同的绿色娱乐层来欣赏展馆与展览。同时，进入太空舱的入口也设计在这一层。弧形顶棚的主体包括了一个400平方米的圆形大厅。表现城市发展主体的展览即是沿着大厅的边缘布置，包括了四个区域。

第三层次位于四个太空舱之中。其中一个太空舱建在了弧形顶棚里面，其余三个舱则悬空布置。在太空舱中的展览主要是关于体育、教育、艺术、文化和作为主题的家庭价值等话题。

2010年世博会俄罗斯馆的设计方案意在创造具有吸引力的国家形象的同时反映世博会的主题——展示未来城市的图景。

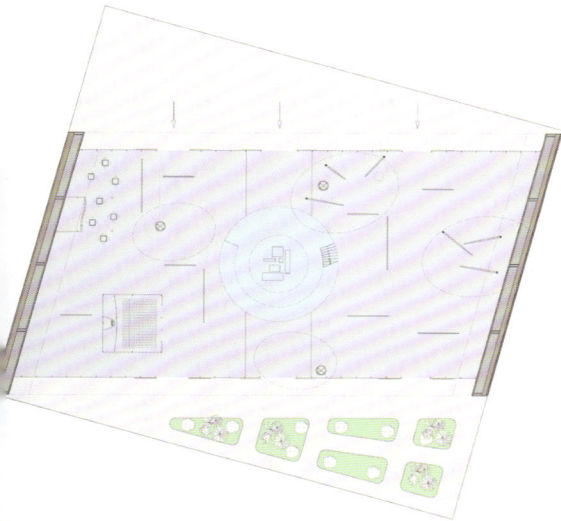

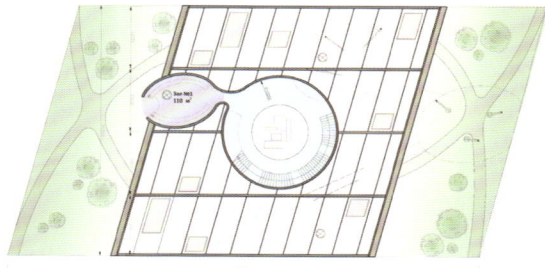

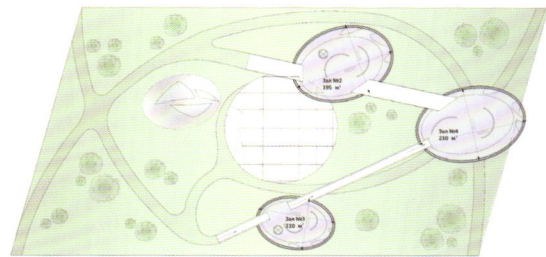

Aachen

亚琛

Firm: studioMDA
Location: Aachen, Germany
Area: 4,645m²

The Centre of Advanced Mobility is a groundbreaking symbiosis between nature and technology, functionality and dynamic form, as it combines high-performance laboratories with the urban environment. Wind studies were used to develop the dynamic shape of the building which provides an optimal wind flow for the Aachen inner city and a maximization of ventilation for the Centre.

The challenge was to develop a new building with 50,000m² total net area for the departments electrical + information engineering, mechanical engineering, and aerospace engineering for the University of Applied Science. The transparent design of the interior will allow for barrier-free visual communication. The combination of theory and pragmatism will facilitate integration between public and semi-public spaces and will stimulate intercultural and interdisciplinary collaboration. The Campus maximizes the use of green space and creates public and semi public recreation zones with a total of 700m².

wind studie

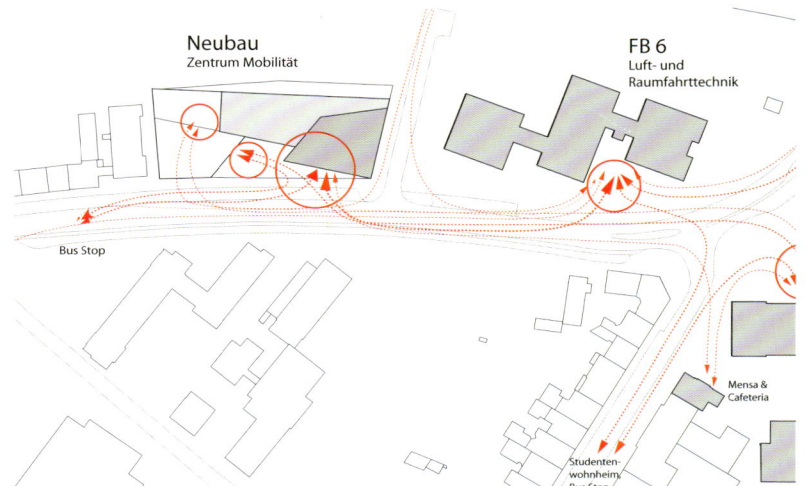

Neubau
Zentrum Mobilität

FB 6
Luft- und
Raumfahrttechnik

Bus Stop

Mensa &
Cafeteria

Studenten-
wohnheim
Bus Stop

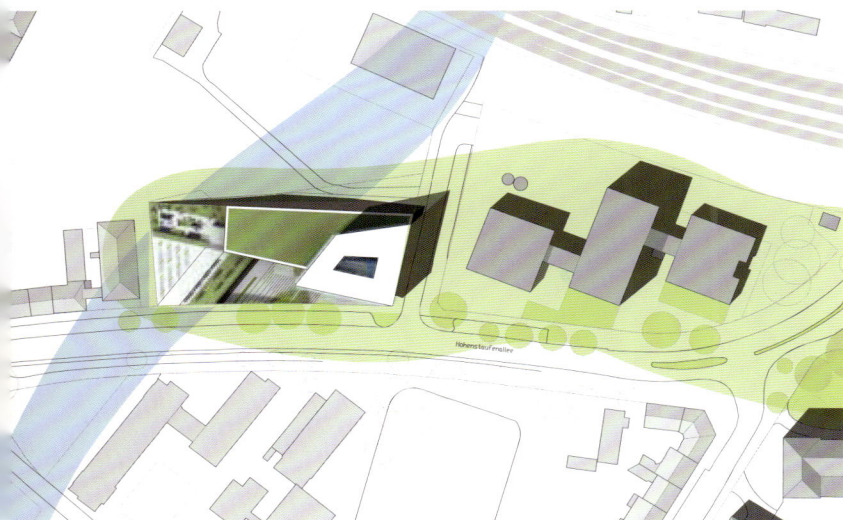

RENDERING . VIEW HOHENSTAUFFENALLEE

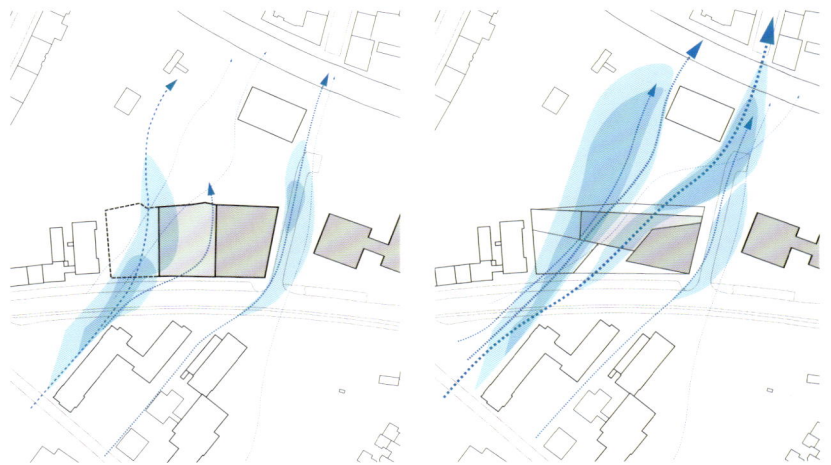

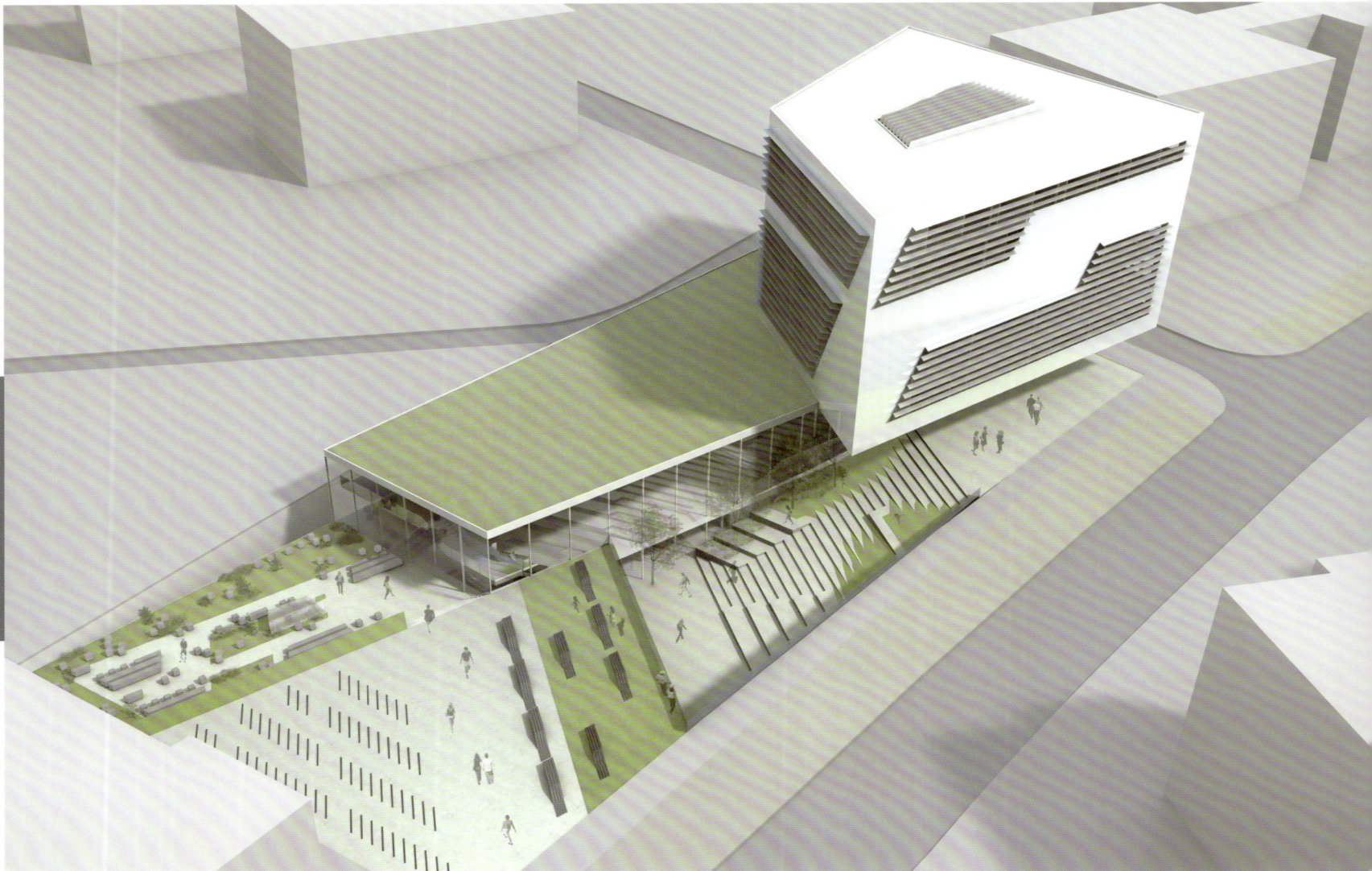

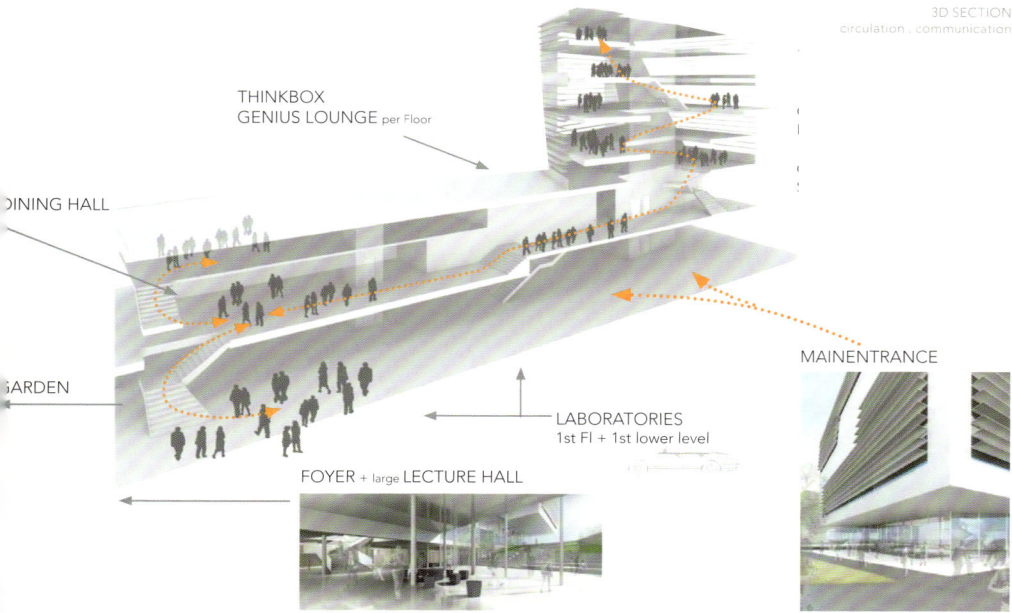

3D SECTION
circulation . communication

THINKBOX
GENIUS LOUNGE per Floor

DINING HALL

GARDEN

MAINENTRANCE

LABORATORIES
1st Fl + 1st lower level

FOYER + large LECTURE HALL

高级移动性能中心创造了自然与科技、功能与动态形式之间开创性的共生关系，它将高性能实验室与城市环境结合起来。风向研究帮助确立建筑的动态造型，为亚琛内城区提供了最佳风流量，并且也为本中心带来最佳的通风。

设计的挑战在于建造一座总净面积为5万平方米的新建筑，用作应用科学大学的电力及信息工程学院、机械工程学院和航空航天学院的新院址。建筑内部的透明设计形成一种无障碍的视觉沟通。理论与实用主义的结合将促使公共与半公共空间的一体化，促进跨文化和跨学科的协作。校园将绿色空间的使用最大化，设计了总面积为700m²的公共与半公共的游乐区。

151

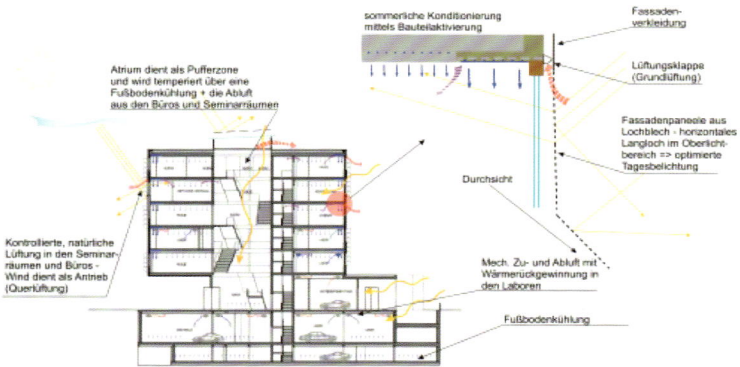

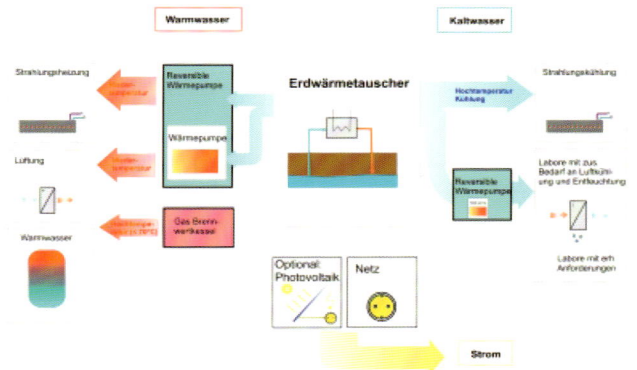

Energy Concept

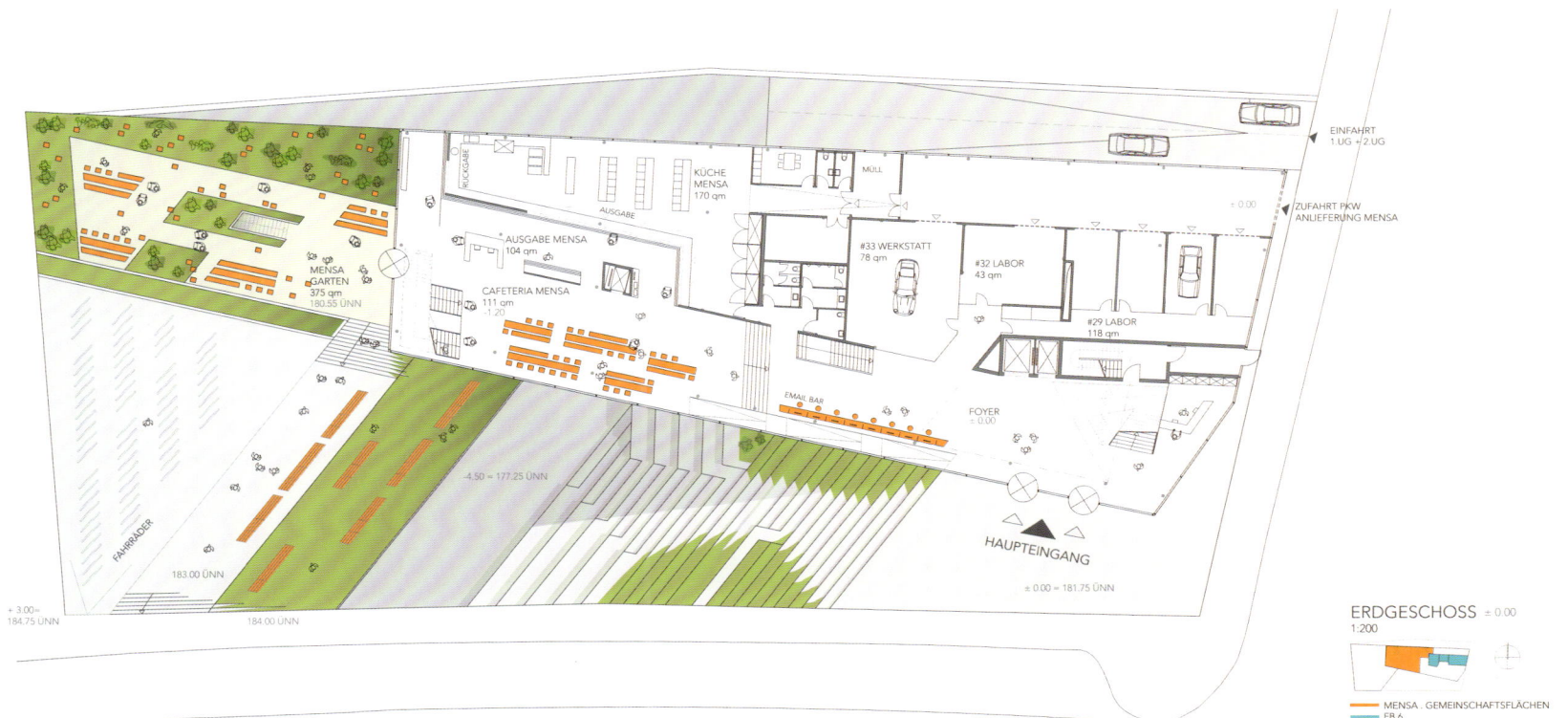

ERDGESCHOSS ±0.00
1:200

MENSA . GEMEINSCHAFTSFLÄCHEN
FB 6

The Danish Pavilion EXPO 2010

丹麦2010世界博览会展馆

Firm: CEBRA
Location: Shanghai, China
Area: 3,200m²

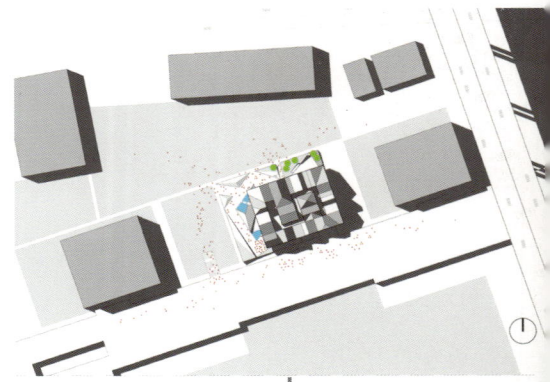

From the general theme "Better City – Better Life" CEBRA has created a strong, narrative concept for both the exhibition and the exhibition space itself, that will boost the global branding of Denmark and the core competences leading up to the UN Climate Conference in Copenhagen (COP15) and beyond.

Emerging from the global climate challenges at hand and the joined responsibilities as individuals and as humanity, the architects actively use experience economy, user orientated interaction technology and dramaturgy tools to interpret the national skills to the global stage. The Pavilions main purpose is to show what Denmark can do for you as an individual, and through that knowledge getting an awareness of the strengths and possibilities in corporation.

The exhibition itself is a journey through innovative Danish products and technologies as well as an introduction to the Danish way to approach sustainability. The exhibition is separated into 3 layers; CO-experience, CO-

creation and CO-existence. Each area has an distinctive flagship area, a area allocated to products and finally an area that is interactive. Startir within the typical Danish home the exhibition then lead the visitors throug a typical Danish city and finally on to Denmark seen in a global context. Th Home area and the City area is inside the pavilion, while the global conte takes place on the rooftop, on the outside looking out towards the enti EXPO area.

The ground floor is designed as archetypes of 9 typical Danish landscape which marks both the entrance and the exit of the Pavilion.

The building has a very small footprint which limits the impact on th landscape as well as it doubles as a shelter for rain and sun.

From each area you will learn more about the common responsibility ar the possibilities in cooperation and get a hands-on knowledge of Danis innovations and creative solutions in regard to sustainable solution sustainable products and the sustainable road forward.

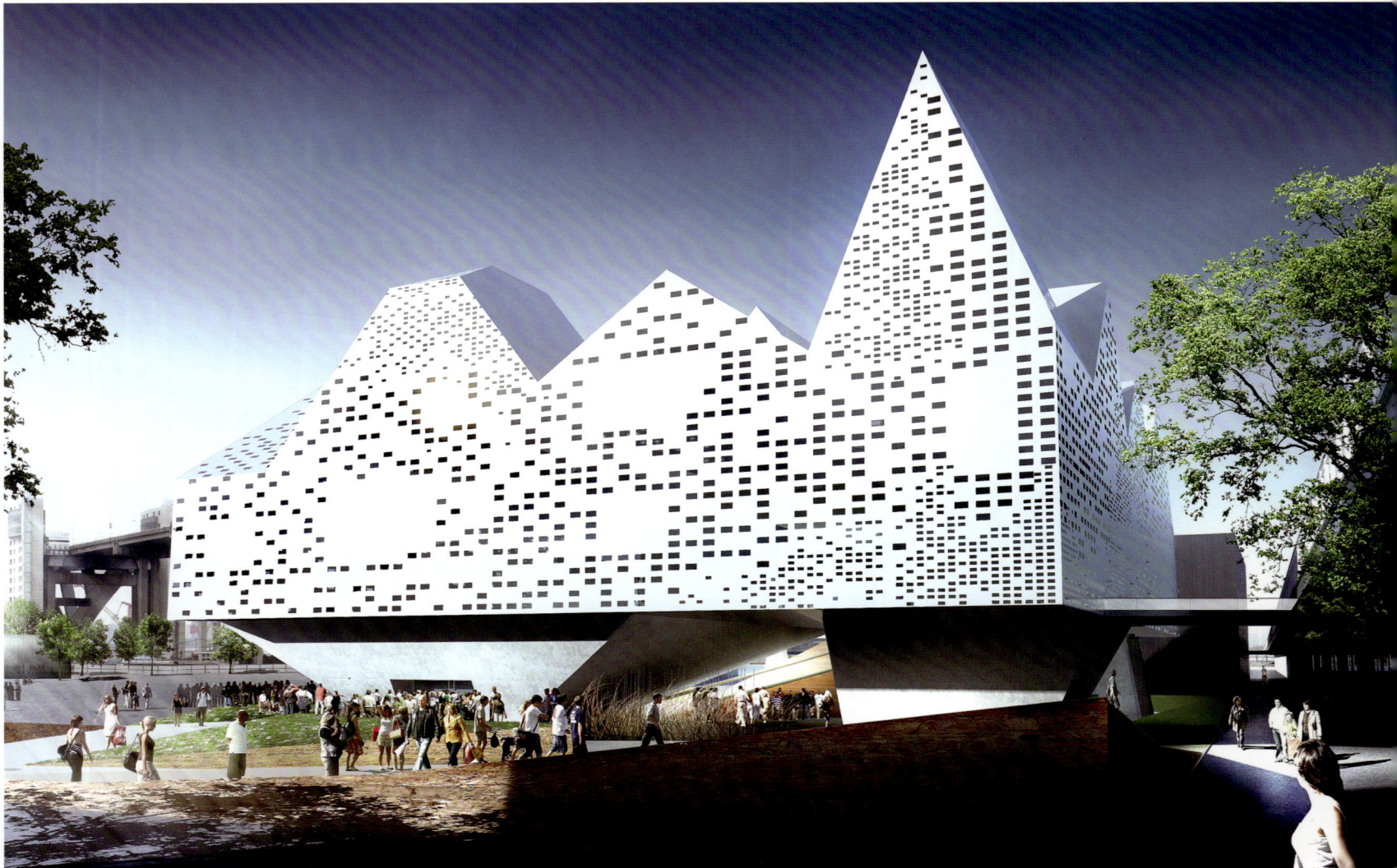

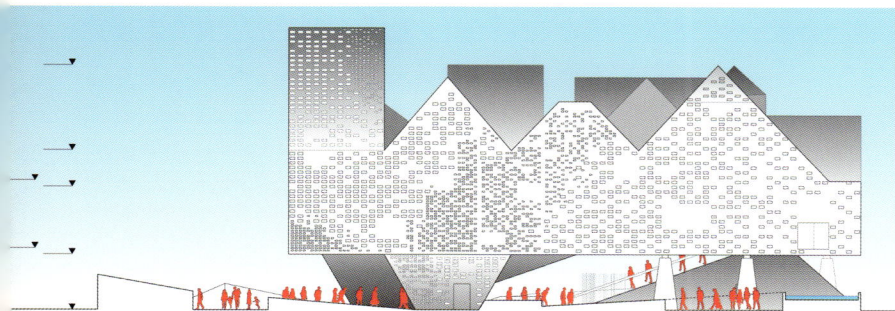

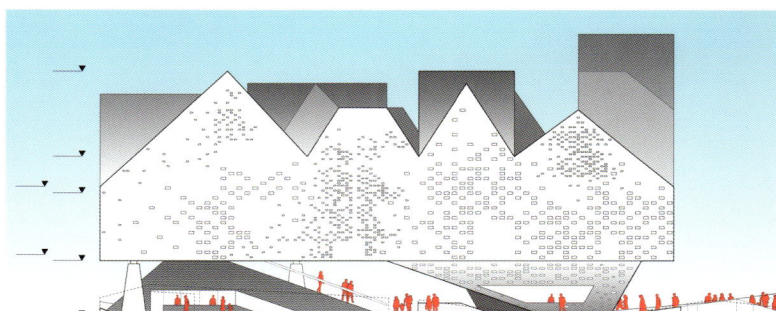

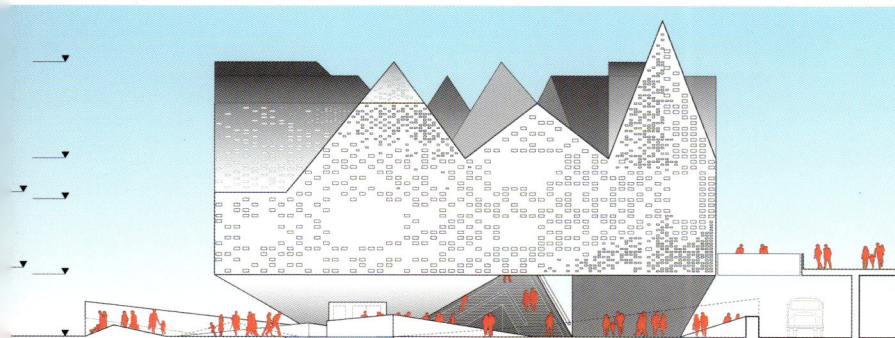

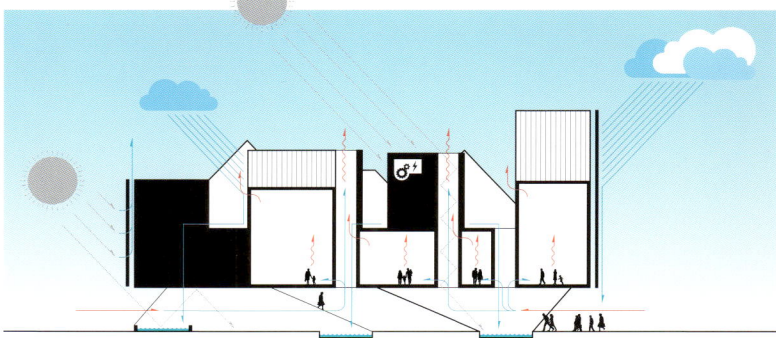

154

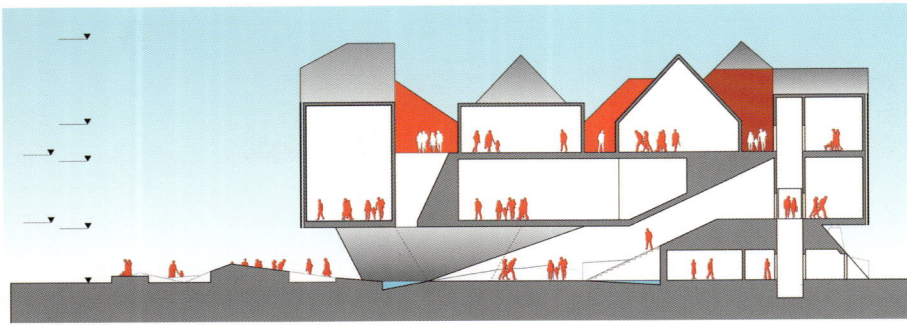

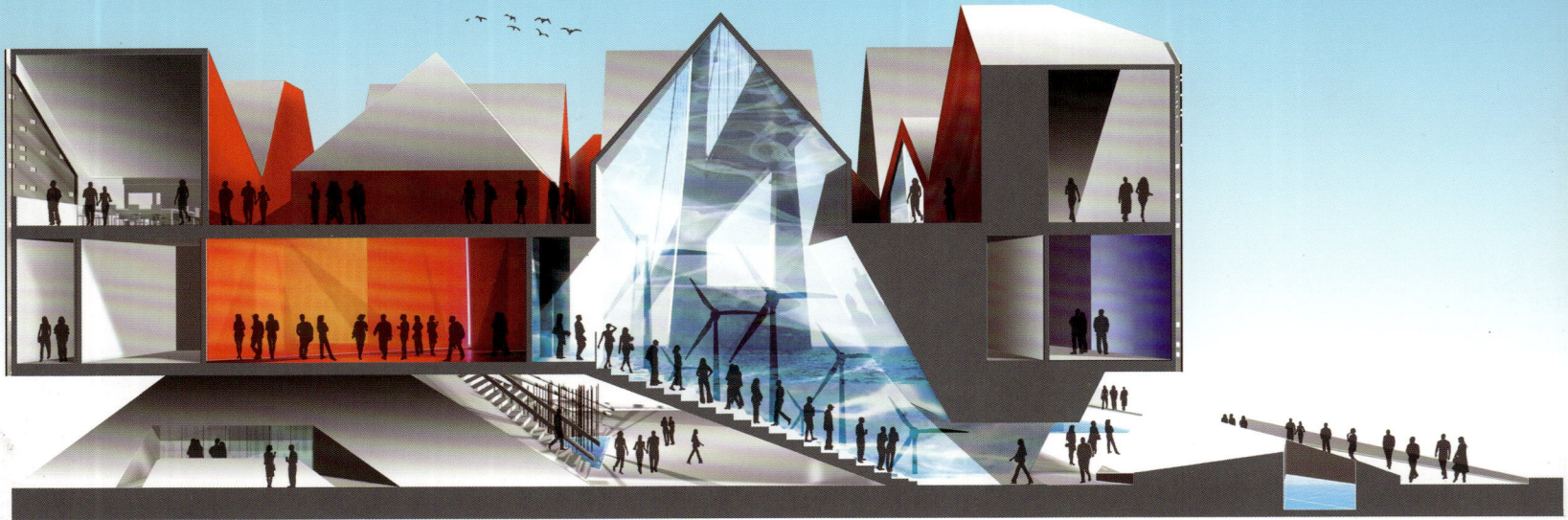

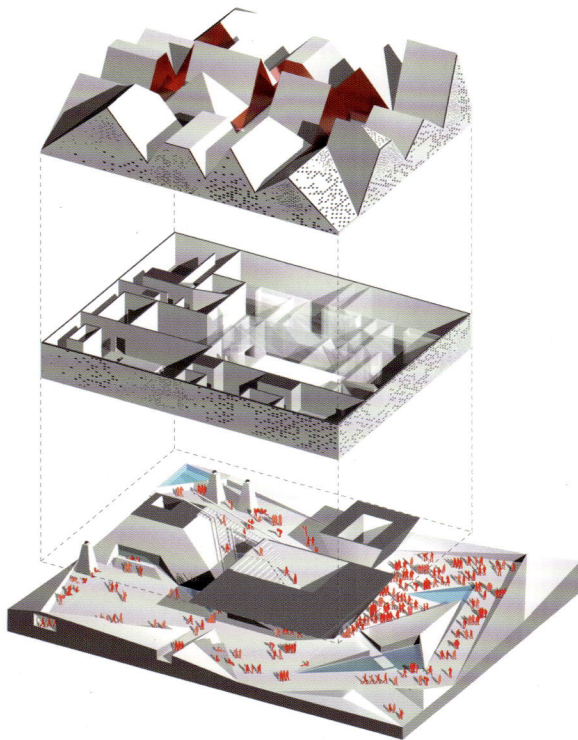

从"城市，让生活更美好"的总主题出发，CEBRA为展览及展览空间本身创造了一种强烈的、描述性的概念，提升了丹麦的全球品牌以及在哥本哈根举行的联合国环境会议中丹麦的主要竞争力及其他。

从眼下全球气候挑战和我们个人作为人类一员的责任出发，建筑师们积极地使用体验经济、面向使用者的交互技术以及戏剧道具在国际舞台上诠释他们的国家技能。展馆的主要目的是展示丹麦能为作为独立个体的参观者做到什么，通过这个知识使人们意识到在合作方面的优势和可能性。

展览本身就是一次体验丹麦创意产品和技术的旅程，也是丹麦式可持续发展方法的介绍。整个展览共分三层，分别为共同体验、共同创造和共同生活。每一个区域都有着独特的旗舰区、展品区和交互体验区。从一个典型的丹麦式家园开始，展览接下来引导参观者通过典型的丹麦城市，最终在全球环境中看到丹麦的身影。家园区和城市区位于展馆内部，而全球环境区位于屋顶外部，在那里可以看到整个的世博园区。

展馆一层设计展示了9种丹麦典型的景观，标明了展馆的入口及出口。

建筑的占地面积很小，不但减小了对景观的影响，而且也兼做了抵挡雨水和阳光的庇护所。

从每一个区域都可以学到更多有关于我们共同责任和合作可能性的知识，了解更多关于丹麦富有创新性和创意性的可持续发展方法、可持续性产品和可持续发展的前进道路。

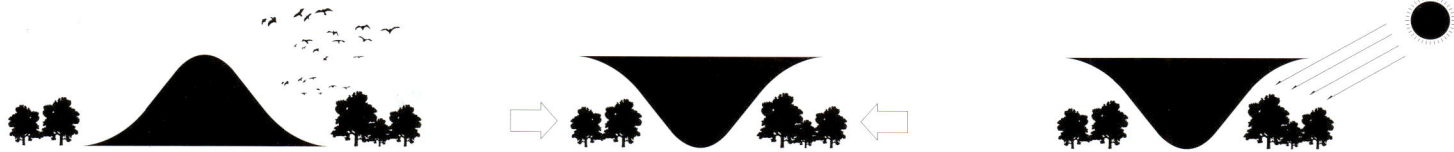

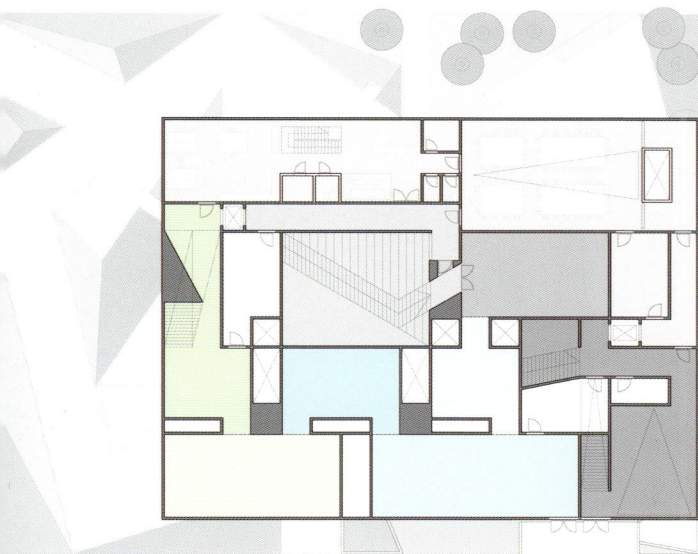

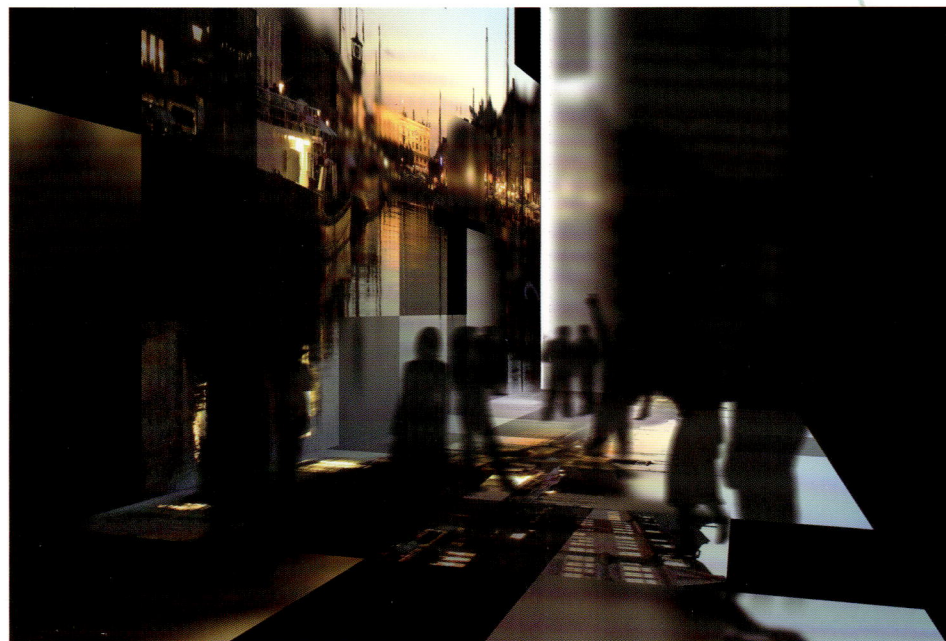

Urban Interlace(Landsbanki Headquarters in Reykjavik, Iceland)

城市编织（冰岛雷克雅维克的冰岛国家银行总部）

Architect: Martin Laursen, Martin Krogh, Anders Lonka, Jonas Smit Andersen, Kasper Svanberg
Firm: ADEPT
Area: 37,500m²
Collaboration: workac (New York) and plusark (Iceland)

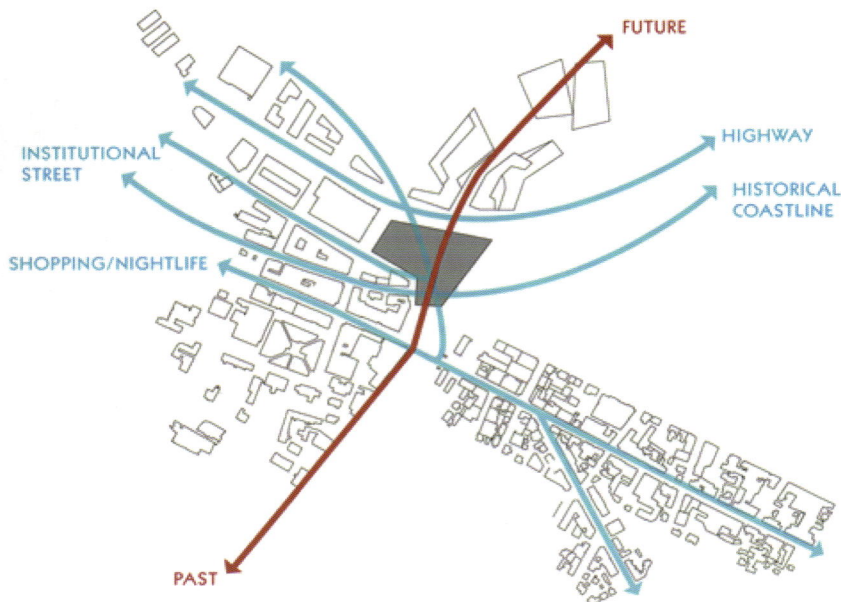

With renovations in the old heart of the city and the recent competition to rebuild the historical Kvosin district, Reykjavik is finally embracing the richness of its past. At the same time, a new Reykjavik is being born – one of great musical performances, international focus and architectural invention. The new headquarters will be located exactly at this intersection between the old center of the city and the new.

The architects propose a building that embodies this state of transformation: a historical, urban, architectural and programmatic connector that knits together the old fabric of the city center with the new scale of the East Harbor. Inspired by patterns of weaving in both the artistic and literary history of Iceland, the project Urban Interlace is a building that facilitates connections and uses an urban vitality to create a diversity of spaces and experiences.

对城市旧中心地带的修复与最近重建历史性的**Kvosin**地区的竞赛让雷克雅维克最终有机会拥抱其丰富的历史。同时，雷克雅维克也作为一座被世界瞩目的伟大的音乐、建筑发明之都得以重生。新的银行总部大楼正坐落在城市的旧中心与新中心的交汇处。

建筑师们所提出的建筑设计方案体现了这种转变的情形：这个作品正是作为一个具有历史感、带有城市特色和充满了建筑学特点的经过规划的连接点将城市中心古老的网络与东港新的区域连接到了一起。受到了冰岛文艺历史中的编织形式的启发，"城市编织"这个设计作品正促进了这种联系并加强了城市的活力，创造出了多样化的空间与生活体验。

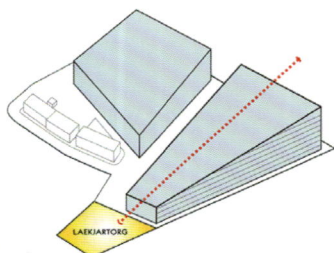

Landsbanki forbinder skala /
Landsbanki scale connector

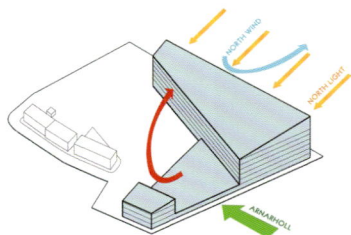

Arnarholl scene og max facade /
Arnarholl scene and max facade

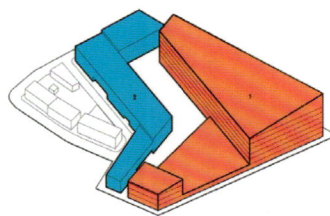

Tillægsbygning /
Second building

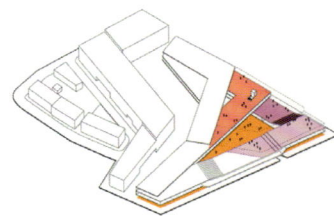

Offentligt eksploderet atrium /
Public exploded atrium

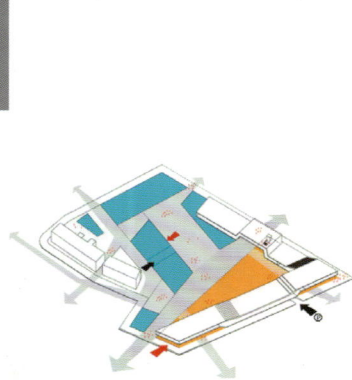

Bymæssige sammenhænge /
Urban connections

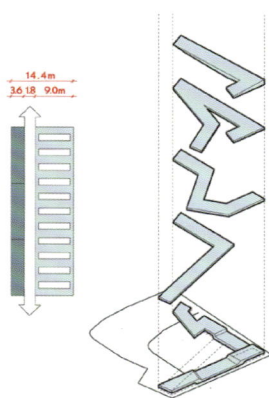

Arbejdsstationer - perfekt kontor bygning /
Workplaces - perfect office slab

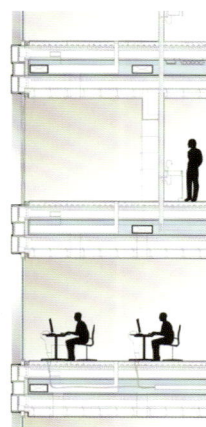

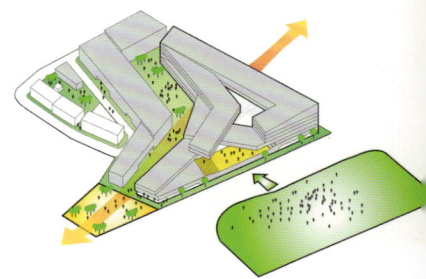

Landbanki oplevelsen /
Landsbanki experience

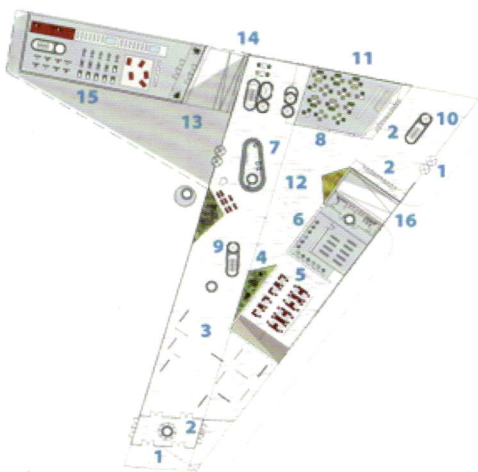

Mestia Airport

Mestia机场

Architect: Juergen Mayer H., Jesko Malkolm Johnsson-Zahn, Hugo
Reis, Mehrdad Mashaie, Max Reinhardt
Firm: J. MAYER H. Architects
Location: Mestia,Georgia
Area: 250m²
Architect On Site: Beka PK hakadze
Constructor: ANAGI ltd.
Photographer: J. MAYER H.
Text: Andres Rusy

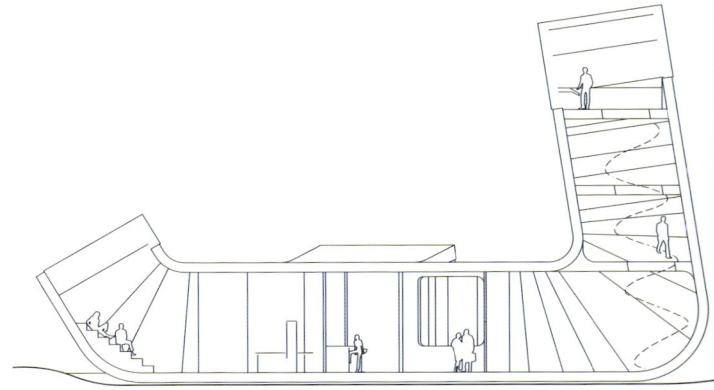

160

It is a V-effect worthy of its name: you hear "airport" and think of something grand and imposing. But then you see this small, alien particle, like debris from an ex-Soviet space station that was spared combustion in the ideological atmosphere of our era by the grace of history, and you cannot help but feel a kind of affection for it. It is remarkable how this specimen goes against the grain of the common interpretation of its typology, which embodies function and system like hardly any other. This Terminal Folly appears like a hitherto undiscovered, alternative happy ending to "2001: A Space Odyssey", in which Kubrick allows the failure of the space mission to culminate not in Nietzschean heroism, but in prosaic slapstick: a first-generation robot that collapsed on its first attempt at walking has reprogrammed its software so as to be able to forge on regardless of the limitations of its locomotor system. The fact that the prosthetic vehicle also reflects th numerous tower buildings in the vicinity makes it an instant monument situational contextualism. In an era where power-conscious potentates the new and old tiger states love to use architecture as a three-dimension representation of their ambitions, this little trouper provides charming pro of the power of weak form architecture. It is an architecture that is n besotted by its supposed significance (the flip side of whistling in the dark its dreaded irrelevance), but instead capable of playfully working through doubts about its own identity and relevance in an era increasingly grounde in uncertainty, and producing an original and contemporary form of poet from this fragile condition.

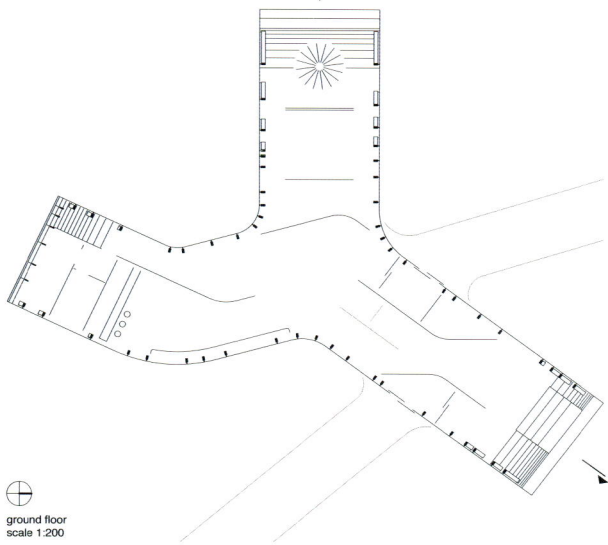

⊕ ground floor
scale 1:200

这是一个配得上它的名字的V形外观设计。当听到"机场"二字便会想到气势宏伟的设计，但是你所看到的却是小巧、陌生的设计，就像是前苏联空间站的残骸部分，因历史的恩典，在穿过我们这个时代意识形态的大气层中幸免燃烧；而且你无法自控地感到受其影响。值得注意的是，这种设计如何违反机场设计类型学的共同原则，几乎没有任何其他设计像它这样体现了功能和系统。这个"愚人码头"像一个至今没有被发现的，电影《2011太空漫游》的另一个快乐结局。片中人物库伯里克接受了太空任务的失败，没有在尼采的英雄主义中结局，而是以平淡的闹剧告终：一个第一代机器人在第一次尝试行走失败后，不顾其运动系统的局限性，重新编程了它的软件以加强它的运动系统。实际上，该建筑的设计手段也反映出附近众多的塔楼，它们使这个设计成为了情境主义瞬间的丰碑。在这个"老虎之国"中有着功耗意识的当权者的时代，他们喜欢用建筑作为三维立体的方式来表达他们的抱负。这样小小的建筑设计为弱建筑的力量提供了迷人的证明。这是一个不被它所谓意义所迷惑的建筑，而是在这个对自身不确定性的态度愈加严重的时代，能够通过对它身份和相关性的疑问态度，在这个脆弱的环境中创造出如同独创的、当代形式的诗歌般的建筑。

Museum of Polish History/Design Initiatives

波兰历史博物馆

Architect: Vlado Valkof
Firm: Design Initiatives
Location: Warsaw, Poland
Area: 43,365m²
Designers: Anne Valkof, Stanislav Christov
Rendering: Assen Balkanski
Status: competition
Date: 2009

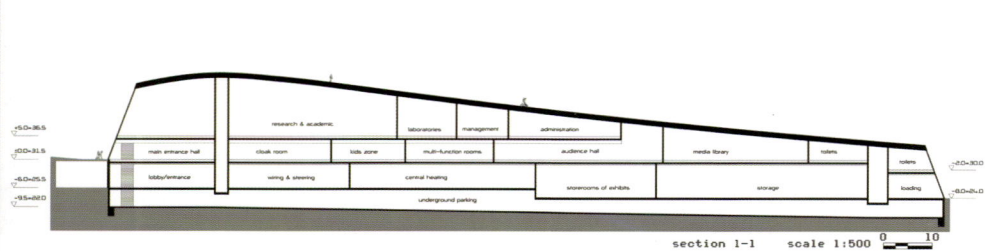

section 1-1   scale 1:500   0   10

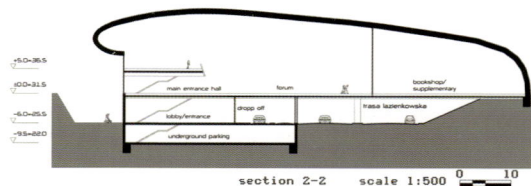

section 2-2   scale 1:500   0   10

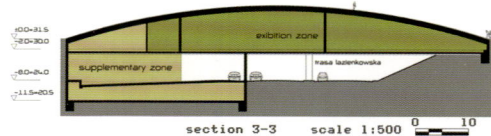

section 3-3   scale 1:500   0   10

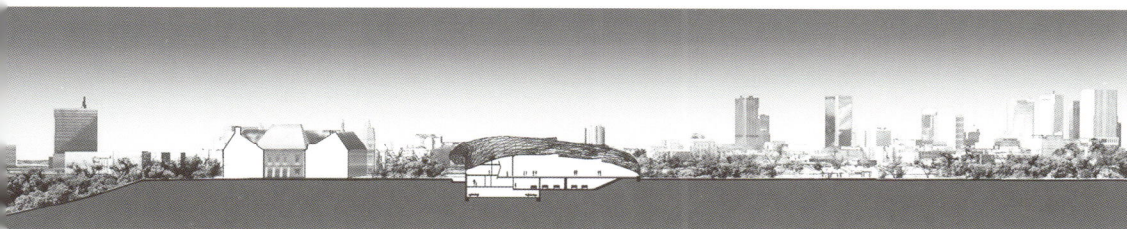

This design restores the integrity of Skarpa Warszawska and improves its slope stability by covering the Trasa Lazienkowska and connecting its two embankments from the foot of the scarp to Plac na Rozdrozu with park terrace and the hybrid structure of the MPH. This hybrid structure is designed to exist independently and still connect the two embankments in case the second stage of extending the park over Trasa Lazienkowska and its complete covering fails or is delayed. The highway underneath the covering shall be mechanically ventilated. In order to cover the Trasa from the foot of the scarp to Plac na Rozdrozu the architects modify the current layout of the Pl. Na Rozdrozu junction by a nearly 90-degree rotation of the right-turn out-lane of Trasa Lazienkowska into Al. Ujazdowskie towards the city centre and Pl. Trzech Krzyzy. The layout of pedestrian passages and biking routes takes into careful account the historic layout of passages and provides connections between the Museum and neighboring facilities. The architects reestablish ul. Jazdow and ul. Lenona as pedestrian passages and emphasize the importance of Os Stanislawowska. A new passage starts from the intersection of the Os Stanislawowska with Al. Ujazdowskie, passing by the MHP facility and connecting with the existing passages and biking routes from the northeast part of the park.

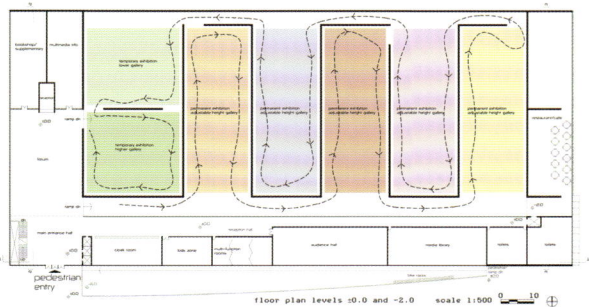

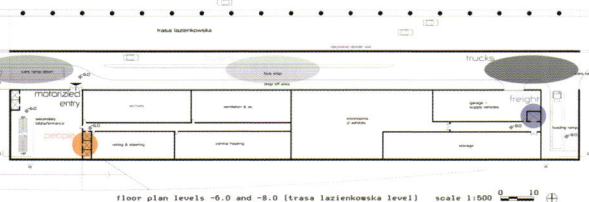

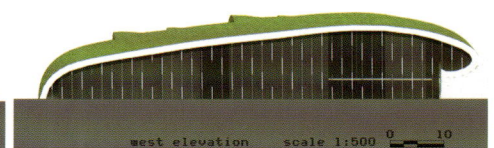

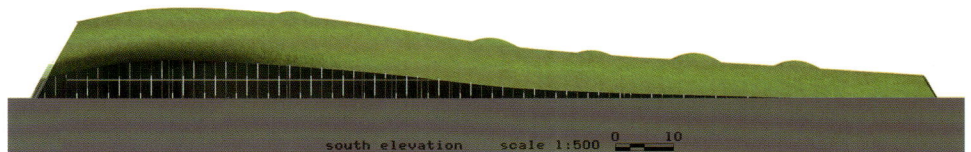

floor plan level +5.0   scale 1:500   0 __ 10

floor plan levels :0.0 and -2.0   scale 1:500   0 __ 10

east elevation   scale 1:500   0 __ 10

west elevation   scale 1:500   0 __ 10

north elevation   scale 1:500   0 __ 10

south elevation   scale 1:500   0 __ 10

floor plan levels -6.0 and -8.0 [trasa lazienkowska level]   scale 1:500   0 __ 10

floor plan levels -9.5 and -11.5   scale 1:500   0 __ 10

　　波兰历史博物馆的设计复原了华沙悬崖的完整性，并通过采用公园平台和混合结构覆盖了Lazienkowska大道，同时从斜坡底部到Plac na Rozdrozu连接其两侧从而改进了斜坡的稳定性。这种混合结构独立存在，同时确保在第二层上延伸在Lazienkowska大道上的公园与其完整的覆盖层出故障或延迟完工的情况下，整个建筑仍能连接大道两侧。在顶棚下面的高速公路将通过机械设备进行通风。为了从斜坡底部到Plac na Rozdrozu覆盖整个Lazienkowska大道，建筑师通过将Lazienkowska大道向右转进入Al. Ujazdowskie通往市中心和Pl. Trzech Krzyzy的出口车道旋转90度来改变了Pl. Na Rozdrozu交通连接点现在的布局。建筑师对于人行步道和自行车道的方案仔细地考虑了以前连接了博物馆和周围设施的人行道和自行车道的历史布局。他们重建了ul. Jazdow 和ul. Lenona作为人行道，并强调了Os Stanislawowska的重要性。一条新的人行道从Os Stanislawowska与Al. Ujazdowskie的结合点出发，通过博物馆设施，并与已有的从公园东北处延伸至此的人行道和自行车道相连。

MOCA Wroclaw

弗洛茨瓦夫当代艺术馆

Architect: Justus Pysall, Peter Ruge, Bartlomiej Kisielewski
Location: Ul. Bernardynska, Wroclaw, Poland
Area: 25,836m²
Structural Engineering: ARUP Krakow, Poland

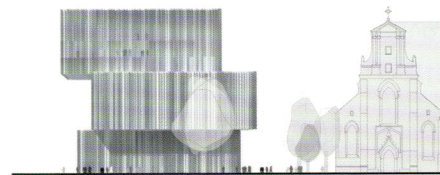

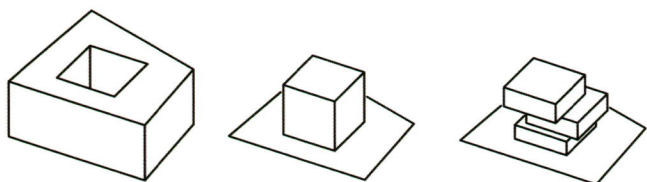

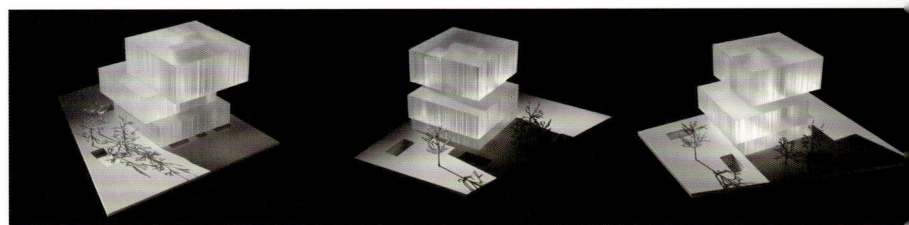

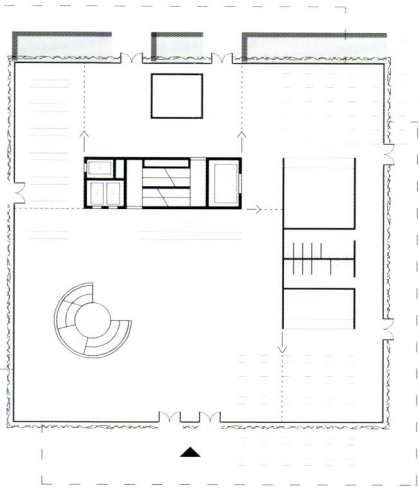

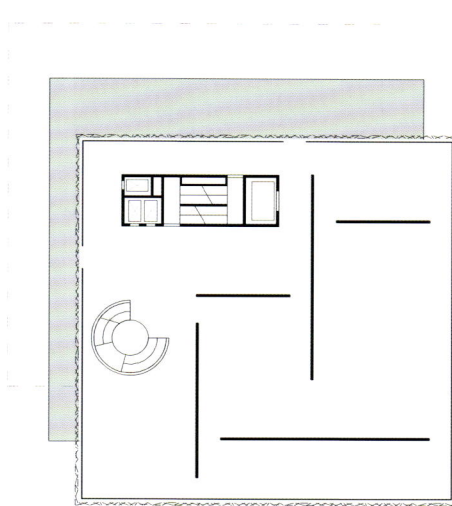

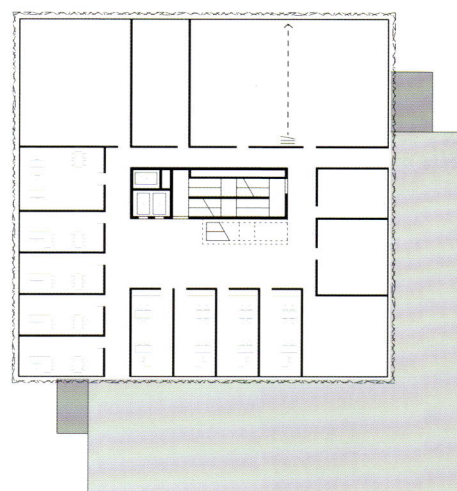

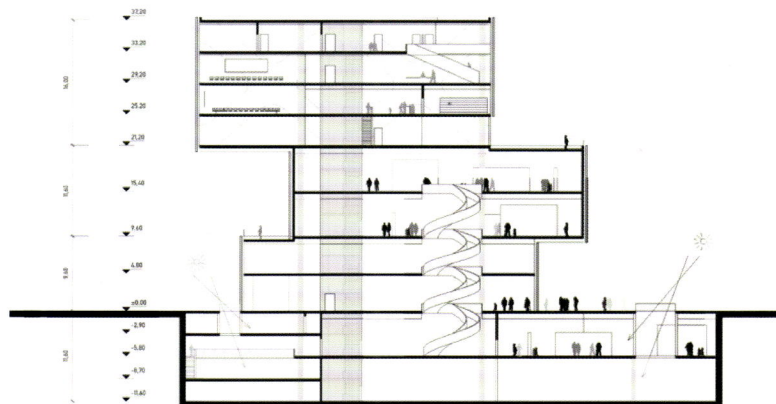

With the revitalisation of the historic Wroclaw, the city is restoring its lost structure of dense traditional city blocks. The unique occupation of a block by a museum of contemporary art demands a new, well-considered urban design solution.

To take the "void" as a built volume, and the vanished building as the open space around it, is the response. Rich and enigmatic spatial qualities for the museum functions are created, as well as exterior spaces which don't disregard the historical context.

The new volume, split horizontally into three offset levels, makes architectural reference to the St. Bernard Church, used today for the faculty of architecture.

The irregular stacking of the levels positions the services core according to the function, either to one side or in the middle. As well, the stacking creates sheltered entrance zones and terraces for open-air exhibitions.

The monochrome sheathing of the building's sculptural form is characterised by the rough, broken edges of semitransparent, textured white cast-glass panels. The rotation and offsetting of the floor-to-ceiling glass facade elements respond to the differing lighting requirements of the different floor areas.

At night the fascinating play of reflections from the facade will be augmented by the gleaming, illuminated soffits of the undersides of the overhanging floors, highlighting the museum and lighting the open spaces at the ground level.

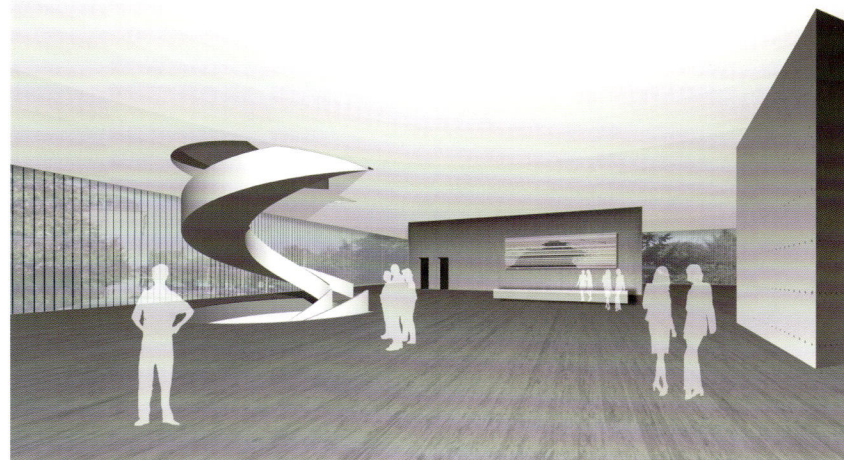

随着历史上著名的弗罗茨瓦夫市的复兴，这座城市曾失去的密集传统城市街区的结构正在恢复。在这样的街区中建设一座当代艺术博物馆就需要一种经过深思熟虑的新的城市设计解决方案。

作为一种解决方式，建筑师们在设计中用"虚空"作为建造的体积，而消失的建筑则是围绕在其周围的开放空间。博物馆在设计上采用了多样而神秘的空间特质来满足其功能要求，而博物馆的外部空间设计也与周围的历史环境相适应。

新建筑被分为三个偏移的层次。这一设计参考了圣伯纳德教堂的建筑风格，在今天被用来作为建筑研究的范例。

不规则的水平层叠结构将服务中心根据功能的需要置于一侧或中间的位置。同时，这种层叠结构还作为入口区的遮蔽结构，并为露天展览提供了平台。

博物馆建筑雕塑外形的单色保护层是用粗糙并不规则边缘的半透明白色质感的玻璃面板构成。地面与天花板之间的玻璃外立面的旋转与偏移的设计考虑了不同楼层区域里不同的光照要求。

在夜里，建筑立面反射的迷人的光线与闪亮发光的悬挂楼层的拱腹交相辉映，突出了博物馆并照亮了一楼的开放空间。

Airbaltic Terminal

波罗的海航空航站楼

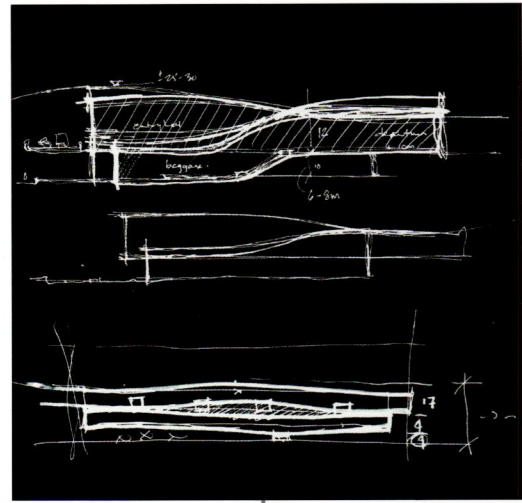

Firm: Zerafa Architecture Studio
Location: Riga, Latvia
Area: 60,000m²
Design Team: Jason Zerafa, Joaquin Boldrini, Katherine Moya.

Inherent in the airBaltic brand is the idea of connectivity. Freedom is the essential association with flight; however, the experience can suffer under the weight of the operational complexities imposed by contemporary air travel. Connectivity and communication are essential components to deal with this conflict.

The movement of passengers between landside and airside is the designer's point of departure and primary design consideration. It can be expressed by a series of parallel virtual corridors, which are created by extruding an inverted u-shaped portal shape in the east/west direction to generate a primary grain clearly describing the landside to airside path. The building is composed of eighteen parallel portals shapes which are each extruded along a unique curve to create an undulating wave composition running from east to west. The effect is to create a volumetric landscape within the greater airport landscape.

The two portal forms at the north and south ends of the building are each extruded out 224m form the departure piers. The volumetric expression of these piers is a clear extension of passenger movement along a landside to airside path. Another portal shape is extruded out to the east towards the city beyond the terminal to provide both a visual link, as well as a physical covered connection from the terminal building to the parking volume. Whe viewed from the air, the building form extends a hand into both landside ar airside.

The use of a repetitive module (spatial unit) to make sequence of direction spaces creates a variety of spatial experiences. Because the extruded port shapes are moving through space along unique curving trajectories, the create areas of openness and compression as they rise up or move downwa towards the floor. The effect is to reduce the scale of the interior volumes b defining a series of varied interior compartments.

As transfer passengers make up approx 70% of the projected passenge capacity, so the movement of these passengers should inform the buildir design in a direct and meaningful way. The internal network of transfe passenger circulation from gate to gate through the required security ar passport controls is consolidated into an L-shaped object to create a distin airBaltic "Transfer Bar". Identified by the airBaltic green, the transfer ba volume is legible at distance to arriving passengers prior to arrival at th contact gates or transfer from remote gates. Passengers can visually identi their unique transfer circulation path and destination before entering th terminal.

AIRBALTIC TERMINAL

EXISTING TERMINAL

SITE PLAN

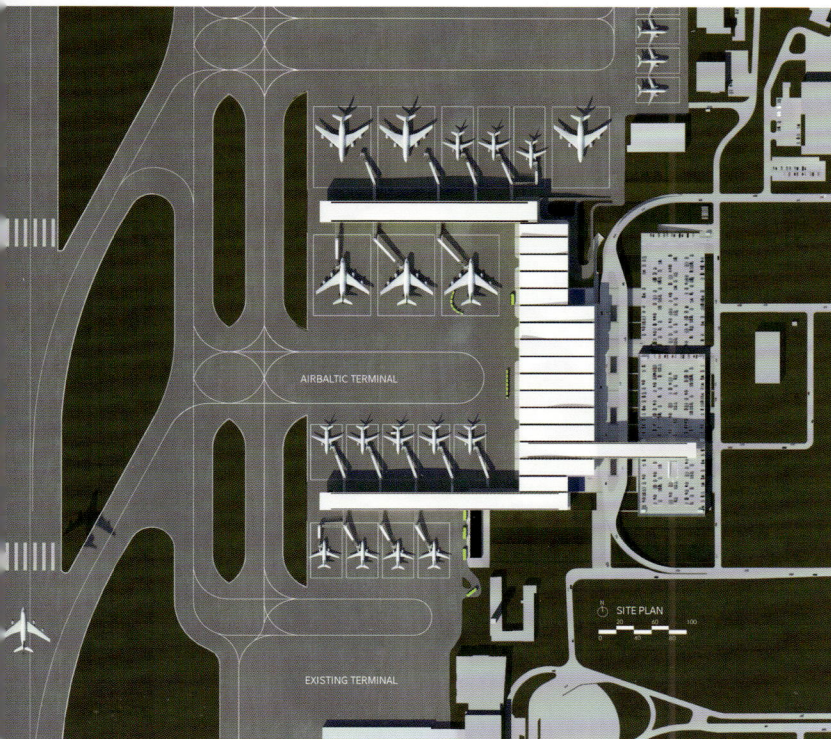

NON SCHENGEN PASSENGER CIRCULATION

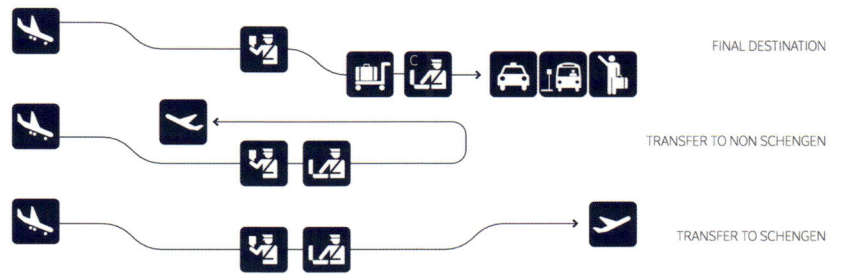

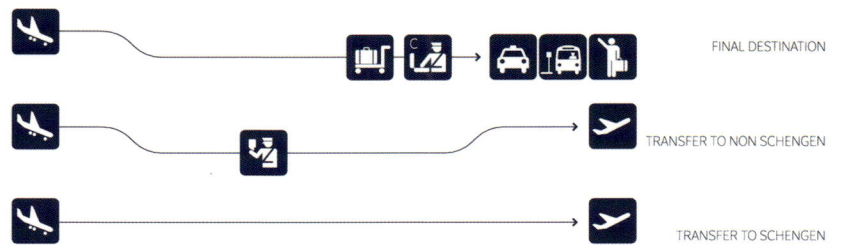

FINAL DESTINATION

TRANSFER TO NON SCHENGEN

TRANSFER TO SCHENGEN

SCHENGEN PASSENGER CIRCULATION

FINAL DESTINATION

TRANSFER TO NON SCHENGEN

TRANSFER TO SCHENGEN

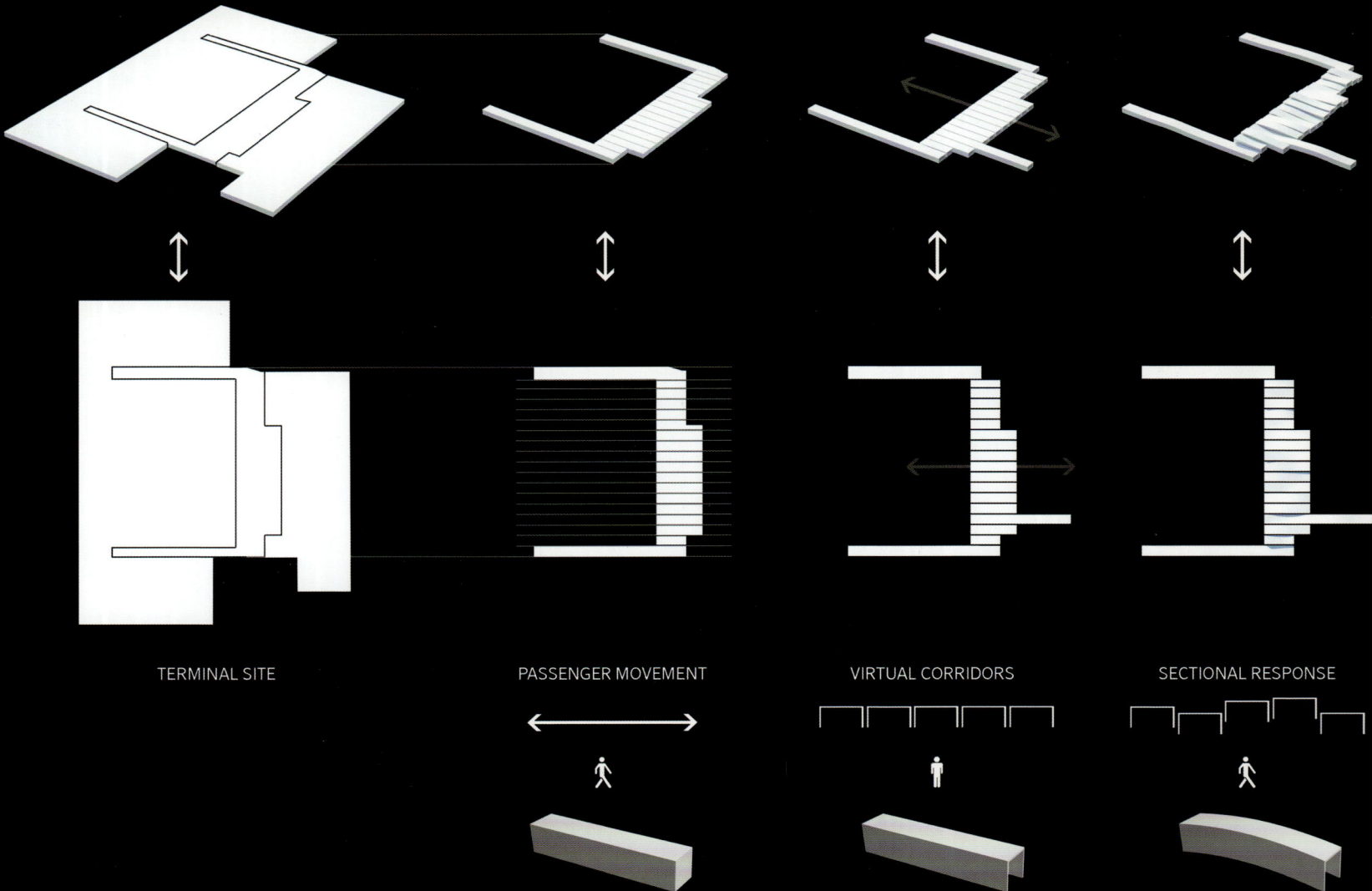

TERMINAL SITE

PASSENGER MOVEMENT

VIRTUAL CORRIDORS

SECTIONAL RESPONSE

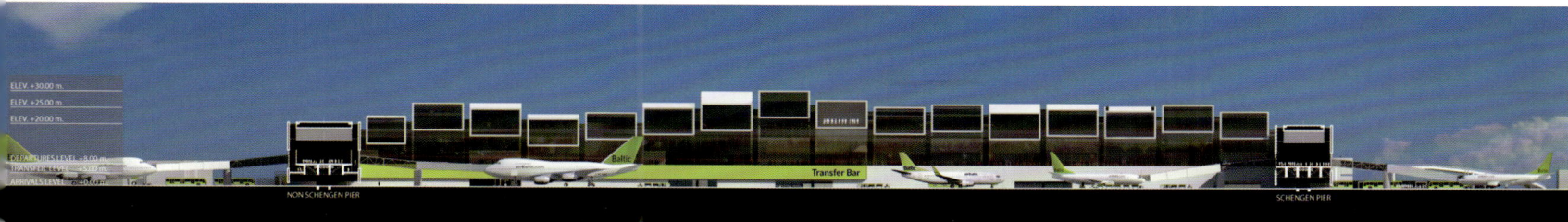

ELEV. +30.00 m.
ELEV. +25.00 m.
ELEV. +20.00 m.

DEPARTURES LEVEL +8.00 m.
TRANSFER LEVEL +5.00 m.
ARRIVALS LEVEL ±0.00 m.

NON SCHENGEN PIER

airBaltic

Transfer Bar

SCHENGEN PIER

airBaltic   airBaltic   airBaltic

ELEV. +30.00 m.
ELEV. +25.00 m.
ELEV. +20.00 m.

DEPARTURES LEVEL +8.00 m.
TRANSFER LEVEL +5.00 m.
ARRIVALS LEVEL ±0.00 m.

airBaltic

TERMINAL

PARKING STRUCTURE

airBaltic

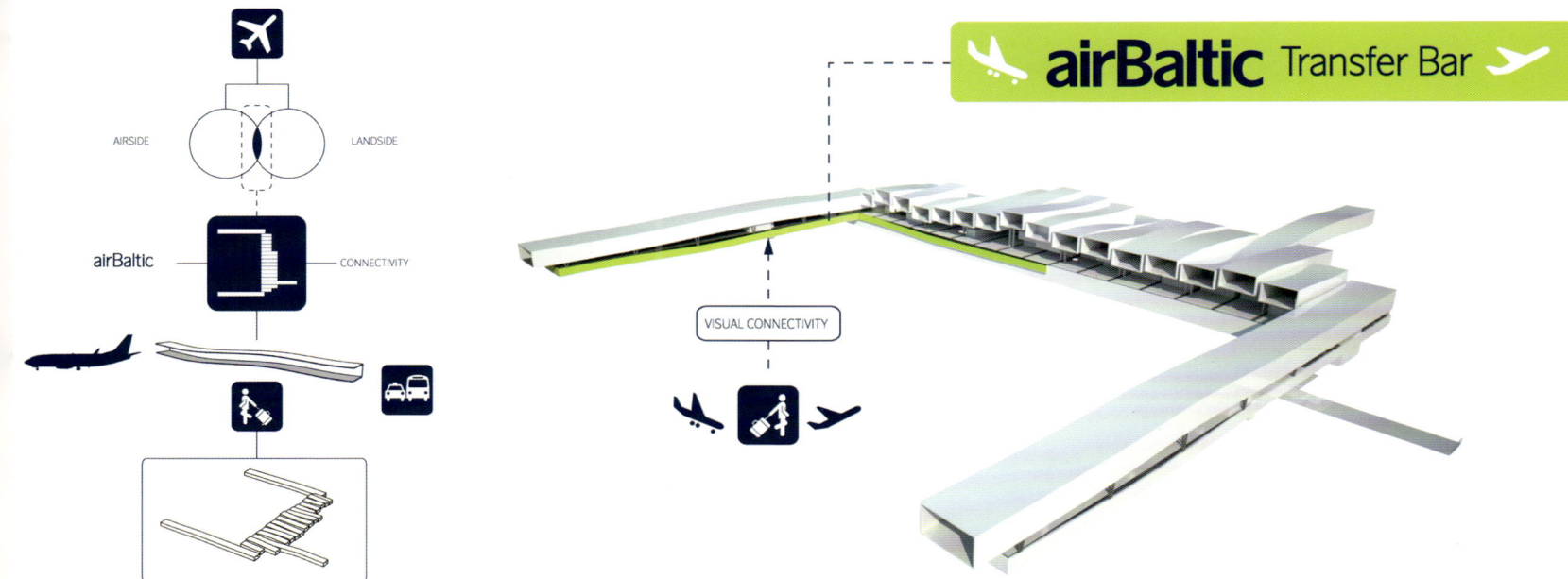

AIRSIDE   LANDSIDE

airBaltic   CONNECTIVITY

airBaltic Transfer Bar

VISUAL CONNECTIVITY

ARRIVALS LEVEL

DEPARTURES LEVEL

TRANSFERS LEVEL

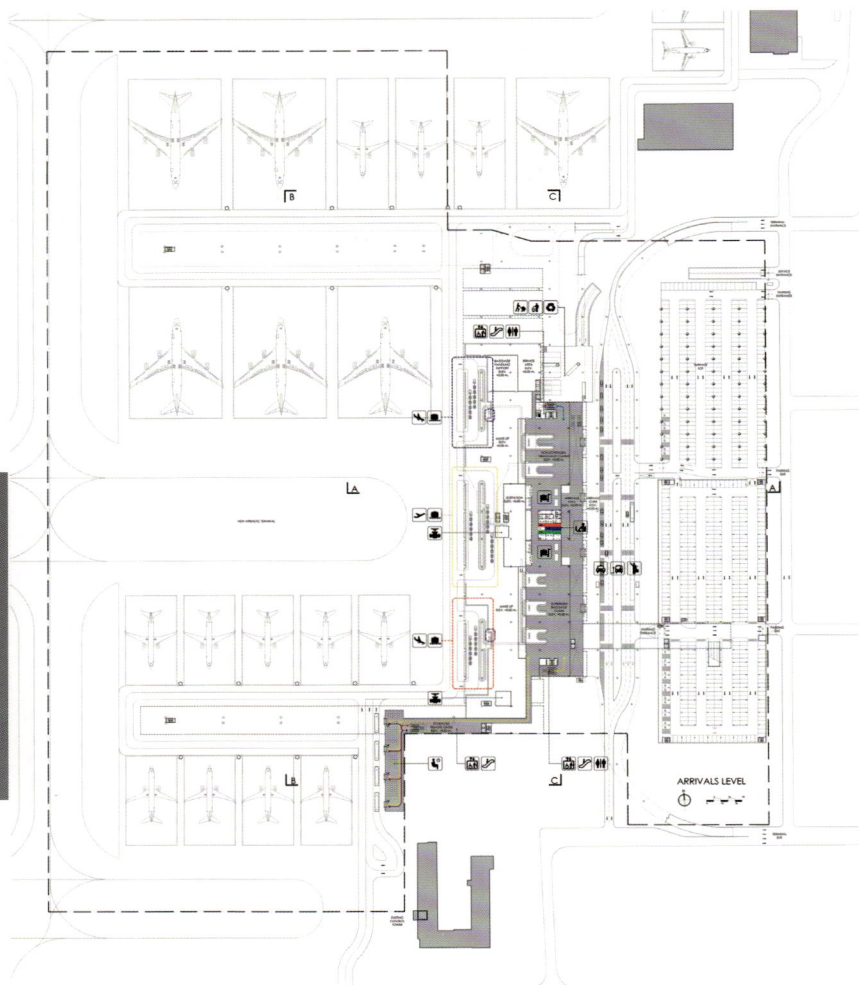

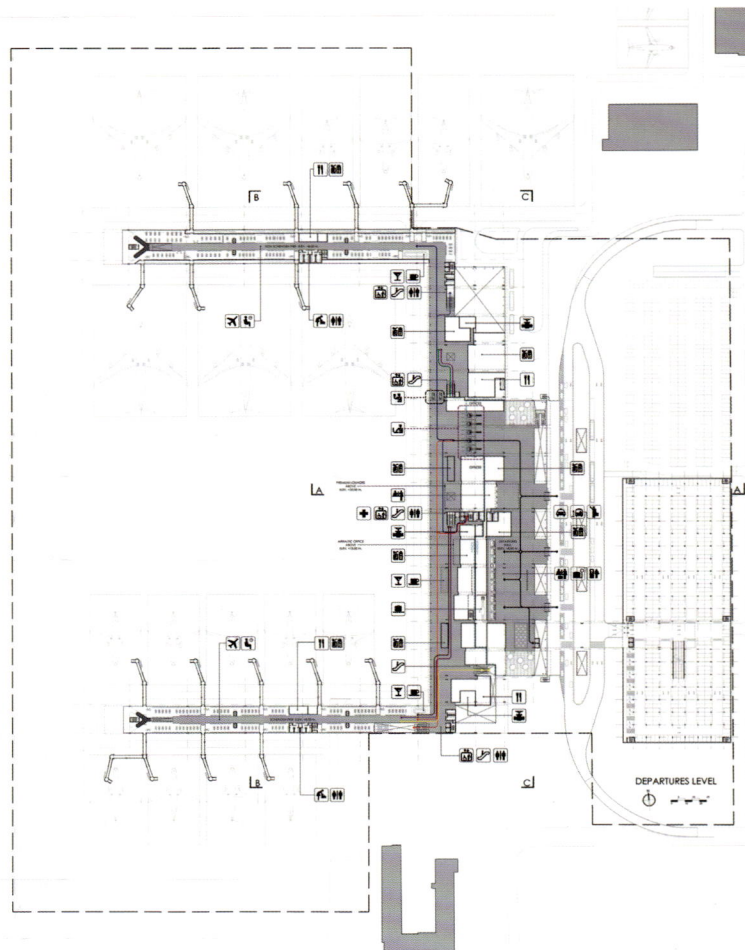

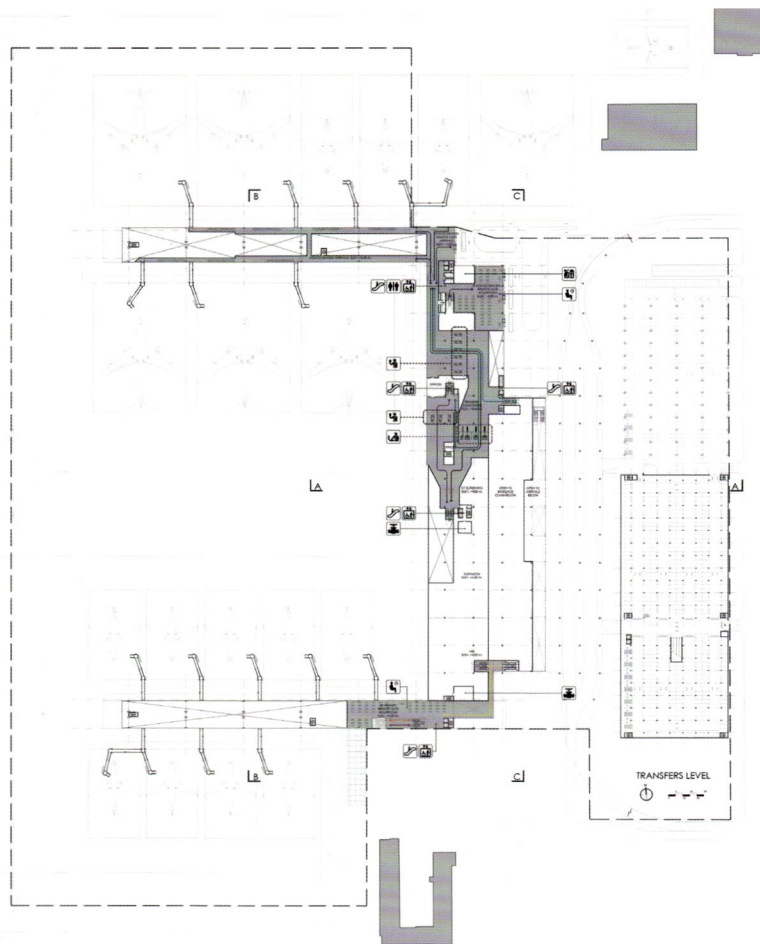

波罗的海航空公司品牌理念中固有的是连通性思想。自由是人们与航行的主要联系，然而由于现代空中旅行操作复杂性的影响，空中旅行体验变得有些痛苦。连通性与交流性是解决整个冲突的主要要素。

旅客在陆侧和空侧间的活动是建筑师的出发点和首要设计依据。设计表示为一系列平行的虚拟走廊，这些走廊由东西向挤出的一个倒U形的入口形状产生，形成了主要的纹理明确区分了陆侧与空侧的路线。建筑由18个平行的入口形状组成，每一个都沿着一个特殊的曲线组合挤压而成，创造了一个起伏的波浪组合由东向西运行。其效果是在更大的机场景观中建立了一个大型的体积景观。

在建筑北端和南端的两个入口形状各自延伸出224m形成了出发港。这些出发港体积性的表达是乘客由路侧向空侧活动的明确扩展。另一个入口形状朝着城市向东向延伸，越过航站楼为城市与航站楼提供了视觉上的连接，同时也是从航站楼到停车场建筑的有形的覆盖连接。当从空中俯瞰，建筑造型向陆侧和空侧二边伸展出来。

重复性模块（空间单元）的使用形成了一系列有指向性的空间，形成了不同的空间体验。因为被挤压出的入口形状沿着独特的曲线轨道移动伸展，在它们上升或者朝向地面下降时，创造出开阔的和压缩的空间。其效果是通过定义一系列不同的室内分隔，缩小了室内体量的规模。

预计载客量中，中转旅客大约占70%。因此这些旅客的活动对建筑设计有着直接和有意义的影响。中转旅客通过必要的安检和护照检查后，从一个登机口到另一个登机口间的内部网络循环被整合入一个L形的物体，形成了明确的波罗的海航空"转机带"。对到达旅客来说，在他们到达连接门或者转乘接驳空桥之前，便可在远处很明显地看到采用了波罗的海航空公司的代表色——绿色的"转机带"。乘客在进入航站楼之前就可以从视觉上识别出他们独特的换乘循环路线以及目的地。

ALL DEPARTING PASSENGERS

SCHENGEN CONTACT GATE

SCHENGEN REMOTE GATE

NON SCHENGEN CONTACT GATE

NON-SCHENGEN REMOTE GATE

TRANSFER NON-SCHENGEN
TO SCHENGEN

TRANSFER NON-SCHENGEN TO
NON-SCHENGEN

TRANSFER SCHENGEN TO
NON-SCHENGEN

NON-SCHENGEN ARRIVALS

SCHENGEN ARRIVALS

DEPARTURES BAGGAGE

TRANSFERS BAGGAGE

SECURITY CONTROL

SCHENGEN ARRIVALS
BAGGAGE

NON-SCHENGEN ARRIVALS
BAGGAGE

IMMIGRATION/PASSPORT
CONTROL

Norwegian Wood

挪威的森林

Architect: Anders Strange, Anders Tyrrestrup and Torben Skovbjerg Larsen
Firm: AART Architects and Studio Ludo
Location: Siriskjær, Stavanger, Norway
Area: 19,500m²

By implementing principles from the see house row "Siriskjær" and "Promenaden" are tied together and made recognizable in relation to the characteristic volume and space situations in the city centre.
The area gives the city recreational spaces at the water front, but also hiding places when necessary. The new buildings works as an enclosure of the outside areas, creates variation and merges the different movements.

Hereby "Promenaden" becomes more than a thoroughfare. It becomes walk through with various experiences and qualities – through the priva or public spheres. The building block is positioned on the area to crea recreational spaces which adapts to changing situations of weather, sun, vie and line of sight.

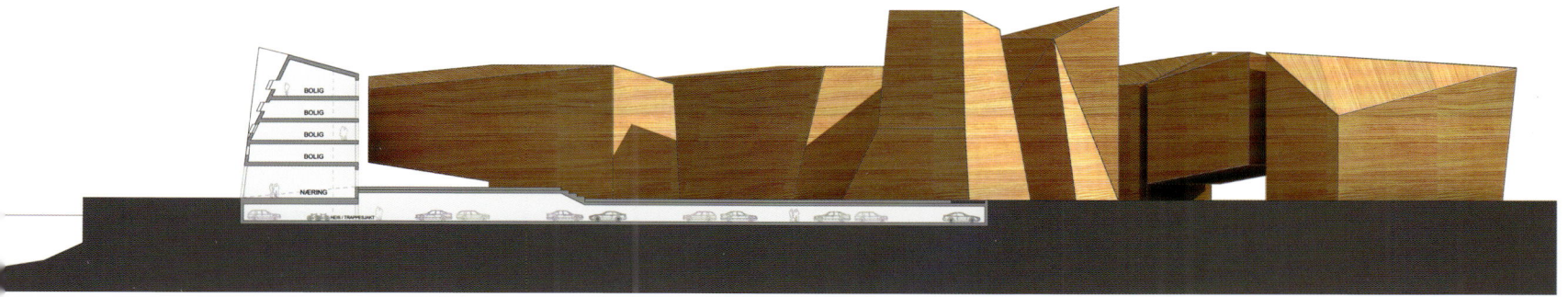

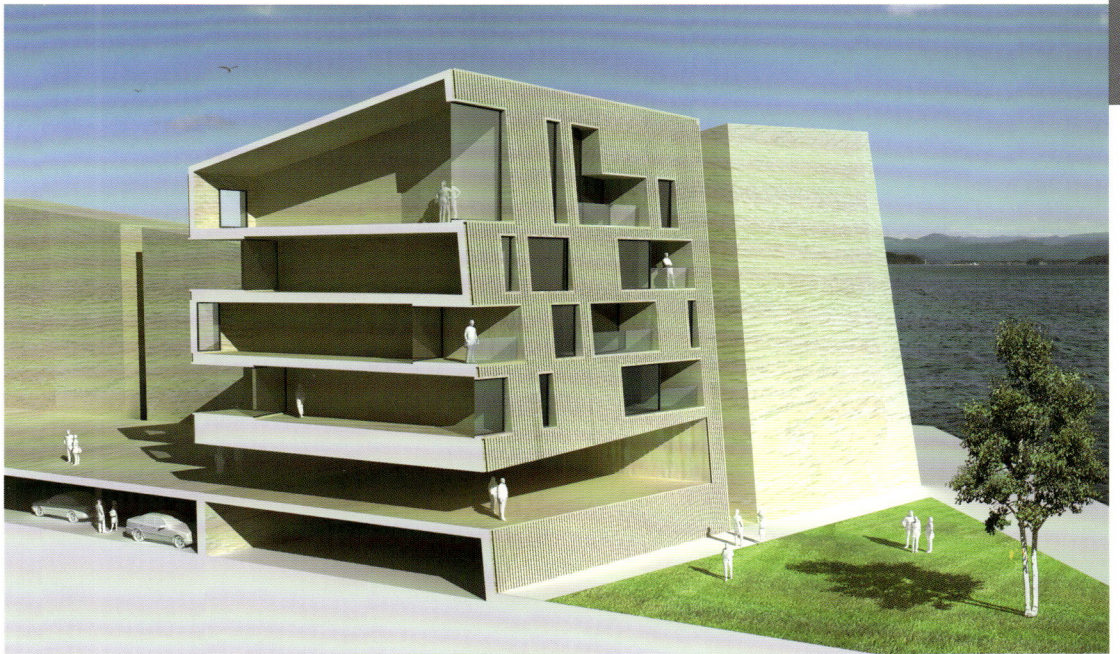

通过采用海边联排住宅的主要特点，"斯里斯克耶尔（Siriskjær）"和"林荫道（Promenaden）"相互连接，并在市中心的外观和空间特点中尤为独特。

该设计的整体环境让城市在水边拥有了娱乐的空间，而且还在需要的时候提供了藏身之处。这些新建筑作为室外空间的附属，充满变幻并融合了不同的动感。

在这里，"林荫道"不仅是一条大街，更成为了穿越多种不同体验和生活的旅程——其中既有私人的空间也有公共的区域。设计该建筑群的目的是力图在这一地区创造能够适应不断变化的天气、阳光、景色和视线的娱乐空间。

THE APEIRON

Apeiron岛大楼

Firm: Sybarite
Building Height: 185m or greater
Building Diamteter: 200m or greater
Total Floors: 28 as designed (max. 46)
Gross Floor Area: 300,000m^2(max.495,000m^2)
Site Footprint Area: 26,500m^2
Wind Load Factor: 350km per hour
Building Use: 5-star deluxe, luxury offices, residential penthouses, high-end retail, culture and leisure spaces

ENVIRONMENTAL RESPONSE

Finding inspiration in nature and embracing ancient techniques common to the Mediterranean region, the design utilizes sea evaporation and cross ventilation to provide natural cooling. The building's shape and materials create a natural shade to deflect the intense midday sun. Louvres contained within the structure's internal facade prevent direct solar gain and are comprised of solar cells, as is the building's ribbon frame, which gives the Apeiron it's name (meaning infinity). Design calculations estimate that the building could generate two thirds of its own energy.

WORLD CLASS DESTINATION

Arrive at the Apeiron by air and enter by the ninth floor bridge, 55 meters above the water. Or arrive by sea, relax on the private beach and have a swim in the crescent lagoon, a tropical seascape filled with corals, plants and fish in a myriad of colours. Imagine enjoying a day at the exclusive spa or a fine meal in a restaurant surrounded by panoramic views of this surreal underwater world. Although originally imagined as an offshore location, the Apeiron design is equally suited to an onshore site.

DRAMATIC SURROUNDINGS

Filled with mature palm trees and fresh water pools of varying temperatures, the outdoor garden on the Floor 11 offers sweeping views out to sea as well as a unique perspective of the 50 meter atrium down to Level 00. This elevated oasis is enveloped by the rising summit of the building's sculptured form. The top floor houses the Butterfly Jungle, a climate controlled sanctuary brimming with tropical plants and exotic butterflies. The contrasting panoramic backdrop of the desert horizon nearly 200 meters below creates a surreal tranquility.

LUXURIOUS SPACES

As well as private clubs, Michelin-rated dining, conference facilities and luxury boutiques, the Apeiron boasts generous space for art galleries and other cultural attractions. The sculpture and art in the gallery space are illuminated by natural light refracting through the pools above.

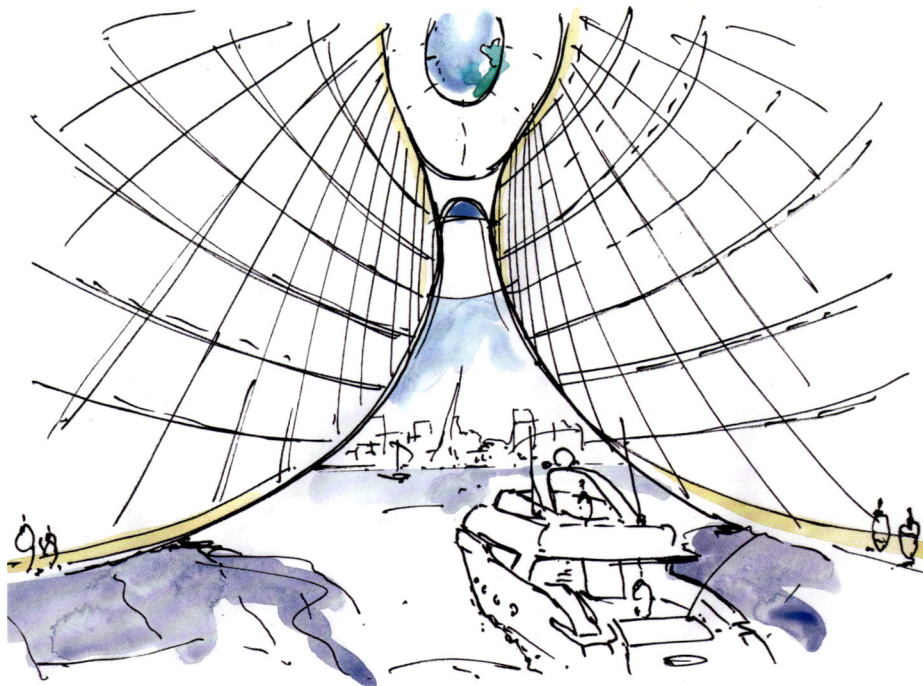

环境反应

　　这个设计从自然中得到启发，采纳了地中海地区常见的传统技术，它运用了海水蒸发和对流通风提供自然的冷却措施。建筑的造型和材料创造了一个自然的阴凉处，以抵抗强烈的正午阳光。建筑内层幕墙中的百叶窗阻止了直接的阳光辐射，并且构成了太阳能电池，是建筑的带状框架，也是这个建筑名字Apeiron（意为无限）的由来。设计计算估计建筑能够供应其自身需要三分之二的电力。

世界级旅游胜地

　　乘坐飞机到达Apeiron岛大楼，从大楼的第九层的桥梁进入，距离水面55m高。或者可以乘坐船只到达，在私人沙滩上稍作休息，也可以在新月形的泻湖中游泳，感受这里充满了珊瑚、植物和各色鱼儿的热带海景。想象在独特的温泉水疗中心享受一天，或者在360度梦幻般水底世界景色围绕的餐厅中享受美食。尽管原本设想建筑在一个海上位置，Apeiron岛大楼的设计同样适用于陆地。

引人注目的环境

　　有着成熟的棕榈树和不同水温的淡水池，位于11层的户外花园可一览无余大海的景色，同时还有着50m高的天井，独特的视角可一直看到1层。这片高高的绿洲由建筑逐渐上升的雕塑式外表所包围。顶层容纳了蝴蝶雨林，一个有着气候控制的圣殿，充满了热带植物和奇异的蝴蝶。下面近200m沙漠地平线突出的全景背景形成了一种超现实的安宁之感。

奢华的空间

　　除了私人俱乐部、米其林星级餐厅、会议设施和奢华宴会厅之外，Apeiron岛大楼还为艺术展览馆和其他的文化经典提供宽敞的空间。展览馆中的雕塑和艺术作品被由上面池水折射的自然光线所照亮。

Waalse Krook

Waalse Krook 大楼

Firm: UNStudio
Location: Gent, Belgium
Building Surface: 19,498.6m²
Programme: Library and Centre for New Media
Status: Competition Entry

The two main aims in the design for the Urban Library of the Future and Centre for New Media in Gent are to create a dynamic, flexible and open knowledge environment, whilst simultaneously strengthening the character of the location with the introduction of a building with a distinct architectural identity.

Sustainability is the guiding factor in the design, based on the conviction that not only must the environmental and user-friendly design of the Urban Library of the Future be able to evolve along with new media, but it must also offer the possibility for future change of use. With an open landscape, spaciousness, extensive views, alternative circulation routes, several meeting areas and a public plaza, the design for the library affords a renewal of its urban context.

The building is both fluid in form and accommodating to its surroundings. This is evidenced by its appearance - which varies according to the orientation - as well as from the decision to lift the building volume above ground level, thereby creating light, transparency and expansive sightlines. However the layered structure and low construction volume ensure that the impact of the design on the urban profile is minimal and that views to the characteristic towers of Gent are preserved. The structure also makes it possible to introduce (green) roof terraces whilst also ensuring low levels of direct sunlight penetration.

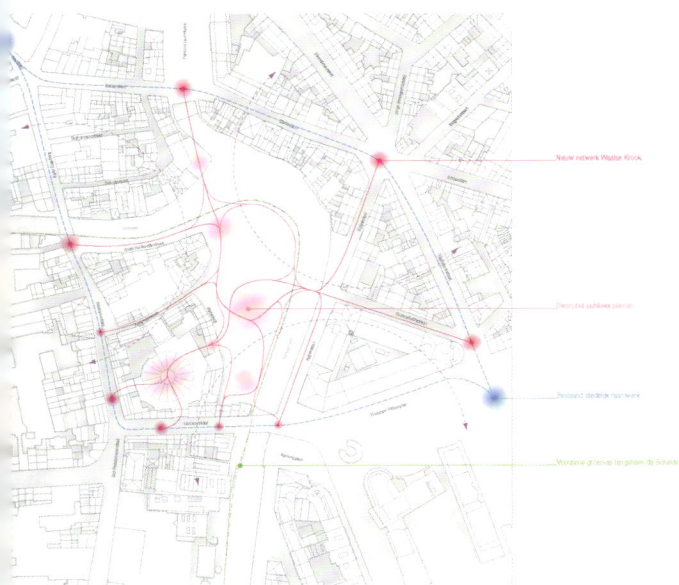

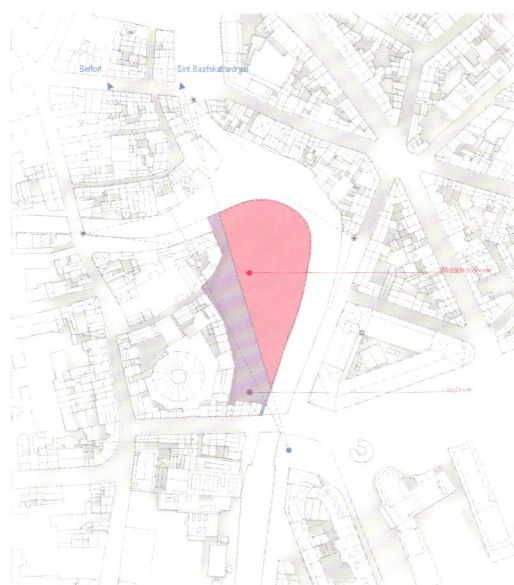

Connection

Based on the functional organisation the volume is lifted in order to create public space around the Library. In conjunction with the promenade along the quays, this results in interaction with the surrounding water and thereby the revival of the Waalse Krook.

The internal organisation of the building is based on an open central void, around which the circulation takes place. This internal void enhances the spatial experience and creates clear orientation through the building.

In addition to providing an extension of the urban context and the junction of the circulation routes, the internal void also functions as a link between the various functional clusters in the design. The void fulfills a bridging function between the city and the Municipal Library, and as such acts as a metaphor for public perception.

External agora   Main agora   Library agora

❶ Public squares   ❷ High restaurant   ❸ Green roofs   ❹ Sheltered access area   ❺ Bicycle parking   ❻ Bridge connections   ❼ Reactivating quays

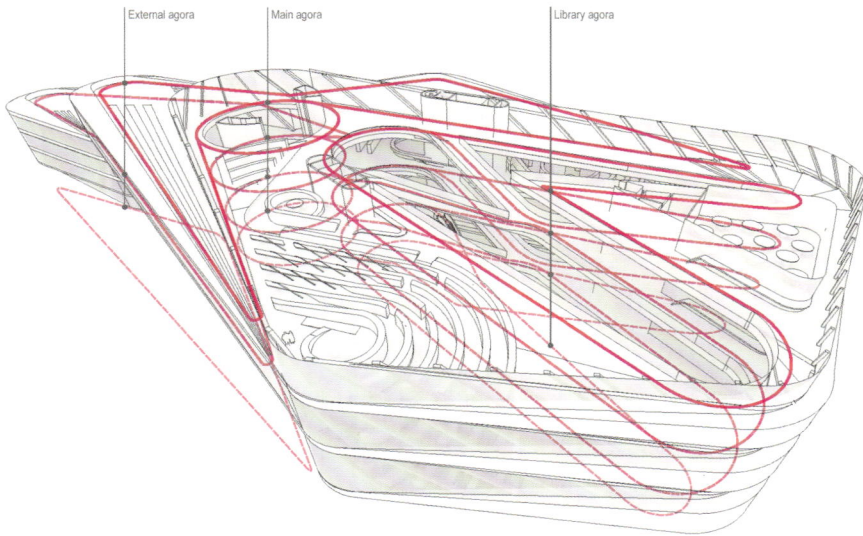

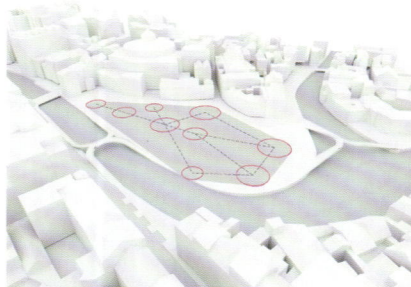

Diagrams Agora

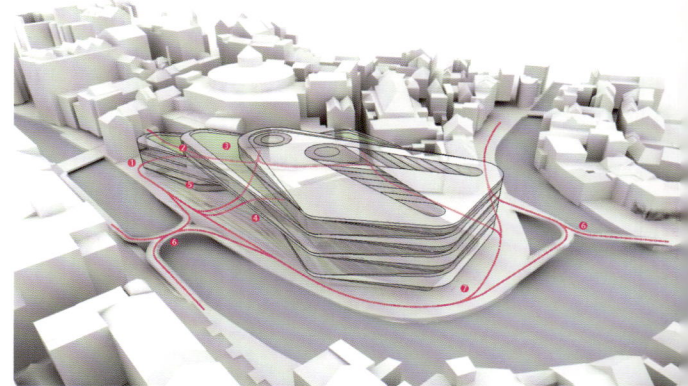

❶ Interconnections and tangents

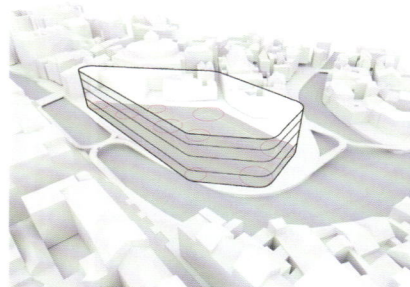

❷ Maximum volume envelope

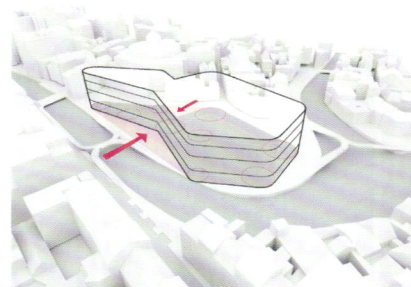

❸ Introduction public squares and accentuation of views to the surroundings

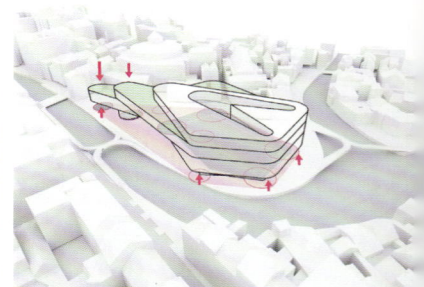

❹ Footprint compensated with green roofs and addition of terraces in order to maintain views from and to the area. Raised corners to enable reactivation of the quay.

Diagrams Volume Manipulation

West view

Aanzicht Oost

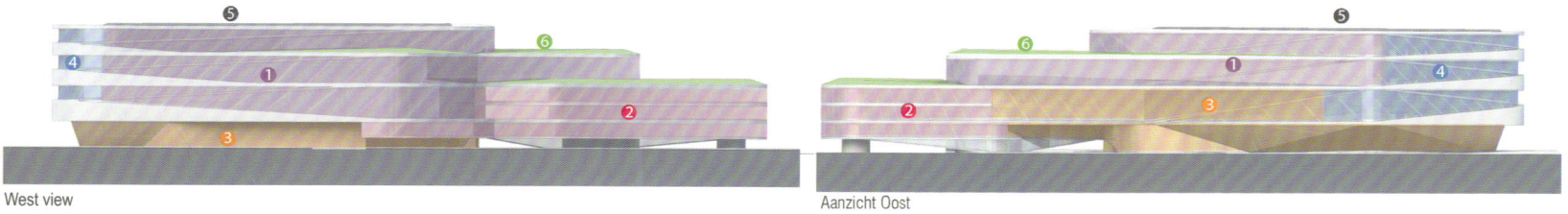

1 Automatic solar control   **2** Solar control glass with fritting   **3** Natural solar control by means of cantilever   **4** Energy efficient glass   **5** Skylight equipped with lamellas   **6** Green roof

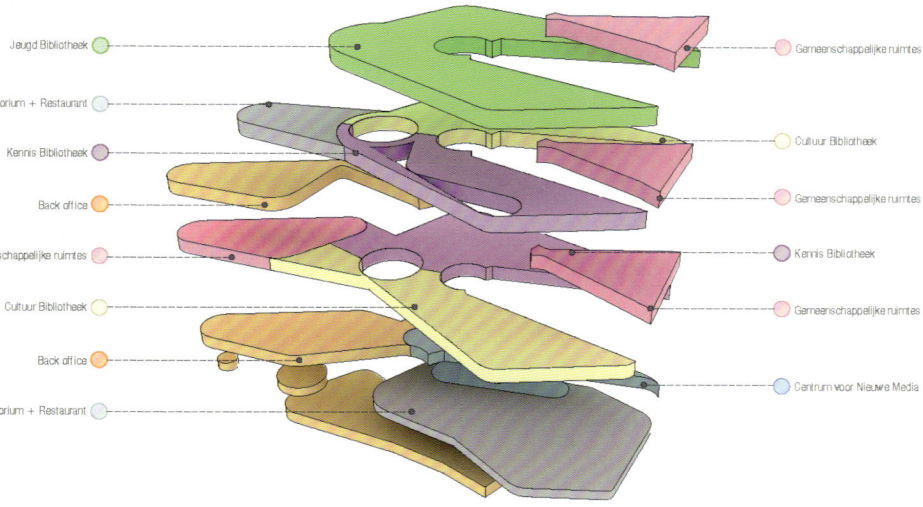

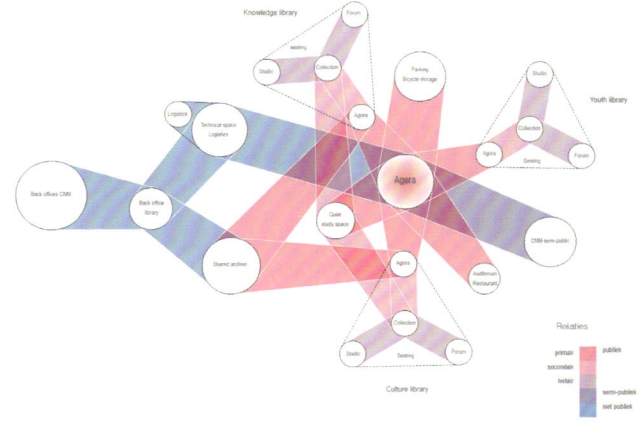

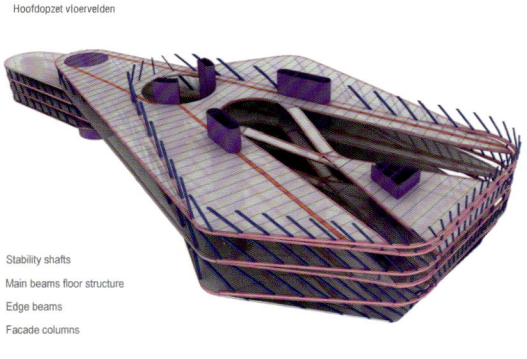

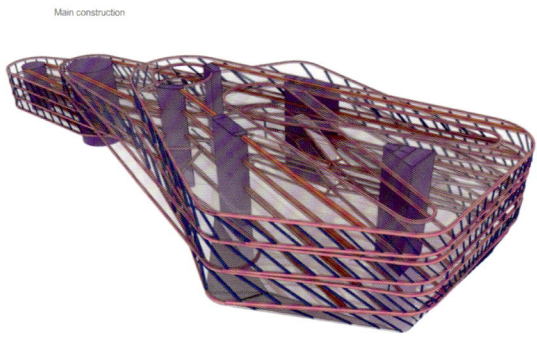

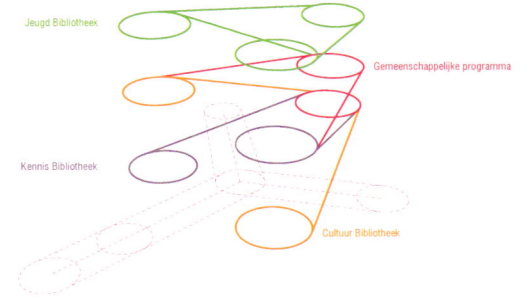

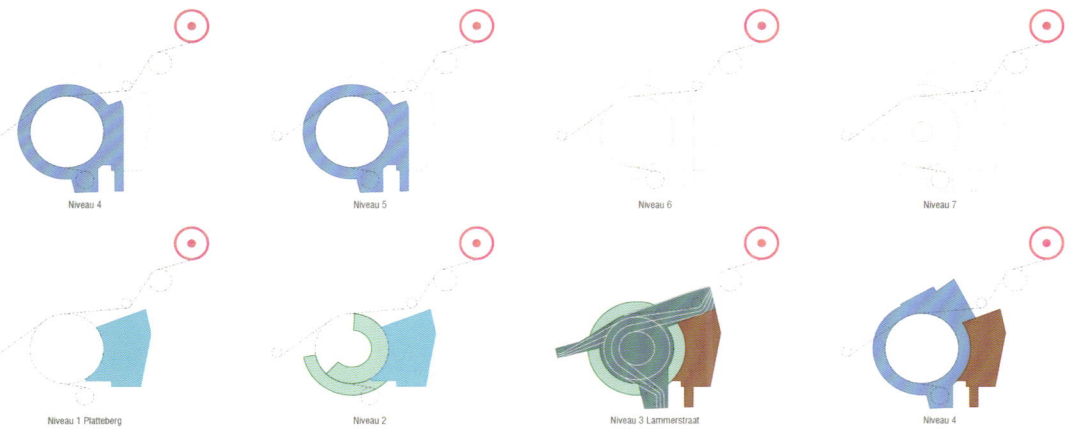

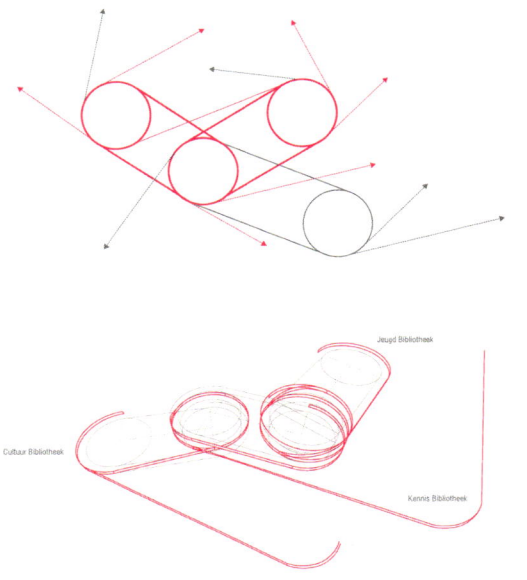

Facade

Jeugd Bibliotheek
Auditorium + Restaurant
Kennis Bibliotheek
Back office
Gemeenschappelijke ruimtes
Cultuur Bibliotheek
Back office
Auditorium + Restaurant

Gemeenschappelijke ruimtes
Cultuur Bibliotheek
Gemeenschappelijke ruimtes
Kennis Bibliotheek
Gemeenschappelijke ruimtes
Centrum voor Nieuwe Media

Hoofdopzet vloervelden

Main construction

- Stability shafts
- Main beams floor structure
- Edge beams
- Facade columns

Jeugd Bibliotheek
Gemeenschappelijke programma
Kennis Bibliotheek
Cultuur Bibliotheek

Niveau 4   Niveau 5   Niveau 6   Niveau 7

Niveau 1 Platteberg   Niveau 2   Niveau 3 Lammerstraat   Niveau 4

Winter Circus

- Vlaams instituut voor archivering
- I-Cubes
- Commerciële ruimte
- Hotel

Jeugd Bibliotheek
Cultuur Bibliotheek
Kennis Bibliotheek

Green roofs with full
sun load

Natural sunlight control
by cantilever

Skylight equipped with lamellas
for controlled daylight penetration

Library

Winter Circus

Support

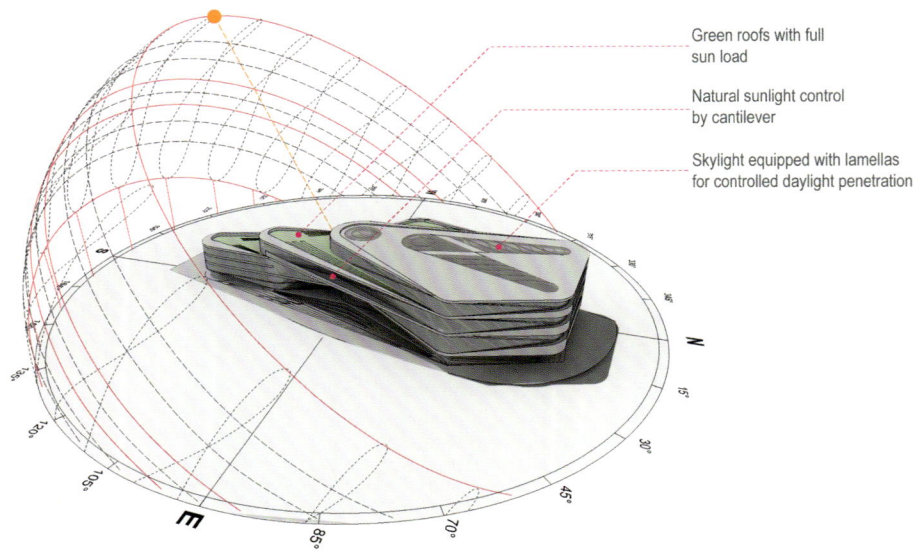

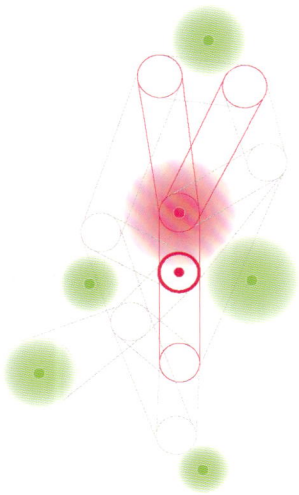

| Agora and public squares | Implementation flight radius | Circulation and orientation | Programme position | Structural grid |

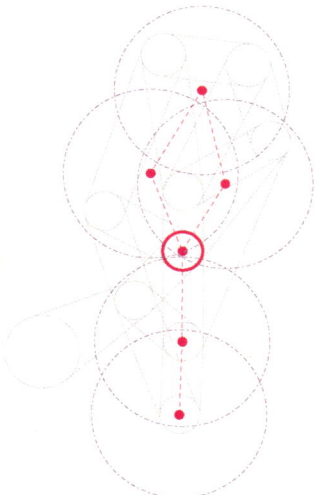

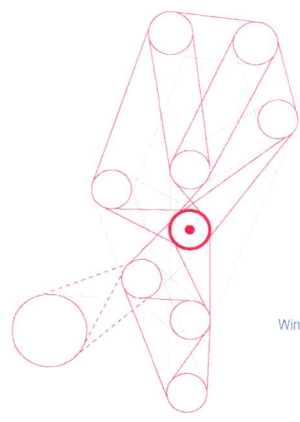

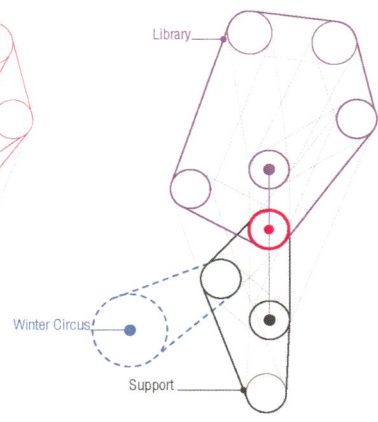

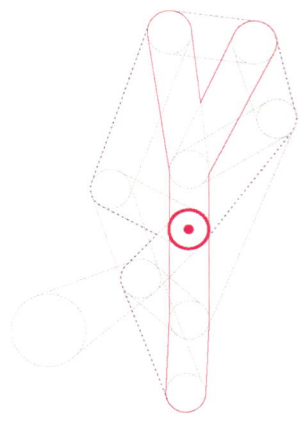

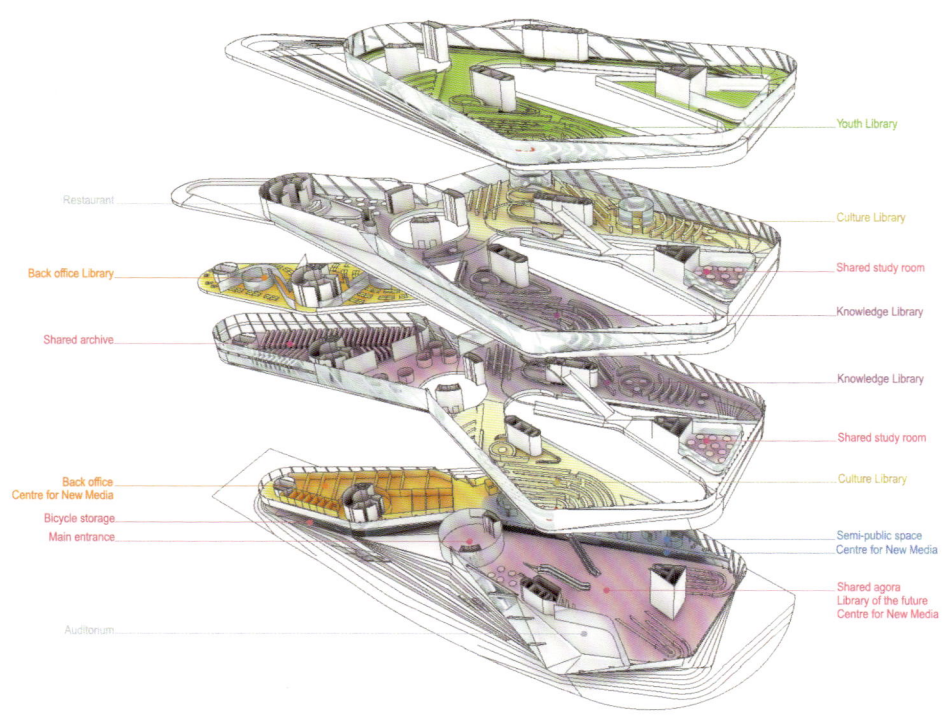

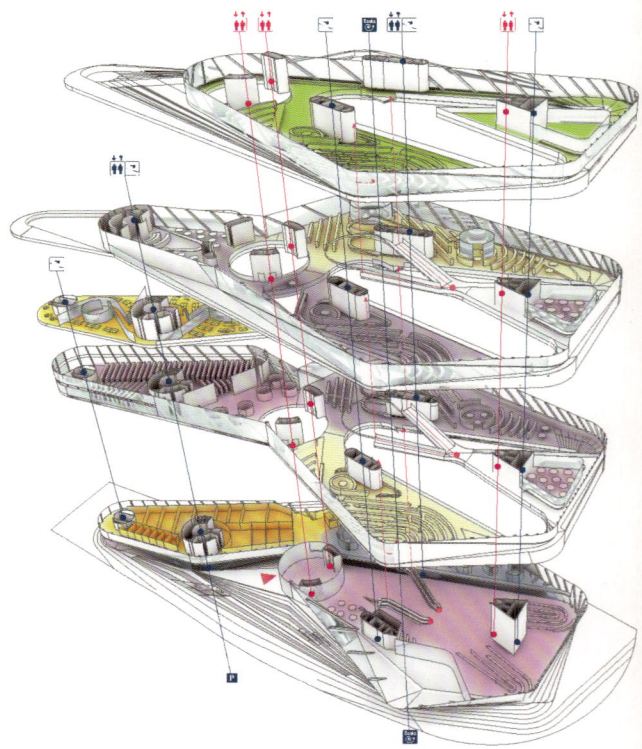

Restaurant

Back office Library

Shared archive

Back office
Centre for New Media

Bicycle storage

Main entrance

Auditorium

Youth Library

Culture Library

Shared study room

Knowledge Library

Knowledge Library

Shared study room

Culture Library

Semi-public space
Centre for New Media

Shared agora
Library of the future
Centre for New Media

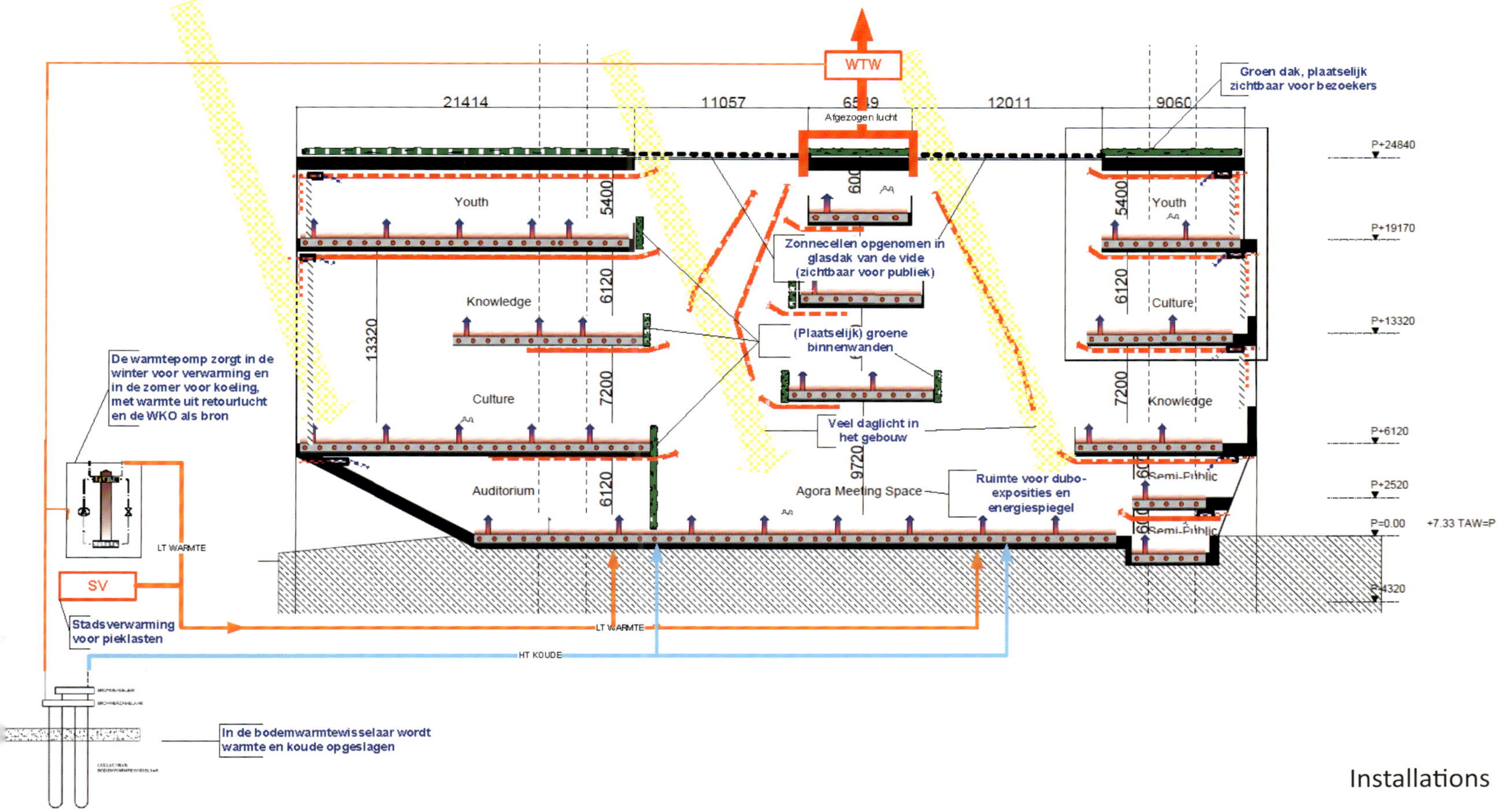

De warmtepomp zorgt in de winter voor verwarming en in de zomer voor koeling, met warmte uit retourlucht en de WKO als bron

LT WARMTE

SV

Stadsverwarming voor pieklasten

LT WARMTE

HT KOUDE

In de bodemwarmtewisselaar wordt warmte en koude opgeslagen

WTW

Afgezogen lucht

Groen dak, plaatselijk zichtbaar voor bezoekers

21414   11057   6549   12011   9060

Youth   5400

Knowledge   6120

13320

Culture   7200

6120

Auditorium

Zonnecellen opgenomen in glasdak van de vide (zichtbaar voor publiek)

(Plaatselijk) groene binnenwanden

Veel daglicht in het gebouw

600

9720

Agora Meeting Space

Ruimte voor dubo-exposities en energiespiegel

Youth   5400

Culture   6120

Knowledge   7200

Semi-Public

Semi-Public

P+24840

P+19170

P+13320

P+6120

P+2520

P=0.00   +7.33 TAW=P

P-4320

Installations

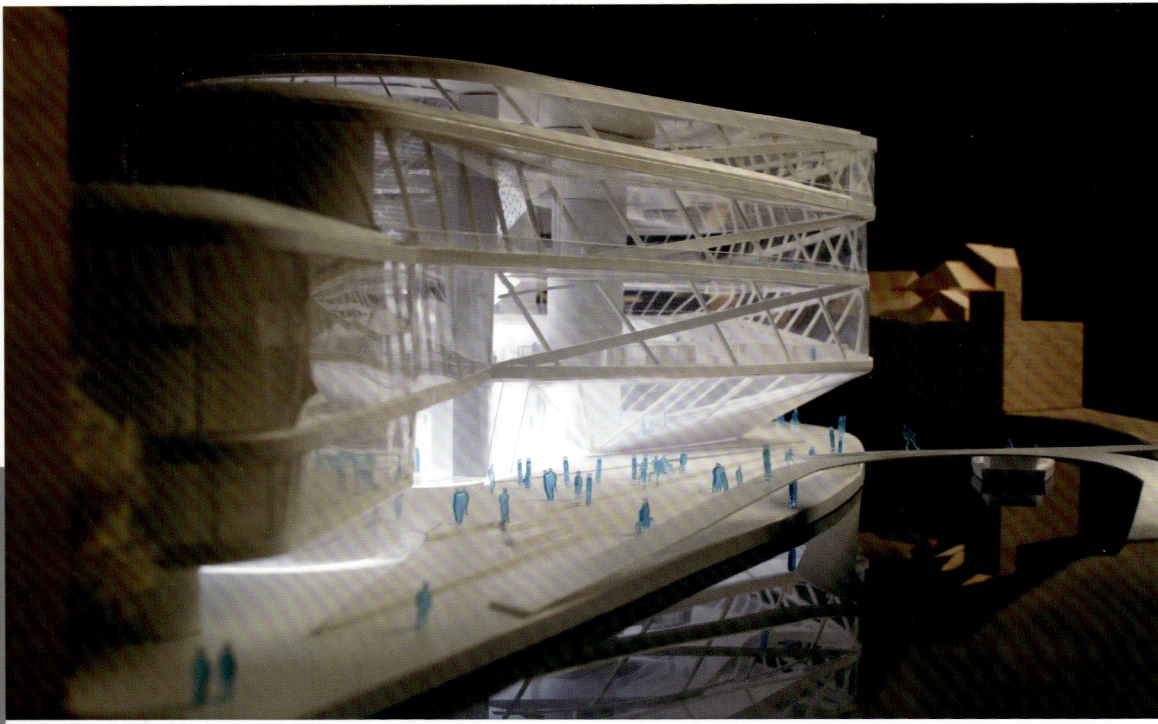

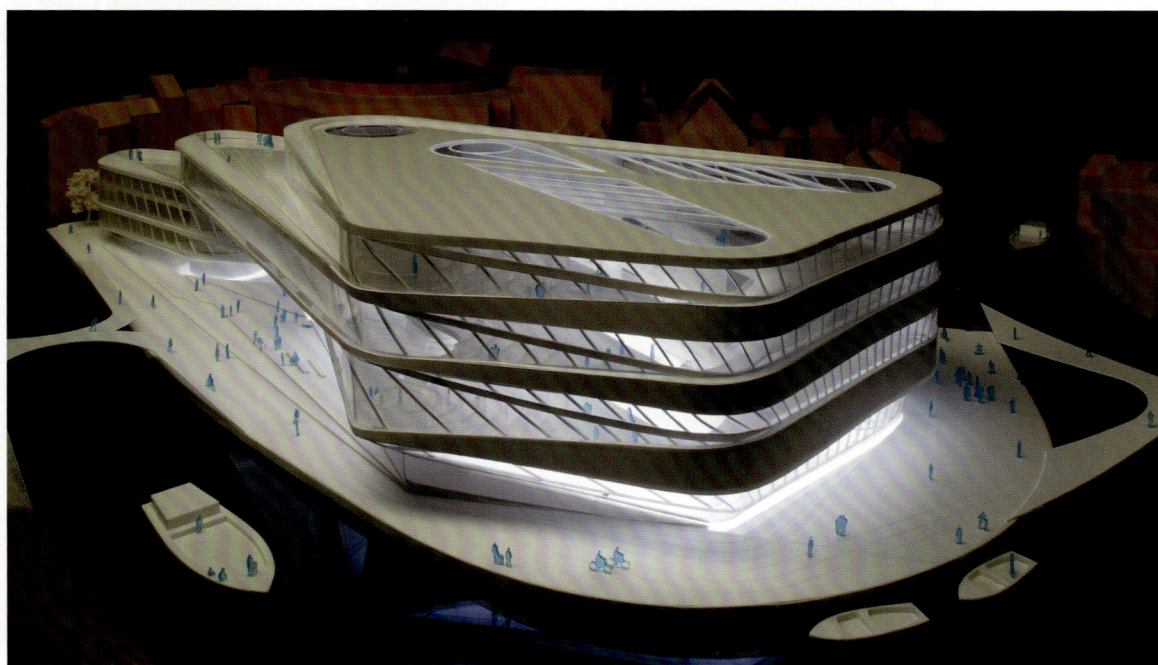

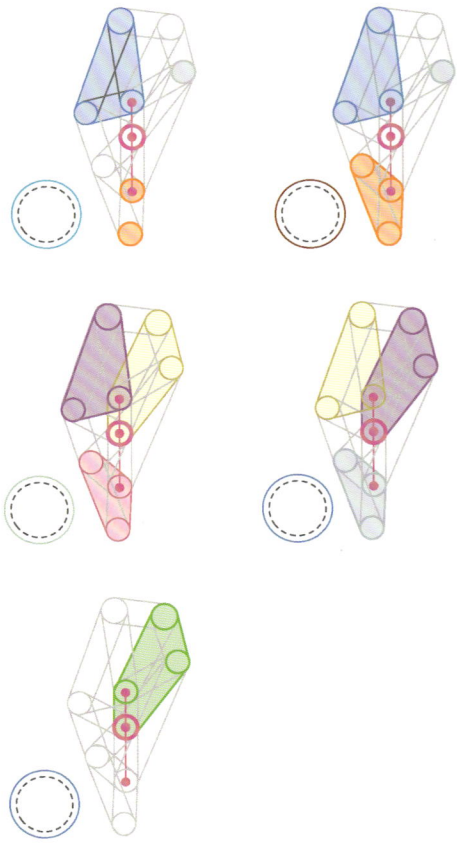

Bibliotheek van de toekomst

- Centrum voor Nieuwe Media
- Back office
- Gemeenschappelijke ruimtes
- Auditorium + Restaurant
- Cultuur Bibliotheek
- Kennis Bibliotheek
- Jeugd Bibliotheek

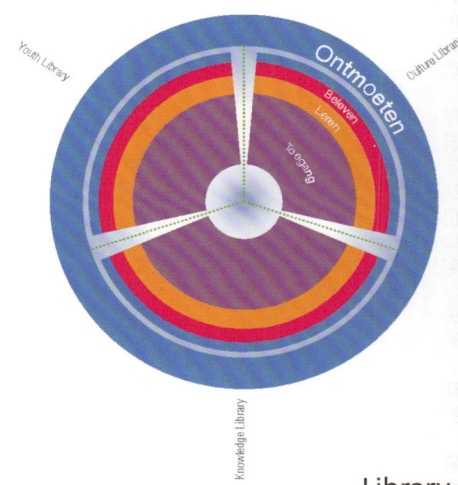

Library

Boekkasten - 500 Meter
Zitgelegenheden - 50 plaatsen

Boekkasten - 450 Meter
Zitgelegenheden - 50 plaatsen

Boekkasten - 350 Meter
Zitgelegenheden - 100 plaatsen

Boekkasten - 300 Meter
Zitgelegenheden - 200 plaatsen

Zitgelegenheden - 100
Moduleerbare lokalen - 4

Zitgelegenheden - 200
Moduleerbare lokalen - 4

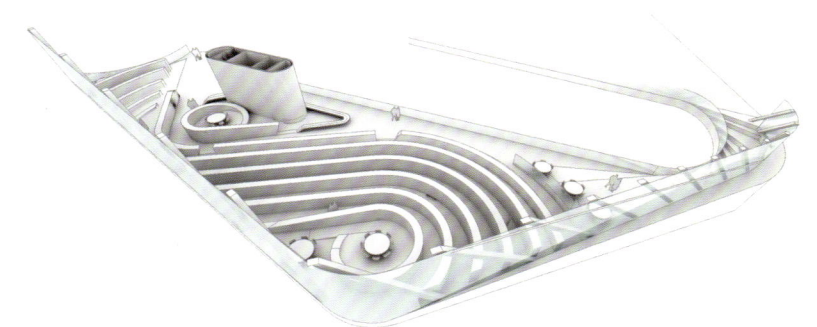

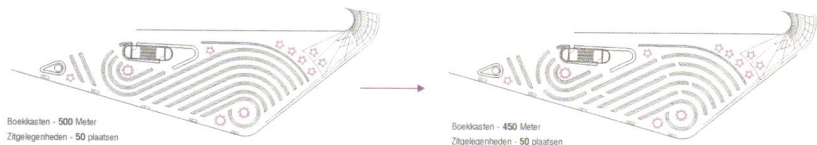

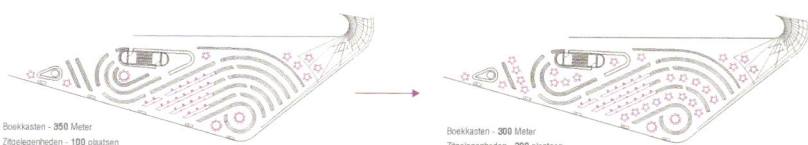

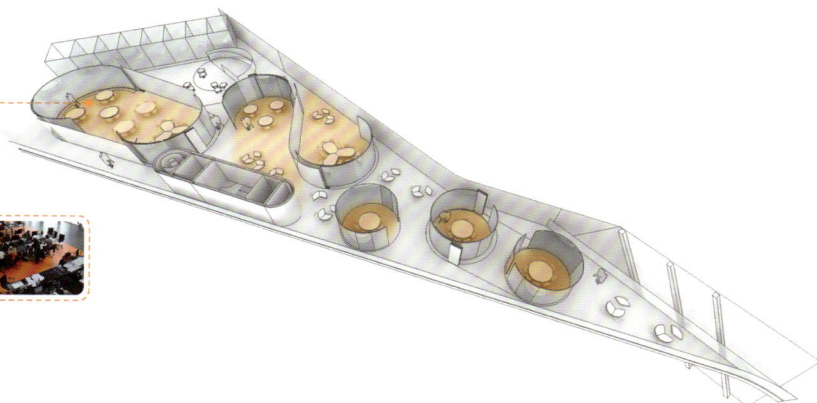

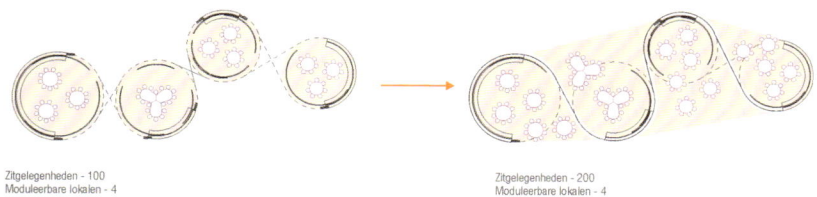

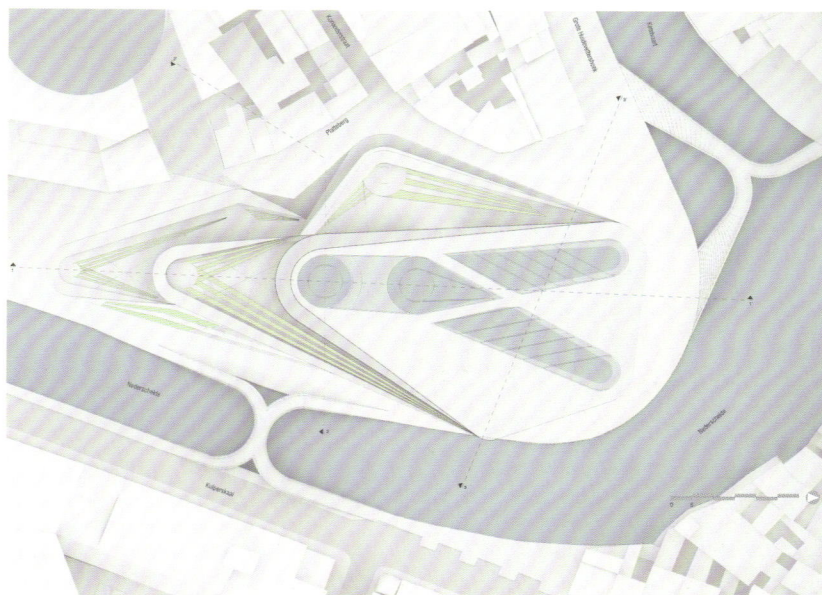

　　为比利时的根特市设计的未来城市图书馆与新媒体中心两主要的目标是创造一个充满活力、灵活多样与共享知识的环，同时又能通过带有鲜明特点的建筑加强当地的特色。

　　可持续性是设计中的指导因素。设计者相信不仅未来城市书馆的环境与使用者友好型的设计要能够与新媒体的发展同，而且设计还要为以后建筑用途的改变提供可能性。通过开的景观、宽敞的视野、灵活的循环路线、众多的会议区域以公共广场等设计，图书馆的设计方案完全可以满足今后其用改变的需要。

　　建筑在形式上十分流畅，并与周围环境相协调。这一特不但从其在各个方向上呈现了不同外形这一方面得以体现，反应在将建筑整体从一层高度升高的决策上——这种设计创了明亮、通透而广阔的视线。但是，分层的结构与低矮的建结构确保了该设计方案对于城区外轮廓线产生最小影响的同仍能使人们从这里看到根特富有特色的塔楼。建筑结构也使绿色）屋顶平台的设计具有可行性，同时还保证了仅有少量阳光会穿过屋顶直射进建筑。

　　基于功能的需要，建筑被提升离地面以便在图书馆周围获公共空间。与沿码头区延伸的滨海步道的连接使得建筑得以周围的水面互动并因此使Waalse Krook得到了重生。

　　建筑内部的布局是基于一个开放的中央天井，在其周围是环的通道。这个中央天井加强了空间感并在建筑中提供了清的定位。

　　在扩展城区环境与连接环形通道之外，中央天井还起到了接建筑里不同的功能组群的作用。天井结构成为了连接城市城市图书馆的桥梁。从这一点来看，图书馆也体现出了对公认知的一种比喻。

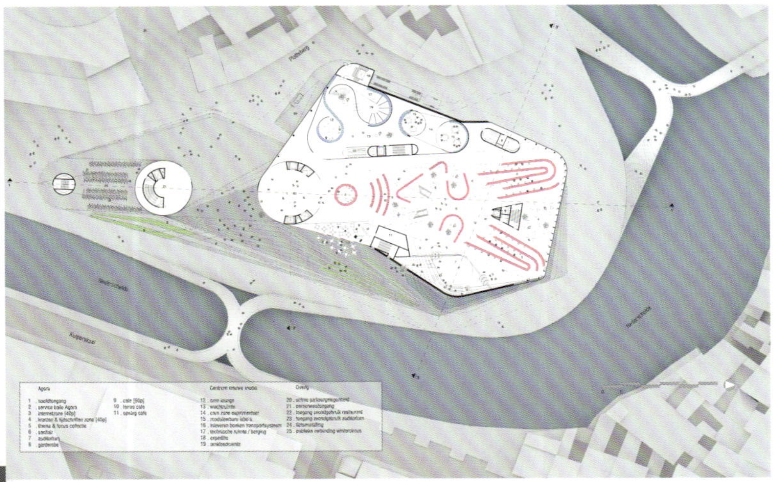

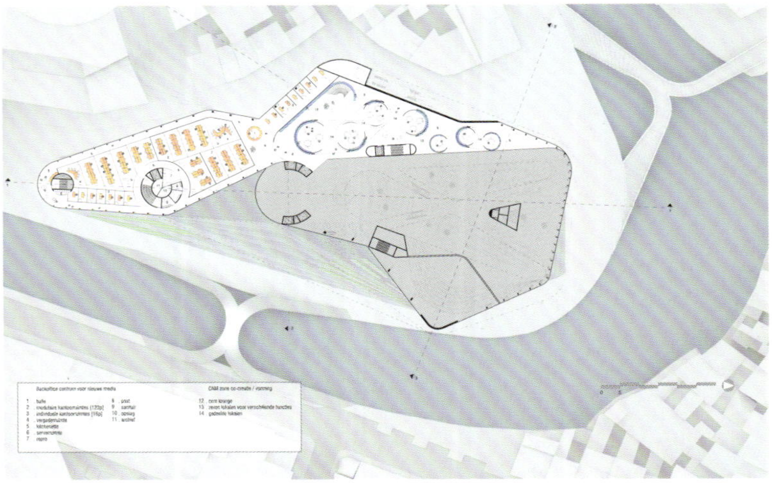

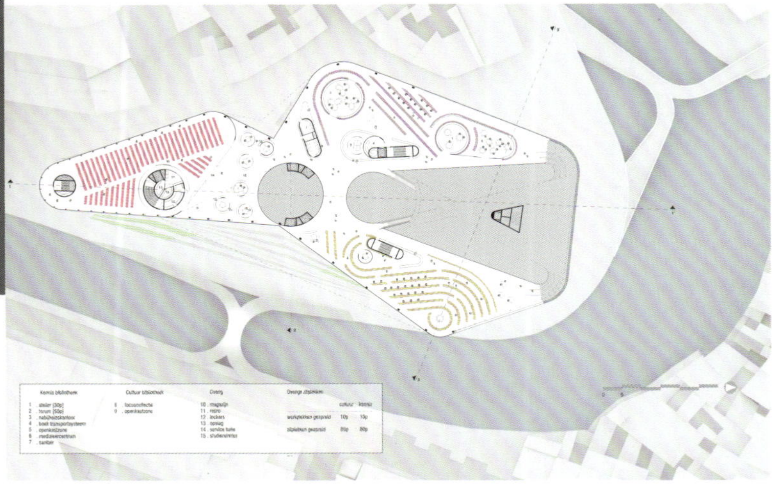

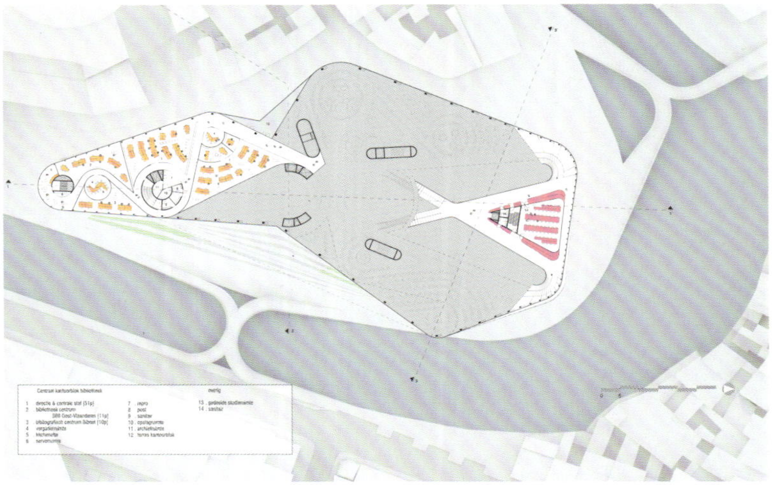

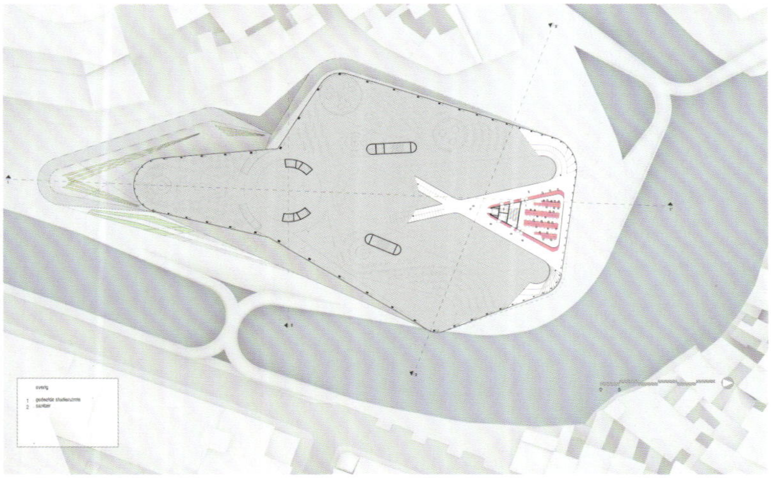

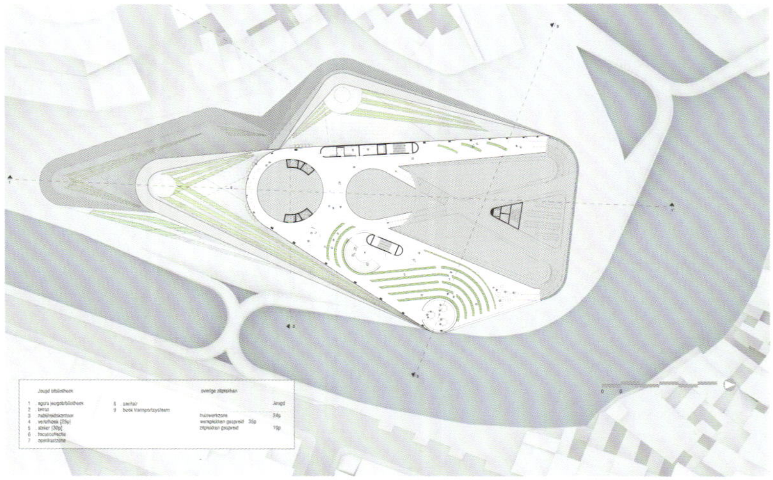

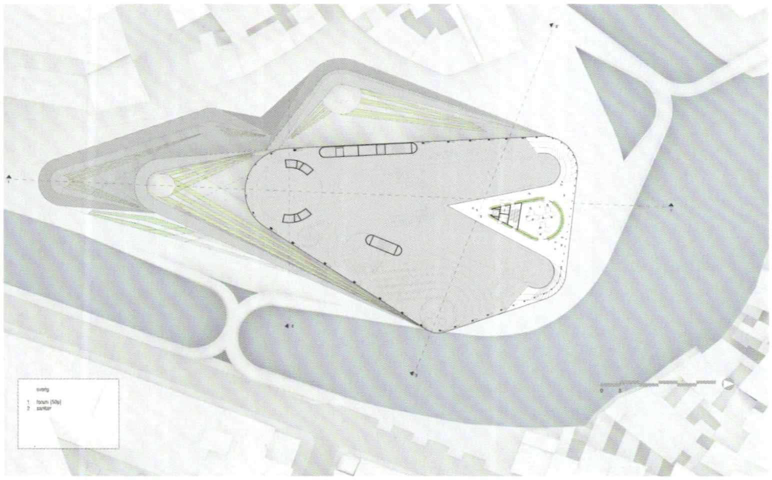

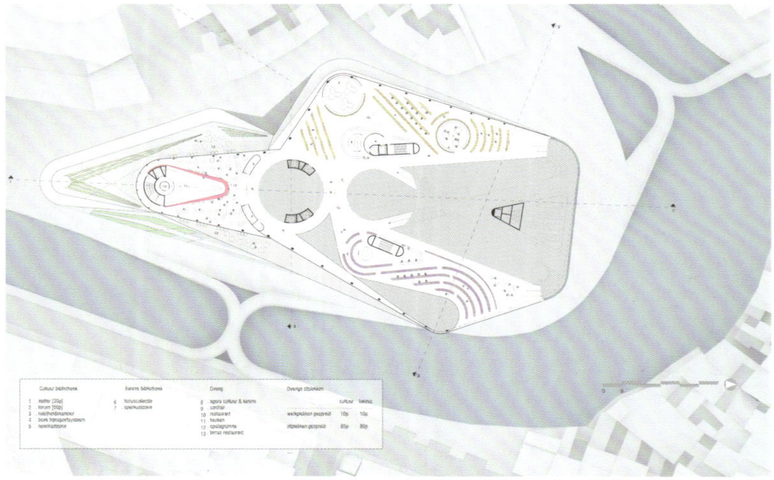

Cybertecture Egg

网络建筑：卵形大厦

Firm:James Law Cybertecture International
Location:Mumbai, India

In the 21st Century, what people will build will not be the same as what people were building in the 20th Century.

Buildings are no longer about concrete, steel and glass, but also the new intangible materials of technology, multimedia, intelligence and interactivity. Only be recognizing this will bring a new form of architecture to light - namely a "Cybertecture".

This project is a "Cybertecture" building and winner of the CNBC Asia Pacific Commercial Property Awards 2009 - The Architecture Award India. It brings together Iconic Architecture, Environmental Design, Intelligent Systems, New Engineering together to create the most innovative building for the city of Mumbai and for India in the 21st Century. This project is conceived together by the combined teams of James Law Cybertecture International and Ove Arup for Wadhwa Developers of India.

The concept was inspired by looking at the world in terms of the planet being an ecosystem that allows life to evolve. The concept for this building is rather like planet earth, where a sustainable ecosystem is derived from an integrated and seamless Cybertecture that is evolving to give the building's inhabitants the very best space to work in.

The analogy to the form of the building is for the beautiful planet form "land" on the site C70 in Mumbai, and creates a new Cybertecture ecosyste for people who will use this building.

The form of the architecture is one that symbolizes with optimism abo the future and of the 21st Century. The symbolic "planet" form is furth stretched to cater for 10 levels of accommodation, deriving an "Egg" shap building. This "Egg" is further orientated and skewed at an angle to crea both a strong visual language as well as to alleviate the solar gain of t building. By using this "Egg" shape, compared to a conventional buildir this building has approximately 10%~20% less surface area. The architectu is sleek and computer designed, with engineering that creates a building extremely high quality and geometric sophistication.

This building will act like a "jewel" for the new Central Business District Mumbai, and a worthy neighbor to the esteemed neighboring buildings the district.

21世纪人类建造的建筑将不会同20世纪所建造的一样。

建筑不仅是有关于混凝土、钢铁和玻璃,而且还包括了很多新的难以形容的科技、多媒体、智能及交互的元素。只有认识到这一点才能产生新的建筑形式"网络建筑(Cybertecture)"。

本项目设计就是这样的一个"网络建筑"并且是2009年亚太财经频道亚太商业地产奖——印度建筑奖——获得者。该设计将形象建筑、环境设计、智能系统及新型工程学结合到一起,为21世纪的印度和孟买市创造出了最具开拓性的建筑。这个建筑是由科健国际有限公司和阿鲁普设计所为印度沃尔德开发公司组成的联合团队设计的。

这个设计的概念源于将世界看成是允许生命进化的生态系统。这座建筑的概念就是我们的行星地球。在这里,一个从无缝整合的"网络建筑"中所衍生出的可持续发展的生态系统不断地进化,以向这座建筑中的居住者提供最好的工作环境。

这座建筑的形式就像是一个降落在位于孟买的C70区域的美丽的星球,并为使用这座建筑的人们创造了新的"网络建筑"生态系统。

建筑的形式代表了对未来与21世纪的美好向往。象征性的"行星"形状进一步地伸展成10层的居住结构,形成了一座"卵"形的建筑。这座"卵"形建筑接着转向东面并倾斜成一个既能产生强烈的视觉冲击又能减少阳光照射的角度。利用这个不同于传统建筑的"卵"形设计,这座建筑大约减少了10%至20%的表面积。这个建筑由计算机设计,拥有光滑的外表,体现了整体高质量的设计水平与精巧的几何结构。

这座新建筑将会成为孟买新的中心商业区的一颗"宝石",也将与其周围其他令人惊叹的建筑相呼应。

TEK– **Technology Entertainment Knowledge Building**

科技娱乐与知识中心

Firm: Bjarke Ingels Group
Location: Taiwan, China
Area: 53,000m²
Partner in Charge: Bjarke Ingels, Jakob Lange
Contributors: Cat Huang, Allyson Hiller, Xi Chen, Esben Vik, Johan Cool, Xu Li, Gaeton Brunet

The Technology Entertainment & Knowledge Center – aka TEK Taipei – a dense urban block of all kinds of activities related to contemporary echnology and media.

The spiraling street of media programs is consolidated in to a 57mx57mx57m ube of program permeated by a public trajectory of people life.

The cube is finished in concrete lamellas serving as solar shading as well s public access. The lamellas recede inwards forming a generous public aircase allowing the public to walk into the facade and all the way to the oof.

EK Taipei will consolidate exhibition spaces, showrooms, retail space, a market place and hotel, offices and conference rooms all related to media in a single superfunctional entity. At the heart of the institution, a big public auditorium will host product presentations, program launches, movie previews and gaming tournaments as well as the biannual TEK Taipei as the reoccurring anchor event for the whole complex.

TEK contains an almost urban mix of programs with no obvious hierarchy. The architects propose to organize the shops and showrooms, offices and hotel rooms, conference rooms and exhibition spaces, restaurants and galleries along an internal extension of the pedestrian street to the south.

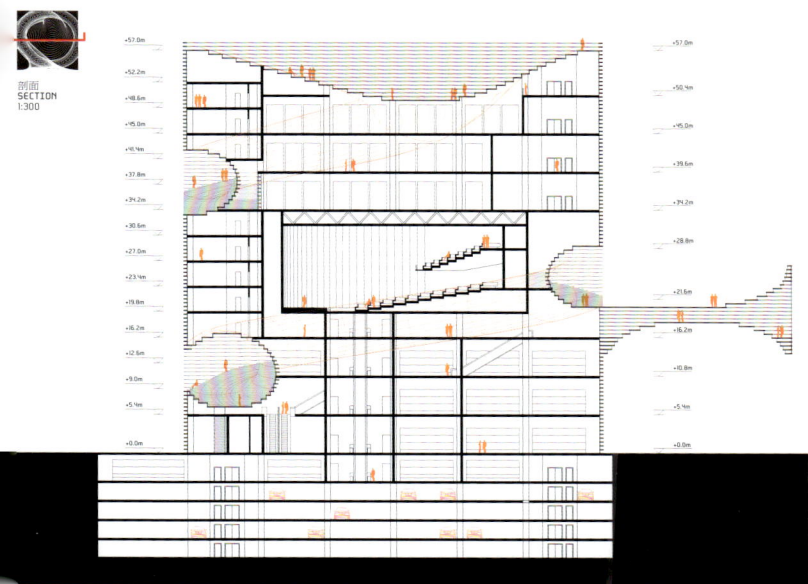

剖面
SECTION
1:300

+57.0m
+52.2m
+48.6m
+45.0m
+41.4m
+37.8m
+34.2m
+30.6m
+27.0m
+23.4m
+19.8m
+16.2m
+12.6m
+9.0m
+5.4m
+0.0m

+57.0m
+50.4m
+45.0m
+39.6m
+34.2m
+28.8m
+21.6m
+16.2m
+10.8m
+5.4m
+0.0m

Section

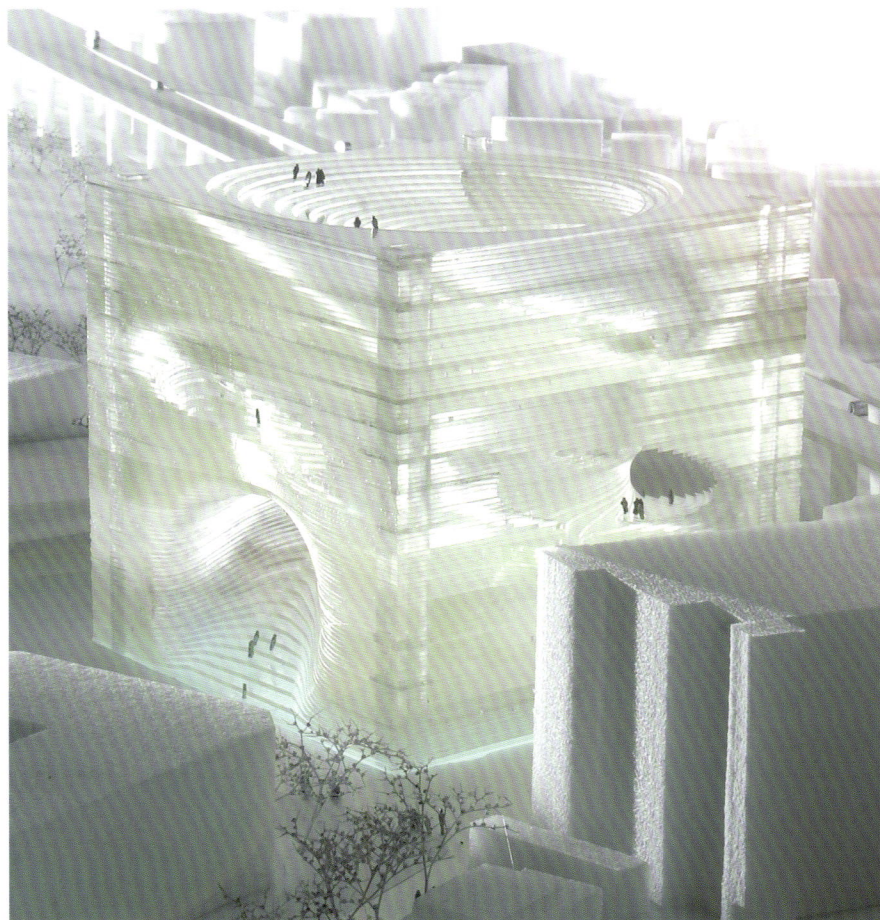

THE COIL : To remain within the site and the maximum building volume, the public street is coiled up in an ascending spiral leading from the ground floor to the roof garden.

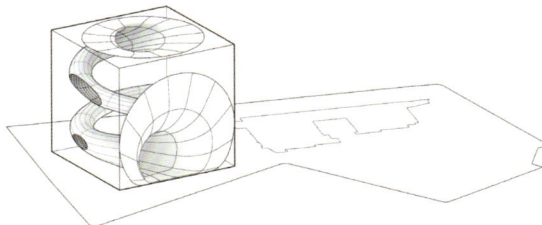

THE CUBE = TEK³ : The spiraling street of media programs is consolidated in to a 57x57x57m3 cube of program permeated by a public trajectory of people life.

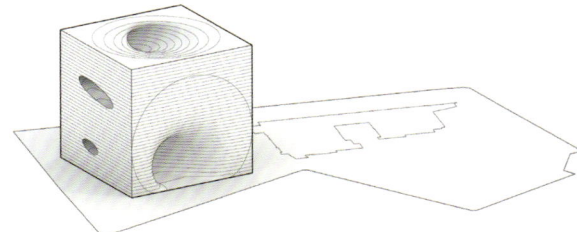

115 STEPS : The cube is finished in concrete lamellas serving as solar shading as well as public access. The lamellas recede inwards forming a generous public staircase allowing the public to walk into the façade and all the way to the roof.

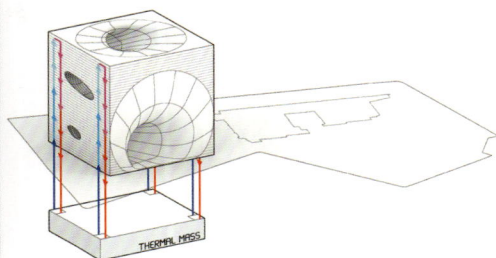

SUSTAINABILITY : THERMAL MASS : Passive cooling from the louvers can be enhanced by the addition of a thermal mass (water tank or concrete) underneath the building.

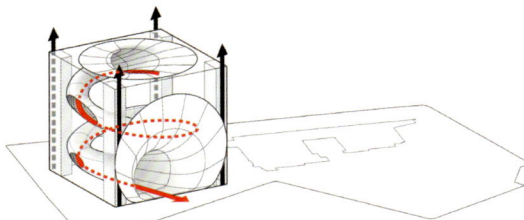

CIRCULATION : The circulation is split in three complementary systems. Primary is the central trajectory that provides seamless continuity from the ground to the roof and back again. Second is a chain link of escalators underneath the trajectory that connects every single floor internally. Circulation happens in parallel inside and outside the building. Exterior circulation is created by the spiral trajectory while an interior path of escalators optimizes the underside of the spiral. Internal circulation is especially provided for times of inclement weather. Finally 4 stair and elevator cores, one in each corner, constitute a rigorous and intuitive means of vertical movement.

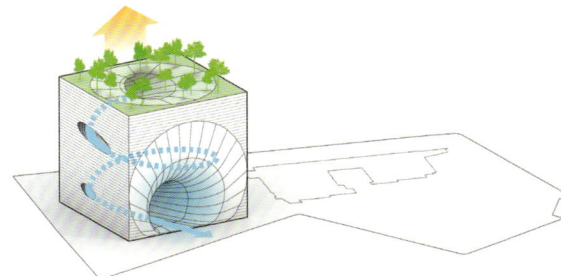

SUSTAINABILITY : ROOF GROVE FOR NATURAL COOLING : A roof grove of trees form a cooling canopy on the roof exploiting natural shade and evaporative cooling to create a local drop of exterior temperature by a couple of degrees C. The relatively cooler air will naturally drop down and drain through the trajectory creating a comfortable breeze throughout the public space.

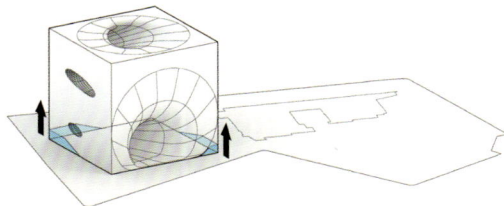

STREET EDGE : Towards the XXX Road and XXX Blvd intersection and the pedestrian street we propose to lift the lamellas to create open storefronts and uninhibited access to the hotel lobby and the ground floor flagship stores.

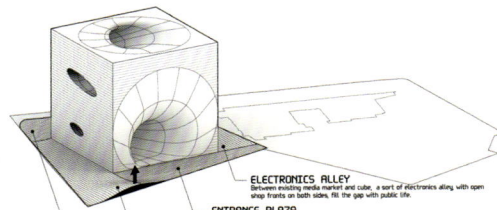

ELECTRONICS ALLEY
Between existing media market and cube, a sort of electronics alley with open shop fronts on both sides, fill the gap with public life.

ENTRANCE PLAZA
Facing XXXX the entrance plaza leads visitors and passersby seamlessly into and up through the public trajectory. Like an irresistible invitation the façade caves in to bring public life to all levels of the cube.

MEDIA PLAZA
From the southwest corner of the cube, scattered benches with integrated audio allow people to rest and enjoy the artwork, information and reoccurring T&D talks projected on the media façade of the cube.

HOTEL PLAZA
Facing XXX Blvd is a drop off and holding area that welcomes visitors to the 150 room boutique hotel.

FOUR PLAZAS : The cube is positioned on the middle of the site, slightly rotated to make space for 4 dedicated plazas.

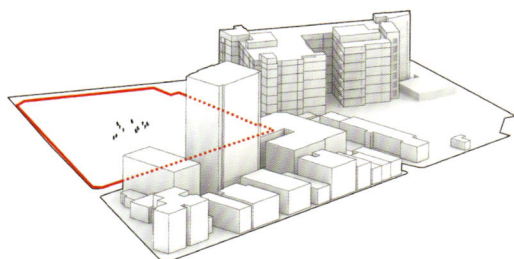

TEK³ EVENTS : TEK³ Taipei will consolidate exhibition spaces, showrooms, retail space, a market place and hotel, offices and conference rooms all related to media in a single superfunctional entity. At the heart of the institution, a big public auditorium will host product presentations, program launches, movie previews and gaming tournaments.

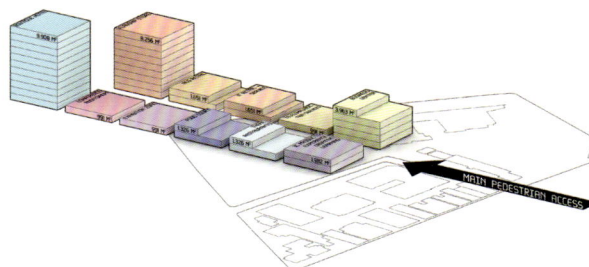

THE STREET : TEK³ contains an almost urban mix of programs with no obvious hierarchy. We propose to organize the shops and showrooms, offices and hotel rooms, conference rooms and inhibition spaces, restaurants and galleries along an internal extension of the pedestrian street to the south.

North wing   West wing

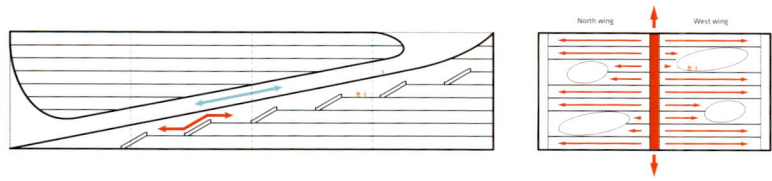

PROGRAM : The Cube is composed of two main organizational diagrams. An L-shaped hotel with two wings facing west and north, and a square stack of generic floors of retail and showrooms sandwiched around a central auditorium for launches and lectures.

EFFICIENCY : The square plan and deep continuous floor plans allow for a very efficient, flexible generic layout capable of accommodating countless conversions. The daylight sensitive programs such as hotel rooms and offices reside on the perimeter, while retail and exhibitions occupy the core.

PARALLEL PATHS : Circulation happens in parallel inside and outside the building. Exterior circulation is created by the spiral trajectory while an interior path of escalators optimizes the underside of the spiral. Internal circulation is especially provided for times of inclement weather.

HOTEL CIRCULATION : The northwest core of the cube is dedicated to hotel circulation. From this central elevator core all hotel rooms are reached. For escape purposes the two adjacent cores as well as the trajectory supplements the main access core.

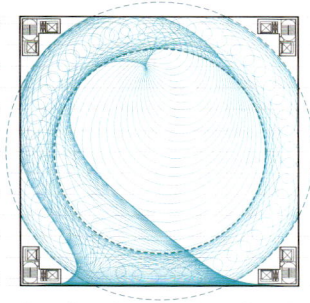

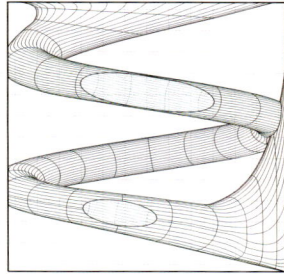

HOTEL

RETAIL
EXHIBITION
FOOD
OFFICE

X.RAY
A perfect circular spiral inscribed within a cube

X.RAY
Consistent slope

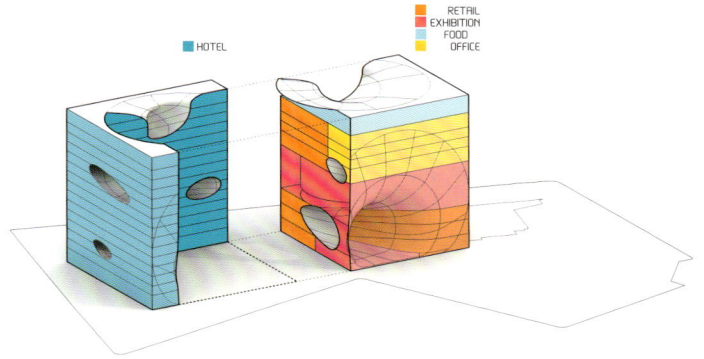

4TH FLOOR
+10.8m

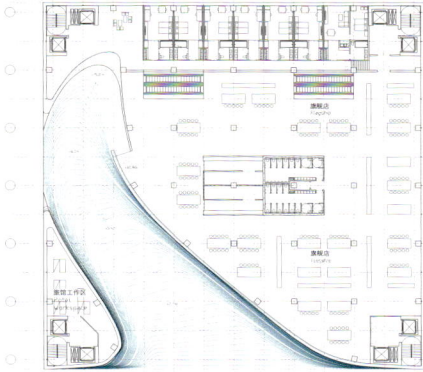

FLAGSHIP STORES
BOUTIQUE HOTEL

5TH FLOOR
+16.2m

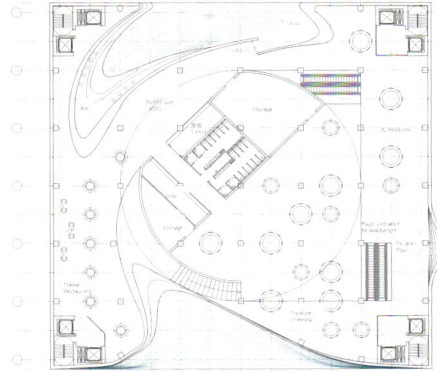

AUDITORIUM
THEME RESTAURANT
EXHIBITION

6TH FLOOR
+21.6m

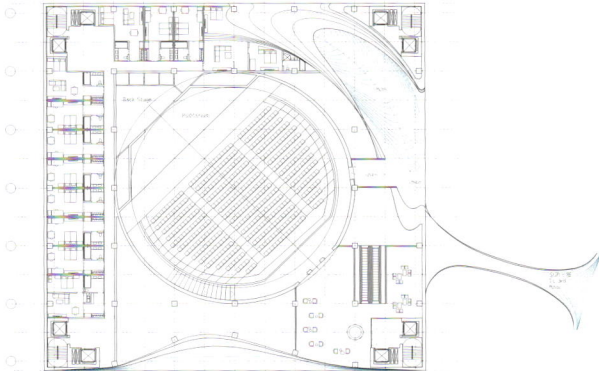

AUDITORIUM
BOUTIQUE HOTEL
3C PRODUCTS

8TH FLOOR
+27.0m

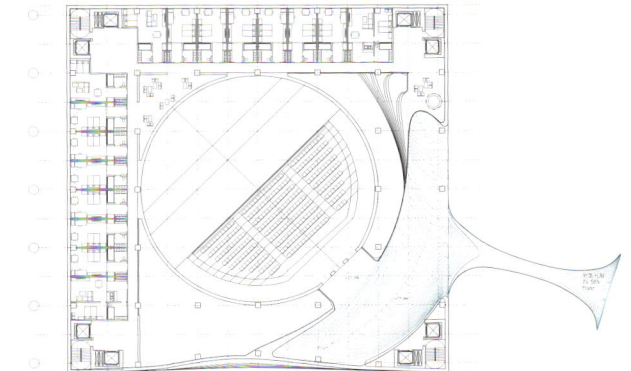

BOUTIQUE HOTEL

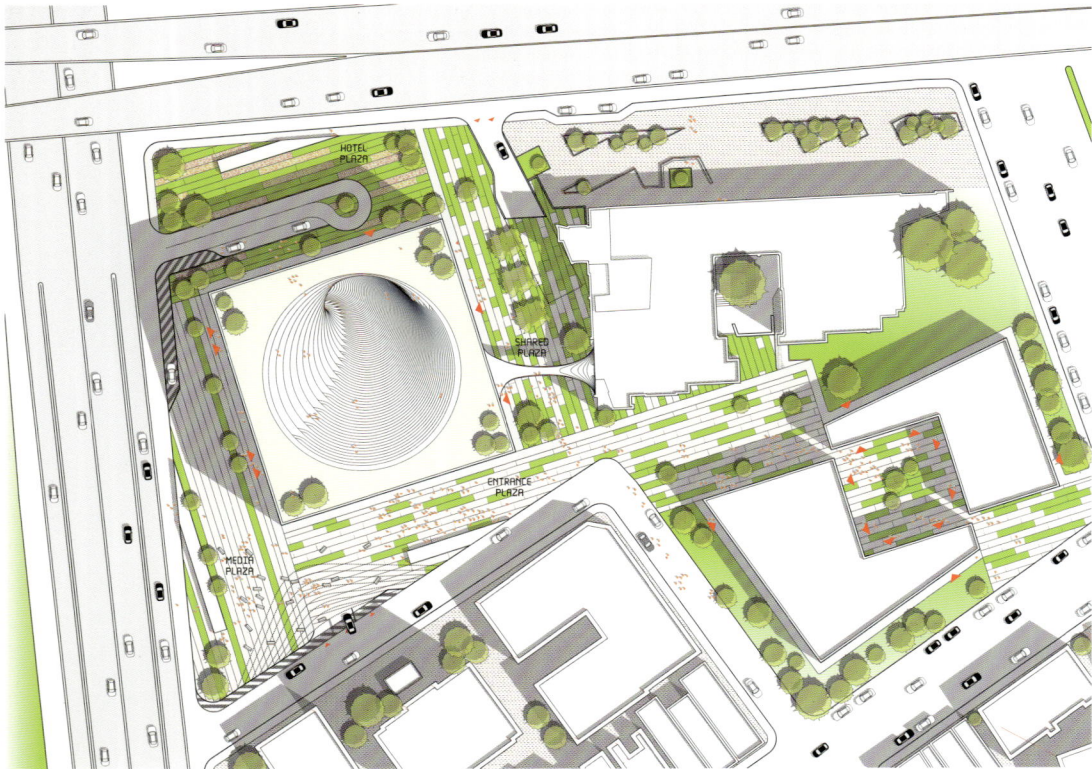

　　科技娱乐与知识中心，即TEK台北，是一座汇集了所有有关当代科技与媒体技术的密集型城市街区。

　　这座拥有各种媒体项目的螺旋形街道被整合在了一座充满了人们生活轨迹的57mx57mx57m的立方体建筑中。

　　这座立方体由多层混凝土结构构成。这些结构既阻挡了强烈的阳光，也作为了公众进入建筑的通道。这些层叠的混凝土阶梯结构逐渐地向内收缩形成了一个公共楼梯，使人们可以直接从建筑的外立面进入并一直走到楼顶。

　　TEK台北在一个单独的多功能建筑中整合了所有与媒体有关的功能，包括了展览空间、陈列室、零售商店、市场及酒店、写字间和会议室。一个大型公共礼堂位于建筑的心脏位置，可以举行产品宣讲会、项目发布会、电影试映会、游戏比赛及TEK台北双年展等活动——该建筑中重复举办的一场重大活动。

　　TEK3几乎囊括了城市中的各种活动，而并没有明显地过于倾向于某一个。设计师提出的这一方案正是作为步行街向南的室内延伸部分，并将商店、展室、写字间、酒店、会议室与展览空间、餐厅、美术馆等功能集中在了这座建筑中。

Bella Sky

贝拉天空

Firm: 3XN
Location: Bella Center, Center Boulevard 5, Denmark
Area: 42,000m²
Award: 1st prize in invited competition2006

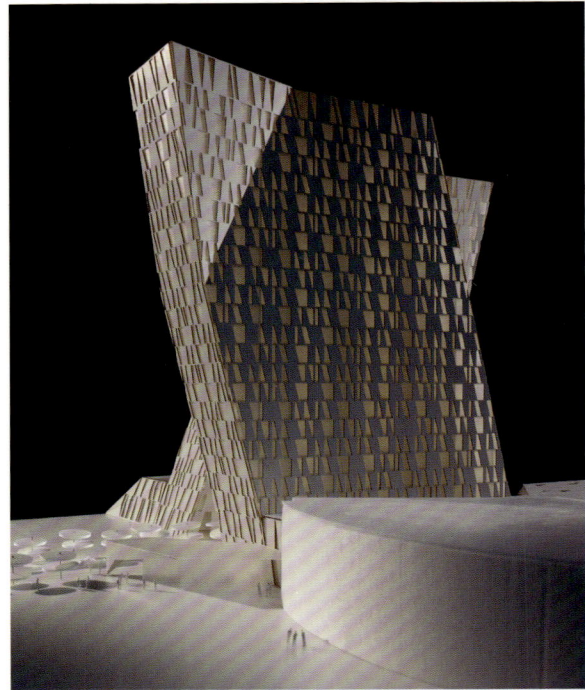

A new landmark hotel and conference center designed by 3XN is the first step of the long-anticipated extension of the Bella Center; Copenhagen's famous Congress Center. The upgrade of Bella Center will add an urban feeling to the place, and the extension will immediately benefit from Bella Center's perfect location: Situated between the old city core and Orestad; Copenhagen's growth center, Bella Center is connected to two important worlds of activities, and furthermore it's close to the Metro as well as the Copenhagen Airport. Initially, 3XN has worked out a master plan for the entire Bella Center area to establish the right place for the two-tower hotel. The master plan is flexible and may be executed in several phases. It draws upon the classical urban weave of rectangular streets and blocks, but leaves each field open to interpretation. The two hotel towers stand close as a pair, yet seem a little shy; the towers incline in opposite directions. The reason for this is to obtain an unobstructed view from all rooms in each tower of the flat landscape. The sky was not the limit in this case; flight safety requires a max tower height of 75m (25 floors) this close to the airport, so one tall tower was not an option. Wind considerations as well as a wish for a landmark signal caused the top twist of one tower, while the wish for a clearly indicated entrance caused the ground floor twist of the other tower. The hotel lobby is merging into the existing entrance lobby of the Bella Center, making the hotel a true integrated part, ready for large events like the Copenhagen International Fashion Fair; this time with rooms for rent.

由3XN公司设计的新的地标性酒店及会议中心是贝拉中心期待已久的发展扩建的第一步。贝拉中心的这次升级将会为这里带来一份城市的现代感，并且扩建的这部分将会马上受益于贝拉中心的绝佳地理位置。贝拉中心坐落于旧城中心与哥本哈根正在发展中的Orestad地区之间。贝拉中心连接着世界上两个重要的活动，此外，它距离地铁及哥本哈根机场都很近。起初，3XN公司为整个贝拉中心地区设计了总体规划，确立了这座双塔式酒店的正确位置。总体规划非常灵活，并且可以分为多期实施。规划建立在传统的直角街道及街区的城市规划基础上，但也为每一个场地留下了可以继续发展的空间。这座双塔式酒店遥相对立，看起来好像还有一点害羞似的朝向相反的方向倾斜。在这里，天空并不是限制了建筑高度的因素，而是飞行安全要求距离机场较近的塔式建筑最高不得超过75米（25层）高。因此，仅仅建造一座塔式建筑是无法满足要求的。考虑到风的因素以及建造一座地标性建筑的愿望，设计将其中一座塔楼的顶部设计成扭曲的造型。同时，设计将另一座塔楼的底层设计为扭曲的造型以清晰地指示出入口的位置。酒店的大堂与贝拉中心现有的入口大厅融为一体，使酒店真正成为贝拉中心的一部分，为哥本哈根国际时装展览会等大型活动的开展做好准备，以满足大量的住宿需求。

Centro Comercial Pedregal

佩德雷加尔商业中心

Architects: Carlos Pascal Wolf, Gerard Pascal Wolf
Firm: Pascal Arquitectos
Location: Mexico
Area: 7,000m²

This project comes to set a new architecture in the Pedregal in Mexico City, this area has been neglected, and nothing new and important has happened since its beginnings in the time it was built Ciudad Universitaria. Today, development pressure and need for services begins to promote change.

How this work relates to the context is through a break with what occurs in the area, which are big houses in big areas surrounded by walls more than three meters where nobody knows what happens inside nor the insiders know what is happening outside.

This is accomplished through a perforated, glazed facade that does exactly the opposite, revealing activity inside and allowing the interior to see and be abroad, so that public social spaces mix and the limits of urban and the private become frontiers.

The project consists of two levels of retail space and a roof garden, which uproots the medium level on the sidewalk and below these two levels of parking.

Include access and spaces for people of different capacities: ramps, special parking spaces, elevators, etc.; zone slowdown in the entry and exit to avoid waiting in cars on public roads, numerous landscaped areas, including the roof; car delivery area located within the basement parking.

This is a project with a sustainable and intelligent automation system and control that provides energy saving assets and liabilities, lighting control, removal, opening and closing the front curtains, air conditioning, security and access control, prevention and protection civil, cctv, all scheduled and synchronized.

The main facade consists of two elements: a zinc sheet lined with large irregular perforations to which a body is embedded laminated glass in shades of yellow and translucent.

4500

2200

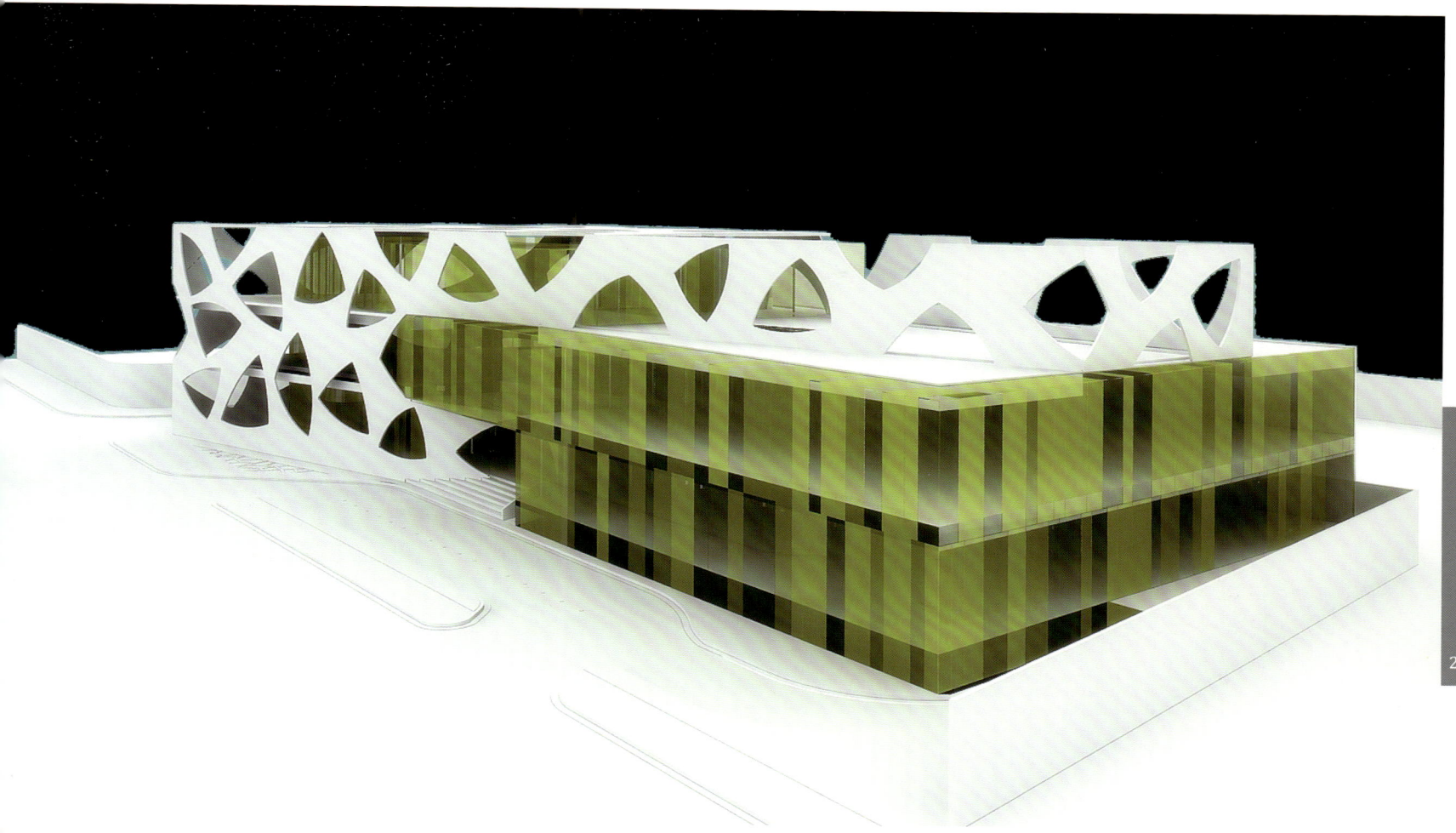

Fachadas

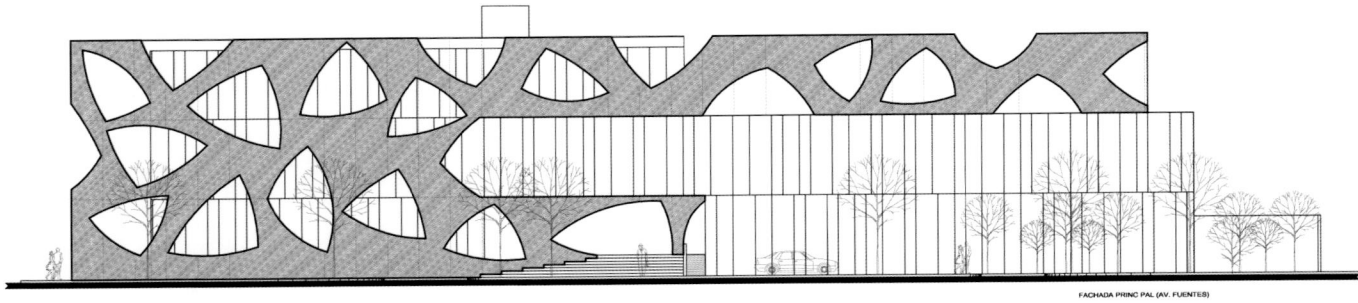

FACHADA PRINC PAL (AV. FUENTES)

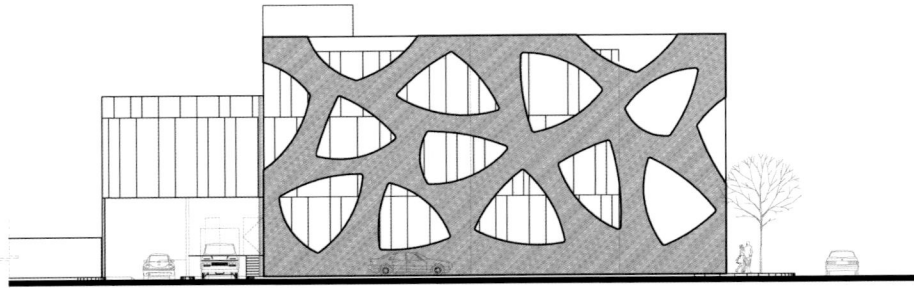

FACHADA LATERAL (AV. AGUA)

ESCALA GRAFICA
0   1   3   6   10
ESC: 1:100

214

Cortes

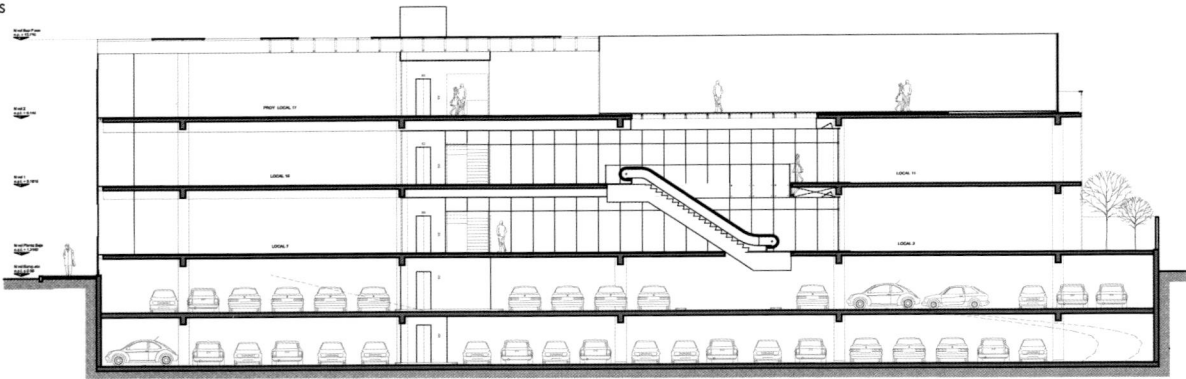

CORTE LONGITUDINAL 01

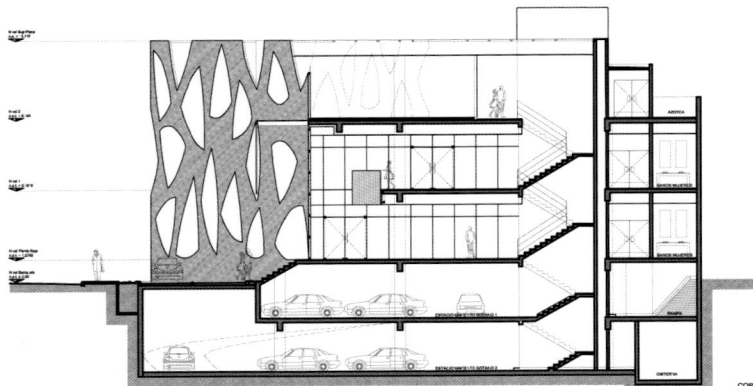

CORTE TRANSVERSAL 01

ESCALA GRAFICA
0   1   3   6   10
ESC: 1:100

这个项目是位于墨西哥城佩加尔的一座新建筑，这个地区去曾经一直受到忽略。从最开始立这个地区开始，除加拉加斯大城以外就没有什么新的和重要的目建立于此。而如今，发展的压和服务的需求推动了变化。

这个设计作品通过打破这一域的隔离达到与周围环境联系的的。这些大型区域里面的大型房被超过三米的高墙所围绕，从而使得人们无法知道高墙里面发生么，也使得里面的人无法知道高外面发生什么。

设计师采用的多孔的玻璃立产生了与上面相反的效果，从而到了使这个建筑与周围环境联系目的。这一设计让人们可以从外看到里面的活动，并让里面的人可以看到外面的情形，并感到生在其中。这种设计达到了整合公生活空间与让城市和私人生活的界成为了新的区域的目的。

整个建筑由两层的零售空间一个屋顶花园组成。此设计将人道的中间层整体抬高，而在这些构下面是两层的停车场。

在建筑中还包括了为不同的们所提供的通道和空间：步道、殊停车空间、电梯等等；为避免车在公共道路上等待而在出入口设的减速带、还有包括屋顶在内众多的景观区域；设在底层停车内的取车区。

该建筑所拥有的可持续性和能自动化与控制系统可以节约能消耗，可以控制灯光，能够移动打开和关闭正面的幕墙，可以管空调系统、保安及进出控制，可进行人群控制和保护，提供闭路视系统。所以这一切都是预设与步进行的。

主立面由两种元素组成：一透明的黄色复合玻璃结构嵌入到有不规则孔洞的锌板。

Planta Conjunto

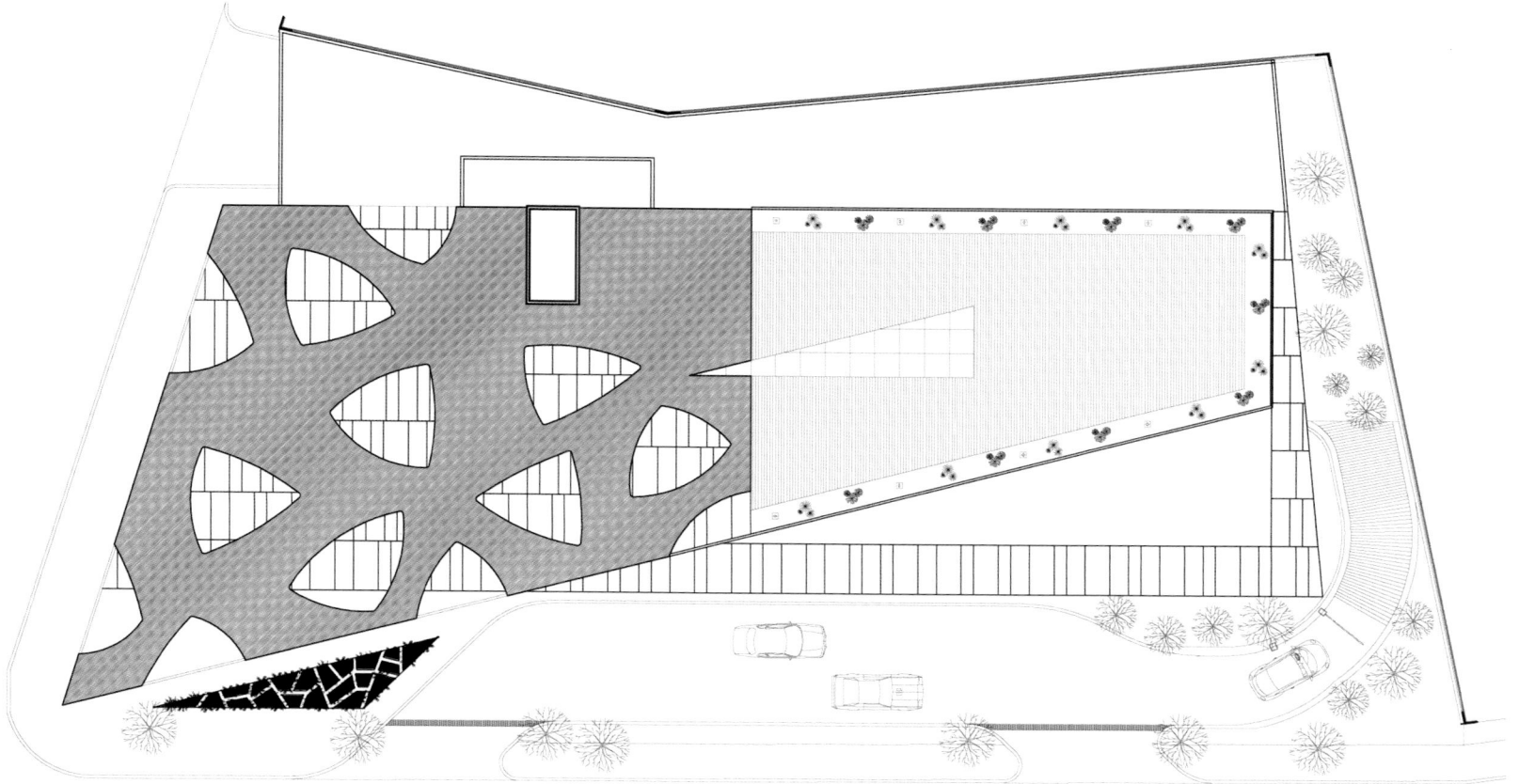

Planta Baja

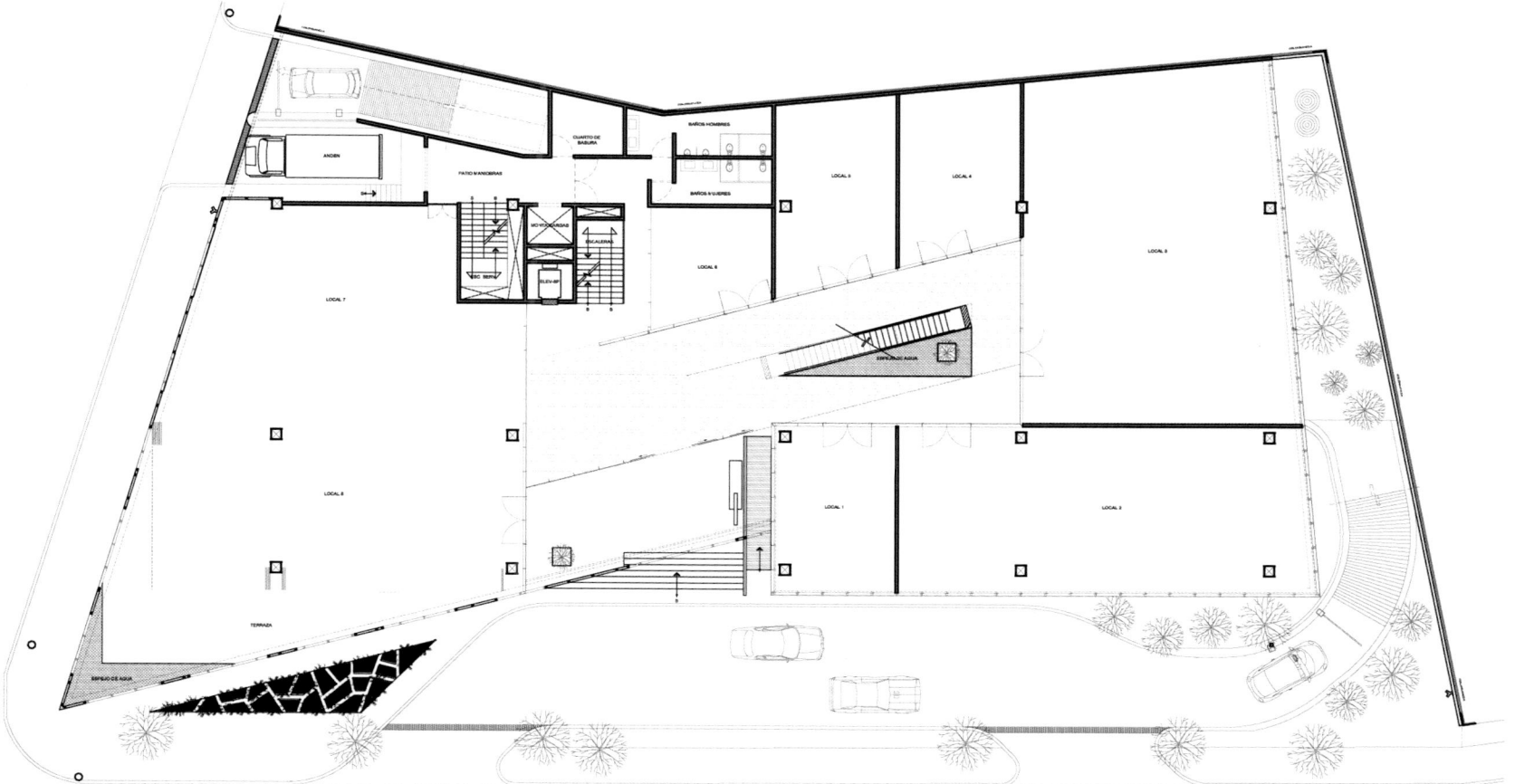

Tornado **Tower**

飓风塔

Architect: visiondivision through Anders Berensson & Ulf Mejergren
Location: Taipei, Taiwan, China

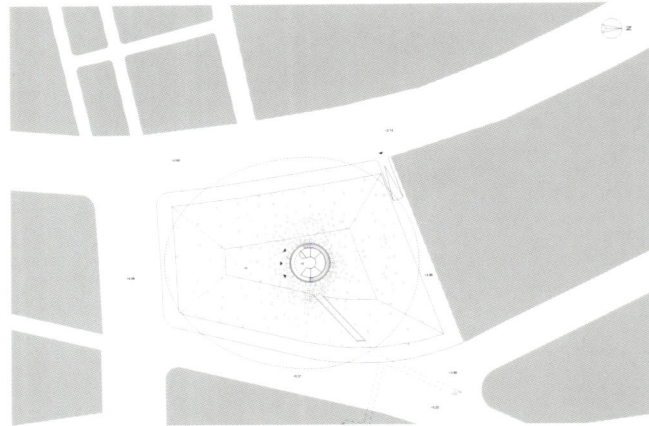

Site Plan 1:1000

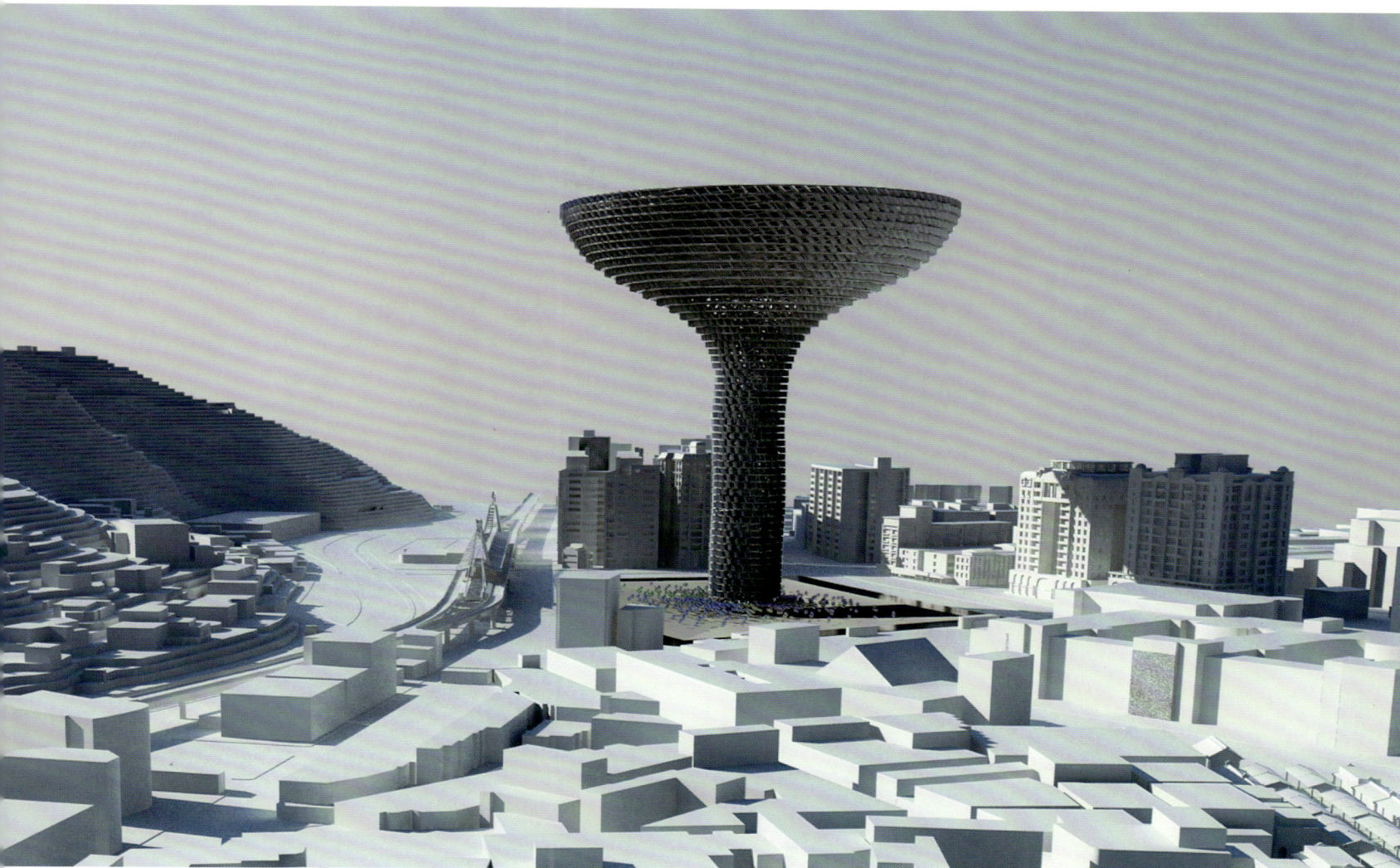

This is a powerful energy generating machine on the outside with a sensual, organic inside that transcends the visitors from the bustling city to a serene world of the performing arts.

One of the positive aspects of raising the building is that it creates a generous public space around it.

The square slopes gently towards the entrance, surrounded by pearls, the visitors descends fadingly into the building like entering an ocean.

A great spiral of pearls is the main focal point of the entrance hall, you can either take the elevator through it or the ramp around it, by foot or with the VIP Train of pearls, taking you all the way into the Grand Theatre.

The Grand Theatre is embedded in pearls, creating an elegant and modern experience for the audience. The semi-transparent pearls are lit from behind and dim the light, creating a glowing sensation. Depending on the performance, the ambience can be set into different modes.

A chandelier of pearls in the middle of the theatre drops seamlessly from the roof like a jewel, radiating an ambient light. The VIP Train is now converted into comfortable seats.

Reaching the roof terrace, the pearls subside into clouds. Walking around among the clouds one can experience panorama views of Taipei.

The building performs on its own for the city, generating culture, urban life and pure sustainable energy for its vicinity.

This is all strongly manifested from the visual effect of its rotating facade.

The facade covered with curved blades is attached on segments that rotate with the wind, generating energy to the building and to the city. The pearls are non-toxic and non-flamable acrylic balls with a tint of reflection.

Pearls in different sizes is combined into various landscapes inside the building to enhance the theatre experience.

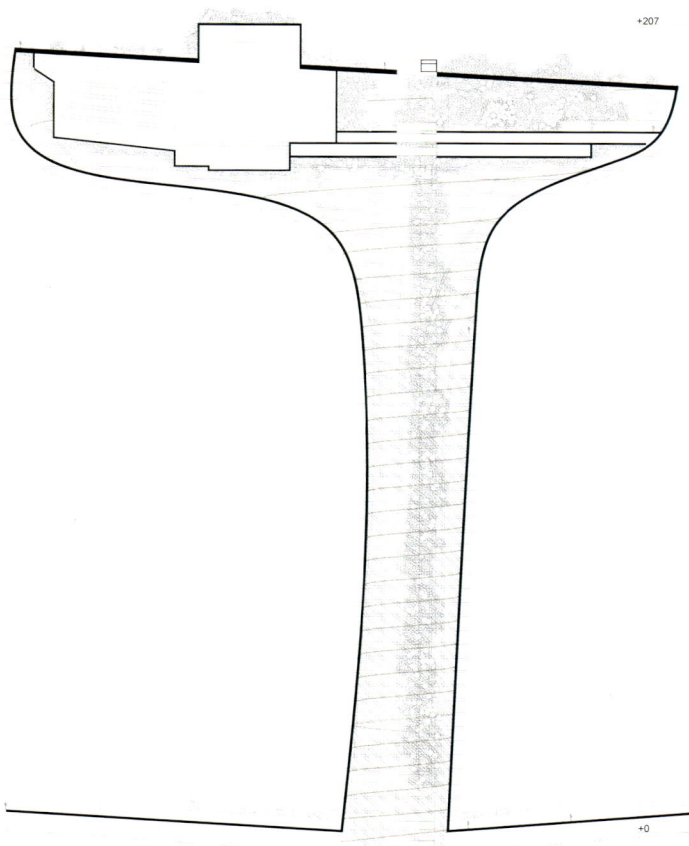

Section

这项的外部是一座能量巨大的发电设备，但内却是一座能将游客从喧闹的都市带到表演艺术的宁静境界的充满浪漫气息的绿色建筑。

建造这样一座塔楼的积极目的之一是它能够为围创造出一个广大的公共空间。

广场缓慢地向建筑的入口处倾斜，并被亮珠所围。游客将逐渐被隐没在向塔楼走去的路上，仿佛进到海洋中一般。

巨大的亮珠组成的螺旋形状是入口大厅的主要点。游客可以选择乘坐电梯通过其中，或者可以沿着周围的通道步行或乘坐贵宾列车，一路进入到大剧中。

大剧院坐落在亮珠中间，让观众感到一种典雅现代的体验。半透明的亮珠被从后面点亮，降低了空的亮度，制造出了热烈的气氛。根据不同的演出，周的环境可以设置为不同的模式。

剧院中央的水晶珠串吊灯好像是一大件珠宝从花板垂下来，向周围放射着光明。这时，游客的贵宾车将变成舒适的座位。

当来到靠近天顶的位置时，游客可以看到珠串散成为了云雾般的状态。在这种云雾里穿行的过程中台北的景色将被游客尽收眼底。

这座建筑将独立地为这座城市提供服务，为周地区提供城市文化生活与完全循环的能源。

这些都将强烈地体现在其可以旋转的外立面中。

这座建筑的外立面被弯曲的叶片覆盖。这些叶分别连接在一个个小单元中，可以随风旋转为这座建和城市提供电能。亮珠则是涂有反光材料的无毒阻燃丙烯酸球体。

不同尺寸的亮珠组合起来形成了不同的景观，建筑内增强了剧场效果的体验。

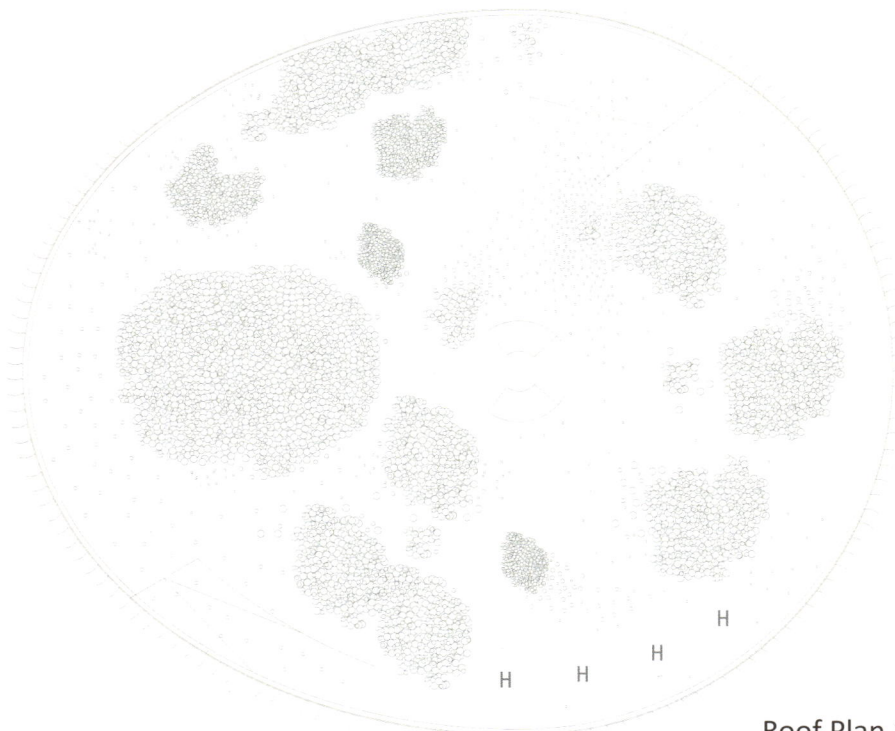

Roof Plan 1:600

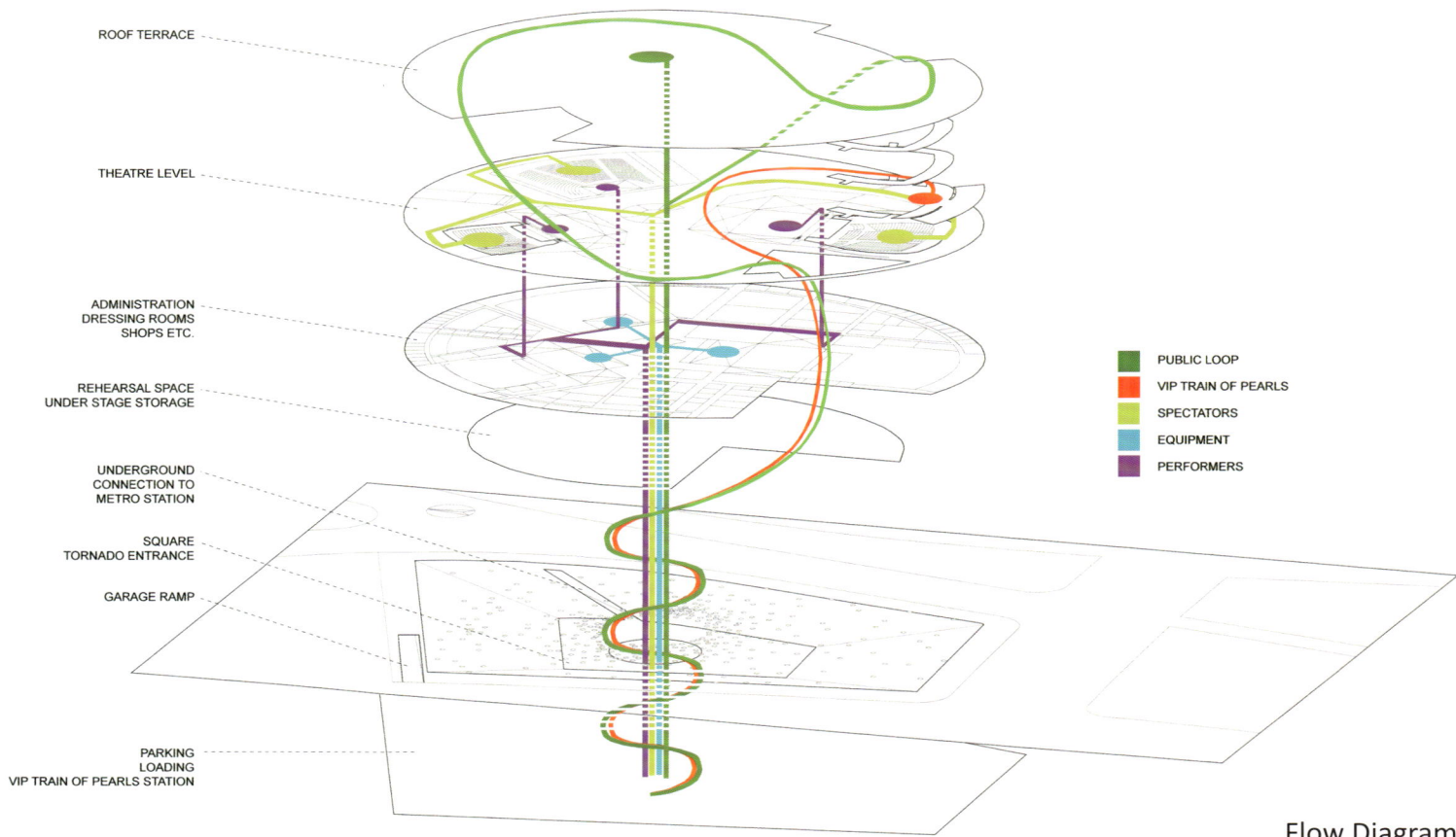

ROOF TERRACE

THEATRE LEVEL

ADMINISTRATION
DRESSING ROOMS
SHOPS ETC.

REHEARSAL SPACE
UNDER STAGE STORAGE

UNDERGROUND
CONNECTION TO
METRO STATION

SQUARE
TORNADO ENTRANCE

GARAGE RAMP

PARKING
LOADING
VIP TRAIN OF PEARLS STATION

PUBLIC LOOP
VIP TRAIN OF PEARLS
SPECTATORS
EQUIPMENT
PERFORMERS

Flow Diagram

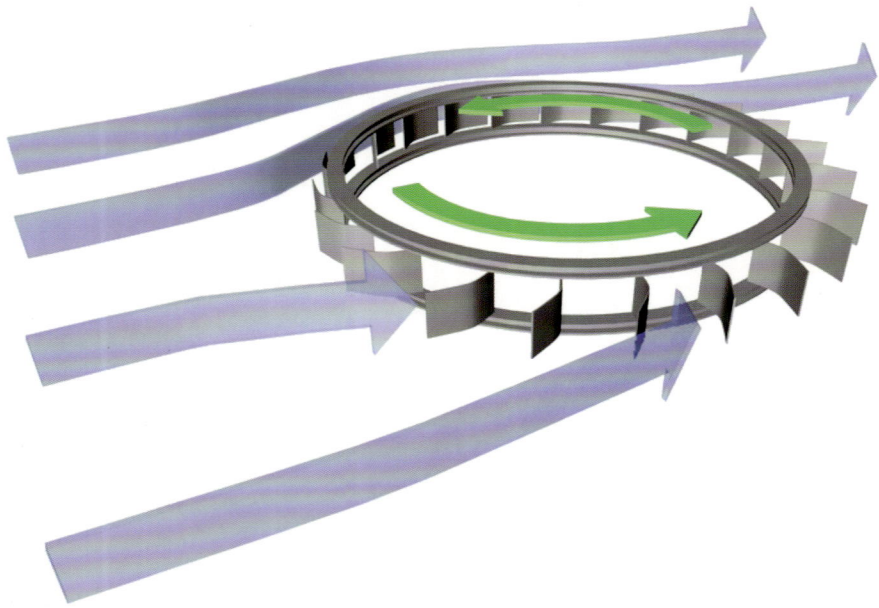

Details

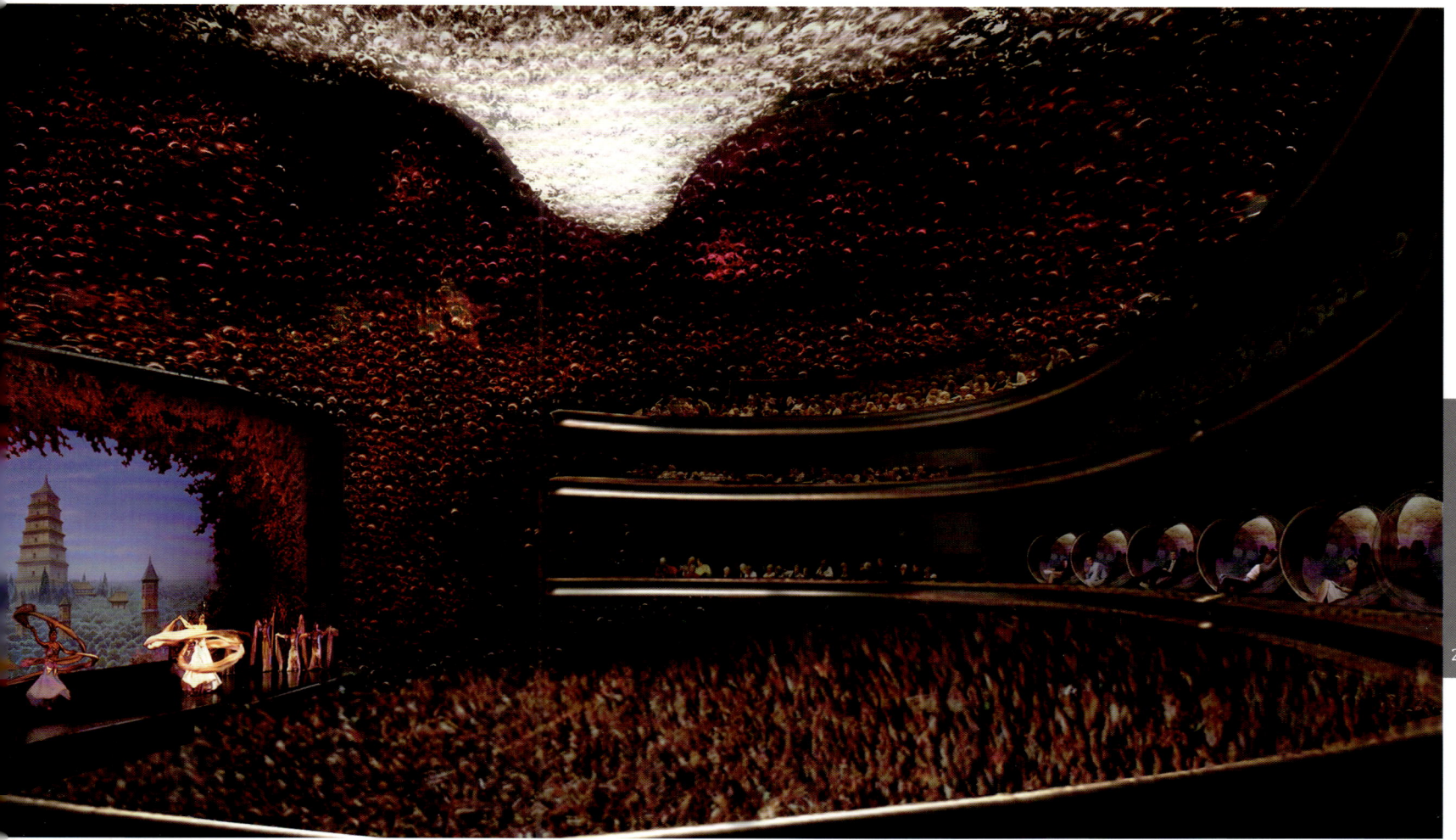

In Remembrance of the Sinan Great Mosque Design

纪念希南大清真寺设计

Firm: Nuvist Architecture and Design

The mosques are not very unique buildings in terms of usage, community relations or reflections of traditional and modern trends. They are significant for sanctity, mystically and figurativeness. Nevertheless, the mosques which are designed recently are not presentable in these features. There are not many examples for that. And there are too many extraordinary traditional examples especially in this historical geography (Turkey). So this situation is a little bit disappointing the architects. In this context, they thought that there is no need to use dome or minaret anymore for mosques buildings. The architects believed that it must be more than these. They have been thinking for a long time that mosque buildings can be discussing its own conditions. In this way, modern thoughts can achieve with these conditions. And also says in competition brief that "there is no competition area for the mosque" because the thing that asked for is the mosque itself. Thus, they can achieve design examples in this subject that has certain strong rules. So the architects thought that they may catch the zeitgeist.

As usual, all the mosque buildings have construction and building elements like domes, arches, minarets, atriums in certain conditions until today. So they thought that they have a chance to improve these things in this competition. Because they believe that there is no need to mosque building elements like domes, arches and minarets. They are all not necessity in this modern age. For example, the architects can create high-tech roofs now, so there is no need to domes anymore or create steel-arch and they don not need old-arches anymore. And specially minarets; minarets were one of the best important invention of its age. Because people has needed to hear sounds of the call to prayer. But now, the architects are using public address system for it. So why they are still using minarets ? In this way, they designed a just symbol for the minaret which represents the religion of Islam in the four main world religions. And they designed skin-shell instead of a dome. Thus, they believed that they created modern dome that they called this "cloak" which represents also connections and fullness in all four main world religions. The architects designed the synthesis of traditional and modern approaches.

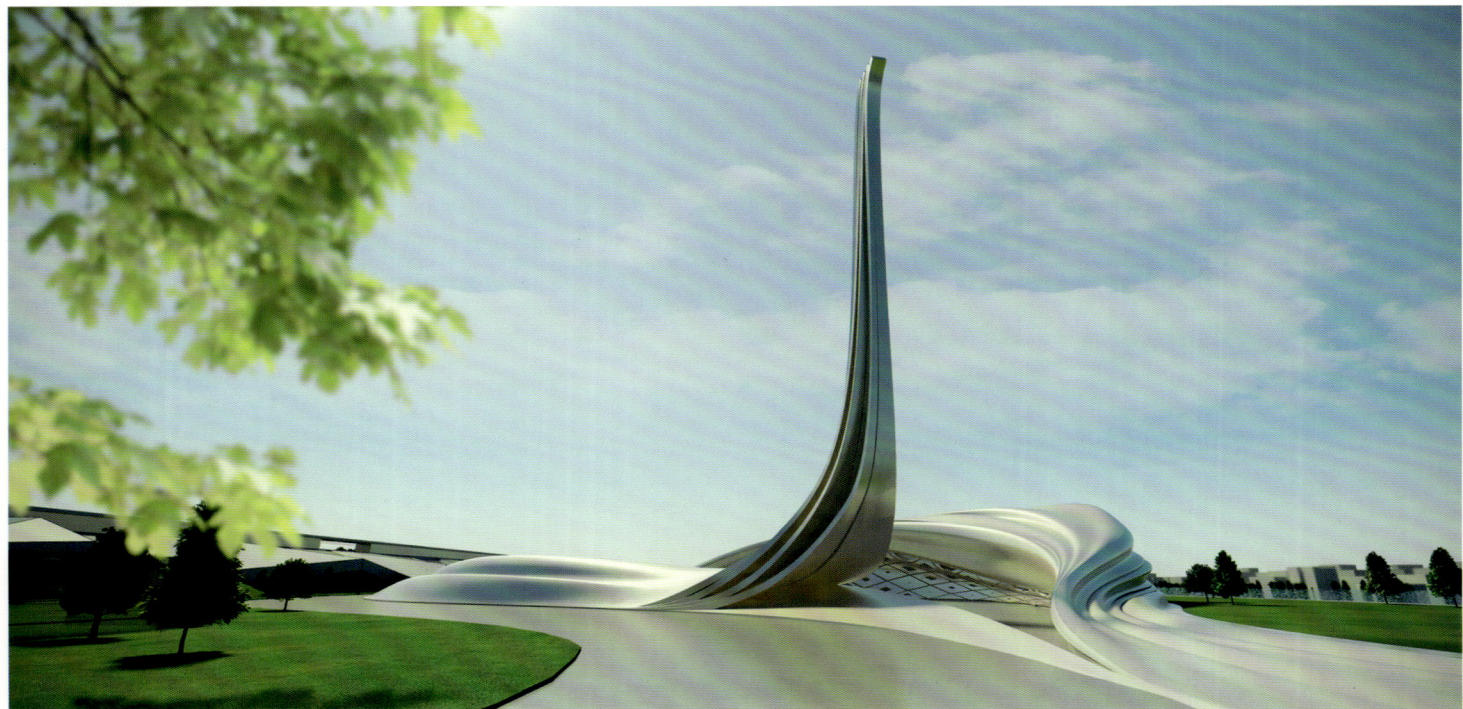

清真寺在使用、社会关系或者反应传统和现代潮流方面来看并不是特别独特的建筑。但是它们有着重要的神圣、神秘和形象性意义。然而，最近设计的清真寺在这几个方面表现得却不很明显。像这样的设计并没有很多的例子，但是在土耳其这片土地上有着过多的不平凡的传统清真寺。因此，这样的情况有点令建筑师们失望。在这种情况下，建筑师们认为没有必要再为清真寺项目设计圆顶或者宣礼塔了。他们相信，最后的设计成果将表现出更多的内涵。他们思考了很长时间，思考着清真寺建筑如何能够表现出自身的状况。就这样，现代的设计思维可以实现这些。竞赛简介中还曾提道"对于清真寺来说没有可竞争的领域"，因为竞赛所要求的东西本身就是清真寺本身。因此，设计师们在这个项目中实现了具有一定强有力的设计规则。他们认为自己能够赶上时代精神。

像往常一样，直到今天，所有的清真寺建筑在某种程度上都有类似于圆顶、拱、天井、中庭等建筑元素。所以建筑师们认为在这次竞赛中他们有机会能完善这些设计。

因为他们相信清真寺中如圆顶、拱和宣礼塔这些元素并不是必要的。在这个现代社会中并不是完全必需的。例如，现在可以创造出高科技的屋顶，所以就不是必须要设计成圆顶或者创造出一个钢拱支架，并且设计也并不需要旧式的拱结构。宣礼塔更是如此，宣礼塔是过去那个时代最重要的创造之一。因为人们要能听得见召唤大家祷告的声音。但是现在，人们使用扩音系统来实现这一功能。所以为何在设计中还要继续使用宣礼塔呢？因此，建筑师们为宣礼塔设计了一个恰当的代表符号，代表了世界四大宗教之一的伊斯兰教。并且他们用外立面覆层代替了圆顶的设计。这样，建筑师们相信他们创造出了现代式的圆顶，他们称之为"斗篷"，代表了世界四大宗教之间的联系。他们设计出了传统与现代手法的综合体。

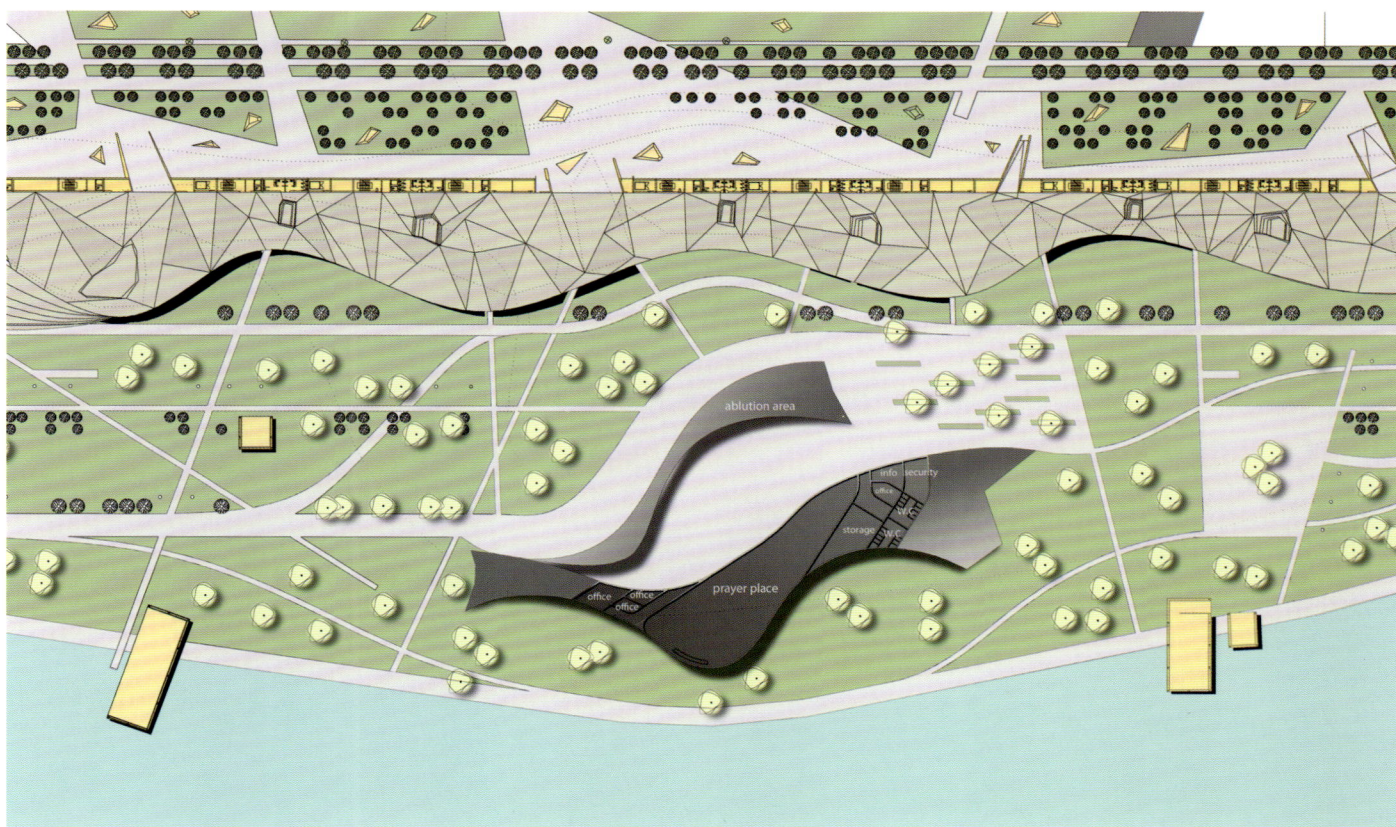

Izmir Opera House

伊兹密尔歌剧院

Architect: Emrah Cetinkaya
Firm: Nüvist Architecture & Design
Location: Izmir, Turkey
Area: 20,000m²
Material: steel and reinforced concrete, low-e double glaze, u-glass, metal panels, fiber reinforced polymer

Art has a great role in showing what happens in human beings' life. Soul, intelligence and the emotions are the reflections of the community, and art forms the social identities of communities. It also shows humans the difference between the various social identities. Global influences effect the types of knowledge acquisitions or the social values in the art, and the artists determine the standards of modern life.

Thus, art centers have many roles beside their fundamental functions that should be participant, shared with society and create a modern level in connection between the citizens and the artists' works. This concept is the key criteria for the Izmir Opera House Project. The project should be reflecting the history, the culture and the modern life in Izmir, furthermore it should be an identity for the city of Izmir.

And competition area also allows the architects to create an identification or a symbol for a city because the area can be seen clearly from near surroundings and specially from sea side areas. This situation came to the forefront. So they considered important in it.

Izmir Opera House has been designed for being a symbol for Izmir. In this way, the architects thought that Izmir will be a center of attraction and also first recognizable preparation place for all domestic and foreign tourists.

They also designed areas and places like playgrounds or amphitheatres for cultural activities, artistic performances, open air concerts, celebration in surrounding areas of opera house. With all these recreation areas the thought that opera house will be meeting and connection point between the citizens, art and the artists' works. These places which support opera house organizations and also can be used separately. In this way, the whole area and functions can be live day and night.

Opera house's functional areas has been designed according to condition and environmental analysis like perceptual data's, vista areas, vehicular and pedestrian areas, functional connections and meteorological factors. In the context, they have started to design an urban art dome that they call Artist Foyer. The urban power effects the functional system, and the function system defines the Artistic Foyer. As a result, all these things together create a parametric topography which forming the dome.

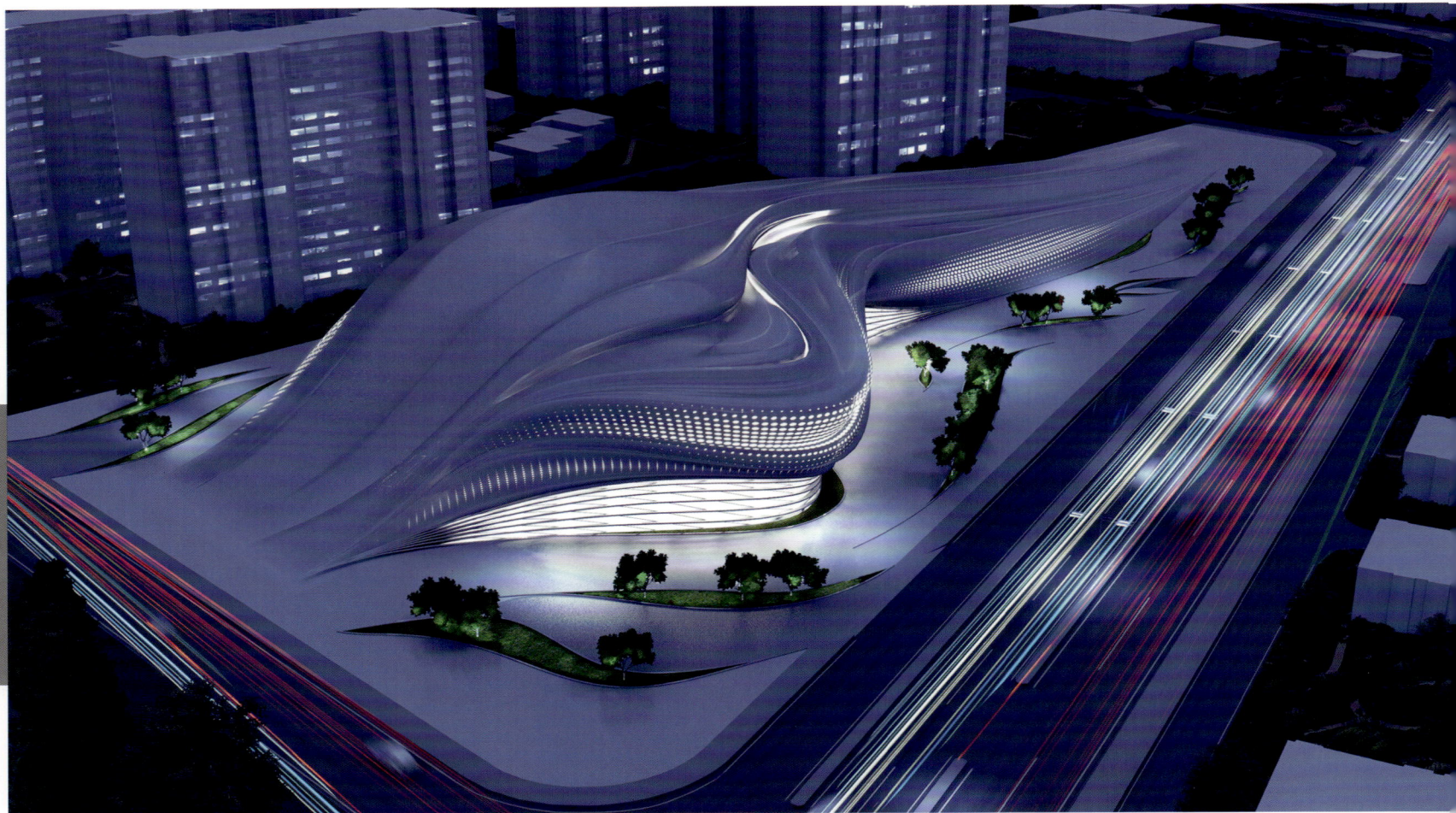

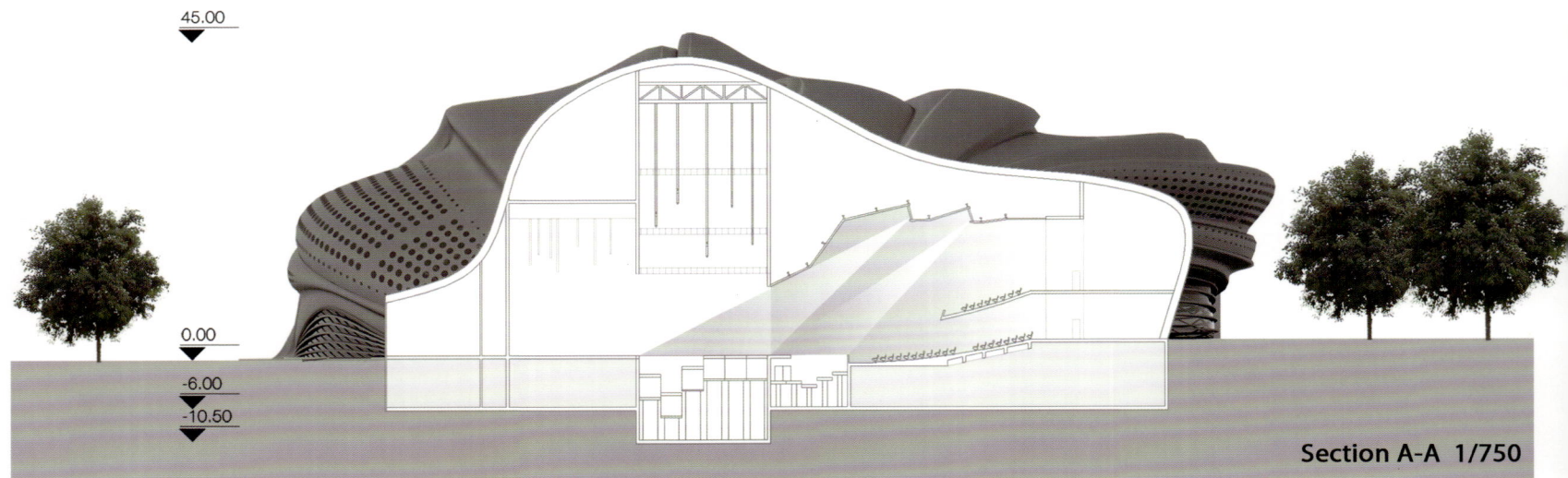

45.00

0.00
-6.00
-10.50

Section A-A 1/750

艺术对于反映人们的生活起到了巨大的作用。灵魂、智慧和情感是社会的反映，而艺术则形成了社会中的各种社会身份。艺术也向人们展示了不同社会身份之间的不同。全球性的影响关系到知识的获取方式或者艺术的社会价值，而艺术家们决定了现代生活的标准。

因此，艺术中心除了其应该具有的基本功能以外还要扮演很多其他的角色，与社会共享并创造一个市民与艺术家作品之间联系的现代化水平。这个概念是伊兹密尔歌剧院项目设计的主要标准。这个项目应该反映出伊兹密尔的历史、文化和现代生活，此外，它还应该成为伊兹密尔的城市形象。

项目场地的设计也使建筑师们能够创造一个城市的标识或者象征，因为从附近的环境，尤其是海边地区都能够清楚地看到这个场地。这个位置是非常重要的，所以他们决定将设计重点放在这里。

伊兹密尔歌剧院的设计是作为伊兹密尔市的象征来设计的。因此，建筑师们认为伊密尔将会成为一个吸引力的中心，并且成为所有国内外游客首先便会辨认出来的地方。

建筑师们还设计了如可举办剧院附近地区的文化活动、艺术表演、露天音乐会、庆活动的运动场或露天剧场。有了这些娱乐区，歌剧院将会成为市民、艺术与艺术家作品间的会议地点和连接点。这些区域既是歌剧院的一部分，也可以单独使用。这样一来，个区域及其功能都可以实现全天候的使用。

歌剧院的功能区基于情况与环境的分析而进行设计，如感知数据、远景区、车辆行人区、功能连接和气象因素等。在这种情况下，建筑师们最先开始设计一个城市艺术穹顶大堂，称之为艺术大堂。城市电网影响了功能系统，而功能系统限定了艺术大堂的计。因此，所有这些共同构成了一个参数地形，基于此建造了穹顶。

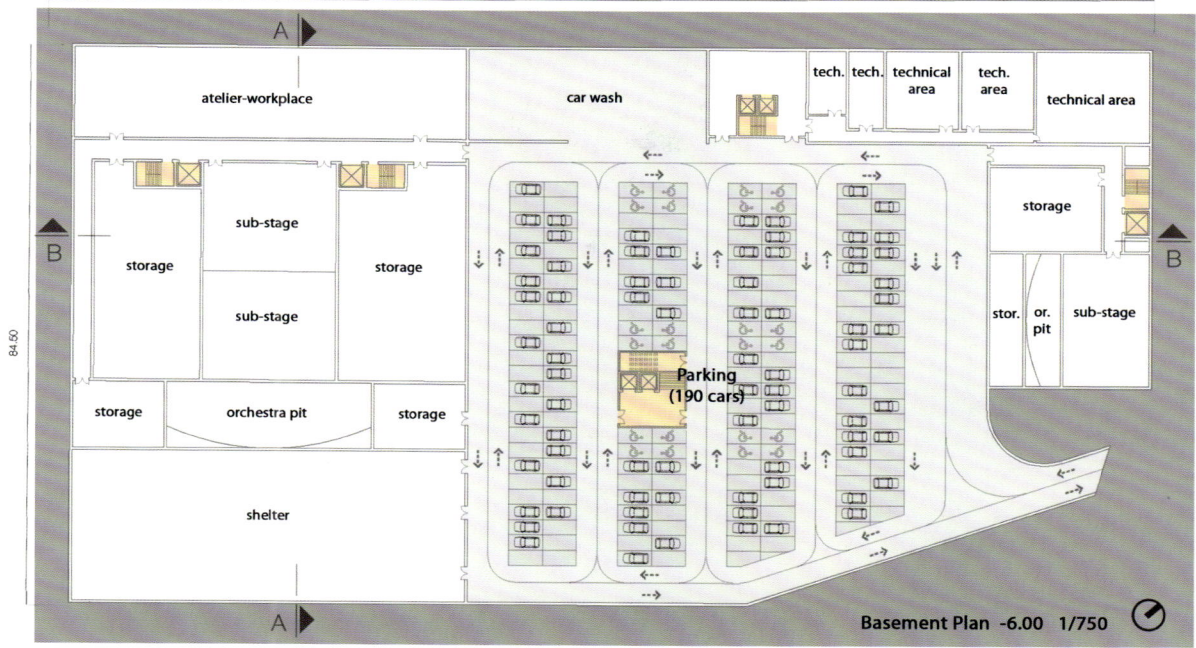

Basement Plan -6.00 1/750

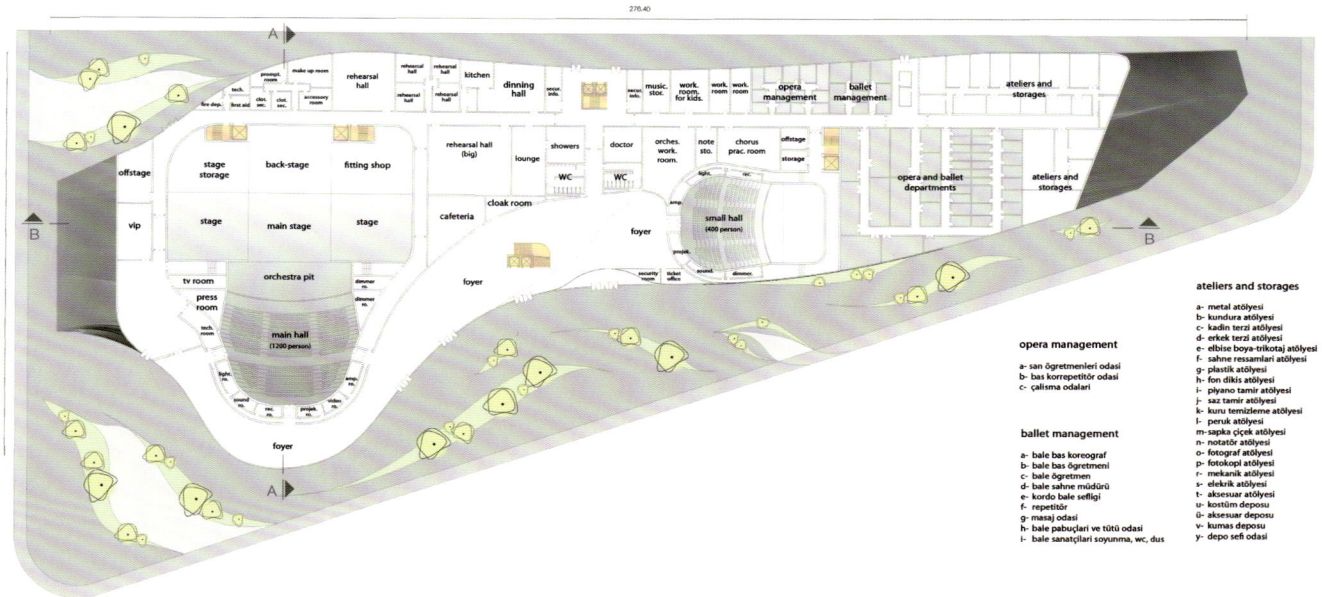

Ground Floor Plan 0.00 1/750

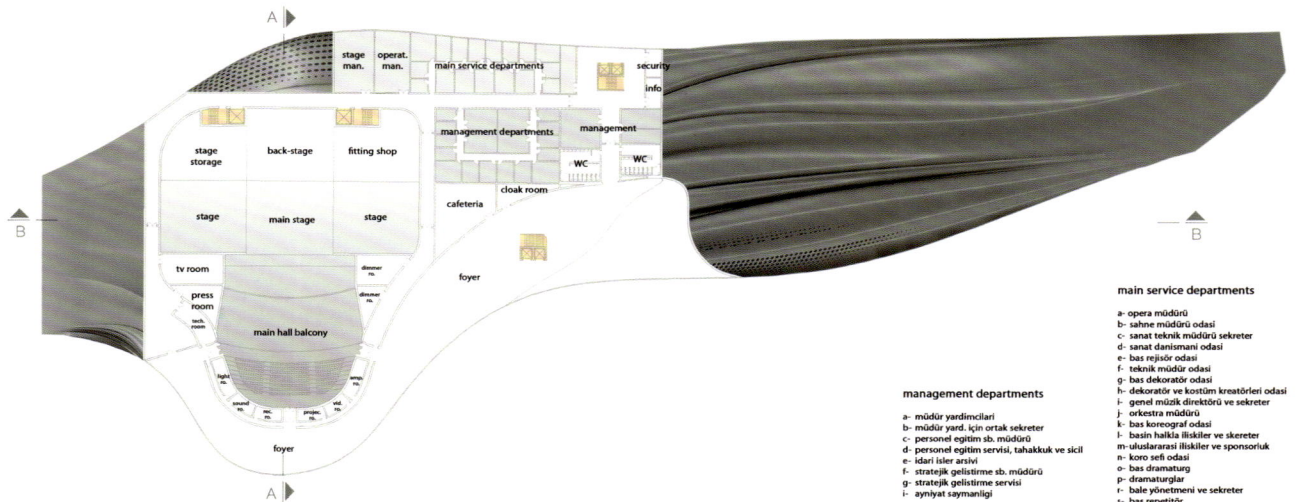

First Floor Plan +6.50 1/750

Rotterdam City Tower

鹿特丹城市之塔

Architect: J.W. van Kuilenburg with A. Chlebinska,
E. Komarzynska, P. Roger Rzepecki, G. Michaud-
Nérard, B. Drogge, M. van Oers, G. Porcu
Firm: Monolab
Location: Maas Harbour, Rotterdam, Netherlands

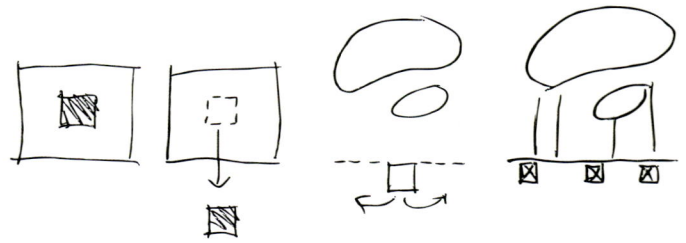

1. city tower
2. metro station Maashaven
3. katendrecht
4. wilhelminapier
5. parking
6. project op parking
7. voetgangers boulevard
   pedestrian boulevard

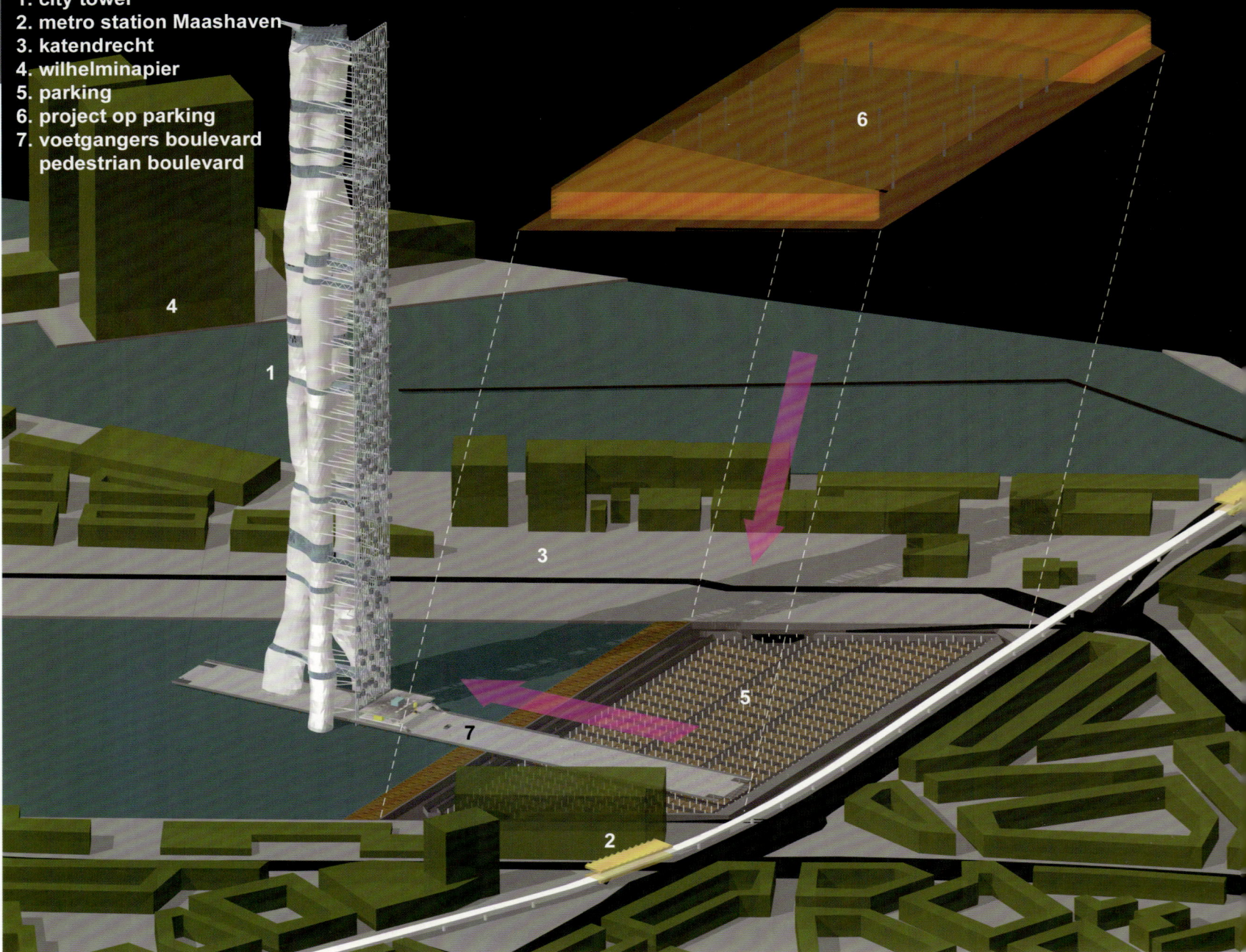

The project consists of four co-operating parts: the tower, the grid, the boulevard, the parking lot. The tower has no traditional core and has a steel structural envelope which is made of triangular elements. Horizontal structures handle forces on the facade every 12m and support the structural envelope sideways. These structures are the floors with "spider structures": beams radiating from center rings. The rings are connected vertically, providing tubes that hold the emergency stairs. The facade is made of approximately 7,500 prefabricated unique triangular panels of Photovoll glass that deliver the tower's electric energy needs. The tower foundations reach very deep to take vertical tension forces and lateral wind forces. The tower has 5 technical floors to handle fire fighting, evacuation, and climatic services.

The grid consists of vertical steel profiles with horizontal members every 16m. Diagonals in the grid serve as passing lanes for gondolas and for lateral stability. The grid is established towards the tower through a forest of steel spacers. To handle possible deflections, hydraulic devices below each vertical member are tuning the verticality of the grid. Sky lobbies connect the grid and the tower, suspended in-between.

The boulevard is spanning the complete site like a bridge. Glass planks with a profiled anti-slipping topside make a semi-transparent walking surface.

All gondolas move individually by their own energy cells and two electric engines. Each engine drives a heads that locks into the steel grid, driving the gondola along the grid vertically and diagonally. In the side window of each gondola an interactive touch screen is embedded in the glass to command the address. Gondolas have two sets of doors: in the front passengers step in and the back gives access to the sky lobbies and the tower. The gondolas are glazed to supply panoramic views while travelling.

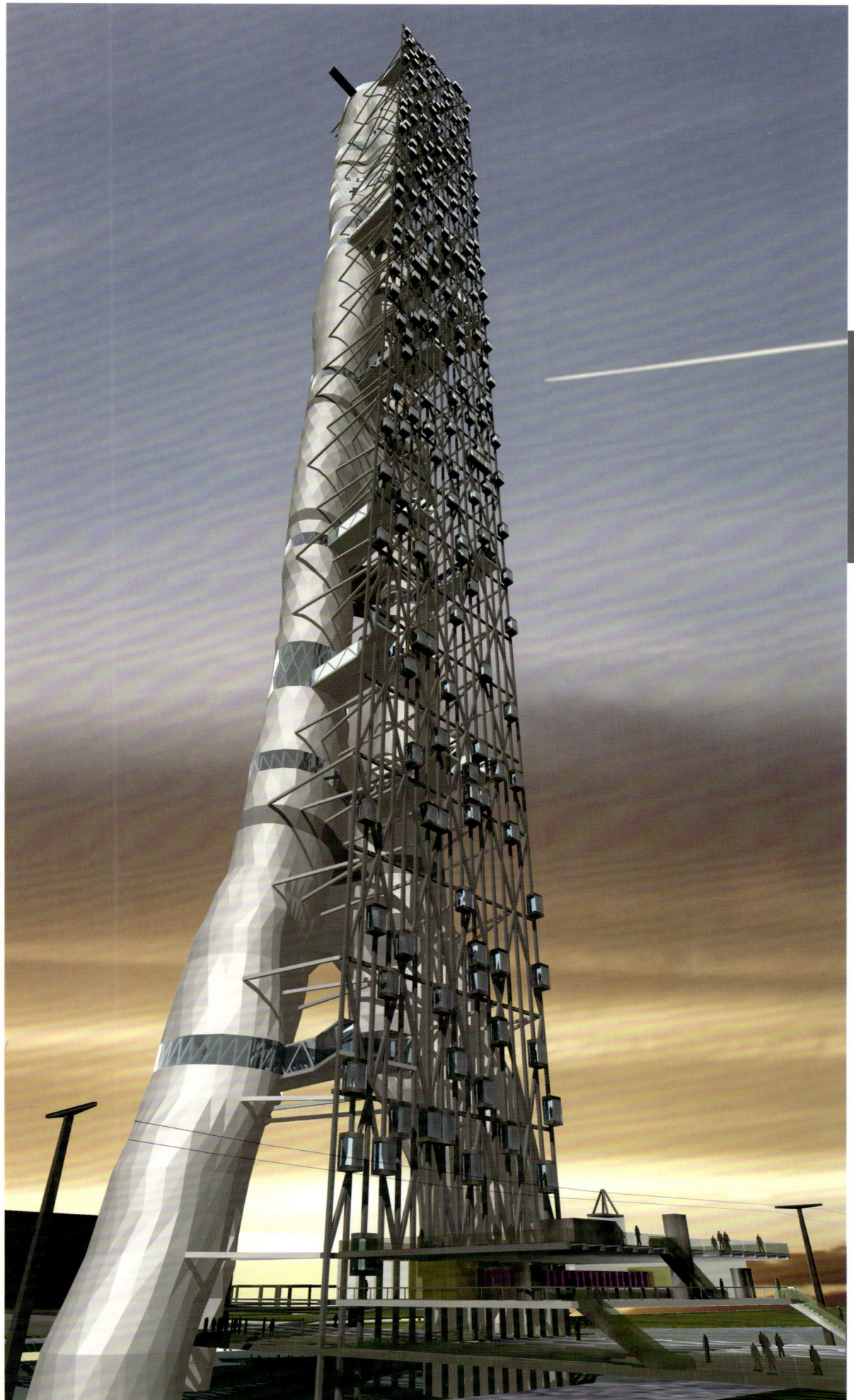

236

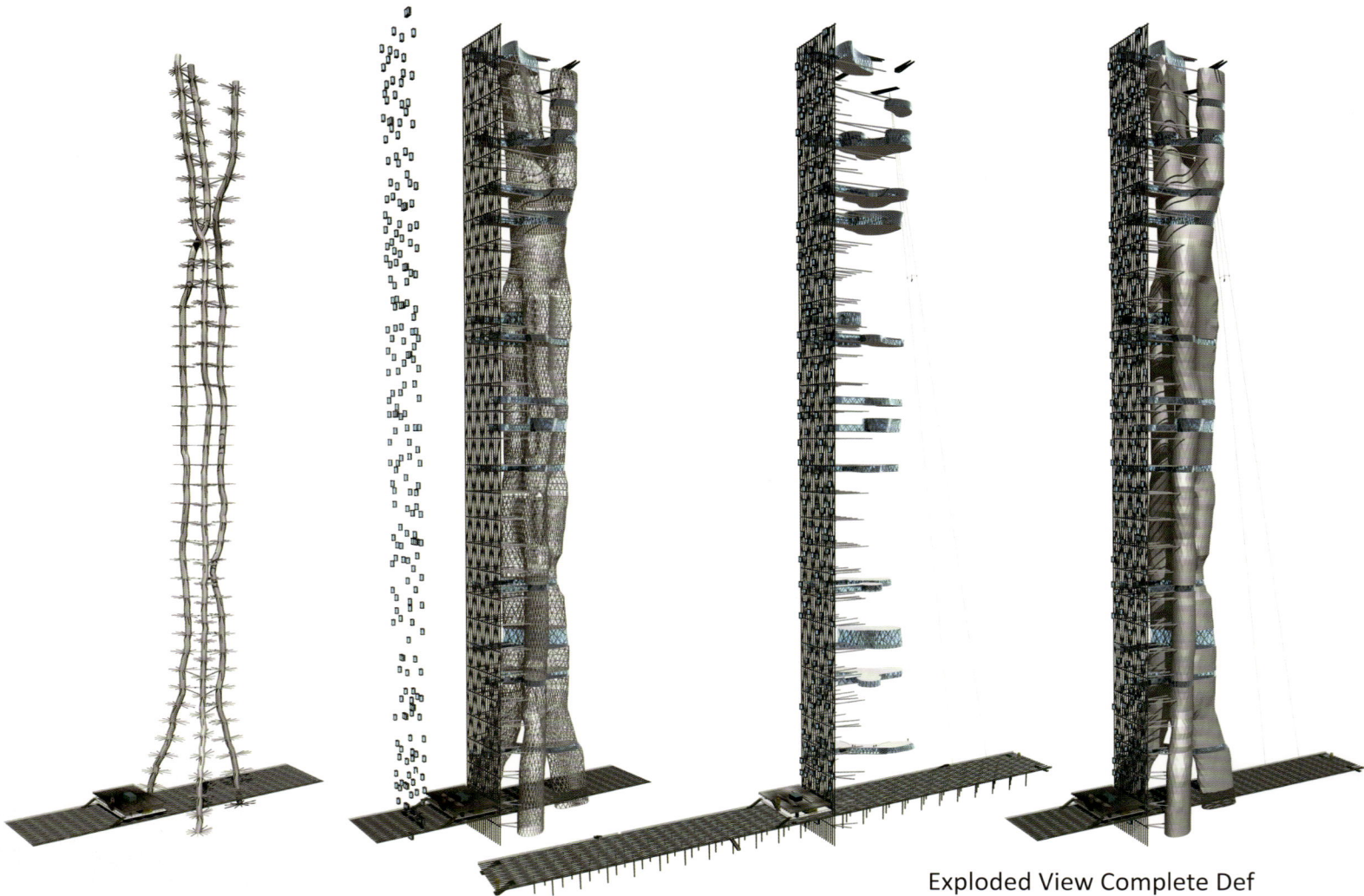

Exploded View Complete Def

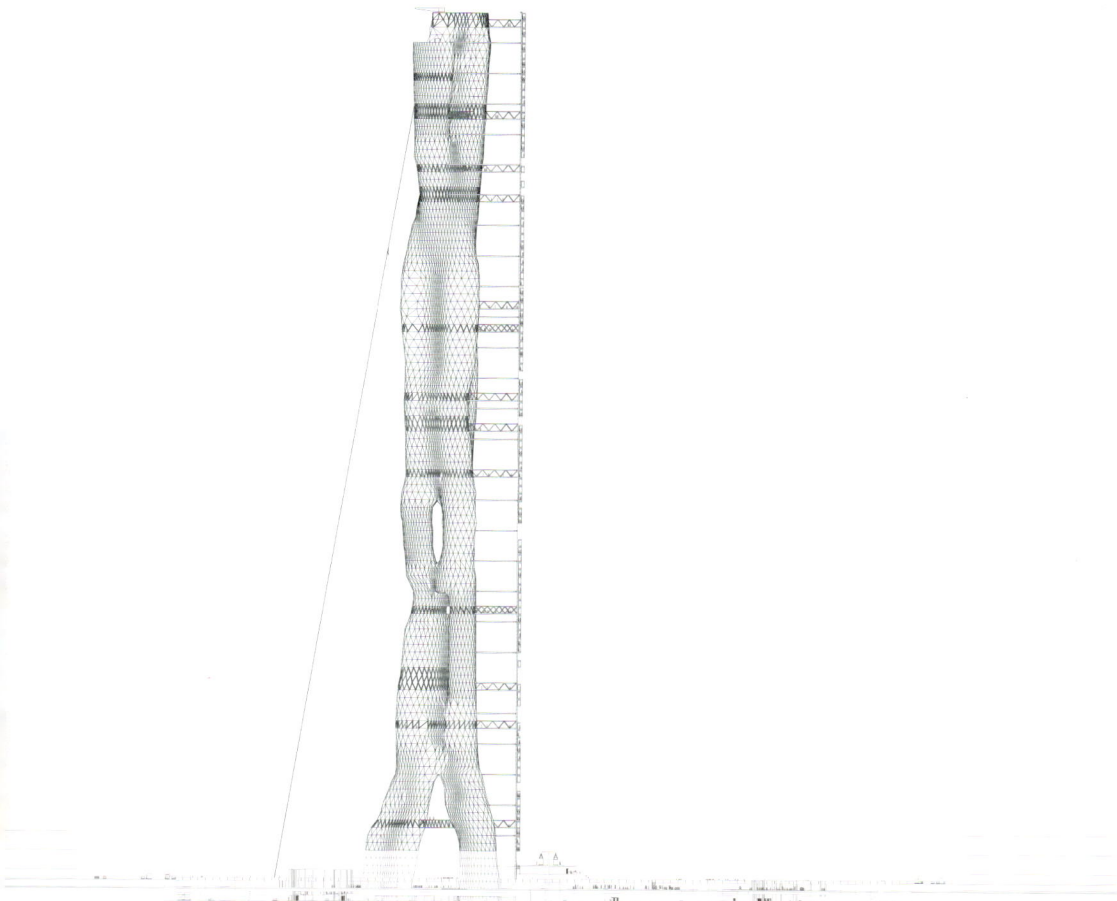
Elevation Side

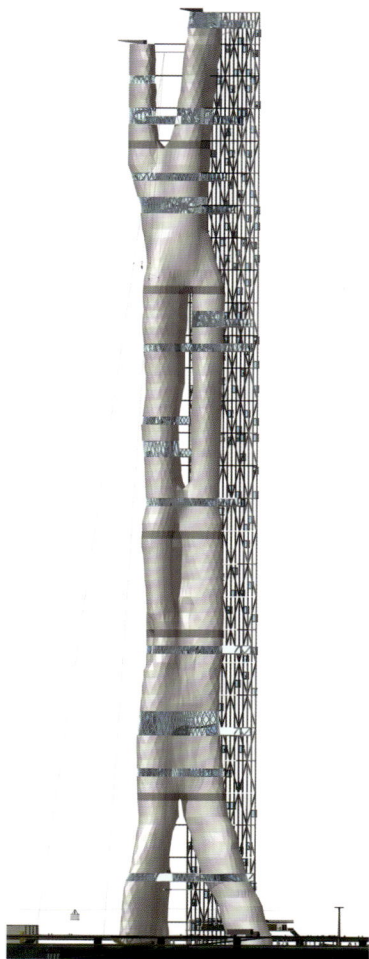
Elevation

Section

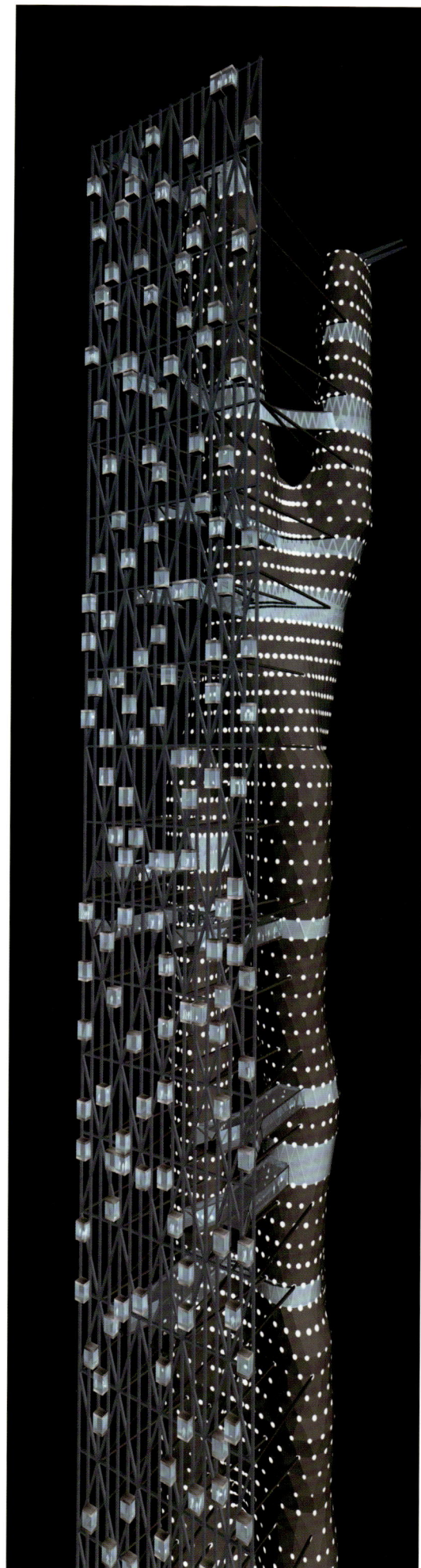

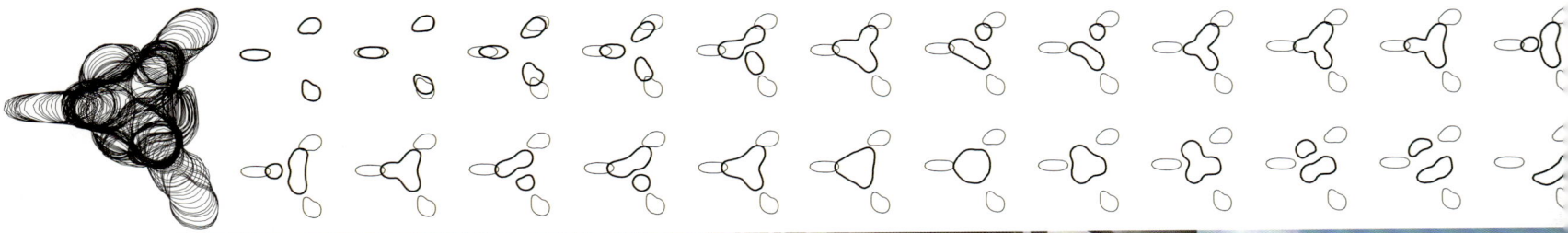

整个项目包括了四个同步运转的部分：塔、网格、大道和停车场。塔的设计并没有传统的核心，而是有一个由三角架构成的钢结构外壳。每12米一个的横向结构将力传递到建筑外立面，并从侧面支持建筑的外表面。在这些横向结构中，支架从中心环放射出去，构成了带有这种"蛛形结构"的地面。环形垂直地相互连接，为紧急楼梯提供了所需的管道。外立面由7500个独特的三角形光电玻璃面板覆盖，并为整座塔提供电能。塔的地基位于非常深的位置，以便能够抵抗垂直的拉伸力与侧向的风力。这座塔专门设计了5层技术楼层来提供消防、疏散与气象服务。

"网格"包括了垂直的钢制框架与每16米一个的水平单元构成。网格中的对角线作为电梯舱通道并提供水平稳定性。"网格"通过密布的钢制连接件与塔相连。为了对抗可能的倾斜，每一个垂直单元下都有液压装置来调整网格的垂直度。空中门廊悬挂在网格与塔之间，起到通道作用。

"大道"如同桥梁一样跨越了整个区域。防滑玻璃板则构成了半透明的走道的表面。

所有的电梯舱都独立依靠自身带有的电池与两部电动机提供动力。每一部电机驱动一个嵌在钢制网格中的引导头，带动电梯舱沿网格垂直或对角线运动。人们可以通过设计在电梯舱侧窗中的触摸屏幕来下达目的地的指令。电梯舱共有两组门：前面的门可供乘客进入电梯，后面的门则可以让乘客进入空中门廊和塔中。电梯舱全部是玻璃表面，以便让人们可以在乘坐时欣赏到周围360度的景观。

Floor Plans

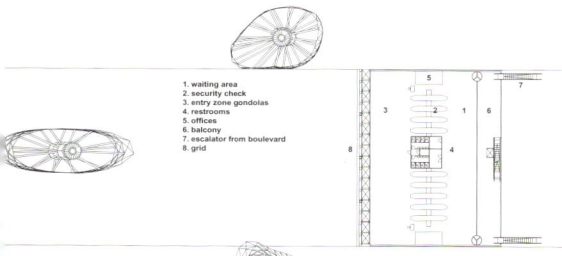

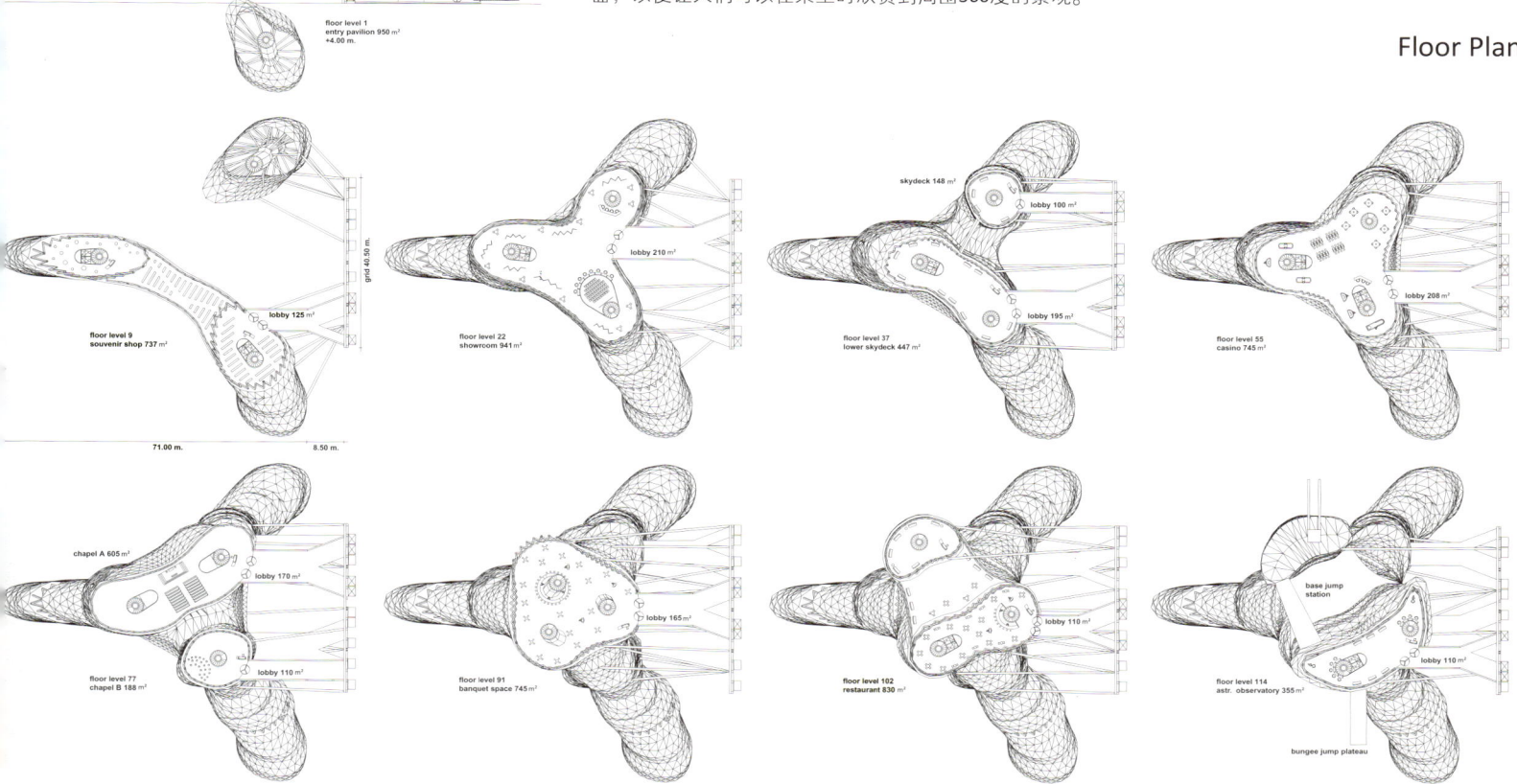

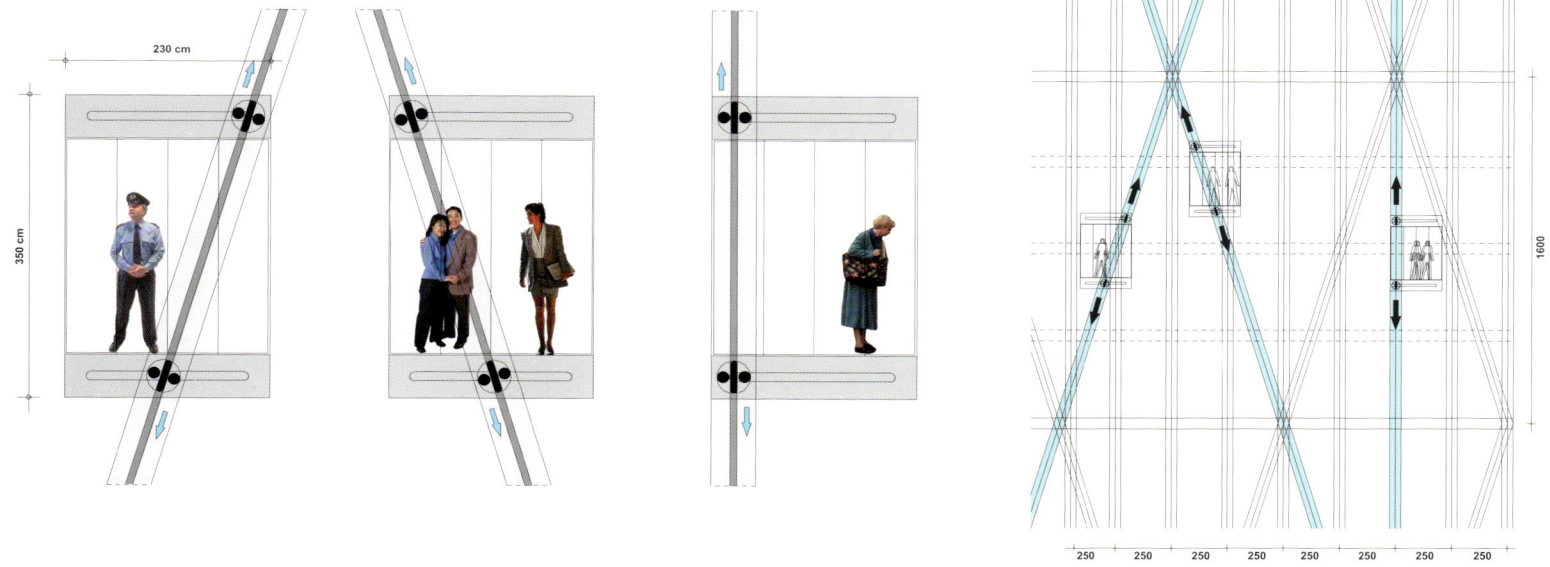

230 cm

350 cm

1600

250 250 250 250 250 250 250 250

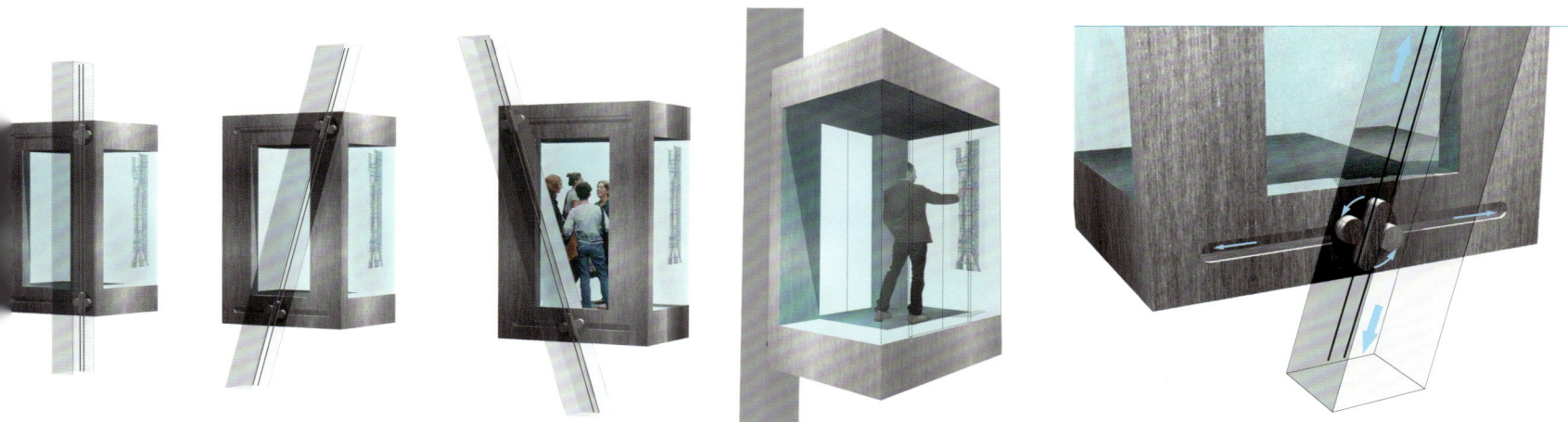

The Lantern Pavilion

灯笼亭

Firm: AWP (leading consultant) /Atelier Oslo
Location: Langgata, Sandnes, Norway
Area: 140m² lantern / 2,500m² public realm
Photographer: AWP/Atelier Oslo

When Sandnes and Stavanger were chosen as cultural capitals of Europe 2008, the Norwegian Wood competitions were launched to promote innovative, timber architecture contributing to make the region an international showcase for it.

Sandnes asked for the design a new square and a sculptural object in wood in pedestrian district Langgata aiming at revitalizing the area, and creating a place where many different activities could take place: a meeting point, markets, informal music concerts and other happenings.

A place able to shelter, invite and encourage more social events and to sustain new practices.

Since the site is visible from afar, and from the railway separating these two distinct areas of the city it was essential to create an object that could be experienced from distance and reveal the square.

Homely but looking at the future, the ambition was to create a manifesto for public space design: not just a decoration, but an urban strategy.

The proposal used the iconographic shape of an old, uplifted wooden house. By the redefinition of its traditional construction and by making it glow in the dark, a new landmark for the city was created, a symbol of the old city upscaled to the new city's dimensions, dealing with the Norwegian wood ancient architecture motives to design a contemporary object. This public Lantern aimed at providing a space where to enjoy light and weather changes if/when nothing else takes place. And if there is always something happening, on account of the Lantern, more will happen: creativity calls for creativity.

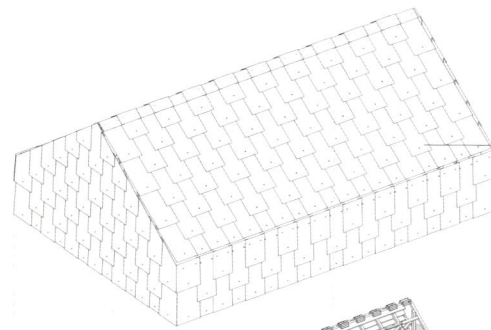

Glass cladding
Page xx

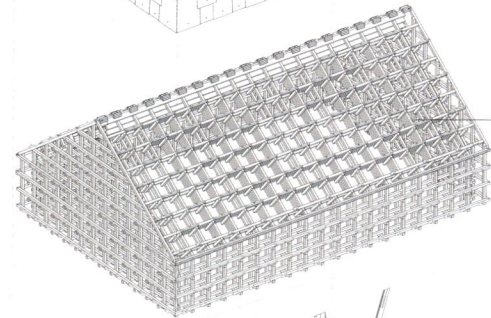

Roof structure
Page xx

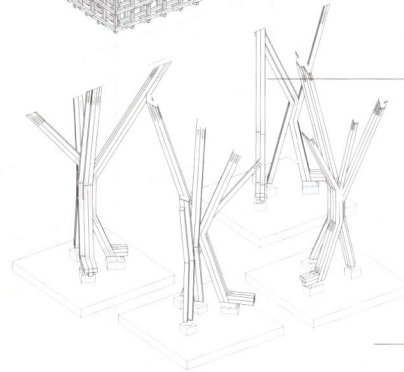

Columns
Page xx

Elements

Concrete base
Page xx

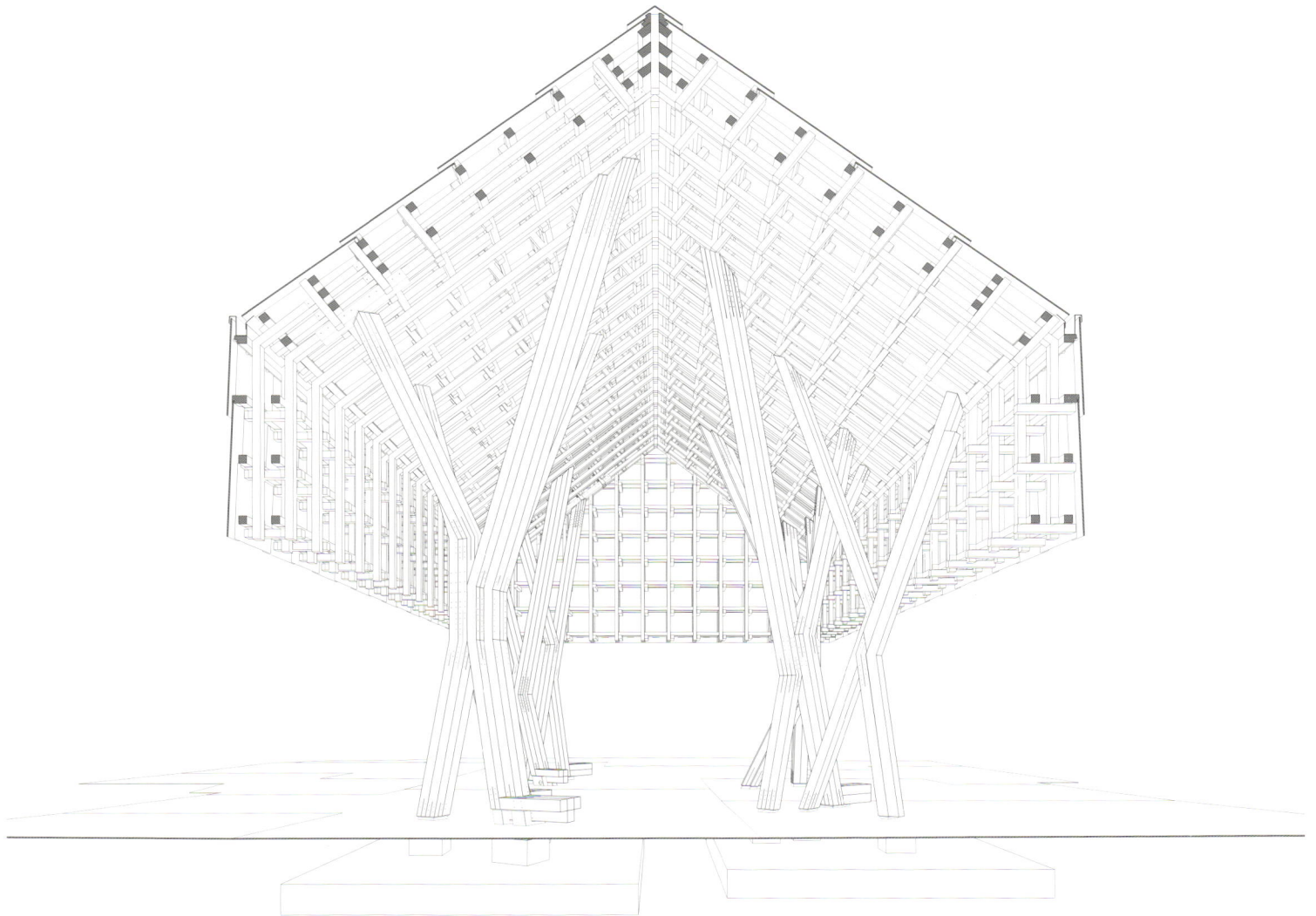

Cross Section Through Whole Structure

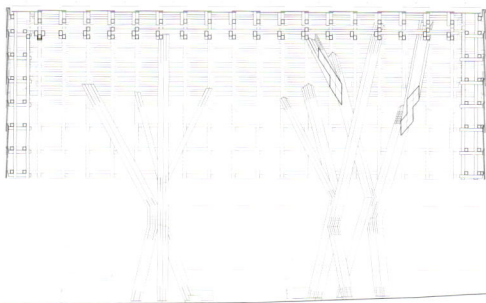

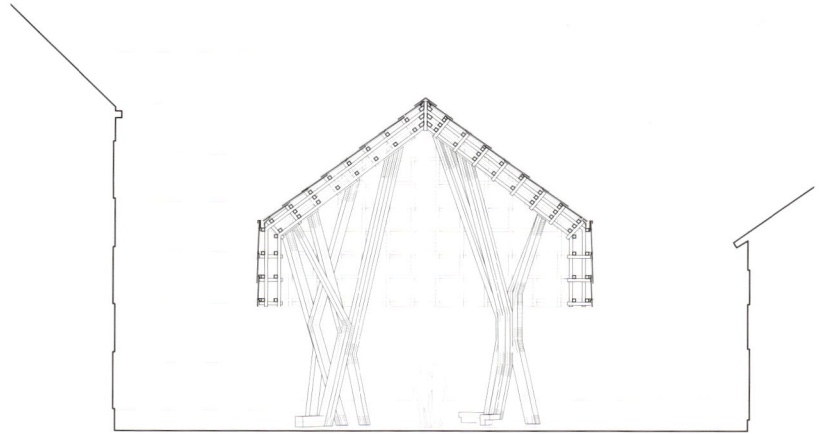

Section

当桑德尼斯和斯塔瓦格被选为2008年度欧洲
文化首都时，"挪威的森林"竞赛开始举行以促
进当地新派木质建筑的发展，并意图使本地区成
为此类建筑风格的代表。

桑德尼斯市征求在朗格塔步行区设计一个新
的广场与木质雕塑式建筑，力求让该地区重现活
力，并以此建设一个可以进行多种活动的地点，
如集会、市场、非正式音乐会和其他活动。

这一地点能够为更多的社会活动提供场所，
并能够促进和鼓励举行各种活动，还能够支持各
种新型的社会活动尝试。

由于这一地区能够从很远被看到，也能够从
将这座城市分成两个区域的铁路上看到；因此该
设计的关键在于建筑一个标志物，使新设计的广
场可以从远处就可以被看到。

该设计既给人以家的温馨也带有未来气息。
建筑师的目的是给公共空间设计带来一种概念，
告诉人们这一设计并非是一种装饰，而是一种新
的城市布局安排。

该设计案采用了一种象征性的老式的高屋顶
木质房屋的形式。通过重新定义其传统建筑概念
并增加了夜晚的灯光设计，建筑师为这座城市带
来了一个新的地标建筑。这个建筑的设计通过用
挪威传统的老式木质建筑风格来建造一座当代建
筑，将一个旧城区的象征扩大为新城市的空间。

这个公共的灯笼亭在没有其他活动的时候可
以提供一个享受灯光与天气变化的场所。如果这
里因为有了这座建筑而发生了很多事情，那么在
以后也将会有更多的事情发生：创造力呼唤着创
造力。

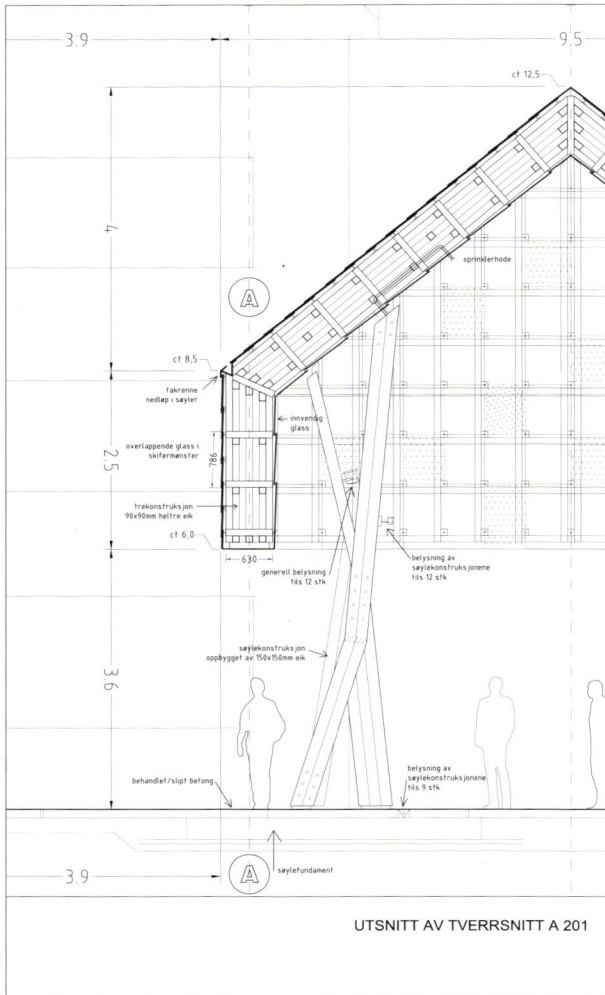

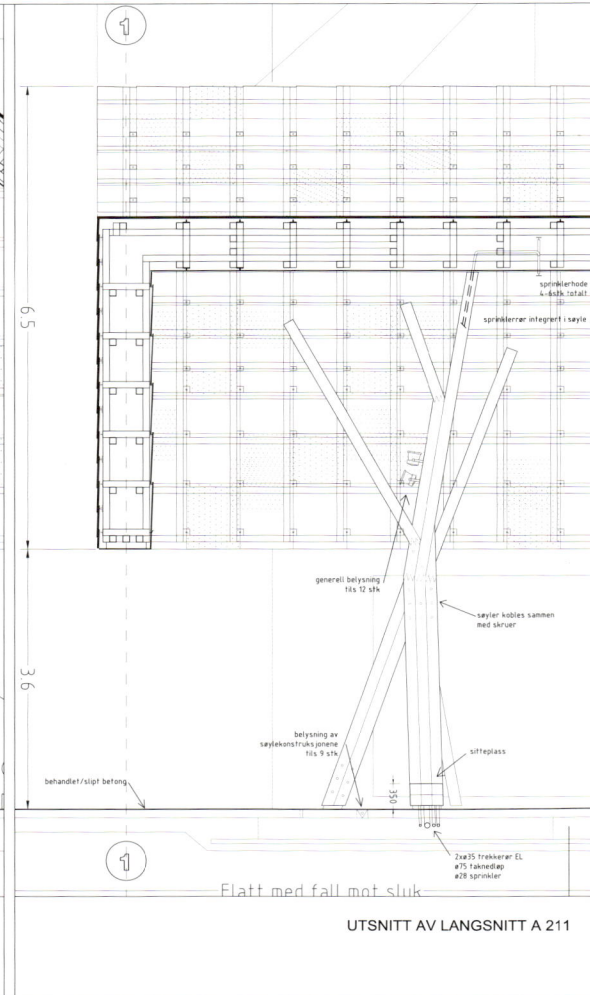

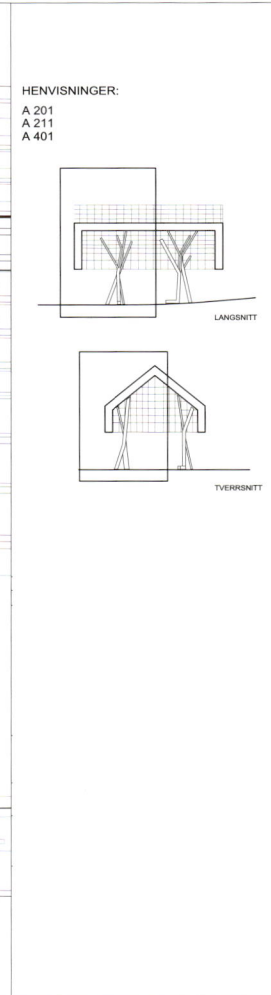

UTSNITT AV TVERRSNITT A 201

UTSNITT AV LANGSNITT A 211

HENVISNINGER:
A 201
A 211
A 401

LANGSNITT

TVERRSNITT

Preliminary Project

City Municipality Ljubljana

卢布尔雅那市政厅

Architect: Rok Oman, Spela Videcnik
Firm: OFIS Arhitekti
Location: Ljubljana, Slovenia
Area: 59,148m^2

gradation of controlled / public areas
- public program - self service terminal
- semi public programs - counters
- controled areas - offices
- reception to controled offices

The site is just on the edge of the Ljubljana city centre, by the river and already occupied by some existing protected buildings. As such, the area represents a chance for unique rearrangement with its identity and to become a sort of symbol and landmark area of contemporary Ljubljana architecture.

Special care was to achieve optimized environmental conditions with a minimized energy demands in three steps: reducing the energy and meeting remaining conditions with high-performance building-integrated systems, and sourcing those systems as much as possible with renewable sources.

One of the core ideas of the sustainability concept is to heat and cool working areas like offices and use the exhaust air of these offices to condition the atrium. The external facades will feature a high performance glazing and an adaptable external shading device to reduce solar gains in summer. The concrete slab imbedded pipe system provides cooling without draft problems and in winter comfortable radiant heating. In summer the cooling of the offices spaces will be achieved for major amount of time just using the slab system. The decentralized ventilation units vent fresh air into the rooms and serve as peak time cooling in summer. During winter the fresh air will be heated inside the units and distributed into the rooms using the displacement ventilation principle. An overflow to the atrium allows the exhaust air to flow from the offices into the atrium and through louvers in the roof to the outside. This means the temperature in the atrium will swing from 15°C in winter to ambient conditions at the upper levels in summer (temperature stratification). To minimize the peak summer conditions, features inside the atrium like the water wall and floor cooling system will condition this space in addition. The water wall is used for cooling and dehumidification purposes in summer, whereas in winter the water will evaporate and humidify the air. Special areas inside the atrium will get this "own" climate. Especially for a front desk this local climate is important to guarantee adequate working conditions. The energy supply system is based on natural resources like river water heat exchanger to provide cooling in summer as well as heating - in combination with a heat pump in winter. Photovoltaic cells integrated in the atrium roof generate power. Another objective of a sustainable building approach thrives to minimize of energy intensive materials instead locally available, sustainable materials will be used.

administration center main hall
level P1 +289about sea level

administration center main hall
level P2 +294about sea level

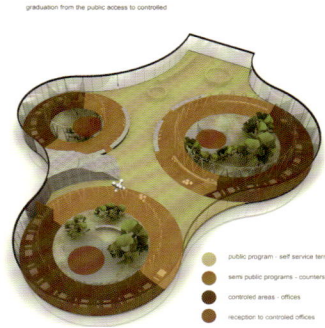
graduation from the public access to controlled

- public program - self service terminal
- semi public programs - counters
- controlled areas - offices
- reception to controlled offices

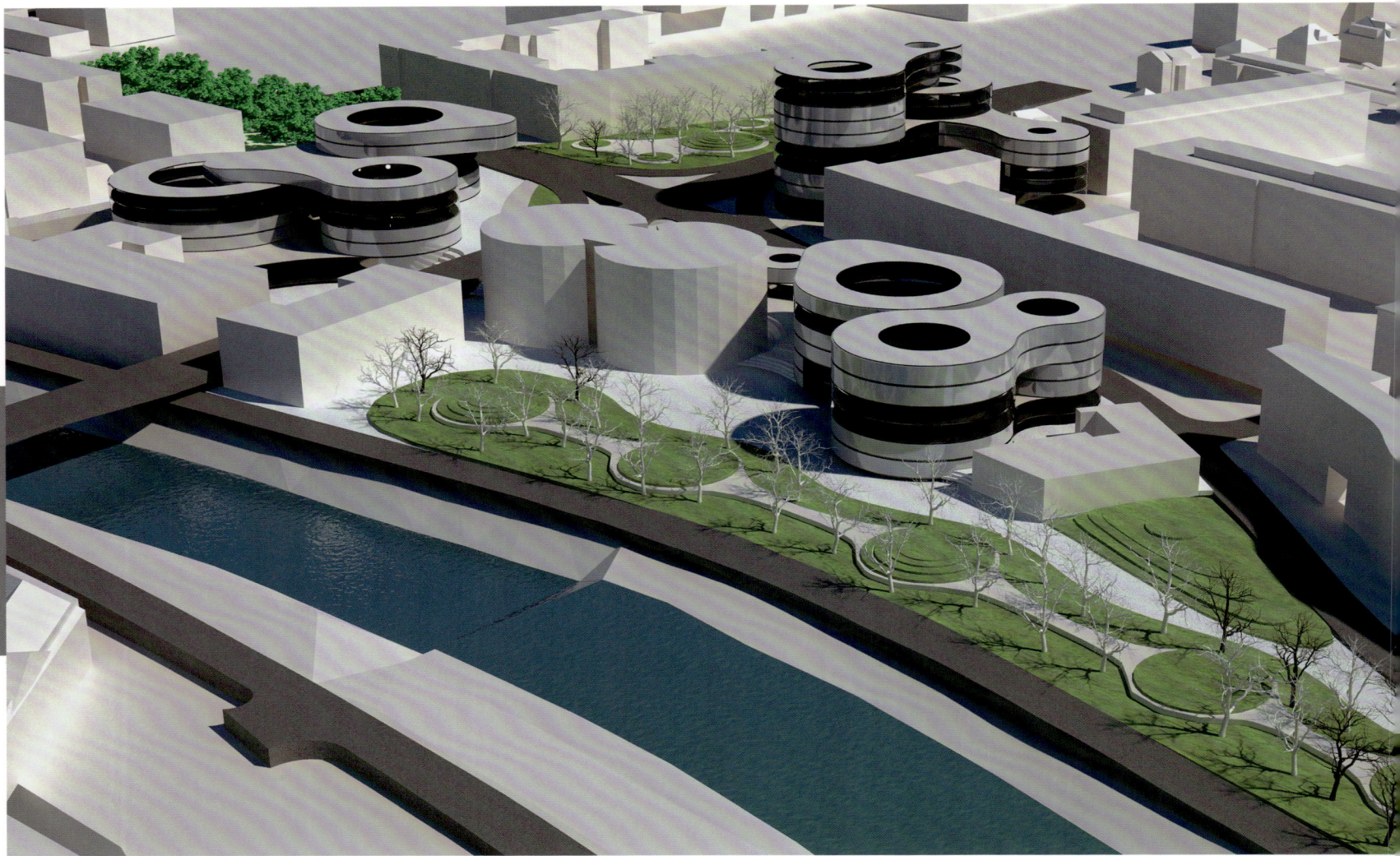
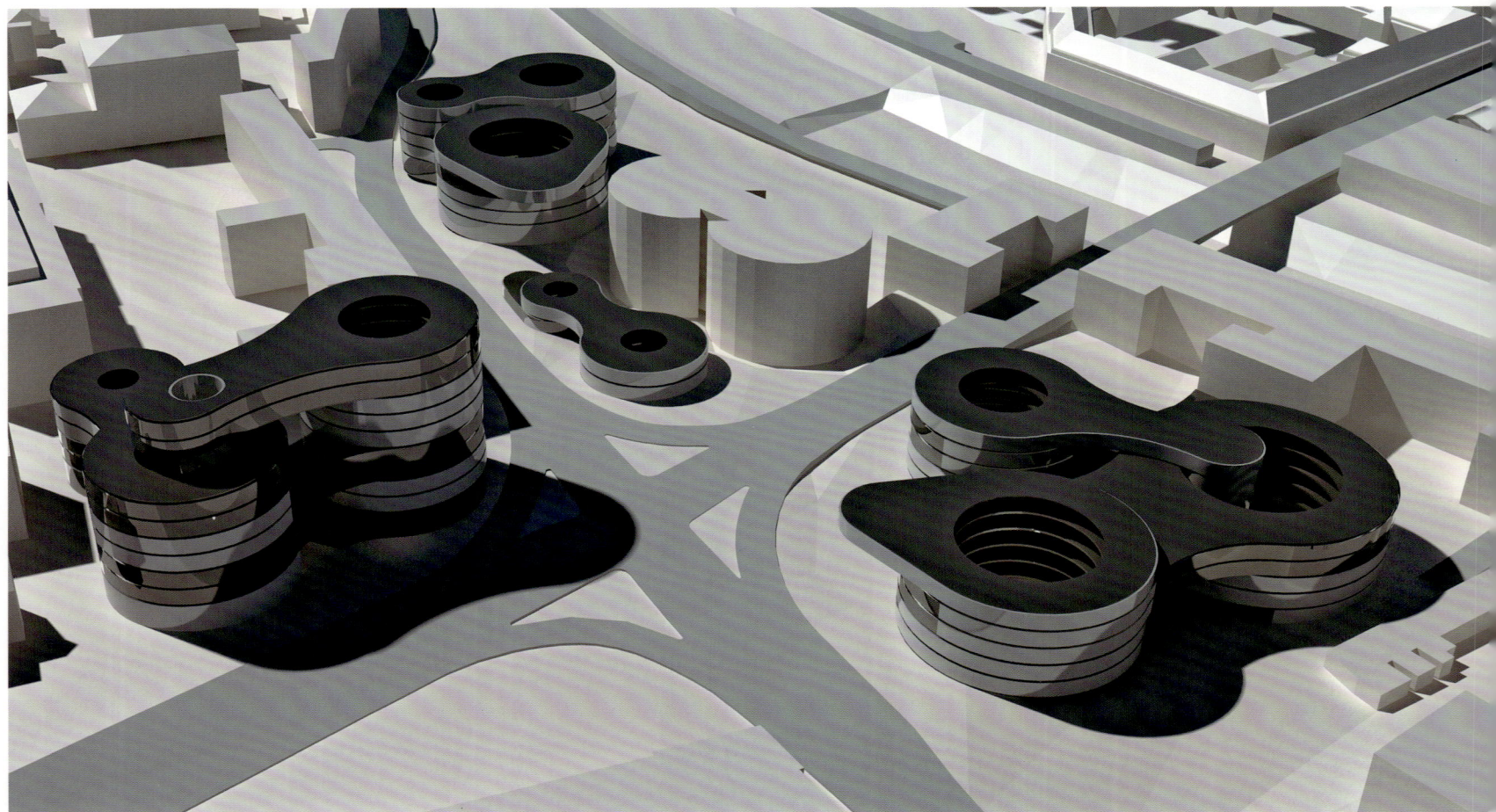

Distribution

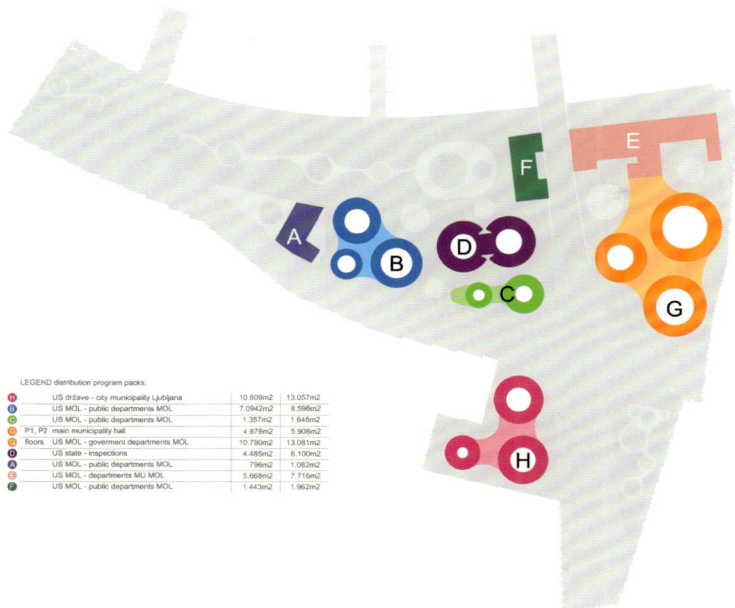

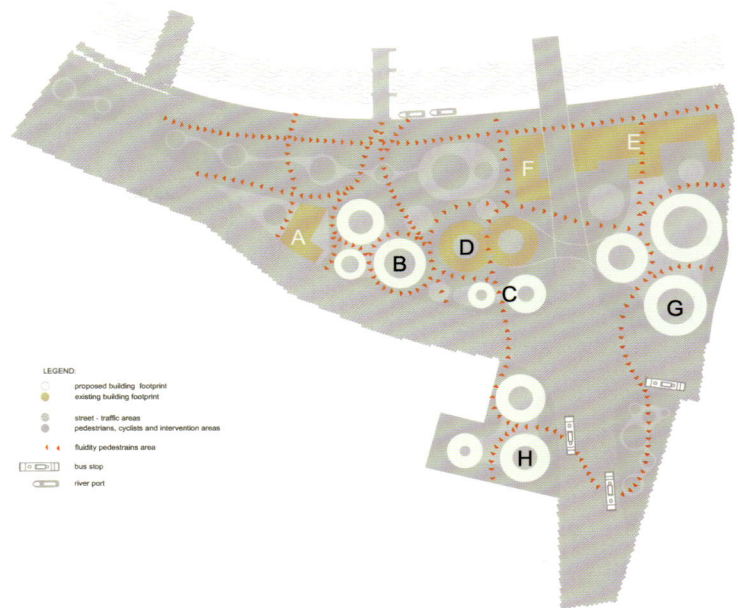

Fluidity

Public Inner Squares

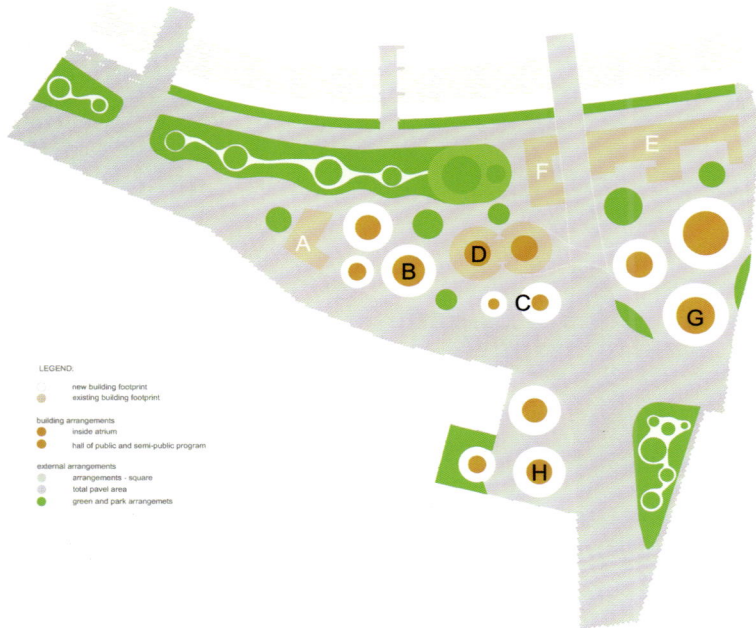

Departments Organisation Diagram

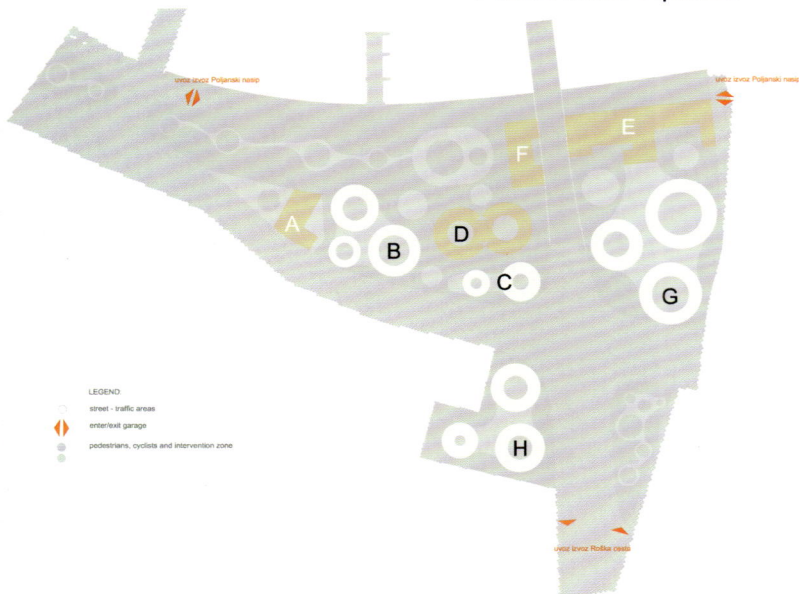

separate offices

1U2U3 offices with shared facilities (meeting space, kitchenette, archives, presentation rooms)

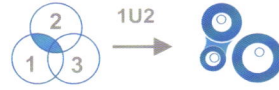

1U3 offices with shared facilities (meeting space, kitchenette, archives, presentation rooms)

1U2 offices with shared facilities (meeting space, kitchenette, archives, presentation rooms)

2U3 offices with shared facilities (meeting space, kitchenette, archives, presentation rooms)

Traffic Areas Organisation

　　该建筑坐落在卢布尔雅那市中心边缘，旁边有河水流过，周围已经有几处受保护建筑。因此，这一区域体现了一种对其特征进行独特的重新组合的机会，也表现出成为体现卢布尔雅那当代建筑的象征或地标的可能。

　　这处建筑的特别之处在于成功地优化了环境状况，而这仅是通过三个步骤来达到最小的能量需求：减少能量需求，通过高效率的建筑一体化系统来满足剩余的要素，并尽量采用可再生能源来为这些系统提供能量。

　　这些可持续性概念中的核心思想之一就是加热与冷却诸如办公室等工作区域并利用这些办公室释放出的空气来调节中央大厅的环境。外部表面采用了高效能玻璃窗以及可调节的外部遮光装置来减少夏天室内所得到的阳光辐射。水泥板中嵌入了管道系统以便为建筑提供降温。同时，这套系统在冬天还能通过散热提供舒适的供暖。在夏天里，办公室空间的降温大多数时间里都将通过使用这套系统来完成。分散的通风装置则为房间提供新鲜的空气并在夏日里的高峰时间起到降温作用。在冬天，新鲜的空气将在装置内被加热后采用置换通风原理分散到房间中。通向中央大厅的溢流结构使得从办公室排放的空气可以流动到大厅中，然后通过天花板上的百叶窗释放到建筑外部。这就使得中央大厅的温度得以在冬季的15℃与上层结构的夏季室温条件之间摆动（温度分层）。为了最大程度地降低夏季最热时的温度，在中央大厅里的设施，如水墙与地板冷却系统，将作为额外措施来调节这一空间的环境。水墙在夏天里可以起到降温与除湿的作用，而在冬季，其中的水则会蒸发并加湿空气。在中央大厅里的一些特殊区域将拥有这一"独有"的气候。特别是对前台工作的人，这个局部的气候对于保证合适的工作环境具有重要作用。能量供应系统是依靠自然资源来运转的，例如，河水热交换设备将在夏天提供制冷，并在冬天与热泵一同起到供暖作用。中央大厅天花板上的光电池将提供电能。可持续建筑法的另一个客观效果就是减少了能量密集型材料的使用，转而促进了本地化、可持续性材料的使用。

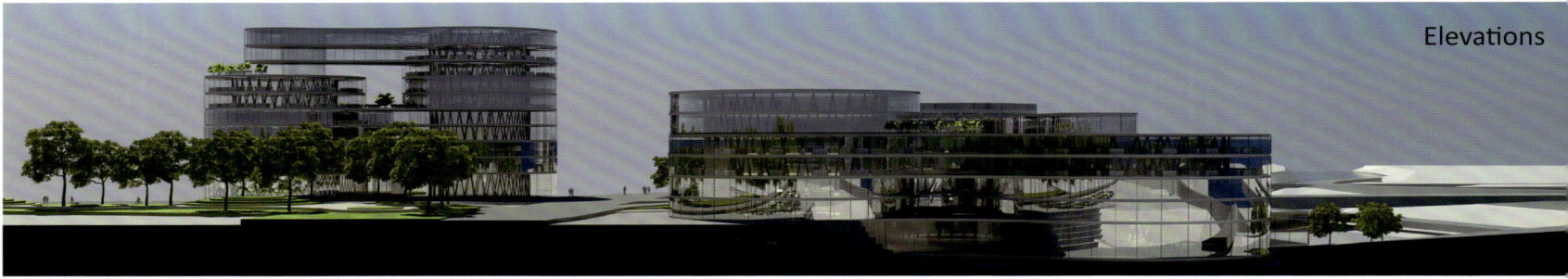

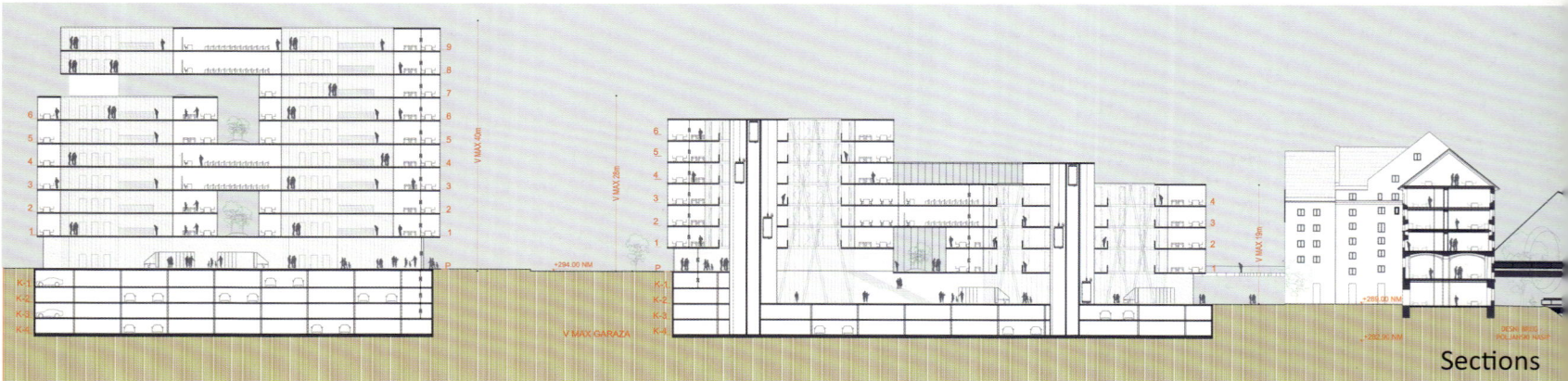

Siteplan

Four-Leaf Clover Kindergarten

四叶草幼儿园

Architect: Rok Oman, Spela Videcnik, Andrej Gregoric, Janez Martincic,
Janja del Linz, Katja Aljaz
Firm: OFIS Arhitekti
Location: Ribnica, Slovenia
Area: 11,500m²

The plot of the building is a four-leaf clover from which each leaf represents one department with associated facilities. Entrance leads to the central part with the common areas, which are positioned and distributed in three floors: ground floor with the main entrance and outdoor central common play area/playgrounds for additional activities, first floor with administration/offices and underground level with kitchen, services and technical areas.

The appearance of a kindergarten through its shape creates distinguishing yet integrated landmark in the middle of a meadow. Each leaf creates its own green bay which functions as intimate atrium for each children department. As such all surrounded landscape is designed as green playground platforms and small meadows. It is located on eastern and southern sunny part of the plot that offers sunlight through all day. Each playroom is connected with external playgrounds through covered wooden terrace.

Bay play-gardens merge in between departments and create "a green effect" atmosphere inside the interior. The design of the garden with islands of green hills in a symbolic way summarizes the characteristics of the loc countryside.

Playrooms are oriented towards the south and have canopies and interi shades. All rooms are naturally ventilated; lateral orientation of departmen allows good air ventilation. The spaces are concentrated, therefore ener losses are small, as well as material consumption during construction ar maintenance. Lobby is combined with changing locks, additional playroor are multipurpose areas.

The material, structure and cladding are local and natural. The structure the building is simple and economical; reinforced concrete cores with bri and wooden fillings. Finalization is mostly natural wood and paint on natu grounds. Selected material and design will allow a pleasant stay. Edited ar rainwater tanks will be able to recycle hot water. The surface of the ro allows the installation of photovoltaic cells.

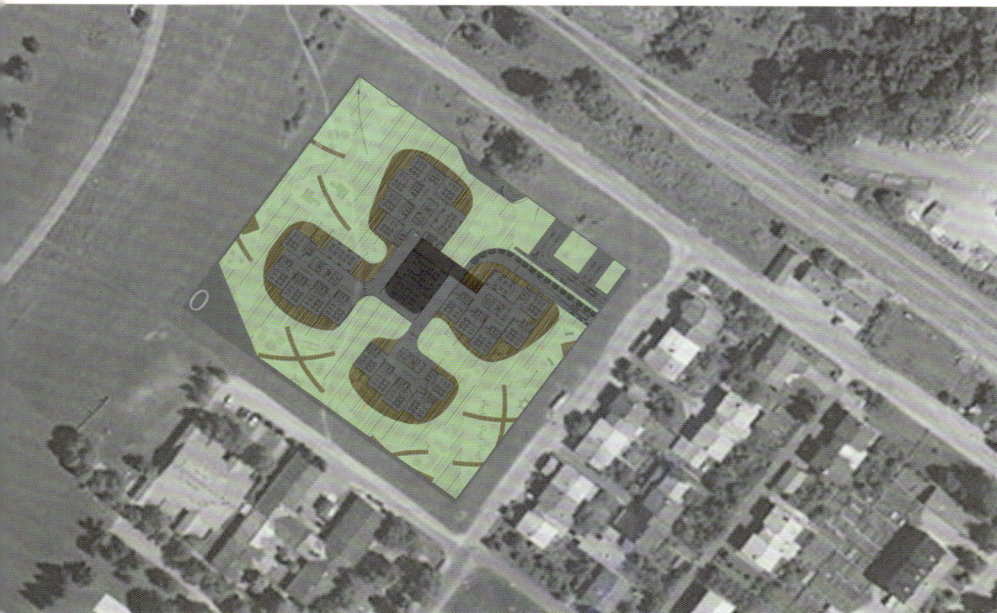

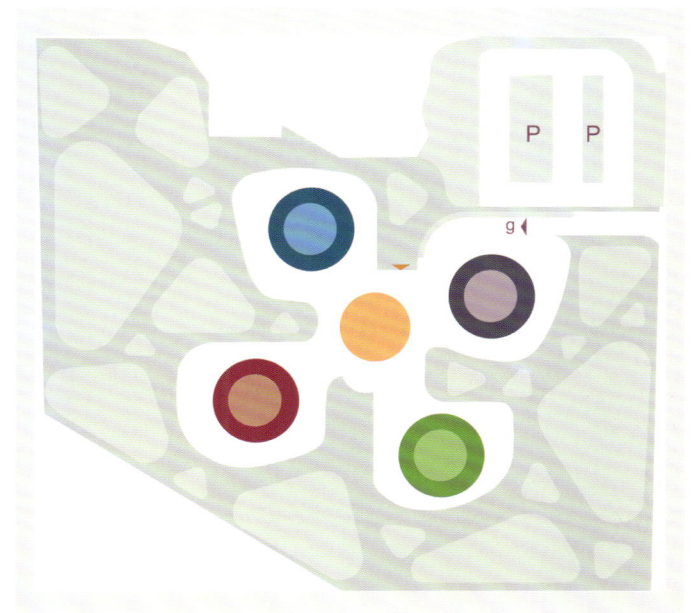

LANDSCAPE

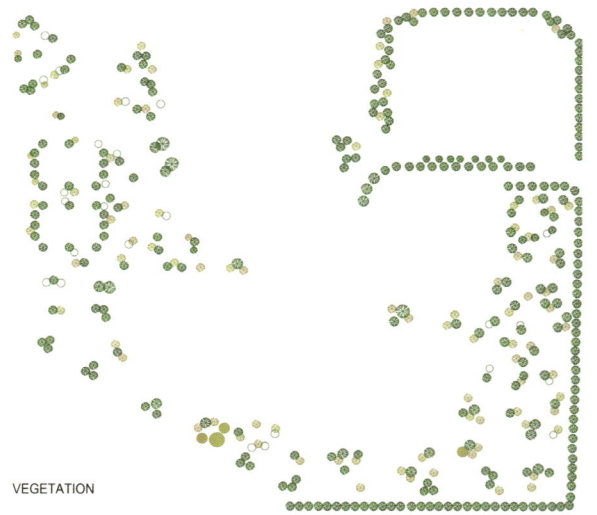

VEGETATION

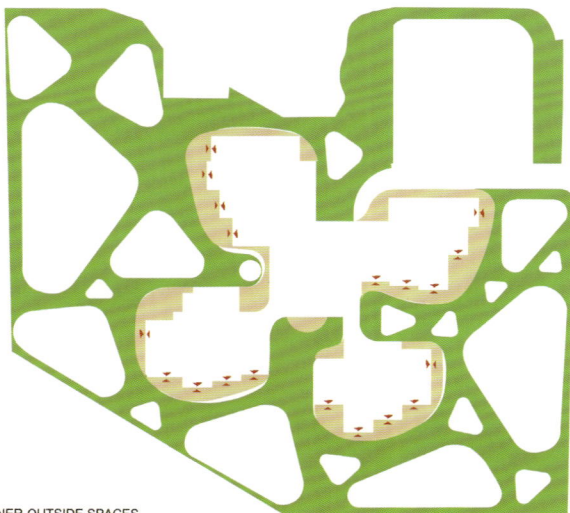

INNER-OUTSIDE SPACES

PLAYGROUNDS

1 2 3 4 5 6 7 8 9 10 11

B-B

0

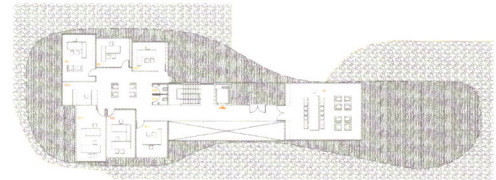

+1

5 6 7 8 9 10 11

-1

A-A

0   10m

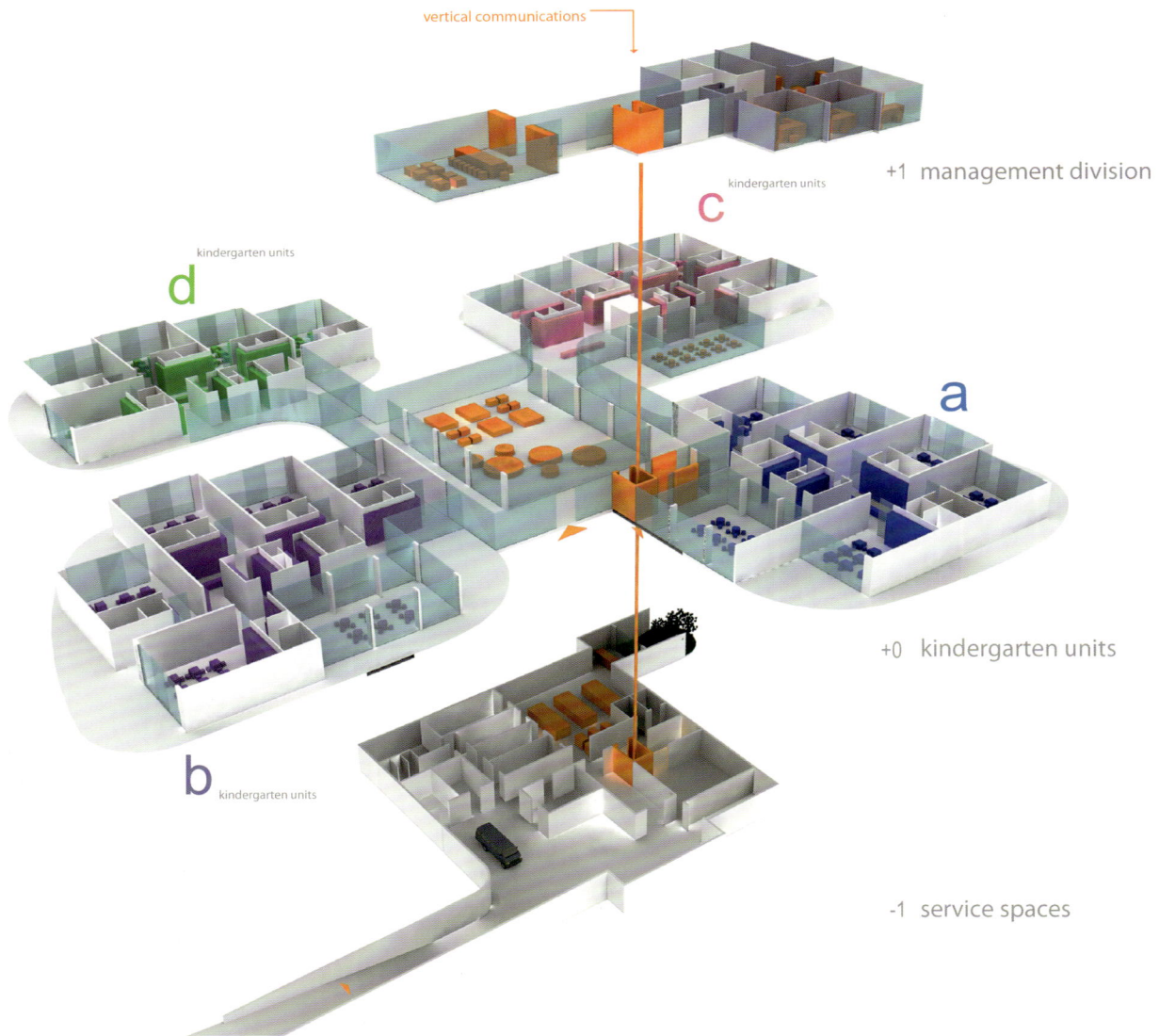

vertical communications

+1 management division

kindergarten units

c

kindergarten units

d

a

+0 kindergarten units

b kindergarten units

-1 service spaces

　　这座建筑的外形呈四叶草状，
中每一片"叶子"都是配备有相
设施的一个部门。人们可以从入
进入位于中心的三层公共区域：
楼是主入口与中心室外公共活动
/提供额外活动的操场，二层是管
区/办公室，而地下层是厨房、服
与技术支持区域。

　　这座幼儿园因其独特又整体化
外形成为了草地上地标。建筑的
一片"叶子"都形成了自己的绿
，并为每一个幼儿部门提供了身
的中庭。通过这种方式，幼儿园
围所有的景观都被设计成为了绿
的活动平台和小块的草坪。这座
筑坐落在整个地块中阳光充足的
部和南部区域，这使得幼儿园可
整天都有充足的阳光。每一个室
活动室都通过了覆盖木质表面的
台与室外操场相连。

　　海湾形的游戏花园在各个部
之间相融合，给室内环境带来了
绿色效应"。在花园的设计中采
了若干小丘形成的"绿岛"象征
当地乡村的地形特点。

　　游戏室都设计为南向，并都
有遮阳篷和内部遮光罩。所有的
间都自然通风，而各部门分支的
向设计产生了良好的通风效果。
为建筑空间的集中形设计，整个
筑损失的能量比较小，而且在建
和维护过程中也会消耗较少的材
。大厅通过改变锁连接在一起，
外的游戏室则作为多功能区域。

　　建筑的材料、结构和覆盖层
是当地和天然材料。建筑的结构
是简洁而经济的增强性混凝土核
填充木料和砖的设计。最后的修
基本采用了天然木料和漆料。精
选择的材料和设计让人们在这里
够有一个愉快的体验。细心设计
雨水收集箱可以循环利用热水，
天棚的表面则可以安装太阳能电

Flowing Gardens

流水花园

Architect: Eva Castro, Holger Kehne, Alfredo Ramirez Galindo, Xiaowei Tong, Mehran Gharleghi, Evan Greenberg, Nicoletta Gerevini, Peter Pichler, Tom Lea, Ying Wang, Katy Barkan, Federico Ruberto, Rui Liu, Danai Sage
Firm: Plasma Studio and Groundlab
Location: Xi'an, China
Area: 12,000m²

Building Elevations

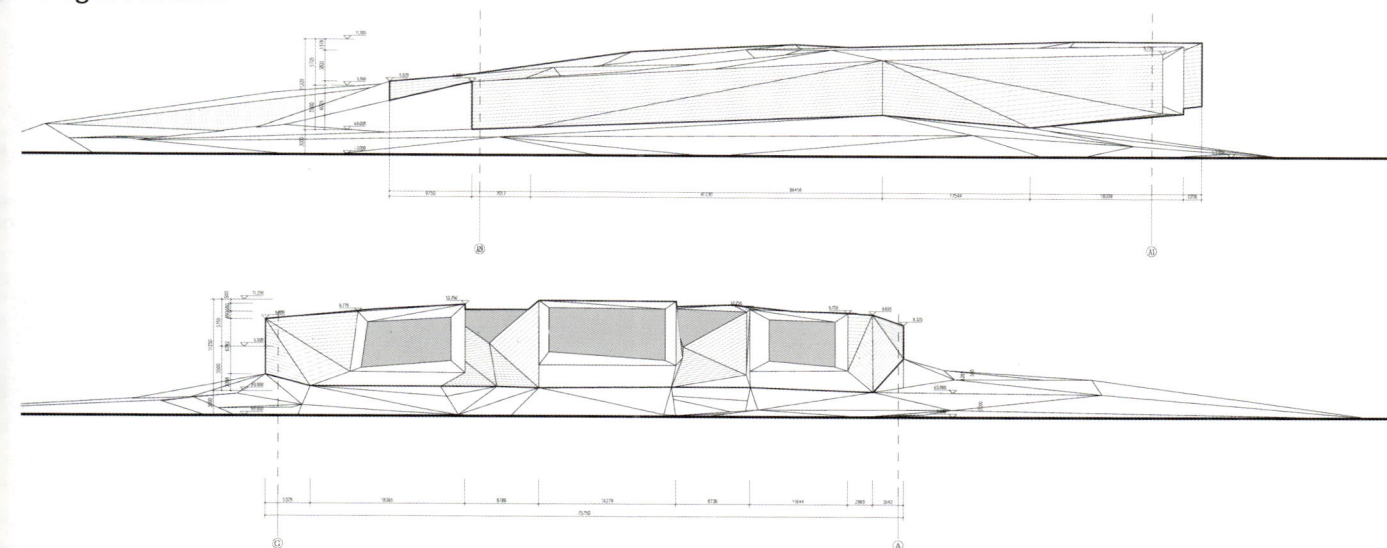

Flowing Gardens begins from a single line—an axis extends from the Gate to the Greenhouse, travelling through the East and West Hills and over the lake, while extending into many sinuous paths, creating a network of intermingling circulation, landscape and water. Much like the legendary Silk Road, Flowing Gardens is connectivity, circulation, rejuvenation, and elegance.

The project proposes a hybrid of both natural and artificial systems. These two opposing systems are brought together in a synergy of waterscapes.

The water cycle begins to become more complex with the introduction of grey and black water treatment. The architects propose to make use of the initial investment and organisation during the exhibition to set up a system which becomes autonomous in function and character. The gardens transform the two conditions of artificial and natural into a sustainable system that becomes more and more maintenance-free once the exhibition is over, allowing the park to become a new model, or paradigm, within the horticultural industry.

The given topography and its existing slopes were used to draw out the paths in a way similar to how roads ribbon around a mountain, negotiating ease of steepness with gradients. These paths vary in width ranging from main

walkways and arteries to towpaths. The patches between these paths become the zones for various planting types and wetland areas, which retain a quality of maintenance.

Flowing Gardens creates a consonant functionality of water, planting, circulation, and architecture into one seamless system. At the major intersections of these pathways lie three buildings; the architecture is an intensification of the ground condition, where each building stands alone as an object yet speaks of the interconnectivity of the landscape.

The Gate is created at the junction of public meeting space, landscape, and circulation; one enters the site through the Gate along the major axis of Flowing Gardens, creating framed views of the gardens. The Exhibition Center is formed at the seam of landscape, circulation and water; one experiences the Exhibition Center's fluid lines as an extension of the landscape with vistas of the lake and the South Hill. The Greenhouse sits at the top of the South Hill, at the connection of various landscape features. The Greenhouse allows one to experience the beauty of Flowing Gardens from across the lake while appreciating plants and flowers from four different climatic zones.

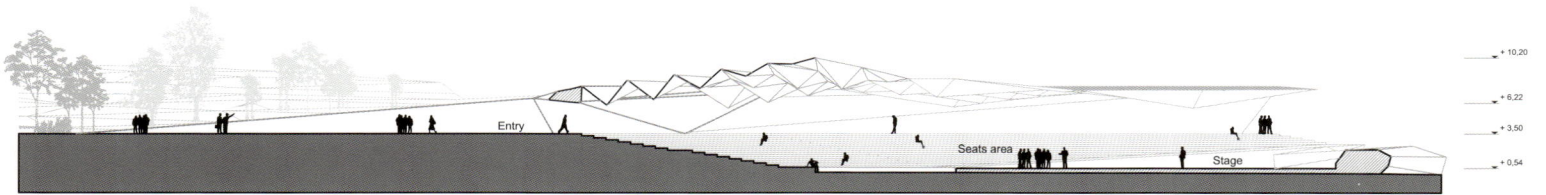

SECTION A-A'

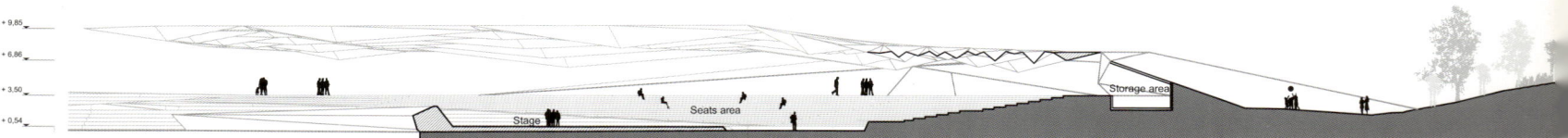

SECTION B-B'

Open Theatre Sections

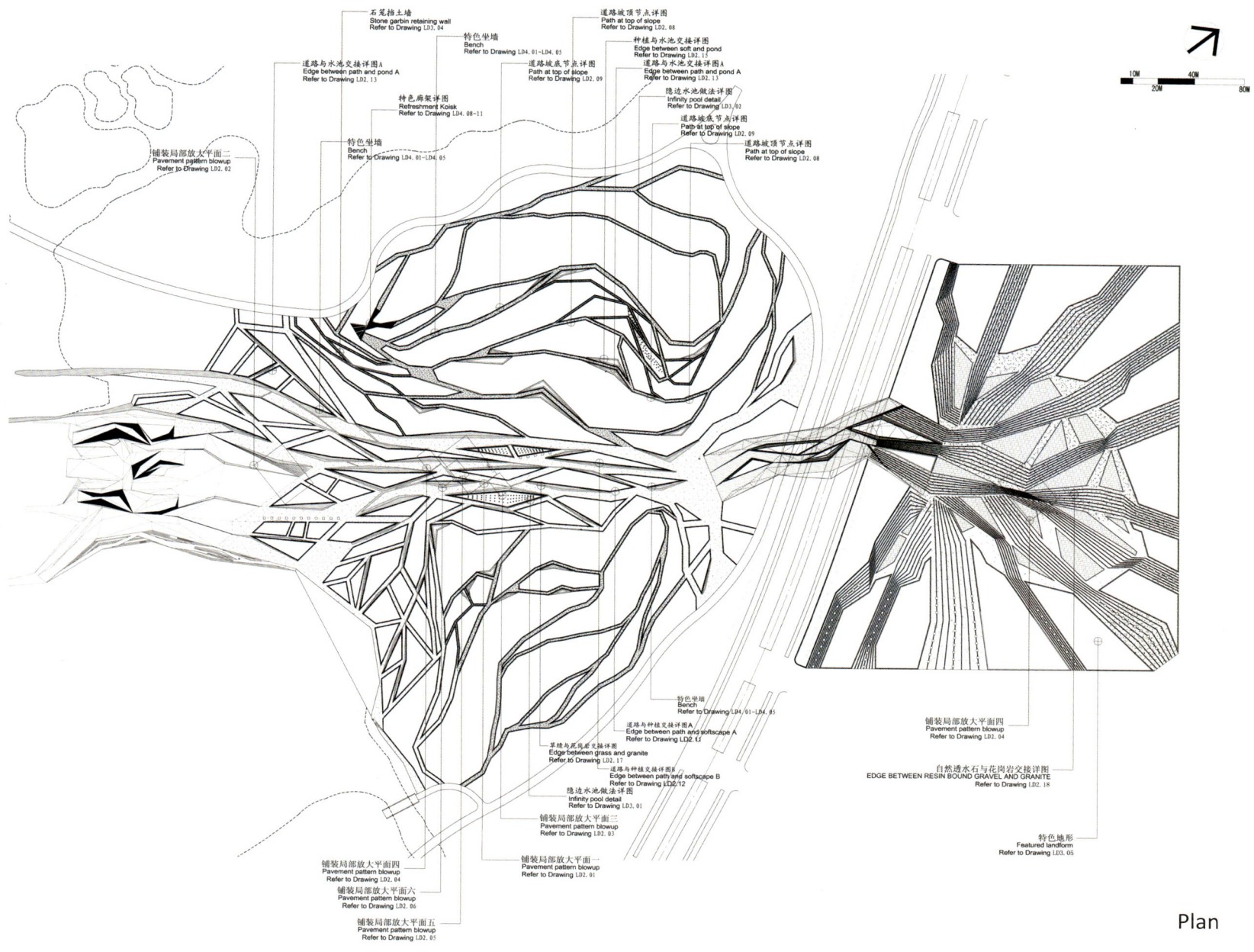

石笼挡土墙
Stone garbin retaining wall
Refer to Drawing LD3. 04

特色坐墙
Bench
Refer to Drawing LD4. 01-LD4. 05

道路坡顶节点详图
Path at top of slope
Refer to Drawing LD2. 08

种植与水池交接详图
Edge between soft and pond
Refer to Drawing LD2. 15

道路与水池交接详图A
Edge between path and pond A
Refer to Drawing LD2. 13

道路与水池交接详图A
Edge between path and pond A
Refer to Drawing LD2. 13

特色廊架详图
Refreshment Koisk
Refer to Drawing LD4. 08-11

道路坡底节点详图
Path at top of slope
Refer to Drawing LD2. 09

隐边水池做法详图
Infinity pool detail
Refer to Drawing LD3. 02

特色坐墙
Bench
Refer to Drawing LD4. 01-LD4. 05

道路坡顶节点详图
Path at top of slope
Refer to Drawing LD2. 09

铺装局部放大平面二
Pavement pattern blowup
Refer to Drawing LD2. 02

道路坡顶节点详图
Path at top of slope
Refer to Drawing LD2. 08

铺装局部放大平面四
Pavement pattern blowup
Refer to Drawing LD2. 04

265

特色坐墙
Bench
Refer to Drawing LD4. 01-LD4. 05

道路与种植交接详图A
Edge between path and softscape A
Refer to Drawing LD2. 13

草缝与花岗岩交接详图
Edge between grass and granite
Refer to Drawing LD2. 17

道路与种植交接详图B
Edge between path and softscape B
Refer to Drawing LD2. 12

隐边水池做法详图
Infinity pool detail
Refer to Drawing LD3. 01

铺装局部放大平面三
Pavement pattern blowup
Refer to Drawing LD2. 03

铺装局部放大平面四
Pavement pattern blowup
Refer to Drawing LD2. 04

自然透水石与花岗岩交接详图
EDGE BETWEEN RESIN BOUND GRAVEL AND GRANITE
Refer to Drawing LD2. 18

特色地形
Featured landform
Refer to Drawing LD3. 05

铺装局部放大平面四
Pavement pattern blowup
Refer to Drawing LD2. 04

铺装局部放大平面一
Pavement pattern blowup
Refer to Drawing LD2. 01

铺装局部放大平面六
Pavement pattern blowup
Refer to Drawing LD2. 06

铺装局部放大平面五
Pavement pattern blowup
Refer to Drawing LD2. 05

Plan

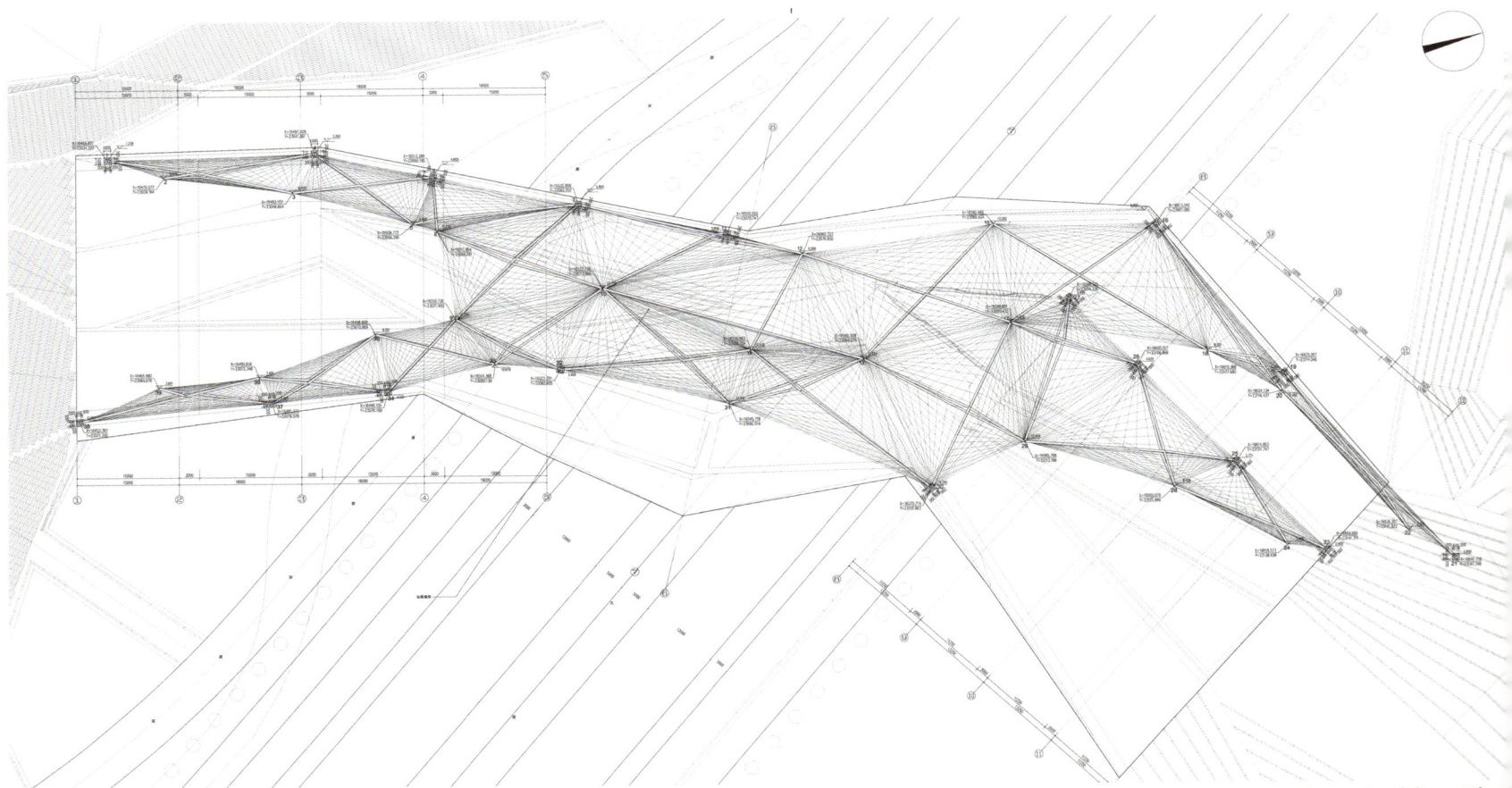

Gate Building Plan

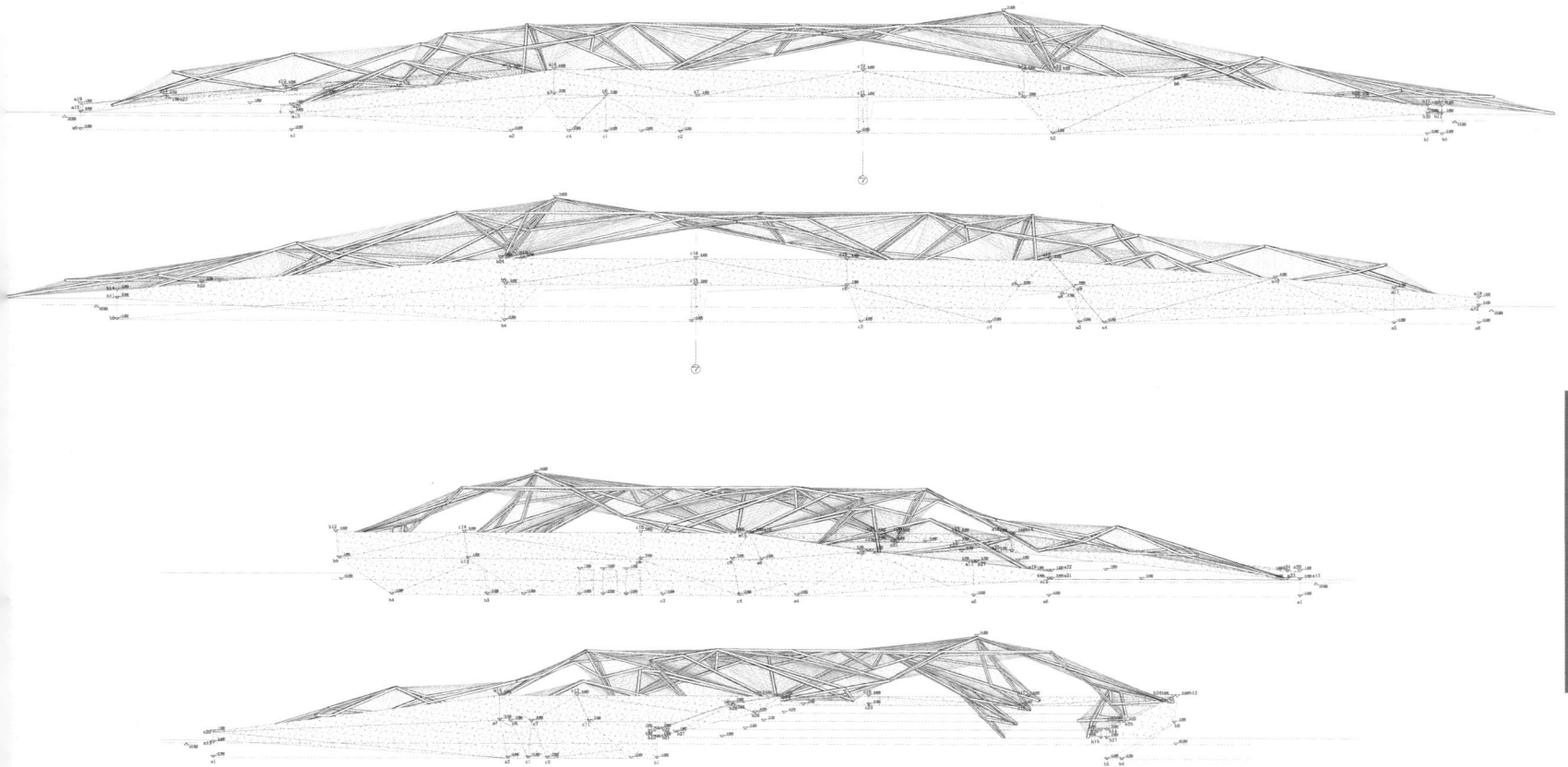

Gate building sections

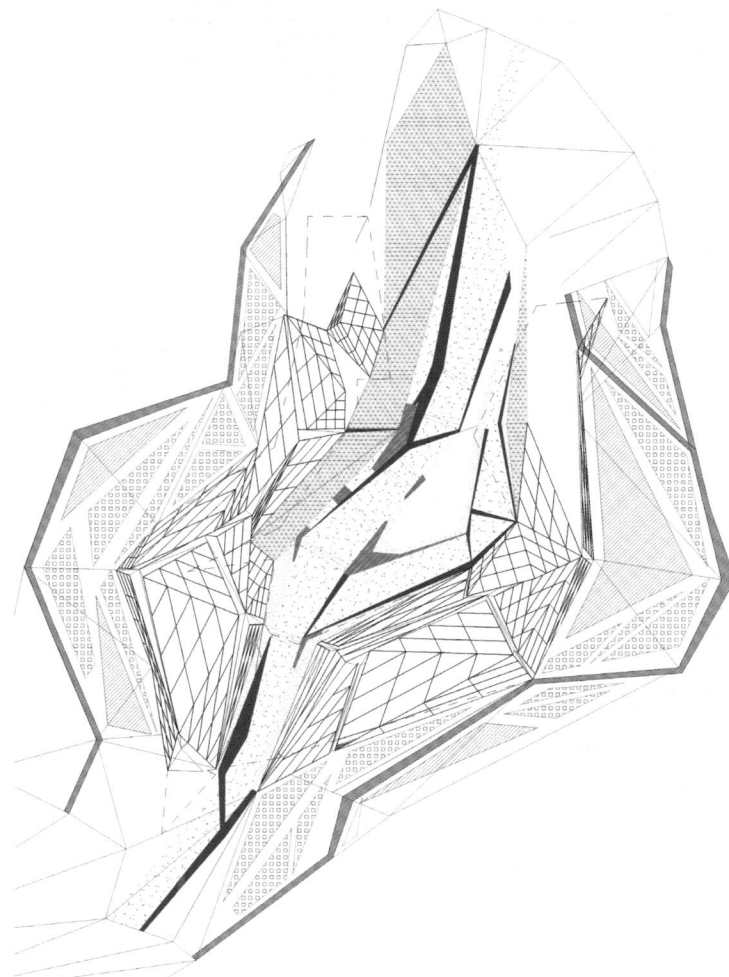

Green House Plan

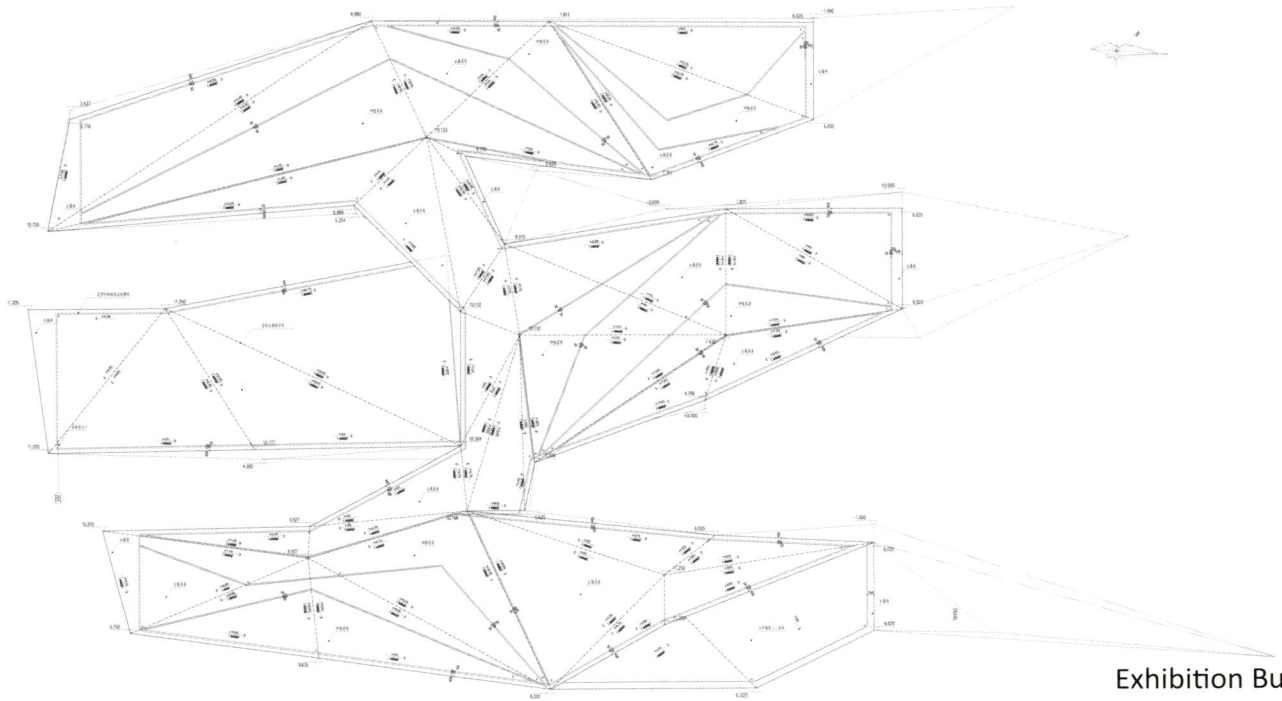

Exhibition Building Plan

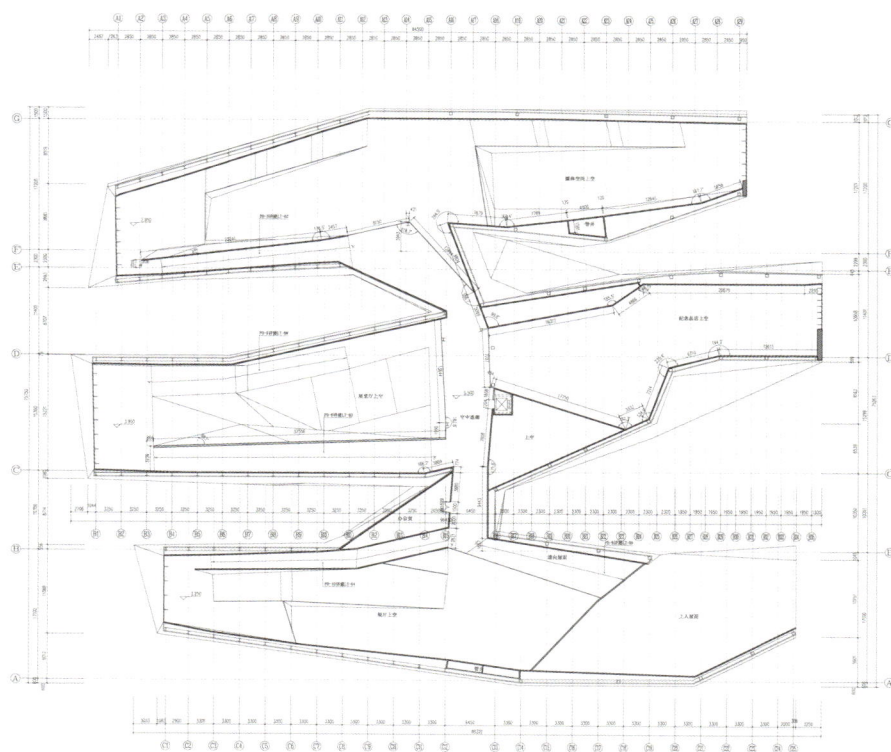

Exhibition Building Plan

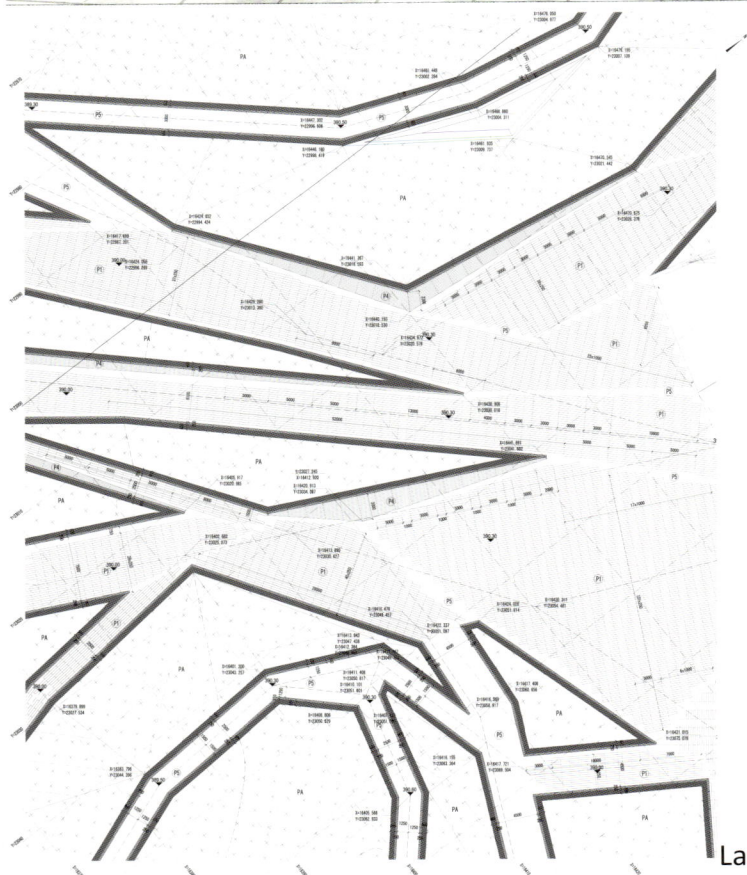

流水花园的设计遵循了一个简单的线路——从温室大门延伸的一条中轴线，穿过了东西山，跨过了湖，并同时延展出许多蜿蜒的小路，形成了一个将景观、水体和活动结合起来的网络。与传奇性的丝绸之路一样，流水花园的设计立足于交流、联系、活力与优雅的理念。

这个项目提出了将自然与人工的系统相混合的概念。这两种相互矛盾的系统被同时用来产生协同化体景观。

水体的循环在引入灰水和污水处理时开始变得更加复杂起来。建筑师们建议在展览的过程中利用最初投入的投资和资源来建立一个可以在功能和性质上独立的系统。这些花园将人工与自然的两种条件变成一种在展览结束后会越来越无需维护的自持性的系统。这使得这座公园成为了园艺领域的新典范。

这个设计方案利用了天然的地形和已有的斜坡来设计了盘山道一般的步道，使地形的倾斜被逐渐缓和。这些步道有多种宽度，从主人行道到小径各不相同。这些步道间的区域为湿地留出了空间并种植了多种不同的植物。这保证了整个系统易于维护的性质。

流水花园的设计方案创制了一个水体、植被、建筑与物质交换相辅相成，完美结合的系统。在这些步道相交的主要路口处有三栋建筑；这些建筑是对地面条件的强化，每一座建筑都是相对独立的，并与周围景观相呼应。

大门建在公共接待区、景观与循环系统的交汇处；人们通过大门并沿着流水花园的主轴线进入园区，通过这种方式可以看到整个花园的轮廓。展览中心则建在景观、水体和循环系统的空隙；游人可以感受到展览中心流动的线条好似狭长的湖泊和南山的景观的延伸。温室正位于南山的顶部，处于各种景观元素的连接点。温室可以使人们在欣赏四个不同气候区的植被和花卉的同时体验到流水花园湖泊两岸的美丽。

Landscape Detail

Auditorium and Library for the University of Amiens

亚眠大学礼堂与图书馆

Architect: David Serero, Yoìchi Ozawa, Ran She, Fabrice Zaìni
Firm: SERERO Architectes
Location: Amiens, France
Area: 1,424m²

Located in the heart of Amiens University, the new library and auditorium constitute a central place for exchanges and meeting around technologies as well as the catalyst of an intense campus social life.

The project sets the two entities, library and auditorium, at the same level of the central garden of the University. The auditorium floor is sloped and follows the natural landscape of the site. In between these two volumes, the reception hall connects the lower level of the site to the garden by long steps of stone.

The architects have placed natural lighting in the heart of the library, thanks to sheds on the roof, which are oriented to the North. With this zenithal lighting, they create a space where exchanges and knowledge are multiple and opened to other discipline or culture.

The auditorium is made of white concrete with a strip of curtain glass window on top of the wall that can be shut by motorized sun shields integrated to the space ceiling.

The library is conceived as a platform, with a unique intermediate post (10.80m range). All spaces are planned on a regular grid of 1.80m x 7.20m, which organizes all architectural and technical devices (lighting system, ventilation, furniture, facade, etc.) of the library in order to offer an opened and flexible interior space. Therefore, the building manages a gradual transition of spaces from the reception hall, for groups and meetings to more individual spaces of silence and high concentration.

A facade of WOODEN scale

The building envelop is designed as a "smart" skin, which controls internal spaces ambiance as well as views.

The facade system is inspired by pine cone scales, which open or close according to the level of humidity in the air. The full height glass curtain wall is protected by wooden sun-shading, whose angles changes along the facade according to the orientation of spaces. This system allows a gradual transition from protected zones to opened and transparent zones. This sun-shading is vertical for a better answer to the main western orientation of the building.

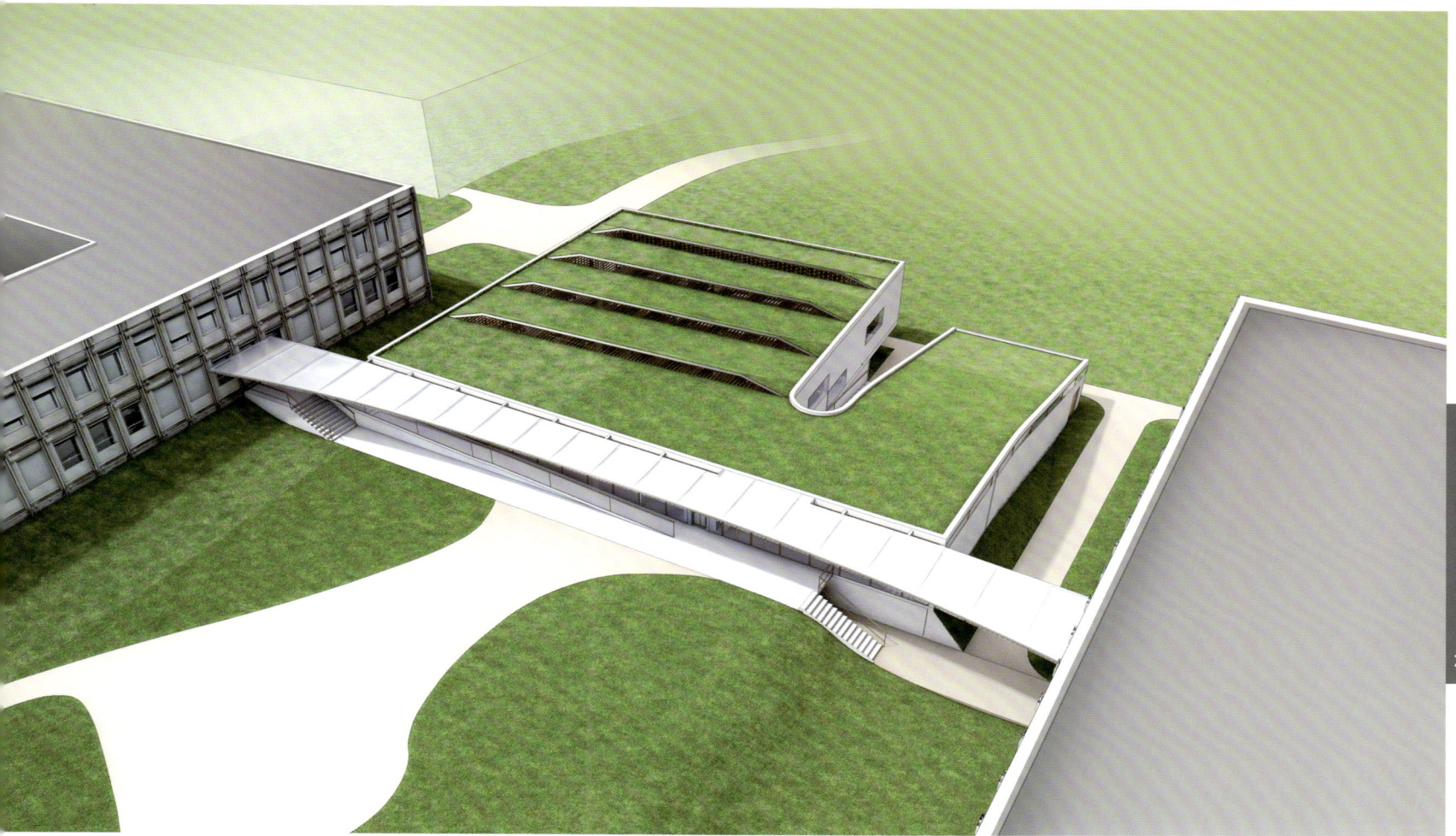

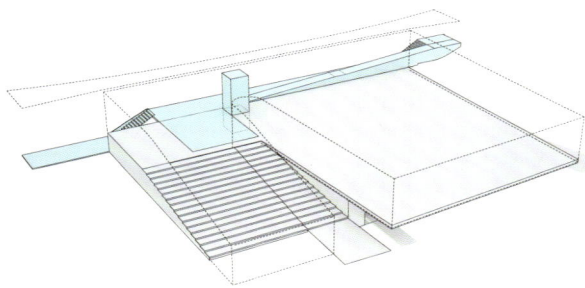

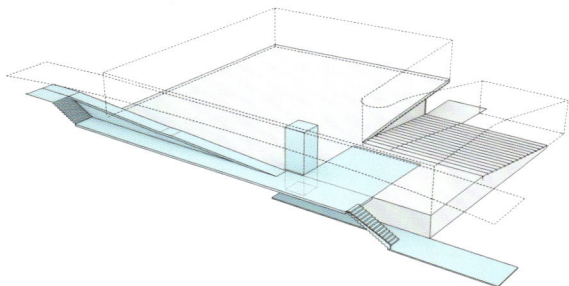

Elevations

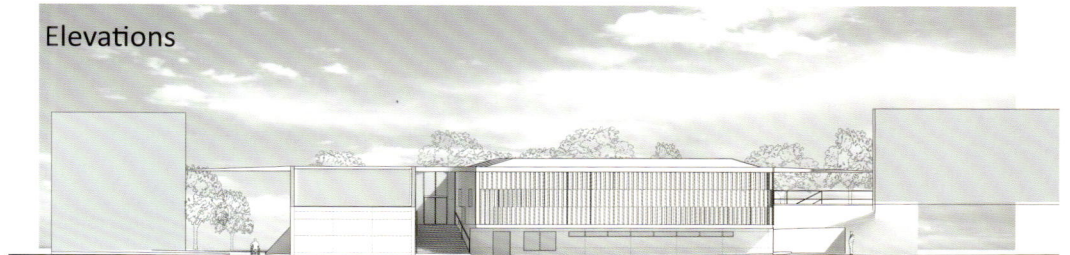

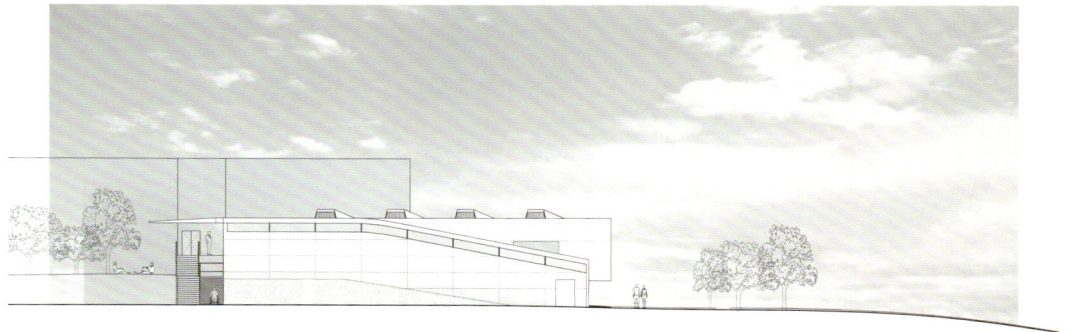

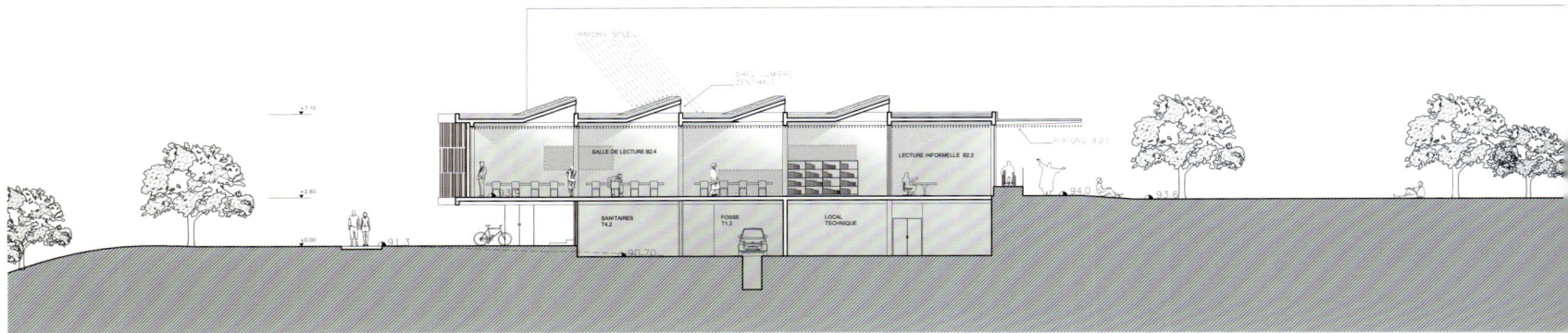

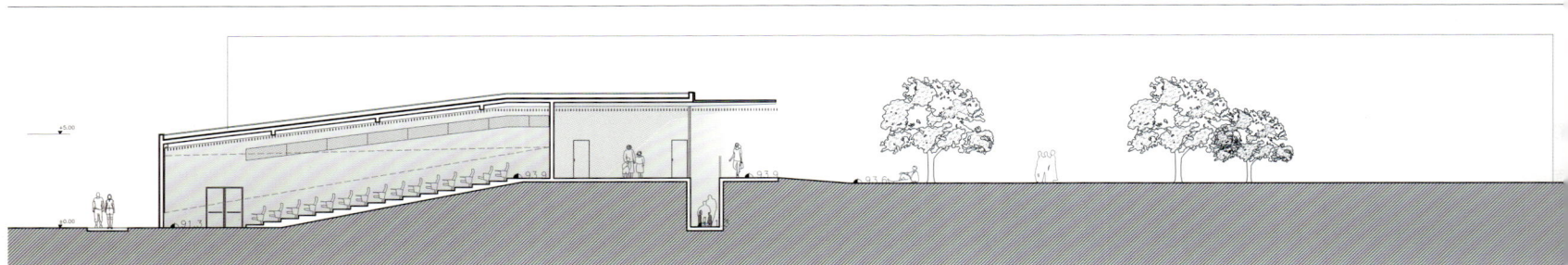

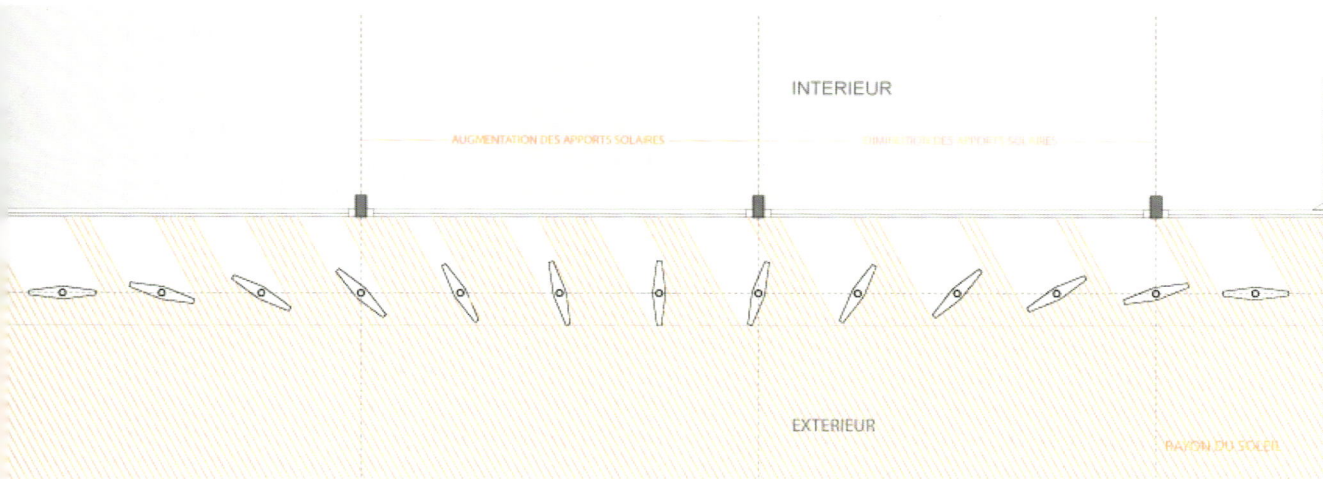

INTERIEUR

AUGMENTATION DES APPORTS SOLARES

EXTERIEUR

RAYON DU SOLEIL

Detail Plan

坐落在亚眠大学核心位置的新图书馆和礼堂是学术交流与会议的中心场所，同时也对彩的校园社交生活起到了促进作用。

这个工程分为两个实体，图书馆和礼堂。它们与大学的花园处于同一水平面。礼堂的面沿着其所在位置自然的地形呈倾斜状态。接待大厅位于两座建筑的中间，连接着这里低的一层，并通过长长的石阶与花园相连。

得益于屋顶上面向北方的小屋，设计可以将自然光源引入图书馆的中心。利用这种天的自然光照，创造了一个拥有多种多样的知识与交流并向其他领域与文化开放的空间。

礼堂由白色混凝土构成。在墙壁顶部是条形玻璃窗，这些玻璃窗带有与天花板整合在起并可自动开关的遮阳板。

图书馆被设定为一个平台，并带有一个10.80m范围内的特别设计的中转站。所有的间都在一个1.80m×7.20m的规矩的方格基础上进行规划，集中了图书馆所有的建筑与技

术设备，包括照明系统、通风、家具、立面等等，提供了一个开阔、灵活的室内空间。因此，建筑完成了一个空间的逐步过渡，从为群体及会议设计的接待大厅，一直到为安静或者需要高度注意力的活动设计的个人空间。

木质鱼鳞瓦立面

建筑的外部覆层被设计为"智能"立面，能够控制内部空间的环境以及视野。

外立面系统受到了松塔的启发，根据空气湿度的不同会自动开合。全高的玻璃幕墙由木质的遮阳板保护，这些遮阳板能够跟随空间的走向沿着立面改变角度。这种系统使空间可以从受保护的区域逐渐过渡到开放的、通透的空间区域。为了更好地贴合建筑的西式设计造型，这种遮阳板都是垂直安置的。

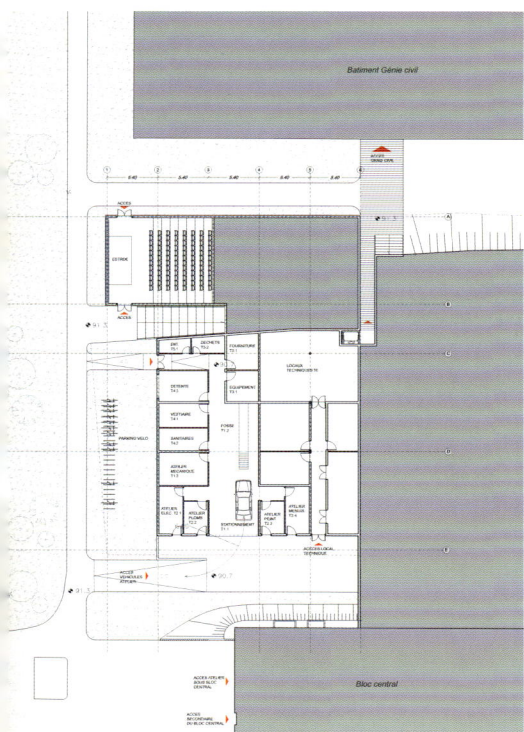

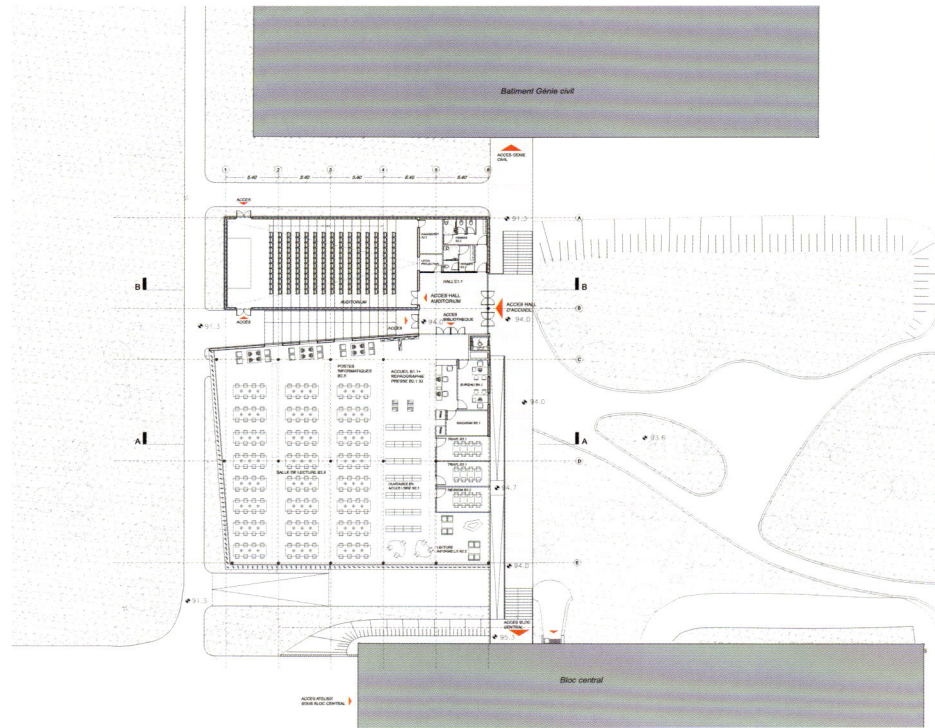

284

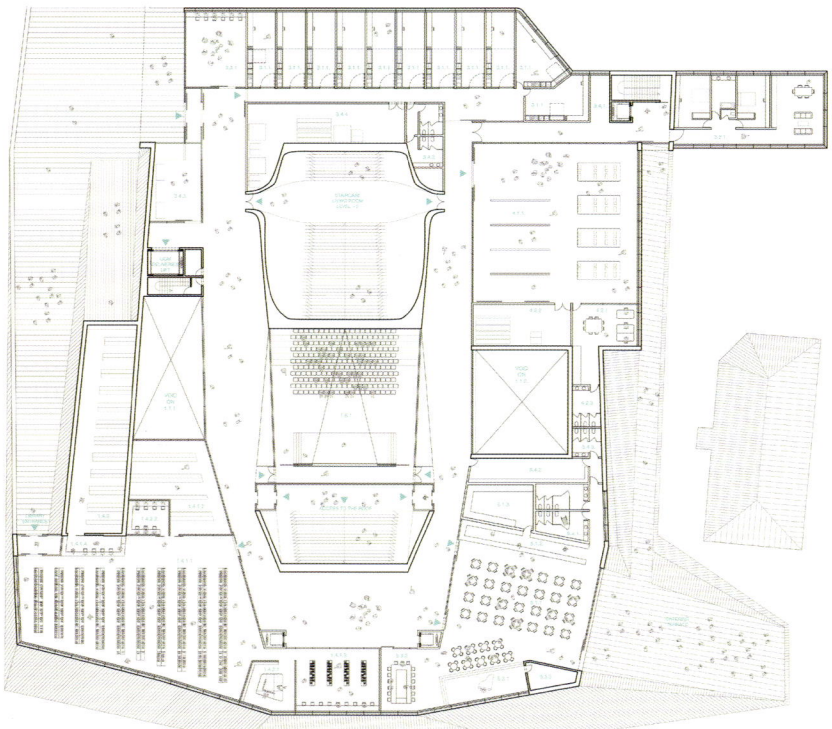

LEVEL +2 SCALE 1:200

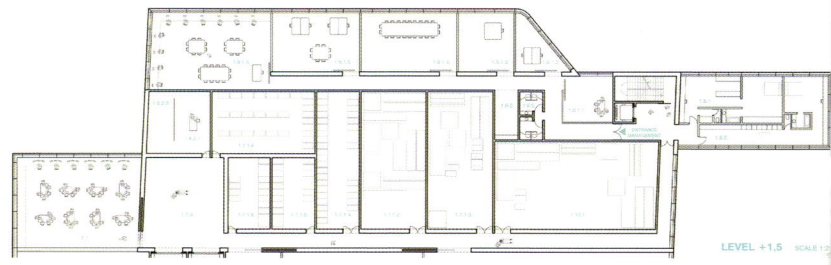

LEVEL +1.5 SCALE 1:200

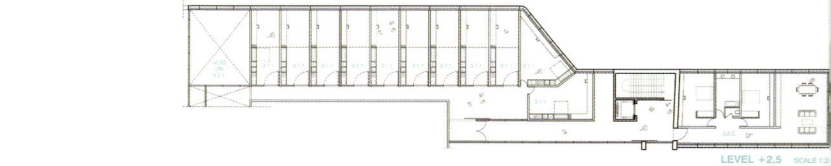

LEVEL +2.5 SCALE 1:200

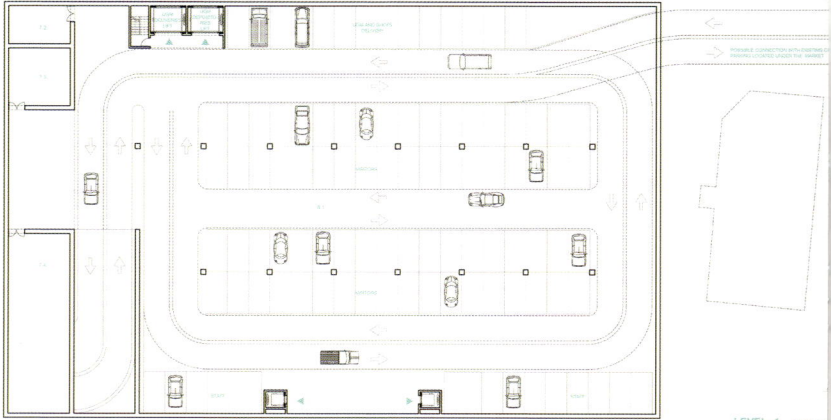

LEVEL +1 SCALE 1:200

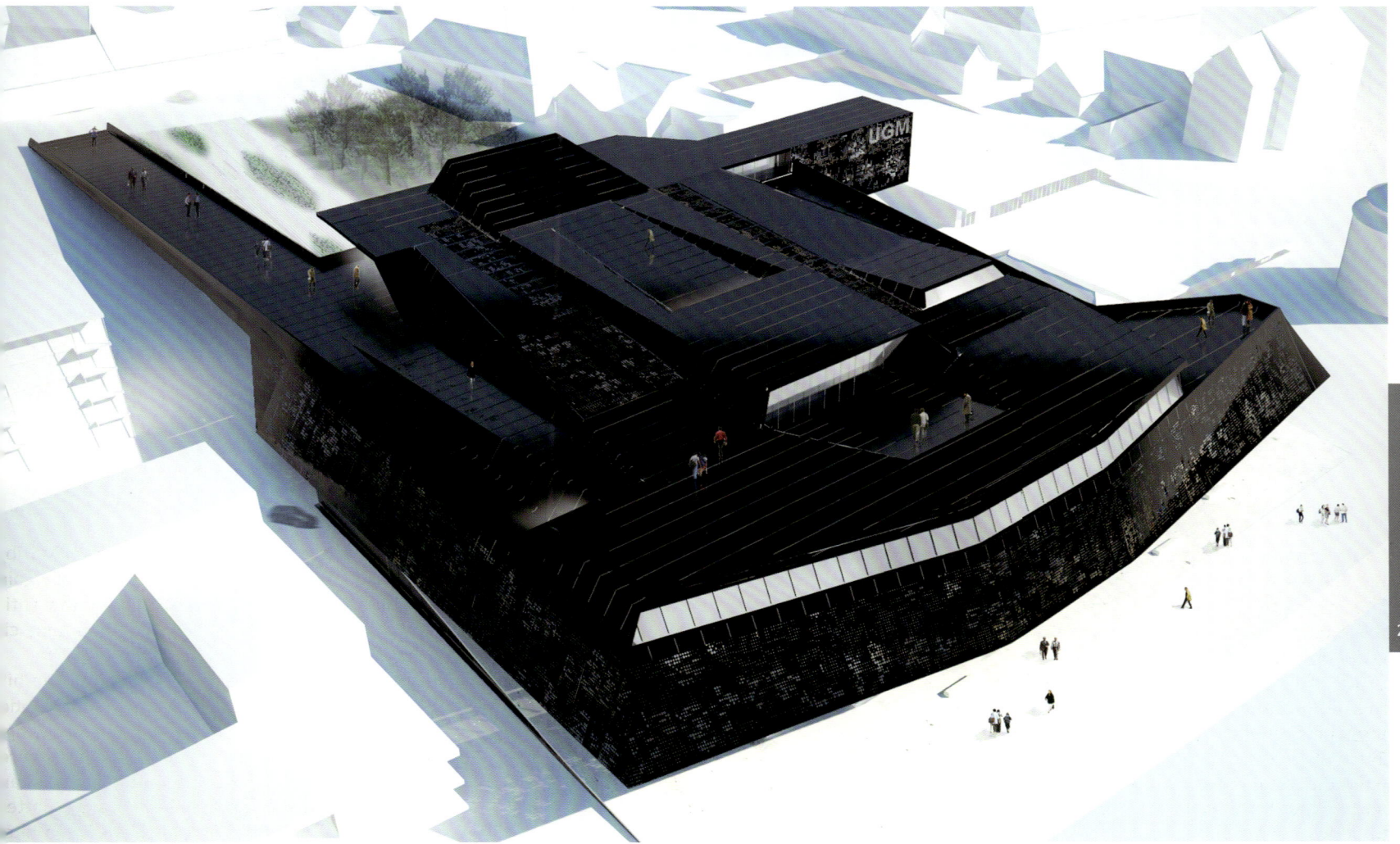

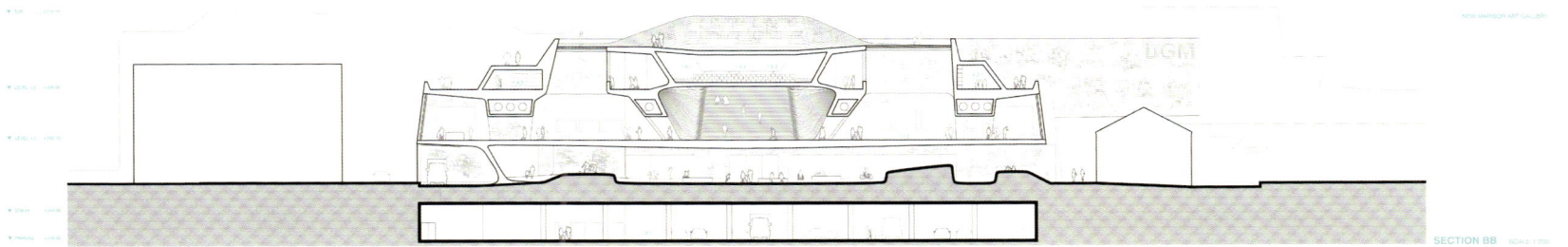

SECTION BB

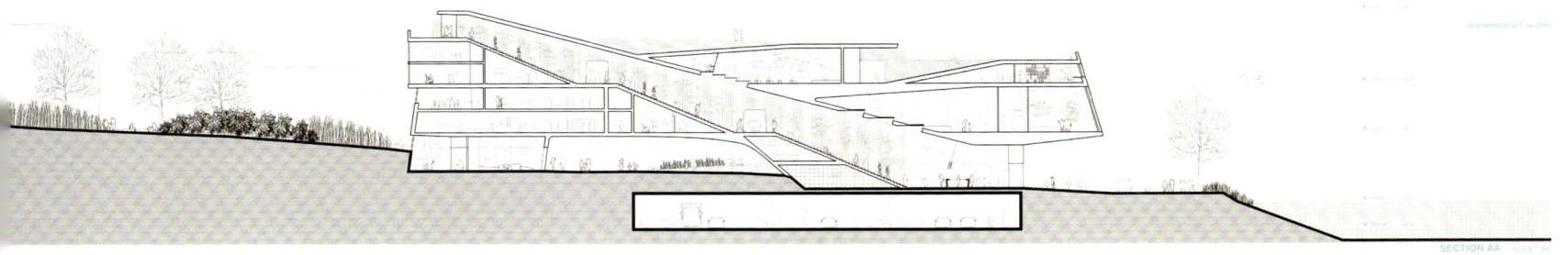

SECTION AA

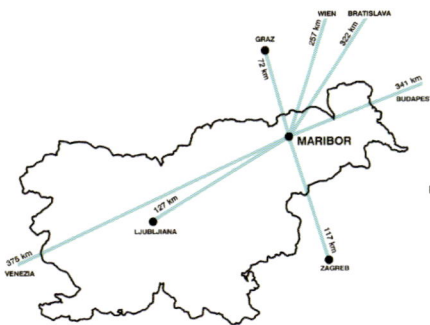

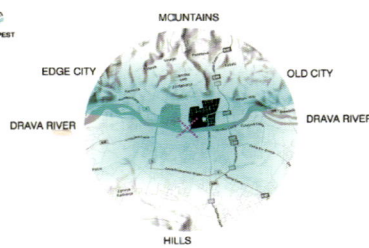

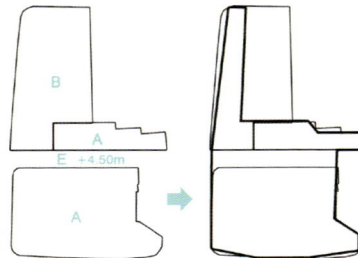

A STRATEGIC SITUATION FOR ARTS
Maribor is well located, at the crosspoint of Alps, Adriatic and the Balkans. Venice, Vienna, Graz or Budapest are all at a close distance by car or by train from Maribor, which means that the New Maribor Art Gallery will be part of an Eastern network of Art cities. Being the second largest city of Slovenia, Maribor will be with the New Museum Gallery institution more attractive, more into the Arts scene. And with the 2012 European Capital label, the short distance from other big museum cities, visitors and tourists will probably stop by Maribor for new reasons.

360° INTERESTING VIEWS
Located along the Drava river, popular way for walks, recreation, relaxation, fun and social events, the site location offers a varied panopticon of sights. On East side, the old city and its judgement tower landmark the museum location. The Museum project seeks to be a smooth transition between the old city and the modern developments located on the west side of the project. From North to South, and all around the city the slopes and mountains are part of the unique panoramic perception of Maribor.

GATHERING ARCHITECTURE, LANDSCAPE AND URBANISM
The project covers the entirety of the proposed area "A", while covering the "E" Street. Detached from the ground, it makes connections with the existing urban fabric, while linking different altitudes on the inclined ground. The "B" area is worked as a landscape element where a ramp connects directly to the roof of the project, and thus to the museum entrances. The project is a porous and furtive mass, in direct visual and morphologic relation to its neighbourhood, from a close or a far distance.

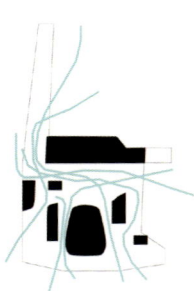

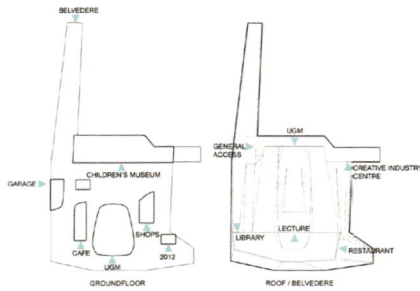

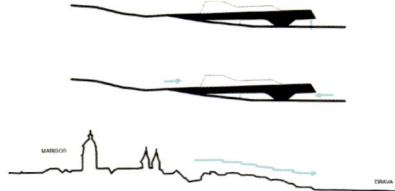

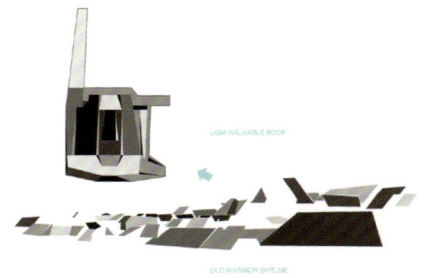

SITE PERMEABILITY
The project frees the existing and inclined ground level, connecting the Drava bank with the upper side of the neighbourhood. People pass through the covered groundfloor, an open public space for outdoor protected activities of the UGM. In this way, the project aims to encourage visual connections between the old and the modern city. Like a Drava bank extension, the ground continues under the museum offering various programmatic-events emerging, like ground distortions.

AUTONOMOUS PROGRAM ELEMENTS
The New Museum Gallery gathers various program elements with different scheduled activities and hours opening. Every elements of program has its own external access. The groundfloor and second floor are freely accessible from outside, permitting direct independant access and different scheduled openings.
Part of the urban potential exploitation, the groundfloor elements contribute to animate this part of the Drava bank, while the second floor located activities contribute to attract people on the roof-belvedere, offering views on Maribor's fifth facade.

STAIRCASE ORGANISATION
The inclined condition of the Museum'site has driven the project implementation and organisation. Elevating the museum galleries at level one offers a covered public space on ground floor, and allows a ramp connection from the upper side to the roof of the project. The museum is thus accessible from the river and the elevated back side of the site. Learning from Maribor urban organistion along the Drava, the project operates by gradually going up or down volumetrically. A staircase working as a daily city room, crossing and supporting the entire museum complex is the key point of the proposal.

FIFTH FACADE
What makes Maribor'specificity is its roofscape. And because of the hilly character of its surroundings, the roofs participate to the emblematic perception of the city, as well as the buildings vertical walls. The project is directly inspired by this perception, showing a fragmented carapace on roofs, facades and ceilings of external covered spaces. Like a stealth, the project is inserted in the city in direct visual and morphologic relation with its context.

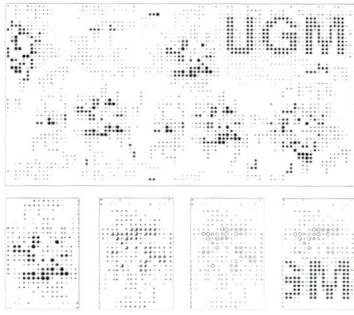

SLOVENIAN LACEWORK TRADITION

The external enveloppe uses a pattern shown in various treatments. Perforated, it works as a brise-soleil and filters natural light for exhibitions spaces. On floors the same pattern is used but in relief, making surfaces non-slippery, especially on the walkable roof. The perforated material is also used in ceilings to hide technical elements like air-conditionning and ventilation systems. The pixelated pattern is also used for signage in the entire museum. Set as a roof cover, the holes stock rainwater, recycled for the building needs.

MUSEOGRAPHIC TOOLS

Exhibition spaces are varied in height, width and length. They all offer different conditions of light, mostly natural light and a system of occultation (see former diagram) allows to convert the spaces into black rooms. In both museum departments, a structural elements contains technical elements like lighting, electricity or air-conditionning. It also contains spaces for stocking materials. Removable partition walls slip from those structural elements to subdivide the exhibition space.

SUSTAINABILITY

The project will be built by slovenian companies using local materials like steel for the structure, aluminium for the perforated pannels and wood for the interior staircase.
The use of perforated surfaces will allow stocking rainwater and filter the sunlight. Black colour for this material will catch solar energy, stocked and re-used to heat the museum spaces. The project is compact, reducing loss of energies and flows. External walls 50 centimeters thick containing technical elements, as well as triple glass windows, ensure a good insulation and no heat loss.
The museum respects the existing ground topography, reducing digging out and the impact on lower grounds like underwaters. The backside of the project shows a landscape using local trees and herbs.

STRUCTURAL INVESTIGATIONS

Circulations in a Museum are the main space where visitors gather, meet or look for informations. The staircase is the starting point of this proposal. From here, a main oblique circulation constitutes the spinal column of the program. Following the site inclination, every room of the museum sitting on both sides of the staircase, climbs gradually to reach the roofs level.

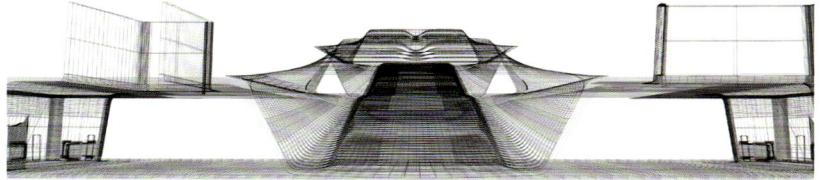

DOUBLE CURVATURE

The staircase as a main structural element is made of the stairs and its cover-ceiling. This mirrored curved becomes a beam with an evolutive section. Steps become ceiling and ceiling becomes steps endlessly to the top. Steps are created in ceilings too, to let natural light enter in this space, while it is used for exhibits purposes in the exhibition departements.

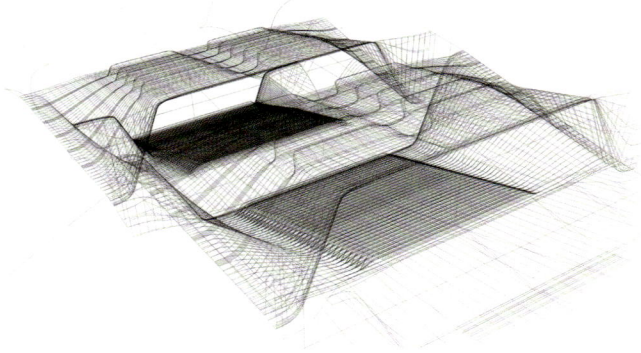

CROSSING OBLIQUE

Every landing in the staircase gives access to a museum department, a stop in this diagonal space. Accessible from outside from the river bank or from the roof, this circulation of big dimensions is a new city living room.

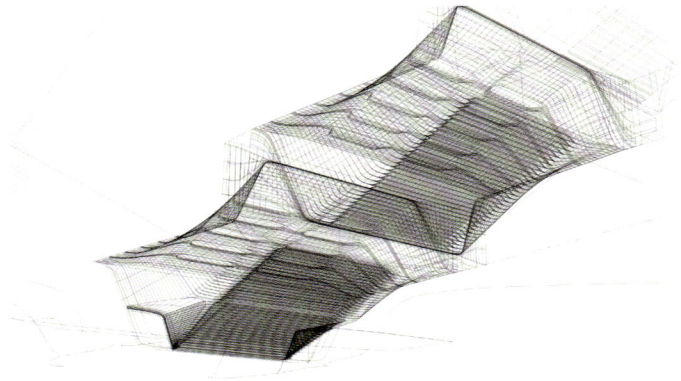

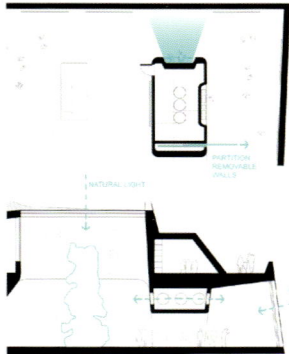

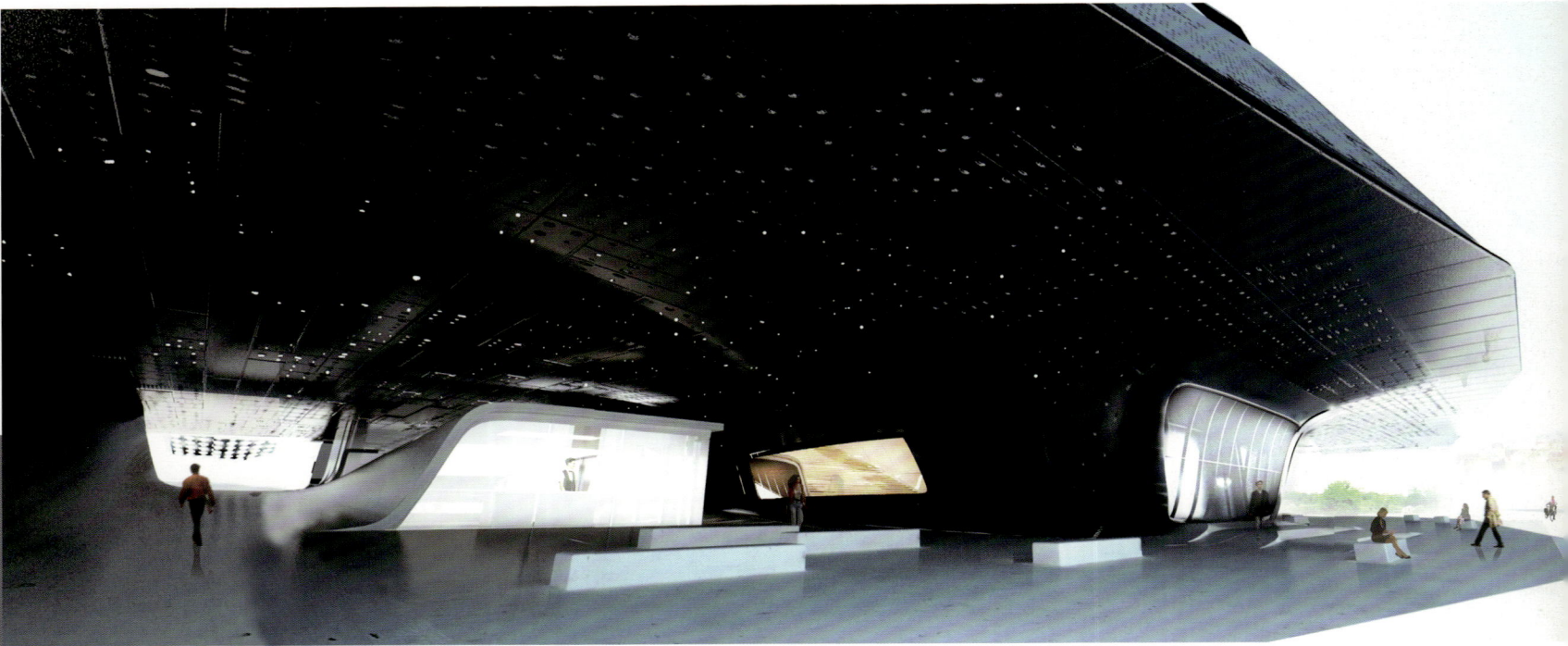

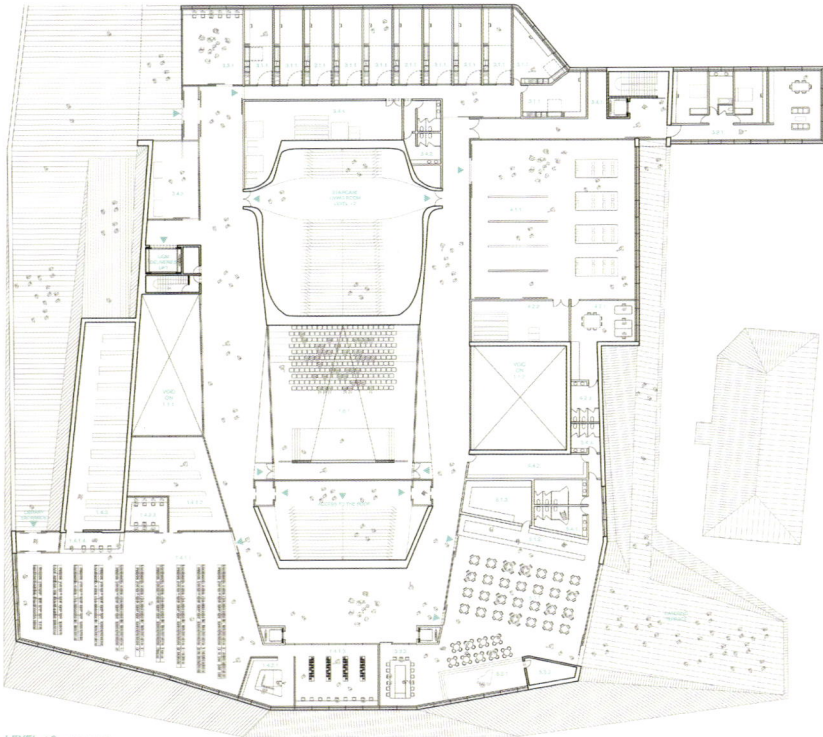

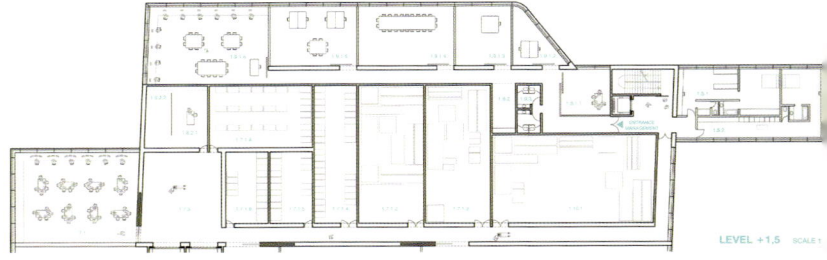

LEVEL +2 SCALE 1:200

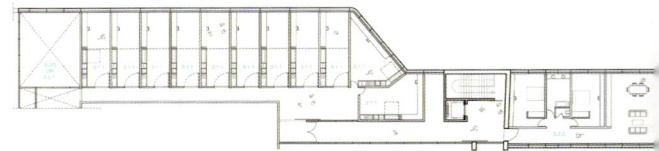

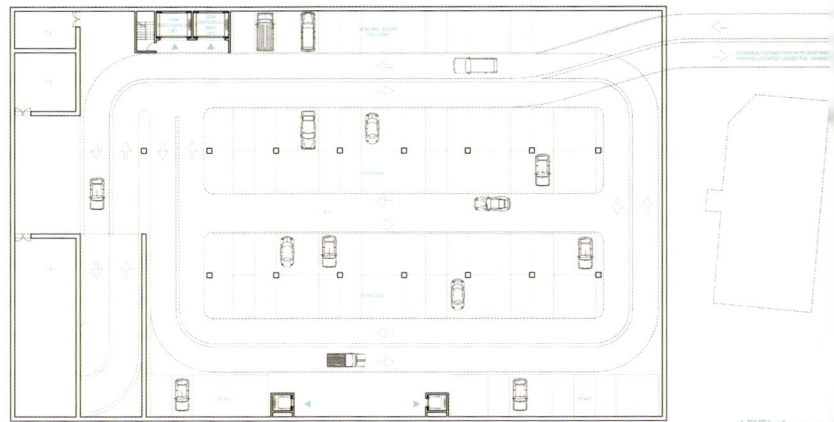

LEVEL +1,5 SCALE 1:

LEVEL +2,5 SCALE 1:

LEVEL -1 SCALE 1:

德拉瓦河是当地人散步、娱乐、放松、玩耍和进行社交活动的场所。这座博物馆就坐落在这条河的河□。这一位置为游人提供了不断变化的360度景观。这座建筑的设计将倾斜的地面表现了出来，将河上游的居民与德拉瓦河畔相连接。人们可以来到被覆盖的底层。这个开放的公共空间可以为博物馆画廊的户外活动提供一个受保护的空间。通过这种方式，这个项目的目的在于加强新老城区在视觉上的联系。新博物馆画廊准备了多种多样的节目，包括不同时间表的活动与开放时间。每一个节目单元都有其从外面进入的独立入口。作为市区潜在开发的一部分，底层里的元素激发了德拉瓦河畔的活力，而第二层内的活动则意图在吸引人们到顶层，那里提供了马里博尔第三立面的景观。

博物馆所处位置的倾斜地面引导了整个建筑的组织与设计。在第一层的博物馆画廊采用不断升高的设计产生了在底层的一个被遮盖的公共空间，并提供了一条从升高的一层通往建筑顶层的通道。因此，这座博物馆可以同时从河边与被抬高的后部位置进入。受到马里博尔沿着德拉瓦河的城区规划的启发，这个项目的体量设计采用了逐渐升高或降低的方法。楼梯的设计跨越并连接了整个博物馆建筑群。这一设计正是整个计划的关键之处，起到了白天的观景房的作用。

马里博尔的特点来自于其屋顶景观。由于这座城市多丘陵的地形，屋顶成为了整个城市的总体印象。这个项目的灵感直接来源于这一印象。这个建筑的外部屋顶、立面及天花板都表现出了细碎的贝壳形状。如同一个潜行者，这个项目通过其与周围环境的直接的视觉和地理的关系融入了城市之中。建筑的外部结构采用了在多个设计中多采用的一种形式。设计中带孔的结构为展览区域提供了对自然光的遮挡及透光的功能。在地面上，设计师也采用了同样的结构，但是不同之处是利用了浮雕的结构。这为建筑表面提供了防滑设计，特别是可以供人通行的顶部。多孔的结构也用在了天花板的设计上来隐藏人造光源与技术设备，如空调和通风系统。这种醒目的结构也成为了整个博物馆的标志化形象。

Golf Dots—Golf Resort at Herning

赫宁高尔夫度假村

Firm: CEBRA
Location: Lovbakkerne, Herning, Denmark
Area: 18,500m²

xury golf hotel situated at Herning. The hotel is 100m high and holds 200 ites, conference centre, wellness centre, restaurants, golf club and sky bar th 360 degrees view of the golf course. The building can be described as olded plate or an "L" consisting of "tower" and base. The functions are ided accordingly with the high-rise block containing the "introverted" and ivate rooms, whereas the base contains the extrovert and more official ilities.

rooms are situated in the high-rise block, where they are placed like arls on a string-only interrupted by the circular cuttings which can either elements of pure experience or they can have a function as bar, meeting om or fitness room. On the top floor is placed a "sky bar" which with its ique view of the whole city offers a different experience.

e base which holds 2 and in some places 3 floors contains the facilities which are open for everybody—irrespective of whether or not you are staying in the hotel overnight. In the ground floor, apart from the absolutely necessary functions as reception and staff facilities, you will find a golf store, a golf club and a bar. On the 1st floor you will find an assembly hall, a wellness area and a sports area, whereas the 3rd floor which because of the inclining roof covers only one third of the base contains a section for conferences and meetings. All these functions are supplemented by locker rooms, kitchen and offices and are linked together by means of a through patio which connects the kiss'n'ride area with the bar and the enormous outdoor terrace. The facilities in the base are consequently all oriented towards a central entrance hall in which you will often find temporary arrangements as for instance exhibitions, receptions and concerts. This can be considered the heart of the house.

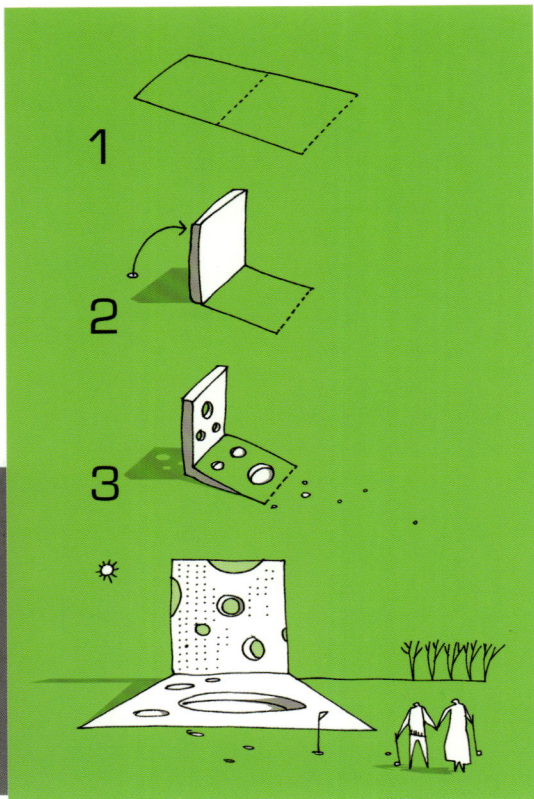

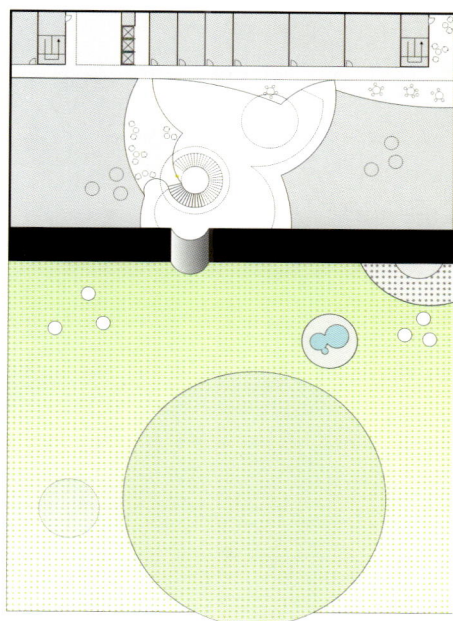

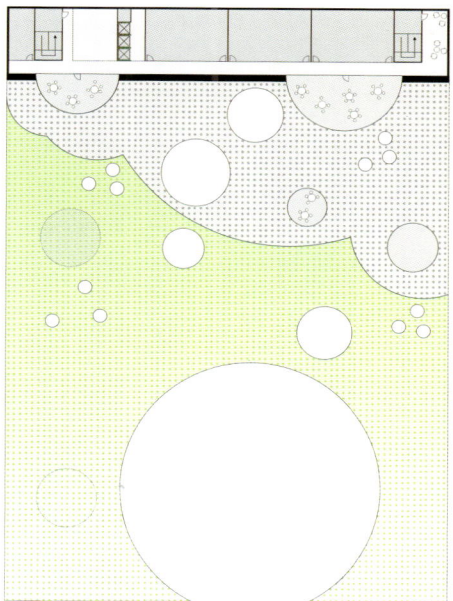

　　本项目位于赫宁，是一个奢华的高尔夫酒店。酒店高100m，包括200个套房、会议中心、健身中心、餐厅、高尔夫球会和可看到高尔夫球场360度全景的空中酒吧。建筑可以被描述成一个折叠的板子或者一个由"塔"和低层组成的L形建筑。功能按照内敛的和有着私人房间的高层大楼和外向的有着更多官方设施的低层。

　　所有的房间都位于高层大楼中，它们就像被穿在绳子上却被圆形的切口分隔开的珍珠般。这些切口有的仅仅是纯粹经验的元素，有的被用做酒吧、会议室或者健身中心。最高层设置有"天空吧"，在那里可以看到整个城市的风景，给客人带来与众不同的体验。

　　低层建筑为二至三层，包括了每个人都可以使用的设施，无论是否是酒店的住客。在一层，除了必不可少的接待处及员工设施，还安排有高尔夫用品商店、高尔夫会所和一个酒吧。在二层，设有一个会议厅、一个健身区和一个运动区，尽管由于倾斜的屋顶只覆盖了三分之一的面积，三层还是容纳了一个会议与集会区。所有这些功能都配有衣帽间、厨房和办公室，并通过天井相互连接起来，同时天井也与停车转乘区和酒吧以及巨大的露天阳台相连通。低层建筑中的设施因此都面向着中央入口大堂，在那里经常会看到临时性的安排，如展览、欢迎宴会和音乐会等。这里可以被看做是整个建筑的中心。

The Iceberg – Isbjerget

冰山

Firm: CEBRA
Location: Aarhus, Denmark
Area: 21,500m²

The Iceberg is a competition first prize project done in collaboration between JDS, SeARCH, Louis Paillard and CEBRA. The building is 21,500m², situated in Aarhus, Denmark – right at the front row harbor – and contains numerous dwelling types as well as smaller commercial facilities at ground level.

Basically the project is a respond to a three dimensional building envelope determined by the municipality. The envelope allowed a building of 21,500m² rising to a maximum height of seven to eight stories. But to allow better views toward the ocean and better daylight conditions, the roofs are pushed up and down to create a mountain like series of buildings. The tops and bottoms of the mountains are constantly shifting so that views between the volumes become possible. This strategy keeps the building at the average height of the allowed seven to eight stories, since the roofs are as much over as under the maximum building height. But it also creates views to future buildings behind the Iceberg site, and this generous feature made it possible to bend the rules and planning regulations.

The varied building shapes are used to create a multitude of different apartment types. At ground level a number of town houses in two leve are integrated into the volume, and obviously the peaks of the buildin contain spectacular pent house apartments also stretching across sever stories. Between these a variety of apartments with different balconie shapes and orientations are found – all to insure an urban environme with a social diversity of people of different ages, incomes and fami relations living together. This supports the radical idea of mixing cond with rental apartments, not only in the same building but around the ve same hallways. One has to imagine the benefits of for instance elder people looking after kids in return for shopping favors or students helpi with the homework or setting up your computer – a community of differe people insuring that the complex is alive around the clock, and that peop who cannot afford to buy a home will have a chance to rent one. Thus t complex becomes a neighborhood instead of just a group of buildings.

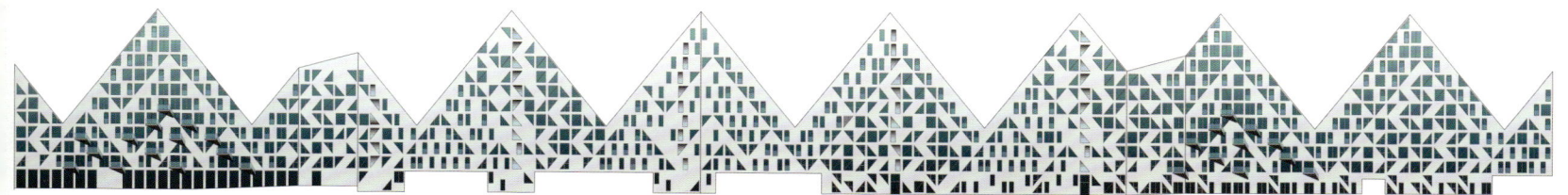

Facade Sketch

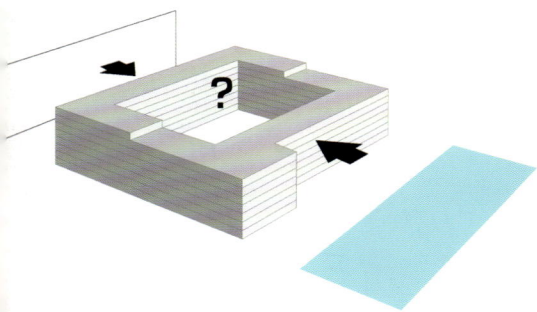

Planning Regulations

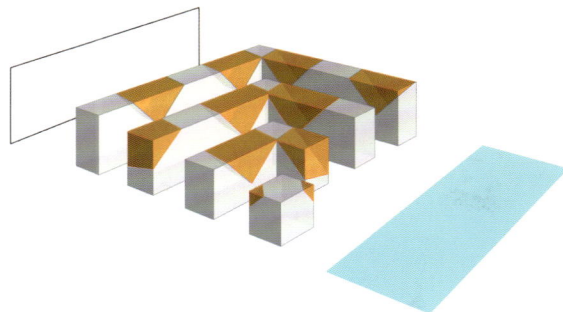

Structure

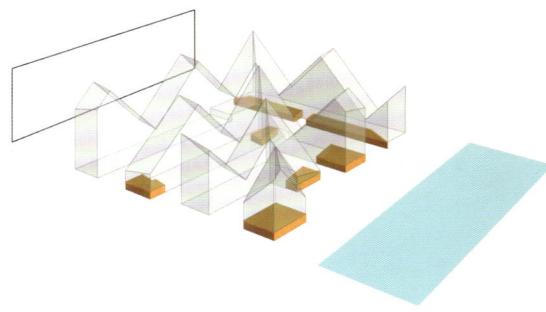

Commercial Space

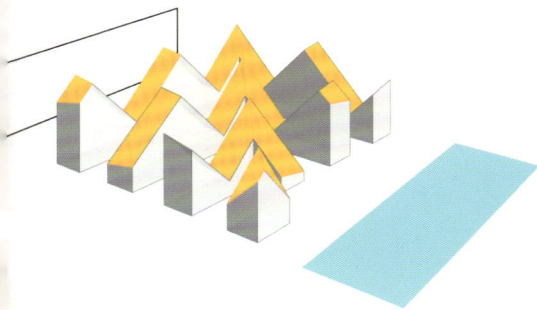

Sun Terraces

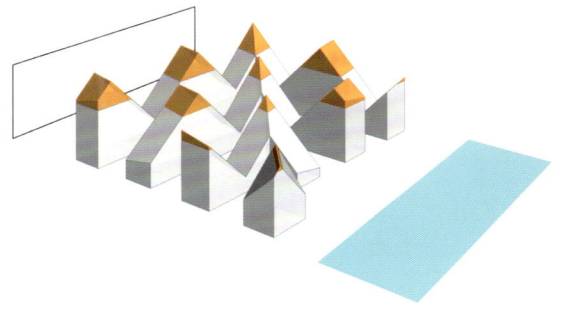

Over Heigh

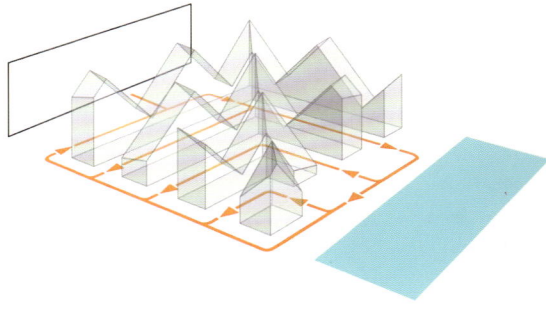

Fire Routes

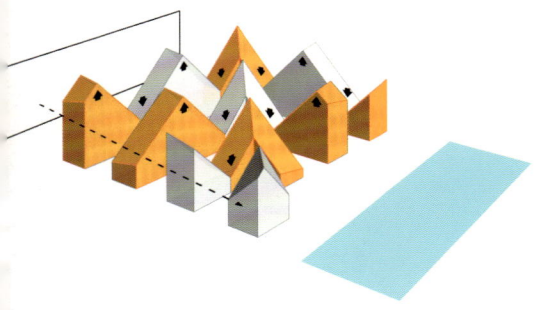

Push Up Down

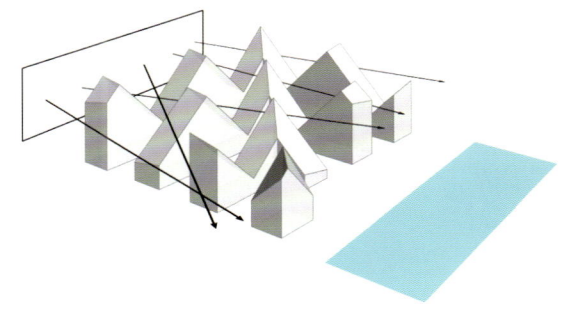

Views Neighbor

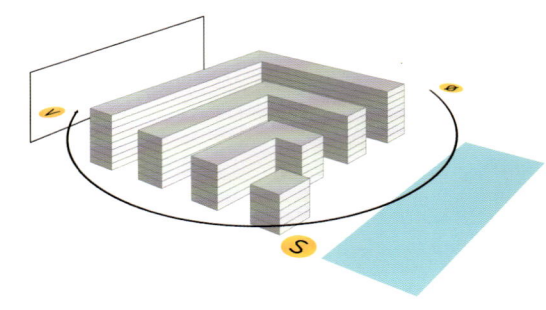

Sun Structure

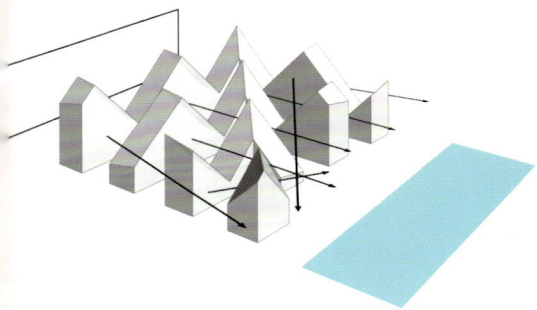

Views

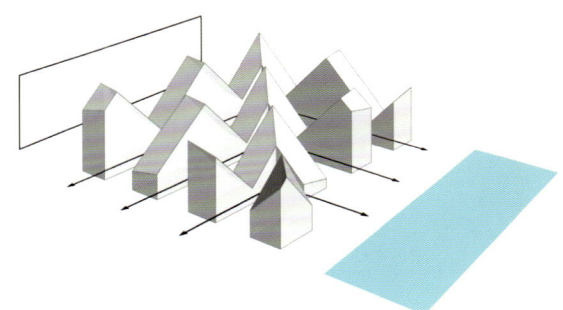

Street Views

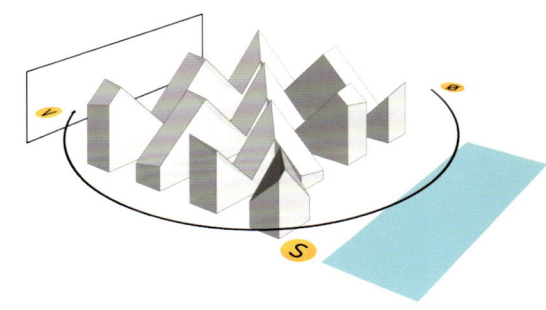

Sun Iceberg

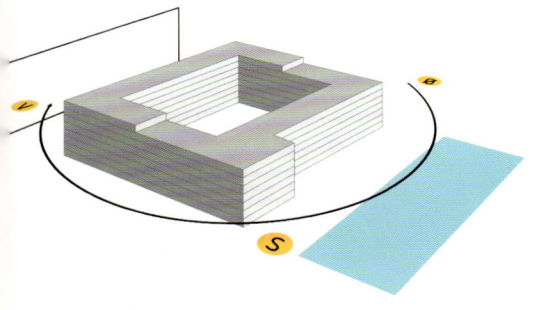

Sun Planning Regulations

 冰山是在一次设计竞赛中的一等奖作品，由JDS、SeARCH、Louis Paillard和CEBRA联合设计。该建筑面积为21500m²，包括多种住宅类型，在一层也有小型的商业设施，它位于丹麦奥尔胡斯，正对着前面的港口。

 基本上，该项目与由市政当局决定的三维立体的建筑围护结构相呼应。这个围护结构为21500m²的建筑留出了最高为七层至八层的高度。但是也留出了更好的海景和更好的日光光照条件。山顶和山脚的不断变化使建筑体量间的景色成为可能。这种策略使建筑保持在可允许的七层至八层的平均高度上，由于屋顶尽可能地保持在最大建筑高度范围内。但这也为冰山项目以后建造的建筑创造了风景，这个慷慨的设计特色使放宽规则和规划条例成为可能。不同的建筑外形用来创造出众多的不同的公寓类型。在一层，两层的联排别墅被穿插到体量中，在建筑的顶层，很明显屋顶阁楼也伸展在几层楼中。在这两种房型之间便是带有不同阳台、外形和方向的公寓，以保证不同年龄、收入和家庭关系的社会多样性的人们可以在这个城市环境中共同生活。这样的设计支持了将产权公寓与租赁公寓结合起来的先锋想法，不仅仅是处于同一栋建筑中，而是围绕着同一个走廊。可以想象到以下的这些益处，例如，老年人帮助照顾儿童，作为回报，喜爱购物的人或者学生可以帮助辅导家庭作业或者安装电脑。一个不同人群组成的社会保证了这个群体全天候的活力，也使那些没有能力买房子的人可以有机会租一间公寓。因此，整个群体成为了邻居而不仅仅是一群建筑群。

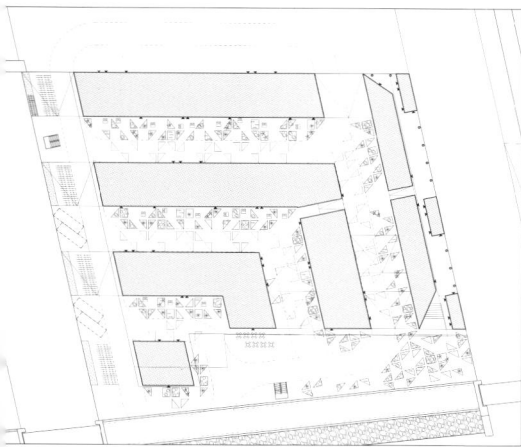

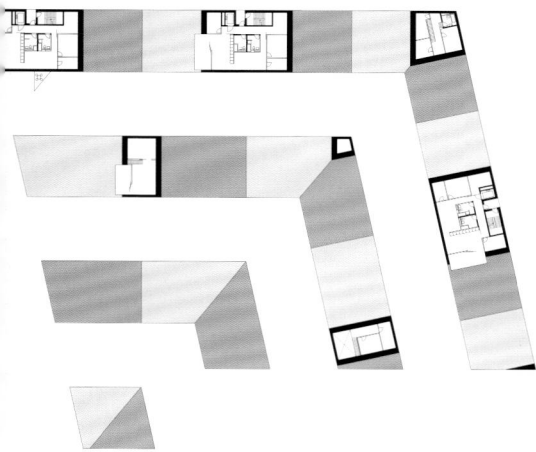
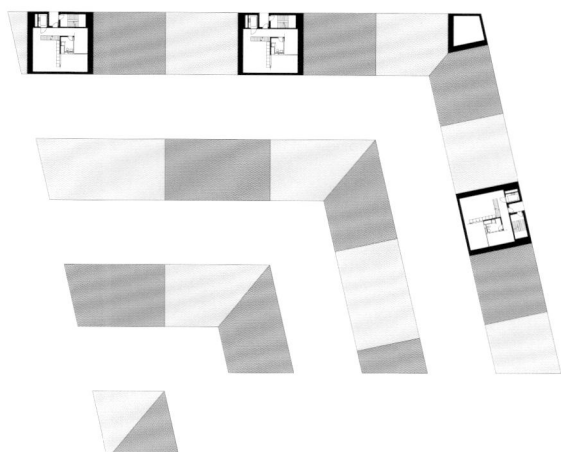
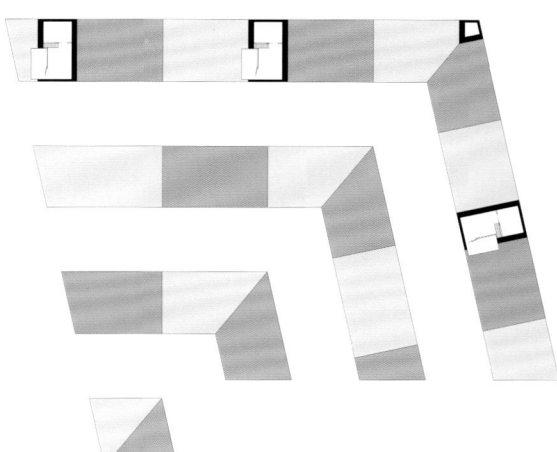

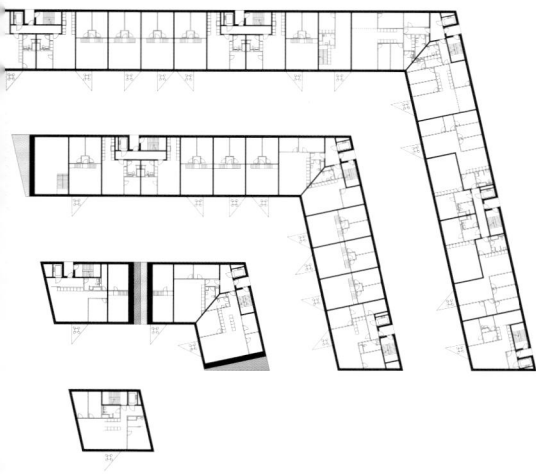
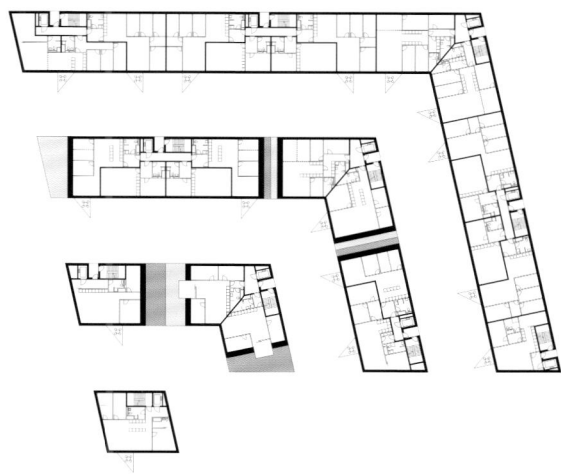
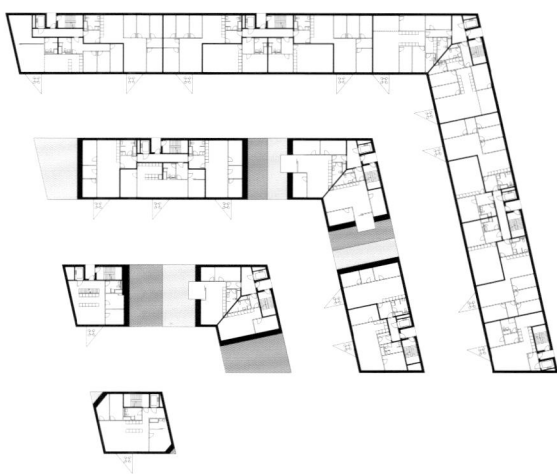

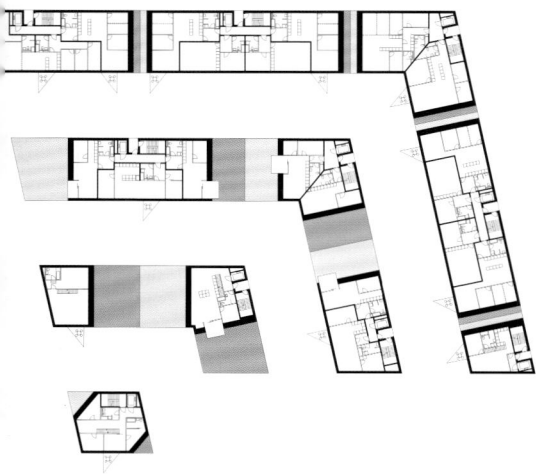
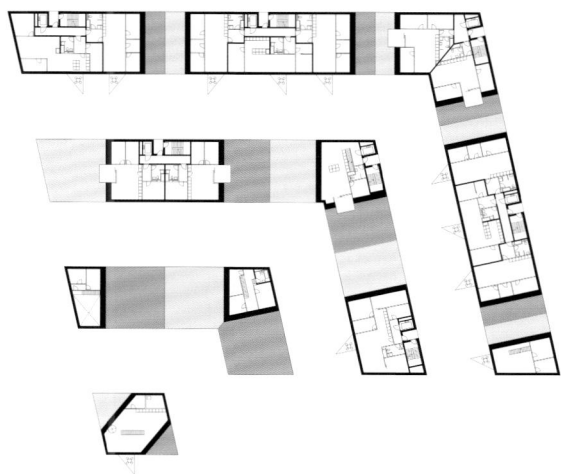

Museum of Polish History/Zerafa Architecture Studio

波兰历史博物馆

Firm: Zerafa Architecture Studio
Location: Warsaw, Poland
Area: 20,000m²
Design Team: Jason Zerafa, Joaquin Boldrini, Pablo Zamorano, Luis Carmona, Katherine Moya
Associate Architect: Gregory T Waugh AIA

For the proposed museum building, the architects have taken a bold step and inverted the typical exhibition space typology. The full exhibit gallery program is conceived as ten, 3-dimensional monolithic objects, a dramatic departure from the gallery defined exclusively as an interior space. Exposing the galleries to be viewed in the round, in space, adds a critical scale dimension to the program, one that is not typically legible in the museum experience. The galleries are given an impressive solidity and monumental scale, and yet many are floating in space—held up on virtual pedestals. It is the complexity of this dual expression of monumentality and lightness which defines the contradiction within the individual museum experience, celebration, judgment and the rejection of sentimentality. The ten gallery objects are juxtaposed to each other both vertically and horizontally to create a 3-dimensional cubic composition within and through a linear circulation volume. The objects are then push-in and pulled-out like drawers to create a series of interior voids and dramatic interstitial spaces.

The six exterior surfaces of the gallery objects, sides, top and underside provide multiple canvases for nonconventional exhibition use. Also, in th configuration, the in between, or accidental spaces can become importa opportunities for surprise temporary exhibits and mobile museum even and program space.

The ten boxes contain the five chronological divisions for the permane exhibit program, the two temporary exhibition spaces, and the exhibitic related educational zones. The free composition of the gallery box does not determine a particular distribution of the permanent gallerie but provides a flexible environment for multiple interpretations of ho the galleries can be allocated and the relationships between them. Th temporary exhibition space is located in two of the upper gallery box clustered together to provide a multi-height venue for temporary exhib when required. The "high" temporary gallery is a 12m tall box which projec out of the eastern facade and through the roof structure to form a dramat tower-like element suspended within the museum.

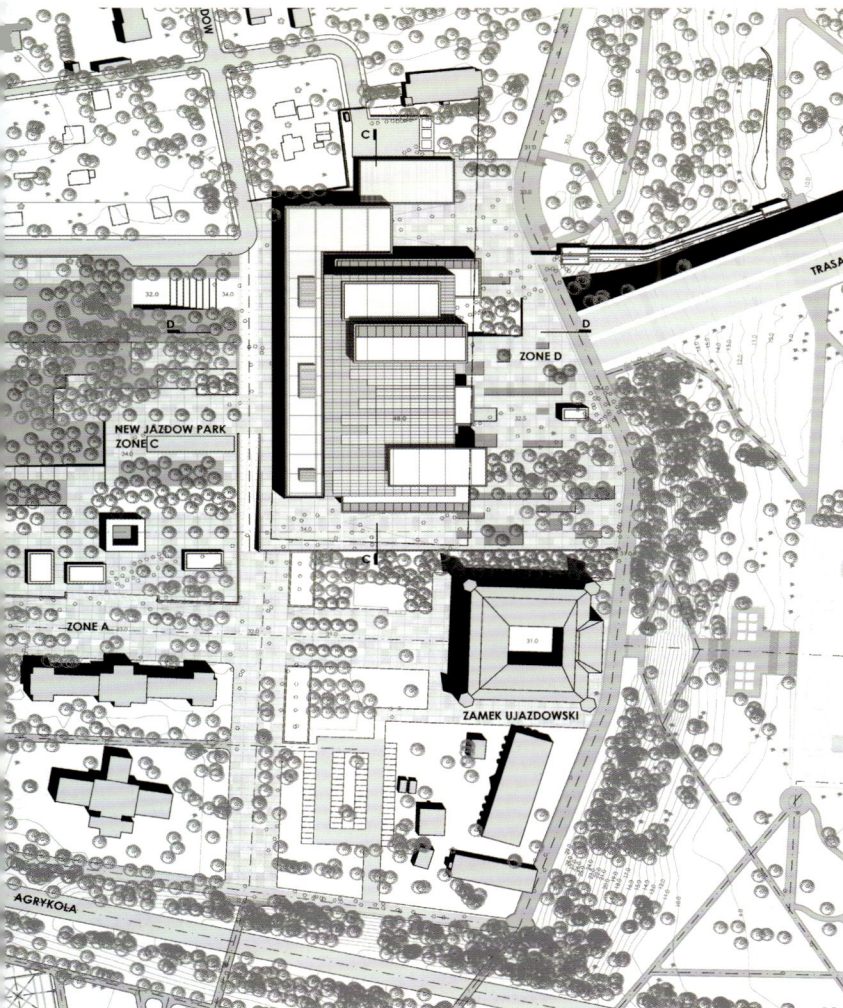

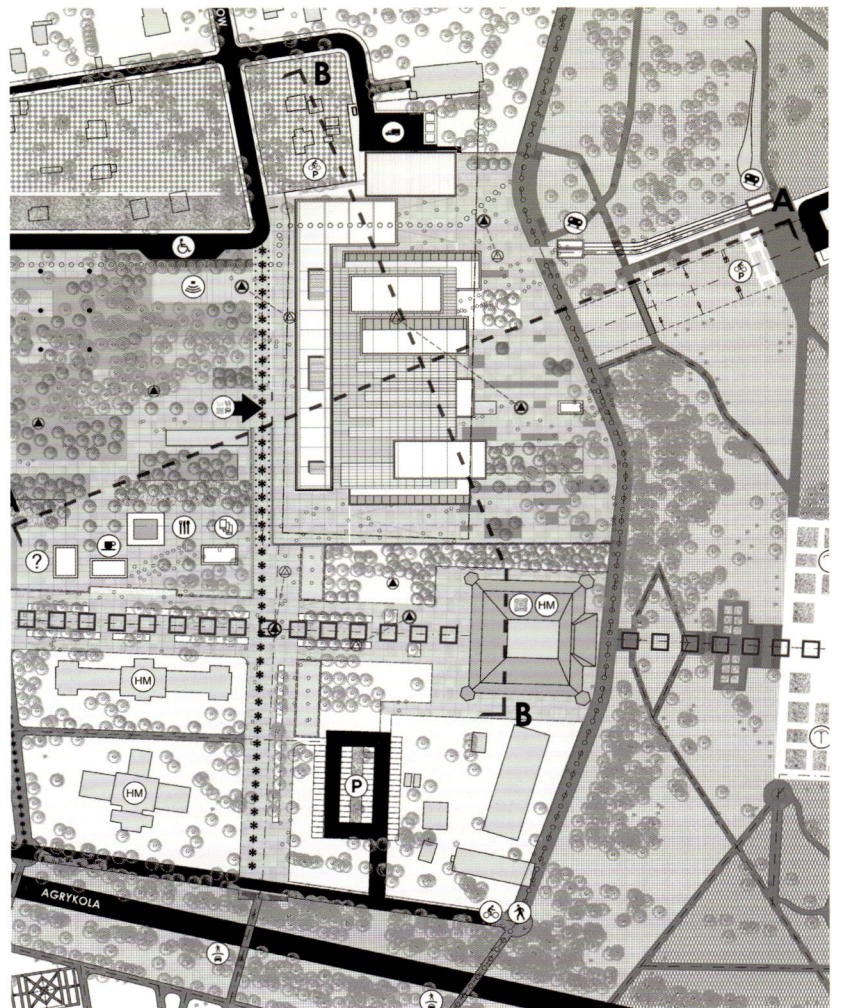

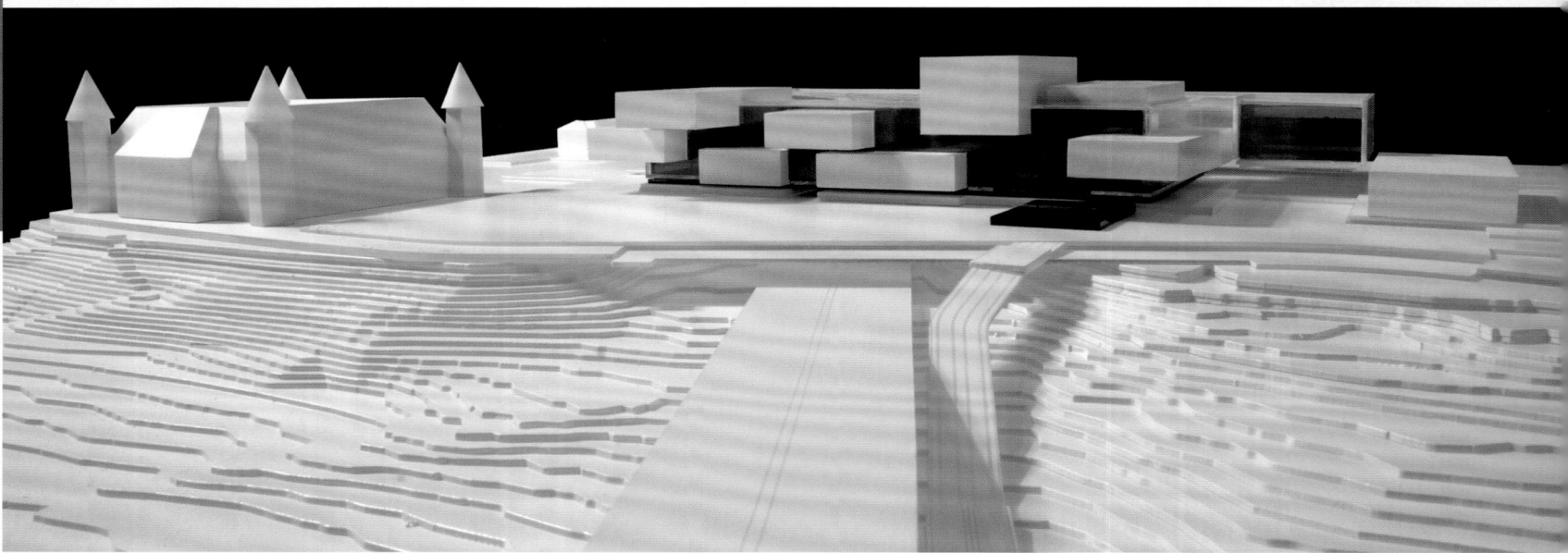

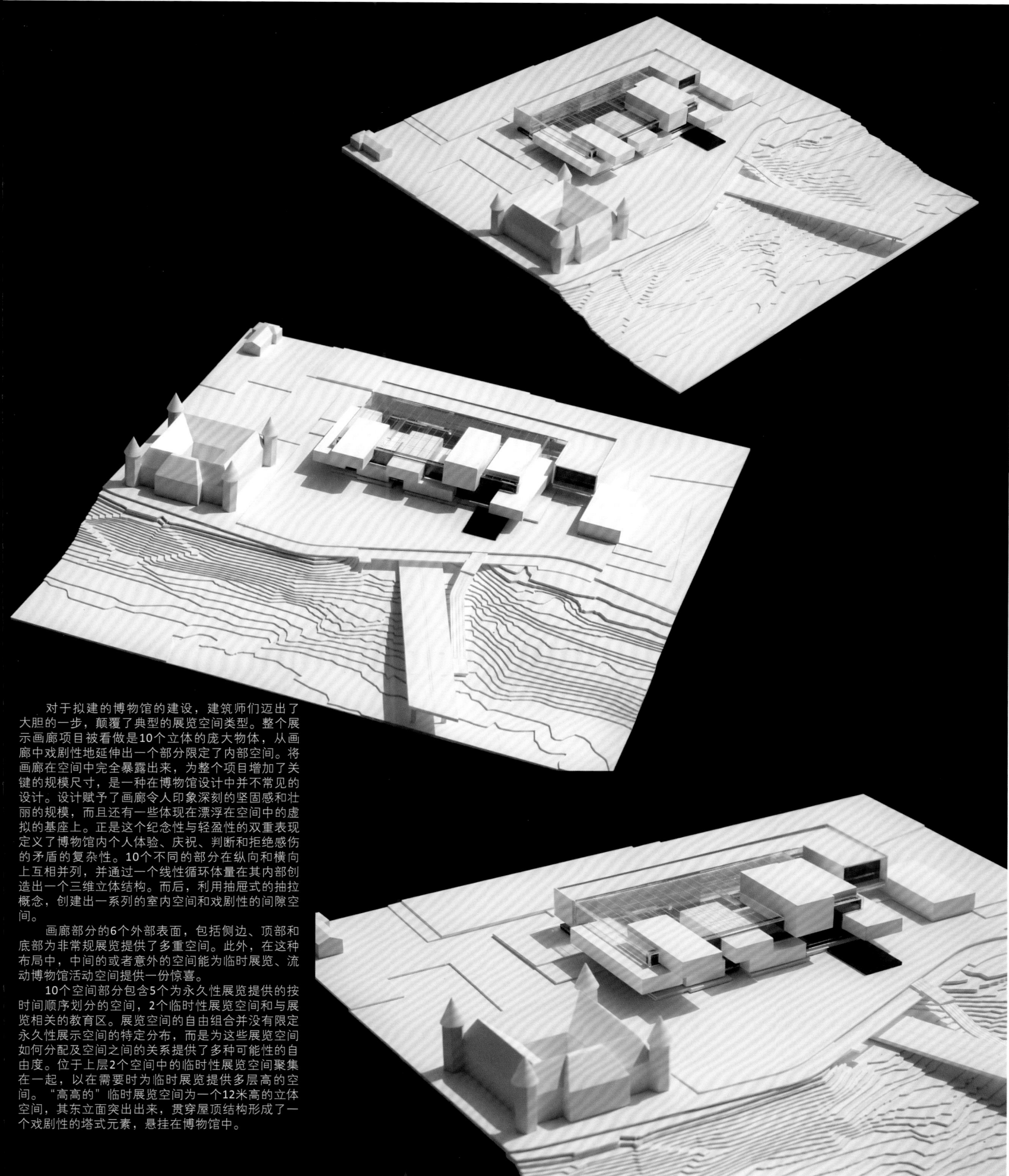

对于拟建的博物馆的建设，建筑师们迈出了大胆的一步，颠覆了典型的展览空间类型。整个展示画廊项目被看做是10个立体的庞大物体，从画廊中戏剧性地延伸出一个部分限定了内部空间。将画廊在空间中完全暴露出来，为整个项目增加了关键的规模尺寸，是一种在博物馆设计中并不常见的设计。设计赋予了画廊令人印象深刻的坚固感和壮丽的规模，而且还有一些体现在漂浮在空间中的虚拟的基座上。正是这个纪念性与轻盈性的双重表现定义了博物馆内个人体验、庆祝、判断和拒绝感伤的矛盾的复杂性。10个不同的部分在纵向和横向上互相并列，并通过一个线性循环体量在其内部创造出一个三维立体结构。而后，利用抽屉式的抽拉概念，创建出一系列的室内空间和戏剧性的间隙空间。

画廊部分的6个外部表面，包括侧边、顶部和底部为非常规展览提供了多重空间。此外，在这种布局中，中间的或者意外的空间能为临时展览、流动博物馆活动空间提供一份惊喜。

10个空间部分包含5个为永久性展览提供的按时间顺序划分的空间，2个临时性展览空间和与展览相关的教育区。展览空间的自由组合并没有限定永久性展示空间的特定分布，而是为这些展览空间如何分配及空间之间的关系提供了多种可能性的自由度。位于上层2个空间中的临时性展览空间聚集在一起，以在需要时为临时展览提供多层高的空间。"高高的"临时展览空间为一个12米高的立体空间，其东立面突出出来，贯穿屋顶结构形成了一个戏剧性的塔式元素，悬挂在博物馆中。

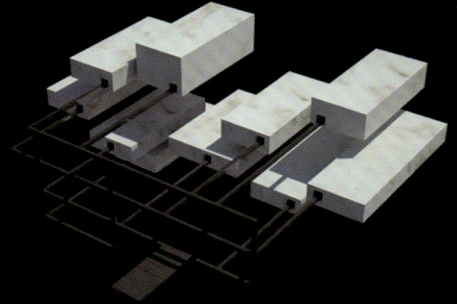

TEMPORARY HIGH GALLERY

GALLERY OF 1914-1915   GALLERY OF MODERN TIMES

GALLERY OF PLR   GALLERY OF MODERN TIMES

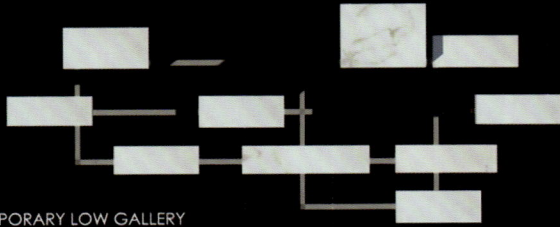

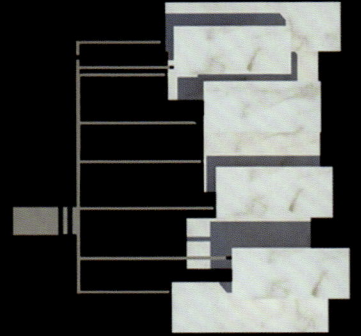

TEMPORARY LOW GALLERY

GALLERY OF 19TH CENTURY   GALLERY OF THE MIDDLE AGES

EXHIBITION EDUCATION ZONE   GALLERY OF 19TH CENTURY

SECTION 01

SECTION 02

MHP

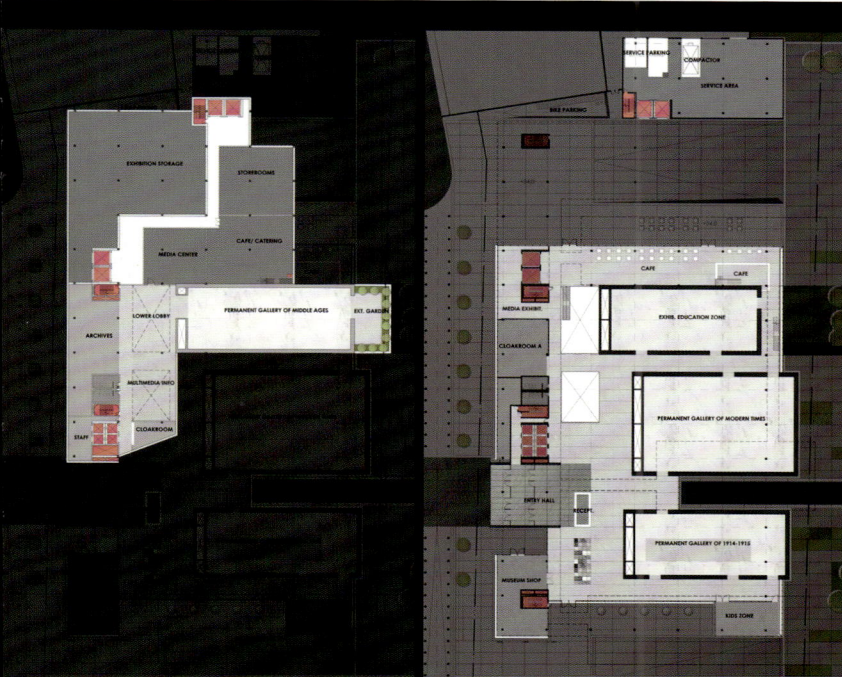

BASEMENT

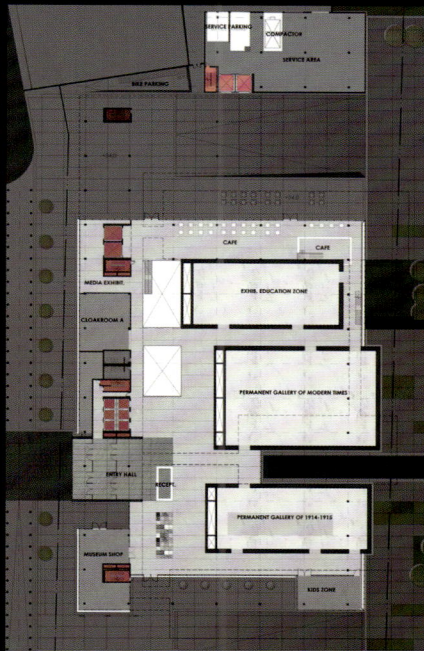

GROUND FLOOR

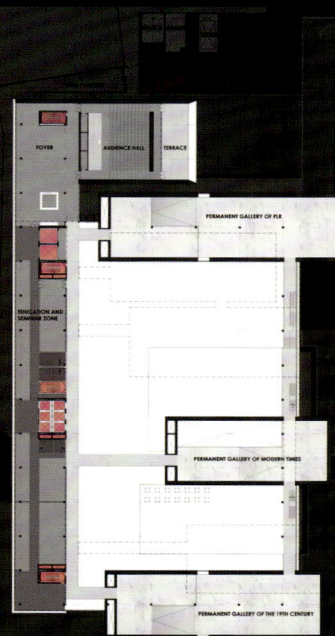

SECOND FLOOR

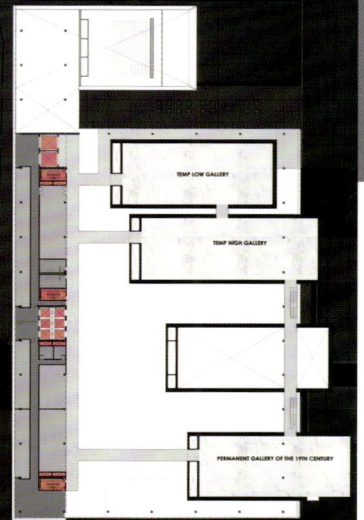

THIRD FLOOR

Quebec Museum

魁北克博物馆

Firm: B.I.G.-Bjarke Ingels Group
Location: Québec City, Canada
Area: 10,000m²
Collaborators: Fugère architectes, Arup Agu, Gustafson Guthrie Nichol
Partner-in-Charge: Bjarke Ingels, Thomas Christoffersen
Contributors: Gabrielle Nadeau, Brian Yang, Daniel Sundlin, Stanley Lung, Alvaro Garcia Mendive, SungMing Lee, Gaëtan Brunet, Malte Kloe

The museum and its new pavilion are enveloped in a context of historical heritage. Listed buildings in a historical park populated by centenary trees constitute a straightjacket of inhibiting concerns. The complexity of the context - the urban requirements, the respect for heritage, the significance of the surroundings, the dispersal of the various museum buildings; the heterogeneity of the historical buildings constitutes a series of paradoxes that the architects had to deal with.

The proposal for the new pavilion has been an exploration of these paradoxes. The latter appears as a simple open exhibition building whose defining feature is that it dives under the George VI street to connect with the with the Charles-Baillargé and Gérard-Morisset pavilions. The blatant gesture creates a natural continuity from Grande-Allée to the existing museum and intuitively brings visitors underneath the park to the museum beyond. Like a Canadian cousin of the tilted tower in Pisa, the new building is familiar to the two existing pavilions in size and volume, but surprising in its submersion to connect to its distant siblings.

The new museum pavilion preserves the perimeter of the former convent, conserving the historical courtyard, while visually opening it up to the surrounding city and landscape. The lifted volume towards Grande-Allée creates an abstract pendant to the church tower, while revealing an inviting public space beneath the galleries. On the other end, the submerged volume minimizes its presence towards the federal territory and the historical context. The submerged volumes open the courtyard towards the park, revealing the church tower from the Parc des Champs-de-Bataille.

The tilted volume encloses the courtyard while opening up to the park. The tilt creates a visual continuity underneath the George VI street. The simple form and choice of materials reflect the Quebecois sensibility, while the gesture of diving under the street transforms the simple stack of galleries into an abstract sculpture. An architecture so well integrated that it stands out.

    魁北克博物馆和它的新馆都被包围在历史文化遗产环境之中。充满历史氛围的公园中排列整齐的建筑之间是一株株百年的古木，这构成了当地人所关心的一切。诸如城市的需求、对遗迹的尊重、周边环境的重要性、各种博物馆分散的布局构成了其周边环境的复杂性；建筑师们也不得不处理历史建筑间的不同而造成的一系列的矛盾。

    他们对于新场馆的设计提案已经成为了对这些矛盾的一种探索。后者则是一座隐藏在乔治六世大街下面并连接了建筑大师查尔斯–贝亚尔热和热拉尔–摩里瑟特的展馆。大胆的外形在格兰林荫大道到现有的博物馆间创造出了连续的自然风景，并在不经意间引导游客从公园下方进入到上层的博物馆中。如同是比萨斜塔的加拿大表亲，新的建筑在规模和体量方面与已有的两座展厅很相似是不足为奇的，但令人惊奇的是建筑还通过下沉部分与它遥远的兄弟建筑连接在一起。

    新博物馆的展厅修建在过去女修道院的旧址上，在保留了具有历史意义的庭院的同时将其与周围的城市与景观融为一体。通向格兰林荫大道的升高的结构不仅形成了连接教堂塔楼的吊桥，还在画廊下方形成了一个吸引人的公共空间。在另一侧，下沉的体量使建筑以最小的方式出现在联邦区域和历史景观中。下沉体量使庭院向公园展开，并使人们可以从战场公园就能看到教堂塔楼。

    倾斜的结构则与庭院相连并向公园展开。倾斜的设计在乔治六世大街下产生了一种视觉的连续感。简单的外形和材料的选择反应了魁北克人的鉴赏力，而下沉到大街下方的设计则将简单的层叠布局的画廊变成了一个抽象的雕塑。整个建筑浑然一体而又不乏特色。

Mercedes Benz Tower, Yerevan

埃里温的梅赛德斯奔驰塔

Architects: Rok Oman, Spela Videcnik, Robert Janez, Janez Martincic, Janja Del Linz, Katja Aljaz, Andrej Gregoric
Firm: OFIS Arhitekti
Location: Yerevan, Armenia
Area: 82,130m²

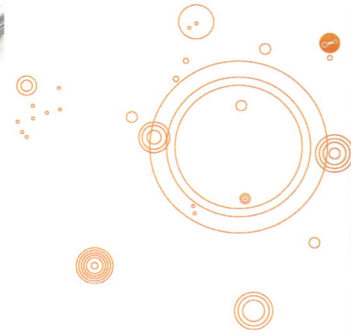

Circle Diagram

The prominent location and dominating position with biblical Mt. Ararat as background "wallpaper" represents a chance for unique rearrangement with its own identity and could become a symbol and landmark of contemporary architecture in the city of Yerevan. The mixture of programs and relations inside the program calls for complex organisation – both inside and outside.

The concept reinstates two terraced cylindrical towers connected in the ground floor embraced with green tent-shape layer.

Higher tower is hotel and business centre (shopping, retail, convention...), lower is occupied with apartment program (apart-business condos, private apartments...).

Public program is connecting the towers, here shopping, exhibition and restaurants are combined with hotel lobby and business entrance.

Parking is mainly in garage under the ground floor, part of the fast parking and taxi/bus drop off are outside at the Plato.

Structural facade skin is covering the terraces of towers, creating openin and full-structural elements.

The facade is a metal mesh that represents landscape element that is risi from the Plato.

With its present also corresponds during the seasons:

In summer the mesh is covered by greenery that is planted at the fence the terraces.

In winter it is white and partly covered with snow. In the garden cafe located with sitting area. With lighted windows glowing from behind it ge festive look.

Circle towers correspond with round structures specific for Yerevan urb tissue.

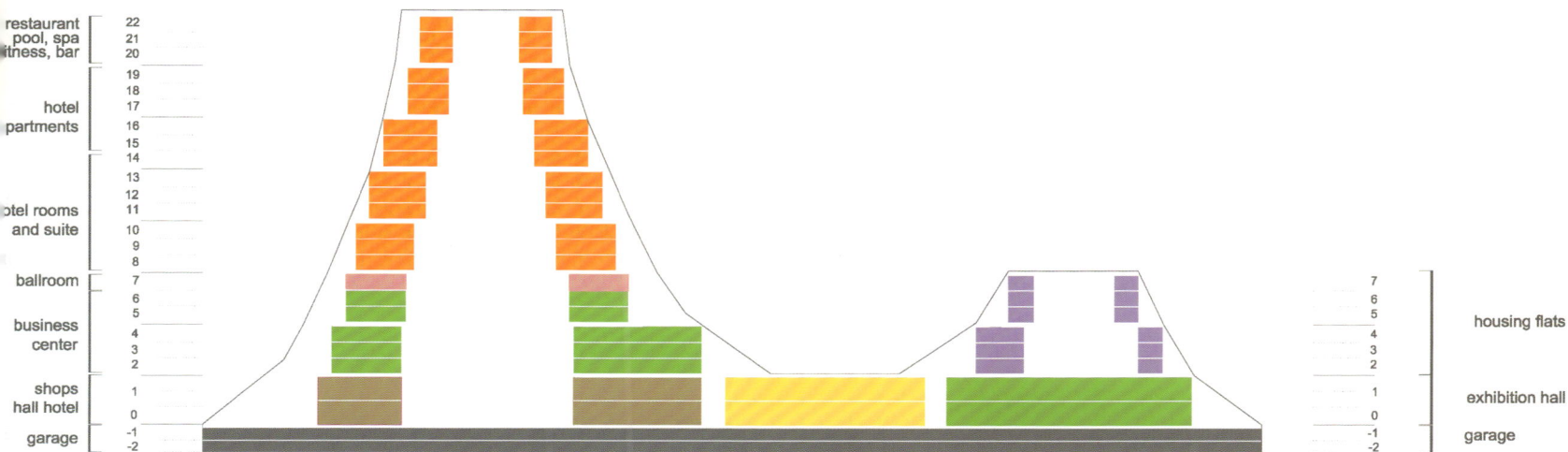

					housing flats
restaurant	22			7	
pool, spa	21			6	
fitness, bar	20			5	
	19			4	
	18			3	
hotel	17			2	
apartments	16			1	
	15			0	
	14				
	13				
	12				
hotel rooms	11				
and suite	10				
	9				
	8				
ballroom	7				
	6				
business	5				
center	4				
	3				
	2				
shops	1			1	exhibition hall
hall hotel	0			0	
garage	-1			-1	garage
	-2			-2	

Site Plan

General plan_connections / access

garage entrance/exit

taxi

A

B

0   10   20 m

South Elevation

0   10   20 m

West Elevation

0   10   20 m

East Elevation

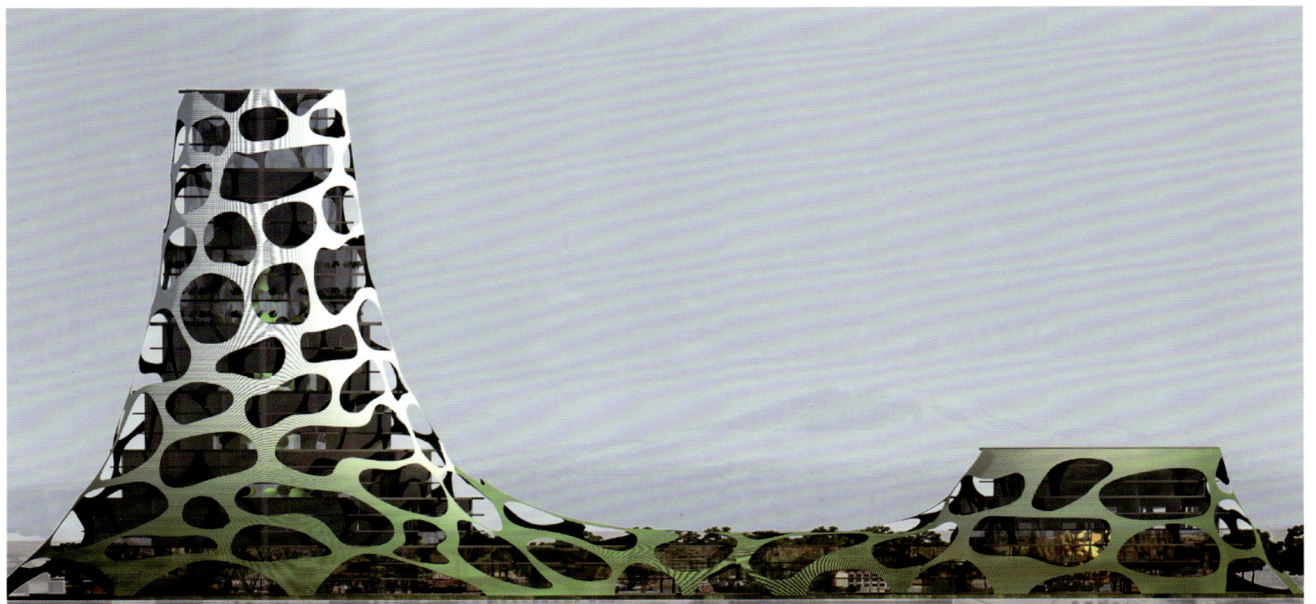
North Facade

0   10   20 m

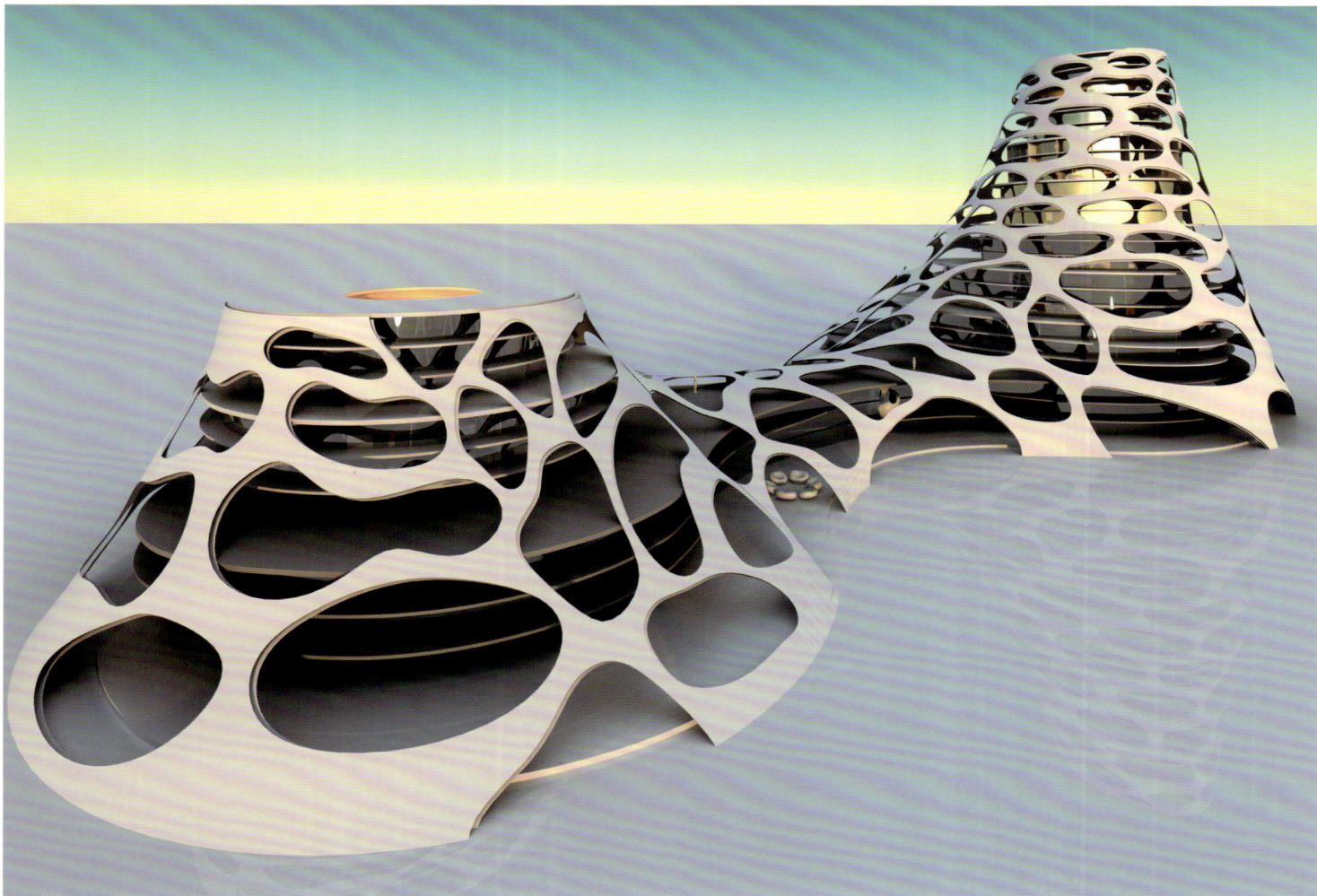

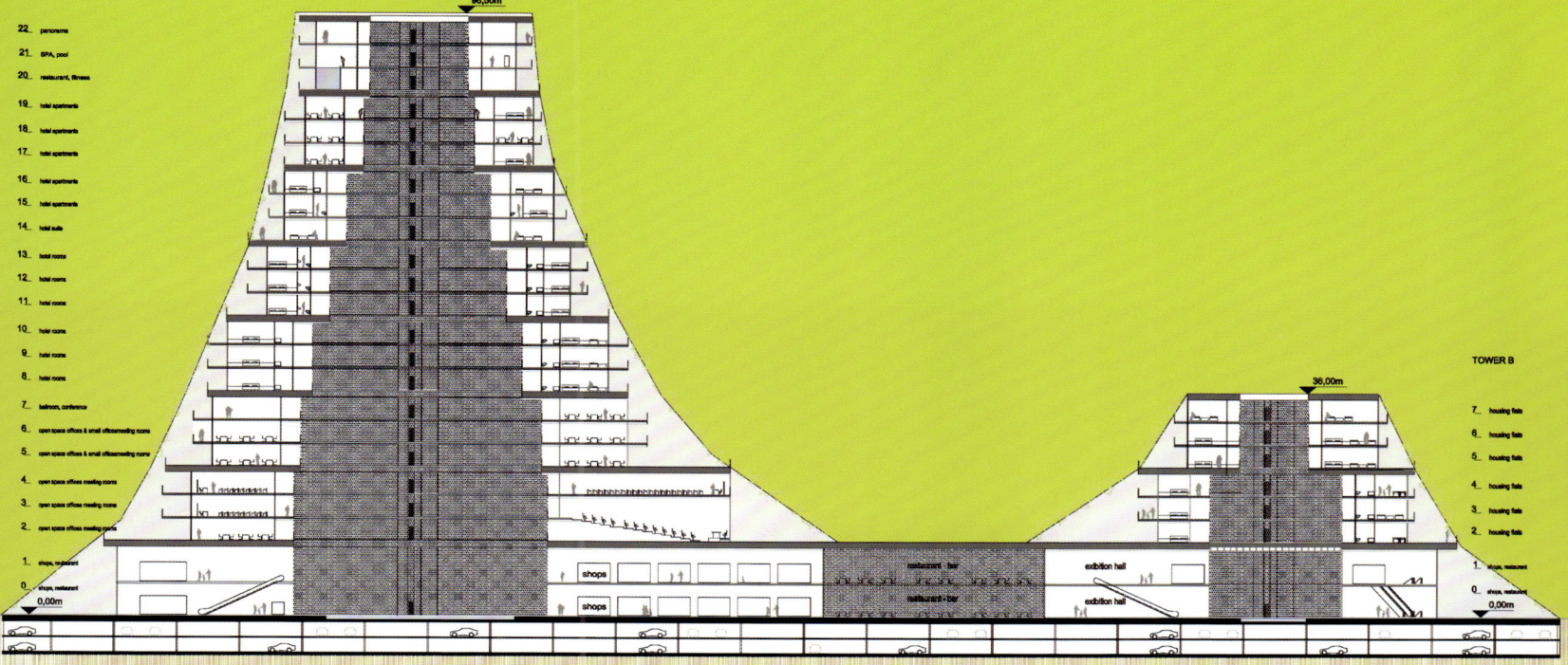

TOWER A

22 panorama
21 SPA, pool
20 restaurant, fitness
19 hotel apartments
18 hotel apartments
17 hotel apartments
16 hotel apartments
15 hotel apartments
14 hotel suite
13 hotel rooms
12 hotel rooms
11 hotel rooms
10 hotel rooms
9 hotel rooms
8 hotel rooms
7 ballroom, conference
6 open space offices & small offices/meeting rooms
5 open space offices & small offices/meeting rooms
4 open space offices meeting rooms
2 open space offices meeting rooms
1 shops, restaurant
0 shops, restaurant

96,50m

TOWER B

36,00m

7 housing flats
6 housing flats
5 housing flats
4 housing flats
3 housing flats
2 housing flats
1 shops, restaurant
0 shops, restaurant
0,00m

313

0   10   20 m

Section

Program Specifications

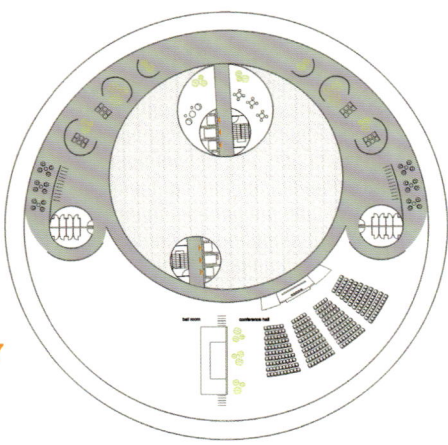

+7

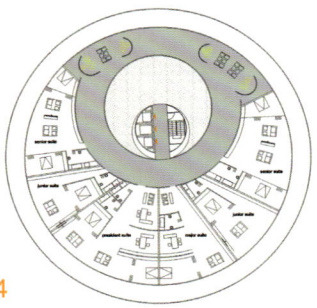

+14

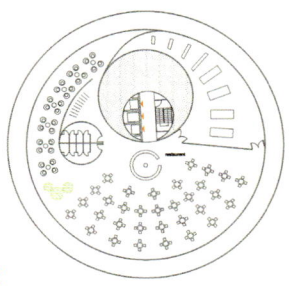

+22

0   10   20 m

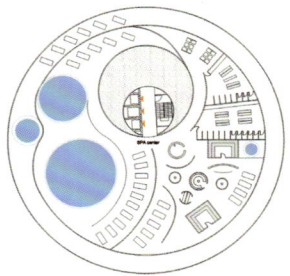

+21

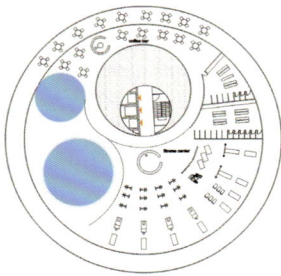

+20

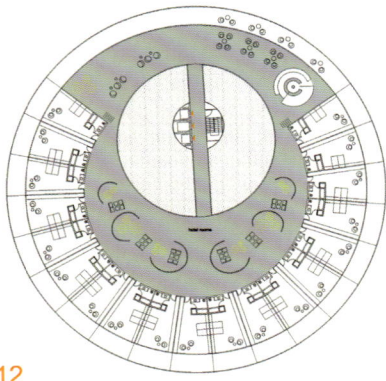

+12

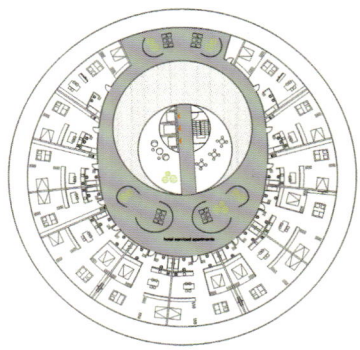

+15

0   10   20 m

Plans

这座建筑将圣经中的亚拉腊山作为了设计的"壁纸"。这一重要的地点与居高临下的位置代表了埃里温城独特的重整的机遇，也成为了这座城市里现代建筑的标志和象征。各种功用的混合以及结构内部的联系在建筑的内部和外部创造出了复杂的形式。

这个设计的概念包括了在底层连接的两个形似绿色帐篷形状的梯形圆柱体塔楼。较高的一座塔楼作为酒店和商业中心（商场、零售店、会议中心等），较矮的一座塔楼作为公寓（商住两用公寓、私人公寓等）。停车场设计在底层之下的地下车库中，另一部分临时停车位和出租/公交的乘降站则设计在大厦的外面。结构立面上覆盖了一层薄膜结构，创造出了开放空间与纯结构感的元素。立面是由金属网状物构成，代表了从当地的地形特点转化出的景观元素。这些金属网状物将随着季节变化而改变：

在夏季，网状结构将会被种植在平台围栏上的绿色植被所覆盖。

在冬季，这些结构将被白雪覆盖而成为白色。在花园里，咖啡厅与休息区相邻。从内部照射出的灯光使得建筑光彩夺目。

圆形的塔楼与埃里温城市圆形的布局风格相呼应。

Science Center Østfold

Østfold科学中心

Architect: Anders Strange, Anders Tyrrestrup and Torben Skovbjerg Larsen
Firm: AART Architects
Location: Grålum, Sarpsborg, Norway
Area: 6,400m²

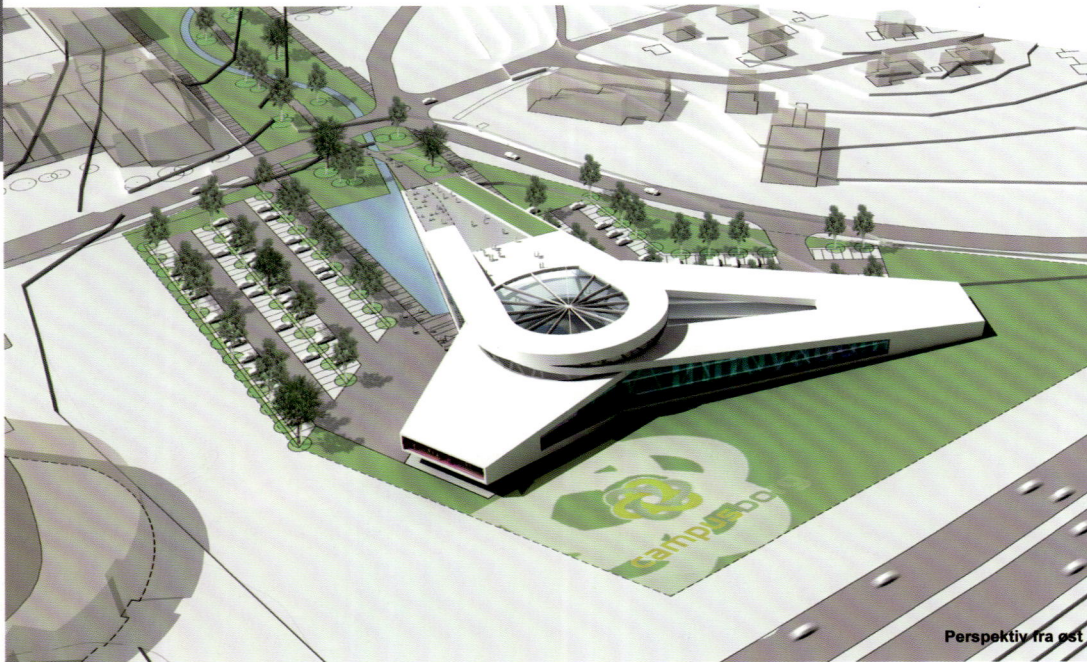

Perspektiv fra øst

Perspektiv fra vest

Science Center Østfold will be the largest and mo spectacular facility for scientific learning and experienc centre for technology and science in Norway. Th centre focuses on Energy, Environment and Healt and as a science and experience centre the audienc includes schools, families and tourists. The centre also going to be a unique venue and event arena.

Science Center Østfold will use art and culture t enhance the communication to the public and ha among other things initiated a collaboration wit Høgskolen in Østfold (Østfold University College) ar Akademi for Scenekunst (Academy of Performing Arts The design of the building is inspired by the cycle cyclic repetitions and spiral forms of nature as well the opportunities offered by technology in terms of th universal power of the circular basic element.

With its interactive exhibitions, working laboratorie classrooms, auditoriums and planetarium the scienc centre creates regional interest in the activities of th centre as well as national interest in the region.

The centre will provide children and youth insight ar knowledge which will help to shape their future choic and their motivation to learn. The communicatic of technology and science embraces not only futu engineers and mathematicians since the technolo of today and tomorrow intervenes and affects parts of society and working life. Basic knowledge ar understanding of science and technology is importa and motivating for all kinds of learning, education ar job.

The science centre is organized with a strong focus the synergy between communication and learning ar thanks to its direct interface between laboratories ar workshops the building supports interactivity and u of models and experiments. In other words the buildi supports and challenges active participation, proble solving and learning — "Fun with meaning"!

The centre which is under construction opens 1 August 2011 and is situated along E6 at the Quali Hotel & Resort in Sarpsborg. It is expected to attra more than 100,000 visitors a year.

Østfold科学中心将成为挪威最大和最引人注目的进行科学技术学习和体验的中心。该中心着重于能源、环境和健康等方面。作为一个科学和体验中心，其参观者包括了学校、家庭和游客。中心也计划成为一处富于特色的场所与大型活动的举办地。

Østfold科学中心将使用艺术与文化来加强与公众之间的联系，除此之外已经与Østfold等学院和表演艺术学院开展合作。

建筑设计受到了圆环、重复的圆环和自然螺旋形的启发，以及循环基础元素的通用能方面所提供的技术支持。

科学中心包含有互动式展览、实验工作室、教室、礼堂和天文馆，在中心的各项活动创造了地区性的利益以及该地区中的国家利益。

科学中心将为儿童和青年人提供智慧和知识，帮助他们塑造他们的未来选择和他们学习的动力。技术和科学的交流不仅仅会塑造出未来的工程师和数学家，现在与未来的科学和技术也将干预和影响社会的方方面面和工作生活。基本知识和对科学技术的理解是非常重要的，对人们各类的学习、教育和工作都有激励的作用。

科学中心强烈关注交流和学习之间的协同作用，得益于它与实验室和工作室的直接联系，建筑支持互动功能，支持使用模型和实验。换句话说，该建筑支持并挑战了积极参与、解决问题和学习——"有意义的乐趣"！

在建的科学中心将于2011年8月1日开放，靠近Sarpsborg优质酒店与度假村的E6区。它将吸引每年超过一百万名参观者。

Perspektiv mot atrium
Innovations Center

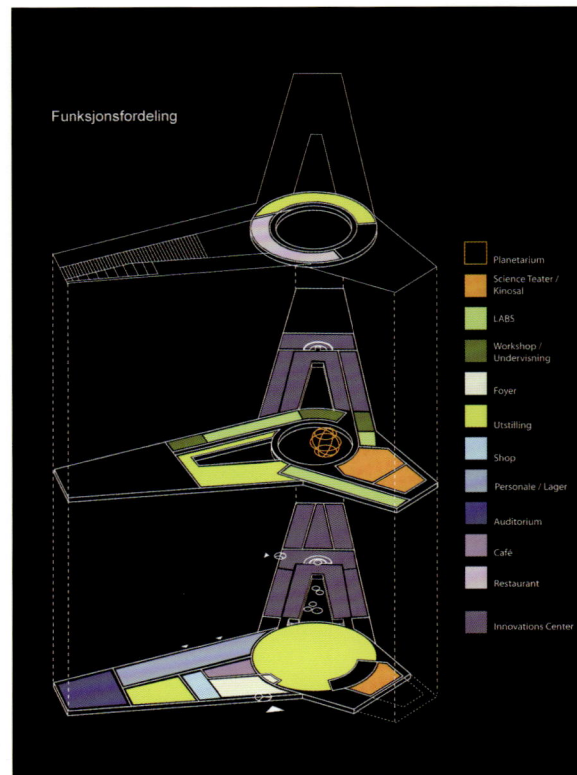

Funksjonsfordeling

Planetarium
Science Teater / Kinosal
LABS
Workshop / Undervisning
Foyer
Utstilling
Shop
Personale / Lager
Auditorium
Café
Restaurant
Innovations Center

House of Culture and Movement

文化活动中心

Firm:MVRDV, ADEPT, SLA landscape, Søren Jensen Engineers, Imitio,
Winnie Ricken, Max Fordham, Ducks Scèno
Area:4,000m²

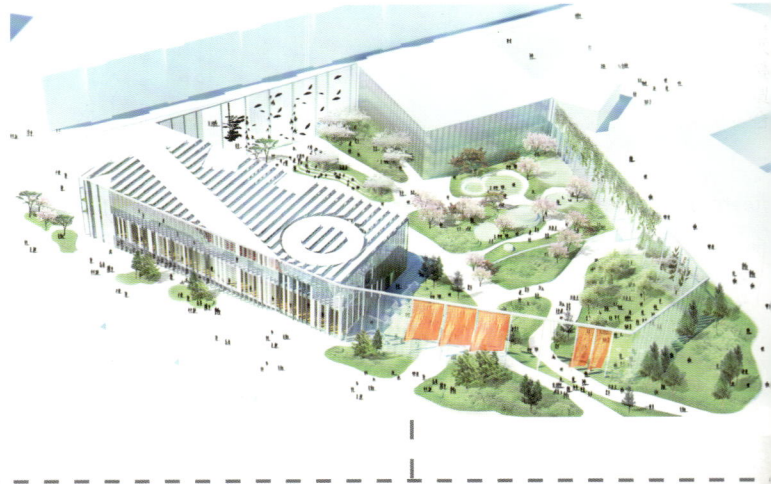

The building is a new urban typology with its mix of community centre, exhibition and performance, playground, park and health centre. The House of Culture and Movement is aimed to engage the population of Frederiksberg in a healthy and active life style.

The main ambition for the House of Culture and Movement is to offer the Flintholm neighborhood a dynamic meeting point for people of all ages taking part in a wide range of activities. Health, culture, leisure and education should smoothly blend together to create a spectacular architectural experience that will become a destination.

The main building, the House of Culture and Movement, or Ku-Be (Kultur- og Bevægelseshus) is a rectangular glass volume containing six stacked ideal programmatic elements. The space in-between can be programmed flexibly as a "play zone" with various activities and main circulation. The stacked elements hold more specific uses: a theatre, a health zone, food zone, a zen area, a study centre and exhibition hall, fitness and activity centre, a wellness centre and an area for the administration. The theatre is flexible and can be used in different stage and audience settings; in addition its large window allows it to be used as an open air theatre where the public stays in the garden. The building is a truly multifunctional public centre which engages its users.

The 3 volumes are wrapped in an "urban curtain" that acts as frame for the garden. It offers great flexibility and can be used for art projects, bicycle parking, water and light installations, performances – and curtains.

The garden is fit for multiple uses acting as an activated, spectacular public space for the area. The landscaping follows the themes of the interior of the House of Culture and Movement with a performance area, health and activity zones, a quiet zone, connecting zones and an empty zone reserved for House of Culture and Movement 2.

The project will be phased. The House of Movement, the garden and the urban curtain will be first realizations. In later stages a commercial building and a second House of Movement will be added. Climate and energy technology is based on reliable technologies such as solar panels, natural ventilation and underground hot and cool storage resulting in a highly efficient low energy building.

该建筑是一种新的城市类型学，将社区活动中心、展览与表演、操场、公园与健身中融为一体。文化与活动中心的目的在于吸引菲特烈堡的人们参与到健康、活跃的生活方式中。

文化与活动中心的主要追求在于为Flintholm地区的各年龄层次的人们提供一个充满活力的集合点，在这里可以参与范围广泛的活动。

文化与活动中心的主要建筑，或者称之为Ku-Be（丹麦语Kultur-og Bevægelseshus的缩写，意为文化与活动，译者按）是一个矩形的玻璃体量，包含有六个堆叠起来的理想的方元素。中间的空间可以灵活规划，成为多种活动"游戏区"同时也是主要的循环动线。

叠起来的元素起到了更多的特殊作用：剧场、健康园地、食品区、禅宗区、学习中心和览大厅、健身活动中心、康体中心和行政区。剧场是一个灵活的区域，可以适应不同的台及观众设置；除此之外，这里的大窗户使这里可以被作为露天剧场，公众在外面的花园中就可以欣赏。这座建筑是真正的多功能公共中心，吸引各种不同的人群。

三个建筑体量犹如城市的屏障般，充当了花园的框架。它有着极大的灵活性，可以举办艺术项目活动、停放自行车、安置水及灯光装置、表演活动，并且也为这些提供了屏障。

花园适合于多种用途，可以视作这个地区有活性的、令人惊叹的公共空间。景观的设计依据文化与活动中心的室内设计主题，包括表演区、健身活动区、安静区、连接区和为文化与活动中心二期保留的一处空地。

该项目将会分阶段实施建造。文化活动中心、花园和城市屏障将会首批建造。在接下来的阶段中，一座商业建筑、文化活动中心二期也将逐渐建造。气候与能源技术是建立在可信的技术基础上的，如太阳能电池板、自然通风和地下冷热储藏等，使这座建筑成为高效率低能耗的建筑。

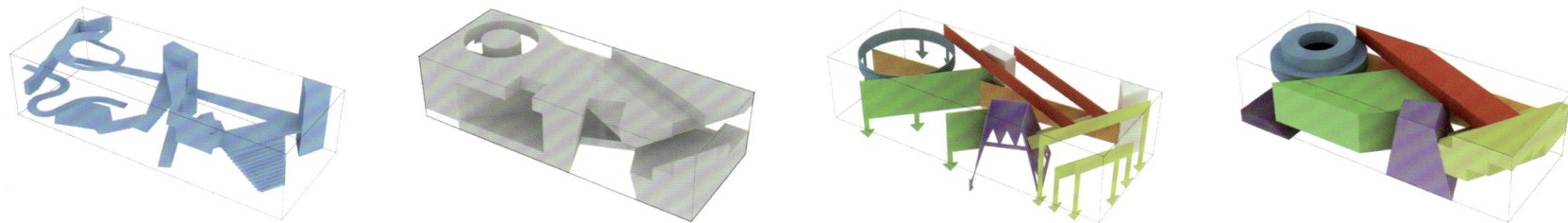

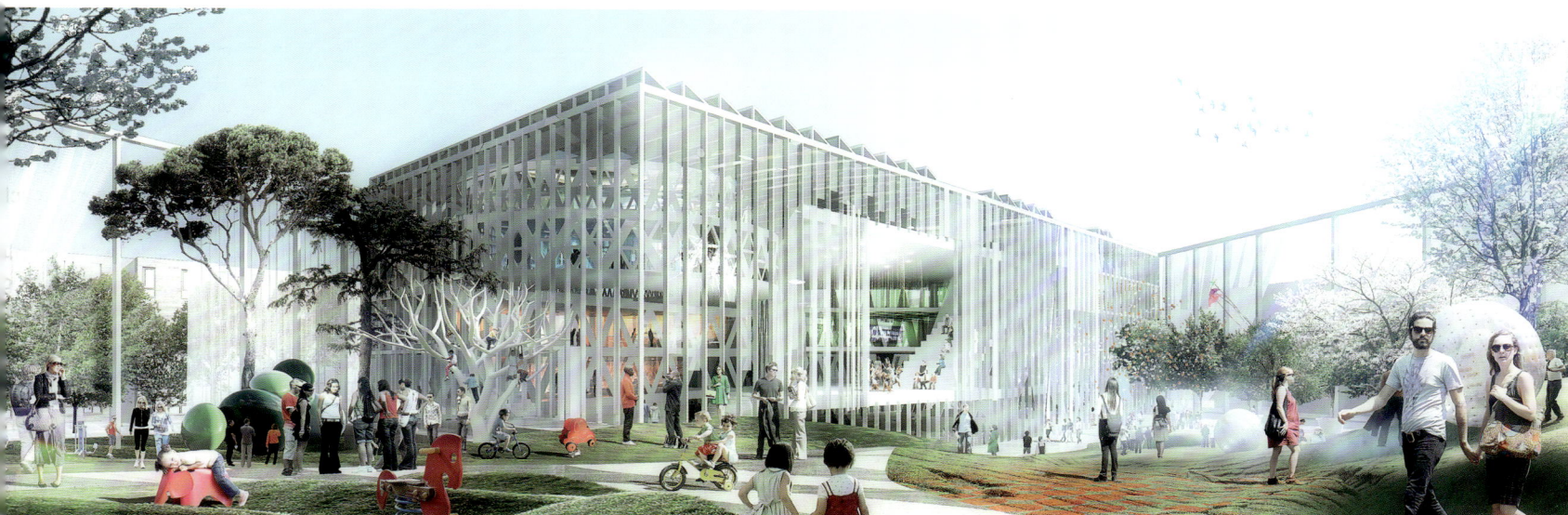

Dalarna Media Arena

达拉纳媒体剧场

SWEDISH MEDIA LIBRARY COMPETITION
Firm: ADEPT, Sou Fujimoto Architects, in collaboration with Rambøll, Topotek1 and Bosch & Fjord
Area: 3,000m²

Dalarna Media Arena matches a new library culture, staging a wealth of opportunities for events and inspiration. The library and plaza at Högskolan Dalarna will be the city's third space – a dynamic meeting point with activities for students, employees and visitors.

The library is organized as a "spiral of knowledge". The sloping terrain continues in a ramp through the building. Wrapping itself the ramp creates a spiral-shaped space— the heart of the building for information seeking and easy orientation. This organization of program creates various learning environments where students can take part in the vibrant life of the library as well as retreat into various study niches. The different sound levels and activities create a diverse and eventful library.

The library has its own spatial character in which library and multimedia functions unite and create synergy with the existing university. Wooden facades integrate the building in the surroundings, refecting the local tradition of using wood as construction material.

Dalarna Media Plaza is created in addition to the library — together they form a new landmark for Högskolan Dalarna. Through a simple reorganizatic the car park becomes a new surface with "islands" of activities. Th programming of the Plaza is fexible and will be developed in the user proces The Plaza, which is a mix of recreational functions and furniture, serves arrival area

and hang-out space for users of Högskolan, the library and visitors in th area.

This project will help to attract people in the neighborhood, there anchoring the library in the local community. In addition the ambition is project strengthen the collaboration between regional and internation educational and research institutions.

With its public functions and activities Dalarna Media Library will becon a dynamo in the area and an attraction for both local inhabitants ar businesses. Activities in the café, exhibitions, lectures, continuing educatio etc. will provide fertile ground for international exchange and interdisciplina collaboration.

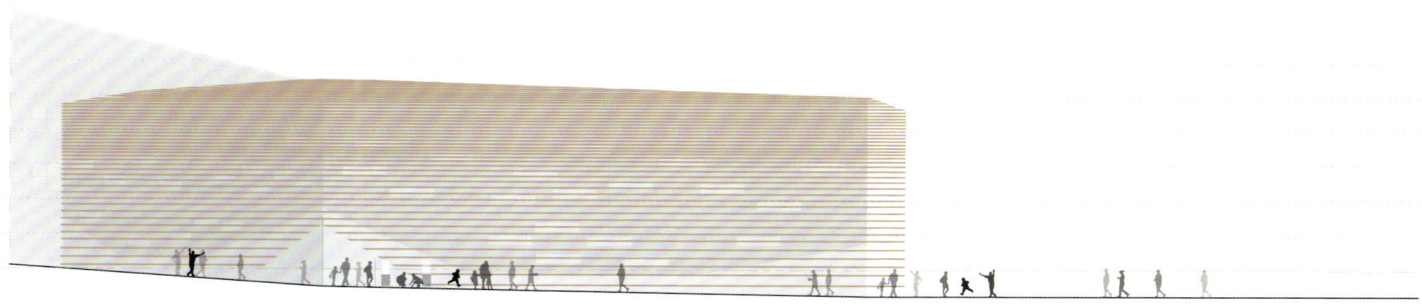

▼ GL +12.00
▼ GL +10.00
▼ GL +8.00
▼ GL +6.00
▼ GL +4.00
▼ GL +2.00
▼ GL 0.00

▼ GL -3.00

SOUTH ELEVATION

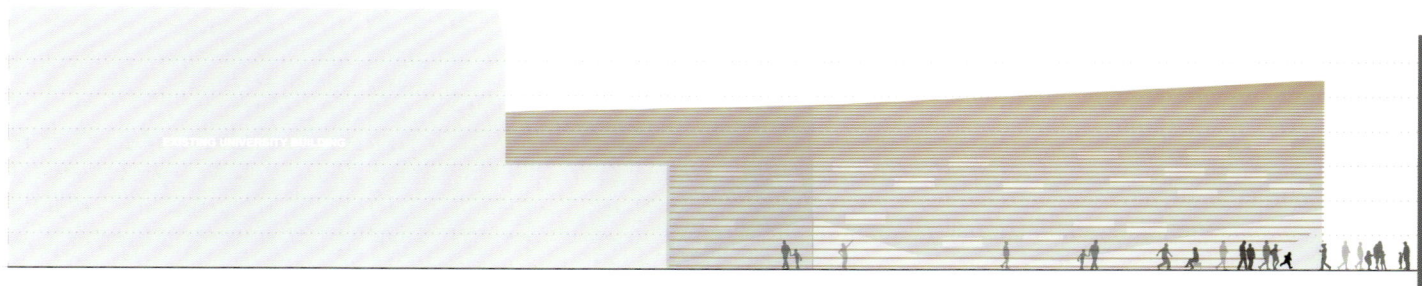

▼ GL +12.00
▼ GL +10.00
▼ GL +8.00
▼ GL +6.00
▼ GL +4.00
▼ GL +2.00
▼ GL 0.00

▼ GL -3.00

WEST ELEVATION

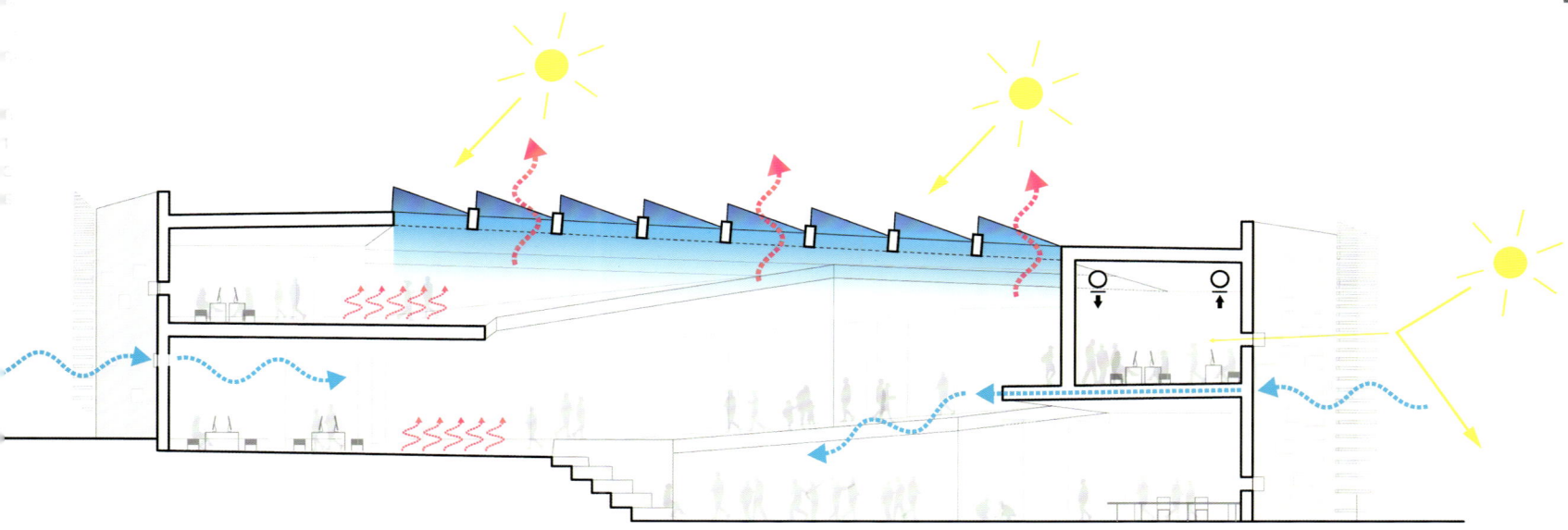

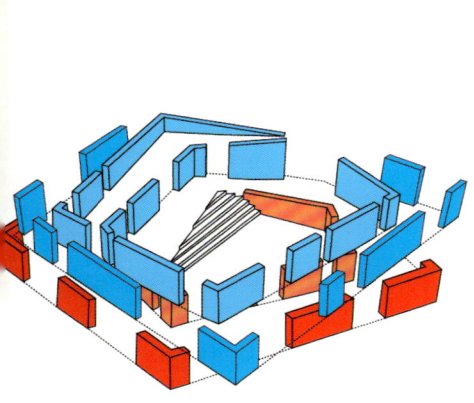

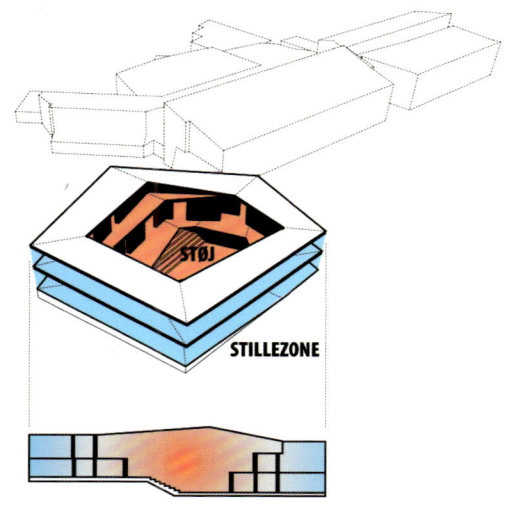

STØJ

STILLEZONE

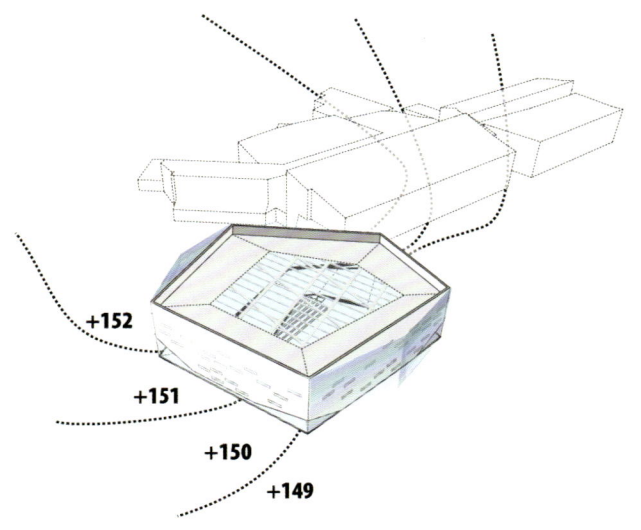

+152

+151

+150

+149

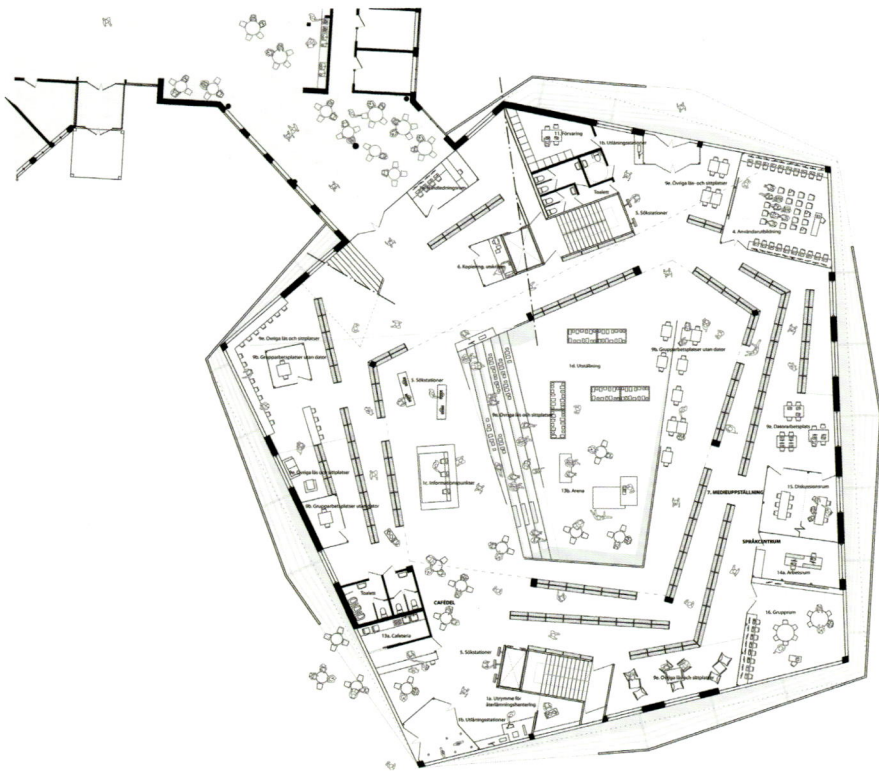

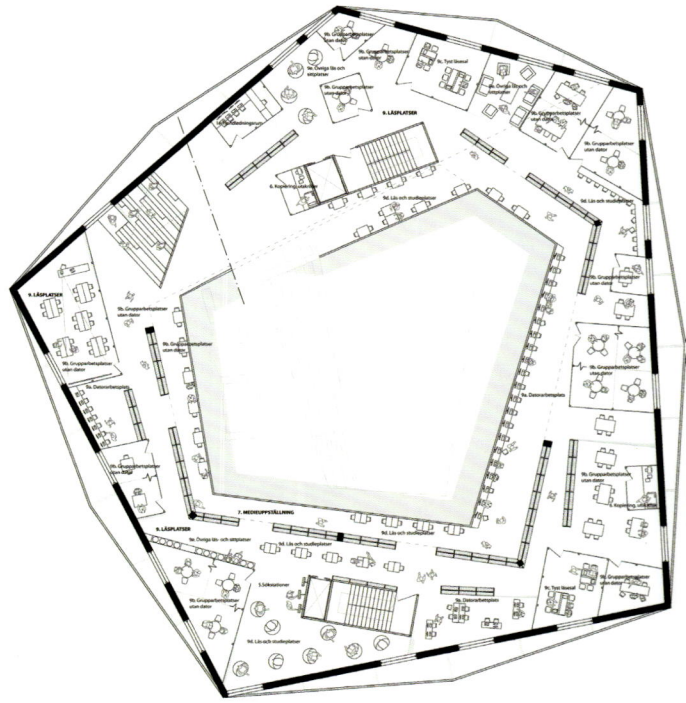

达拉纳媒体剧场与新的图书馆文化相吻合，有丰富的机会举行活动及带给人们灵感。达拉纳学院的图书馆与广场将会成为城市的第三个空间——一个充满活力的聚会点，学员工和参观者都可以举行各式活动。

图书馆被组织成"知识的螺旋体"。倾斜的地形沿着坡道一直延伸到建筑中。坡道将包裹起来，创造出一个螺旋形的空间，这也是搜寻资讯和轻松定向的建筑中心。这样的安排形成了不同的学习环境，在这里学生们可以参与到图书馆充满活力的生活中，也可入到不同的学习小空间中。不同的声级和活动产生了多样化和多种类的图书馆。

图书馆拥有它自己的空间性格，图书馆和多媒体功能单元同现有的学院一起形成协同。木质立面将建筑与周围环境融为一体，反应出当地使用木材作为建筑材料的传统。

除了图书馆之外，建筑师们还设计了达拉纳媒体剧场，它们一起成为达拉纳学院的新地标。经过简单的重新规划后，停车场的地面变成了新的活动"岛屿"。广场的设计非常灵活，是按照使用者的方法来开发的。广场是娱乐功能与设施的混合体，作为学院的使用者、图书馆和这个区域的参观者的下客区和闲逛空间。

这个项目将会有助于吸引周围的人，因此使图书馆扎根于地方社会中。除此之外，项目的追求还在于加强地区与国际教育及研究机构间的合作关系。

因为它的公共功能和热闹活动，达拉纳媒体图书馆将会成为这个地区的主要核心，吸引着当地的居民和商业机构。咖啡馆中的活动、展览、讲座、继续教育等将为国际交换及跨学科协作提供一片沃土。

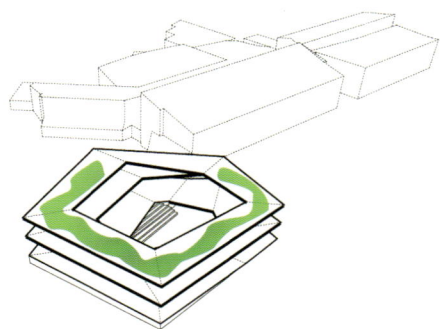

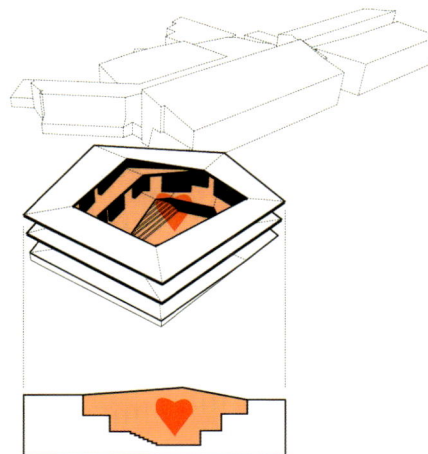

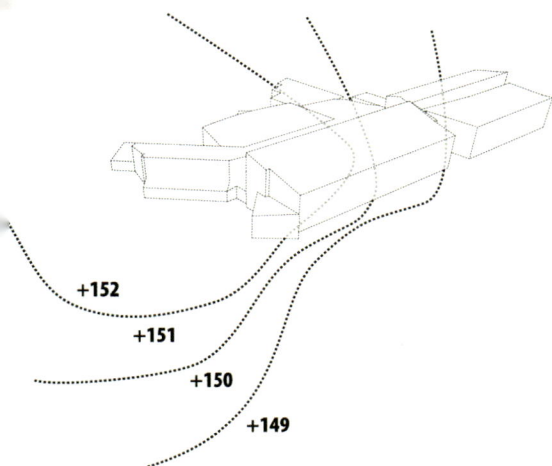

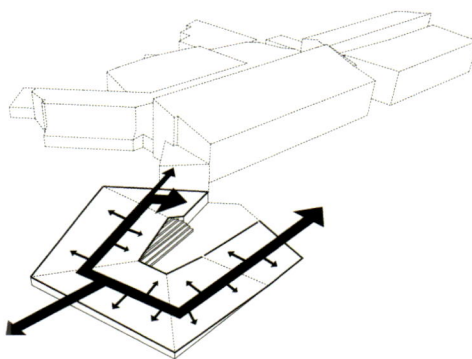

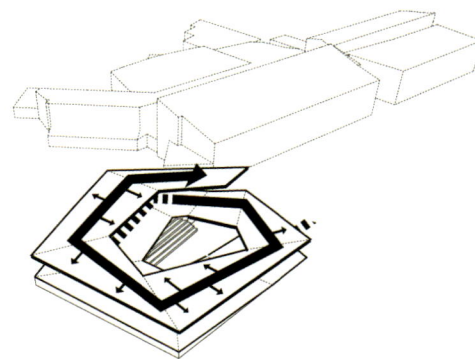

+152
+151
+150
+149

Program Arrangement Dagram

Star Light

星光

Firm: visiondivision
Location: Taipei, China

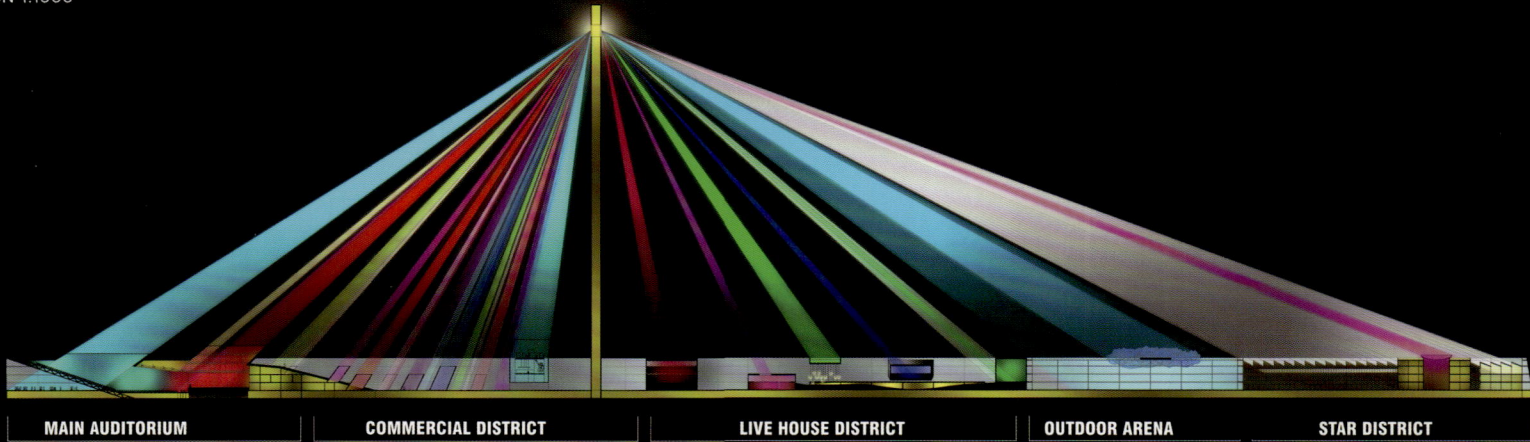

MAIN SECTION 1:1000

| MAIN AUDITORIUM | COMMERCIAL DISTRICT | LIVE HOUSE DISTRICT | OUTDOOR ARENA | STAR DISTRICT |

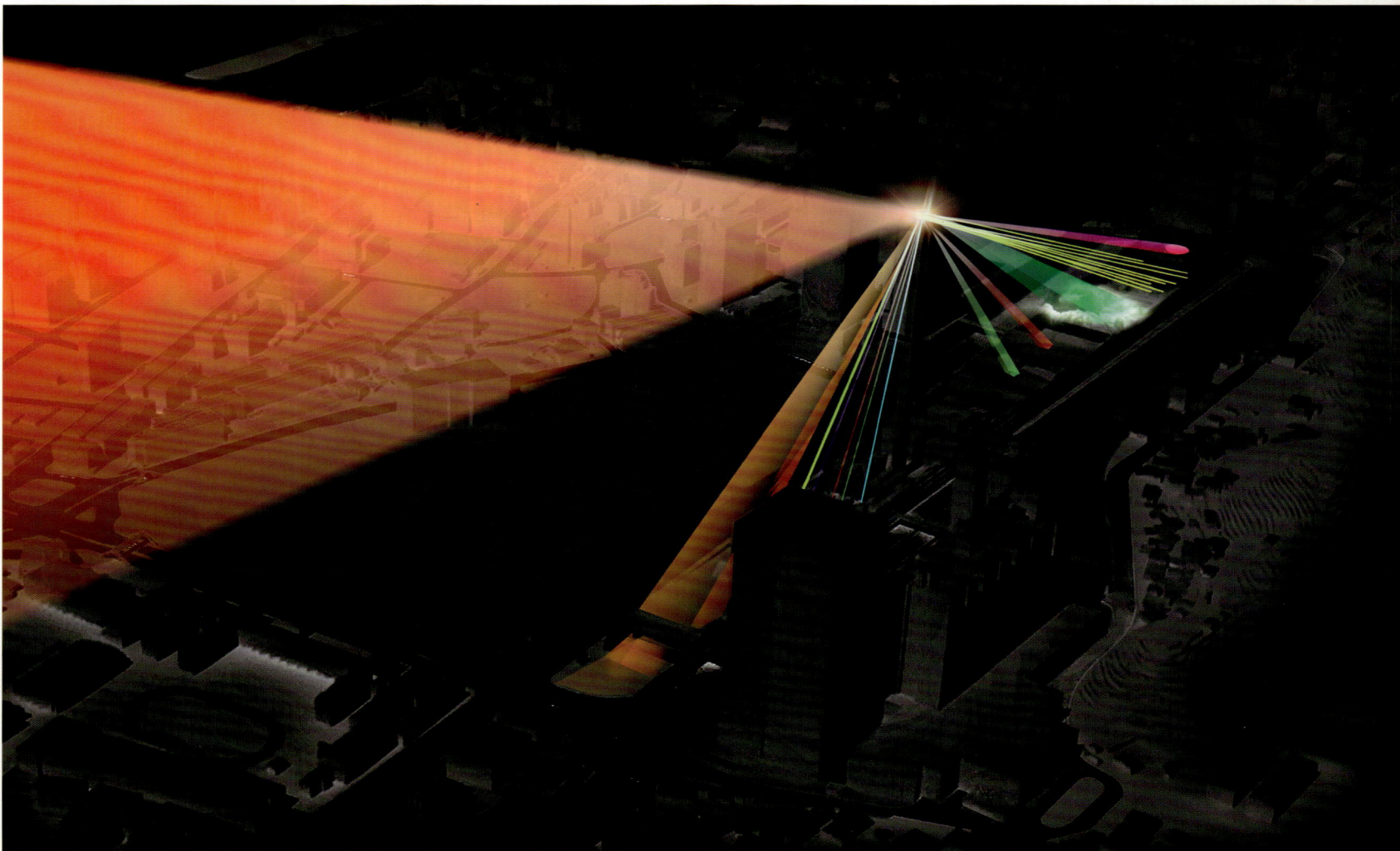

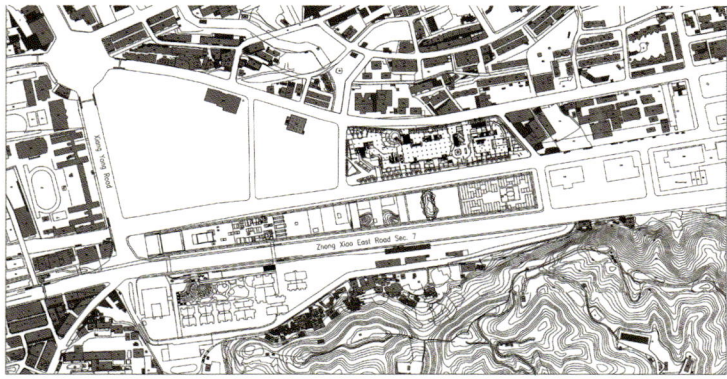
Site and Surrounding Area   0   100   200 Meters

The new Pop centre in Taipei is a city in itself, designed to transcend its visitors into a total escapism of pop. The enclosure makes the experience inside strong and focused and gives the structure's different districts a framework of function and identity. The wall will act as a noise barrier as well as a very straight forward tool of communication.

The tower in its center projects light to the different parts of the town as stage lights or to enhance the setting. Every district uses the effect from the tower in some aspect and is consequently designed according to the angles of light.

The lights can make the whole building expand dramatically in appearance, from rather low key from a distance into a spectacular body of light; forming a unique landmark for this developing part of the city.

The projections also works in a smaller scale, such as high precision laser to light up a visiting Tina Turner's Martini for example.

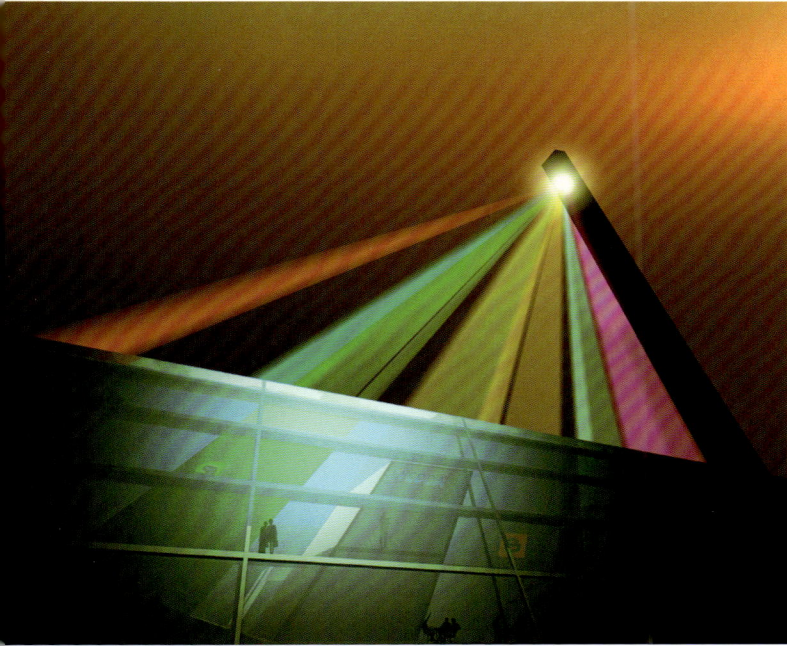

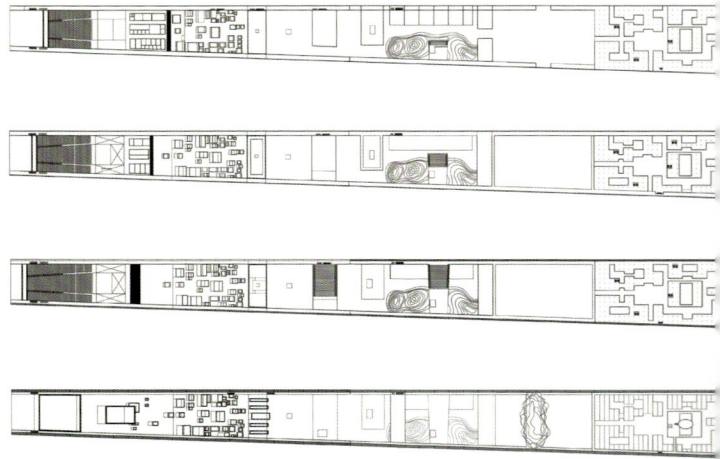

PLANS

台北的新流行文化中心本身就是一座城市，将参观者完全带入了逃避主义的流行艺术中。中心的外围设计强化了内部的体验并将注意力集中在此，同时也将功能和特征的构架带入了结构中的不同区域里。墙壁将作为噪音隔离设备以及交流的直接工具。

在中心部位的塔楼将光线以舞台灯光的形式或加强布景的目的投射到不同部分。每一个区域都在某种方面利用了塔楼，并根据光线的角度加以设计。

光线可以极大地扩展建筑的外形，将其从远处一座低调的建筑变成了一个光线构成的景观；也成为了这座城市中这一正在开发的区域的独特的地标。

光线投影也小规模地使用，例如，用高精度的激光来照亮来访的蒂娜·特纳的马提尼。

Peak Series

山顶系列

Firm: visiondivision
Location: Sweden

This is a series of pre-fabricated summer houses that allows a great portion of social life on a relatively small space. While being a comfortable vacation home for a family it can also easily accommodate a large amount of guests thanks to a sleeping mezzanine floor. The house is pyramidal in its shape with a wood facade with a gap between each board allowing climbing. The house is divided into three floors; in the middle is the sleeping mezzanine located, sandwiched between an open social first floor with a kitchen and a living room, and the upper floor with the master bedroom with bathroom. Each guest bed can be reached through a hatch that connects with the outside, thus giving each guest its private entrance, a fire escape and a nice view from each bed. The upper floor has also hatches from the bedroom and the bathroom. The house will be for sale later this autumn at Sommarnöjen,

an exclusive Swedish summer house manufacturer that sells pre-fabricate houses from the leading contemporary architects in Sweden. The hous can firstly be ordered in two different sizes; 45m² and 90m² but can als be adaptable upon request, perhaps as a vessel or a larger scale, such as summer camp.

The smallest house (45m²) has one master bedroom at the upper level ar can hold six guest beds at the mezzanine level.

When doubled it holds a total of twelve guest beds and comes with a larg bathroom between two master bedrooms.

The house will at first be launched in the Scandinavian countries but is course adaptable to other countries and larger scales as well.

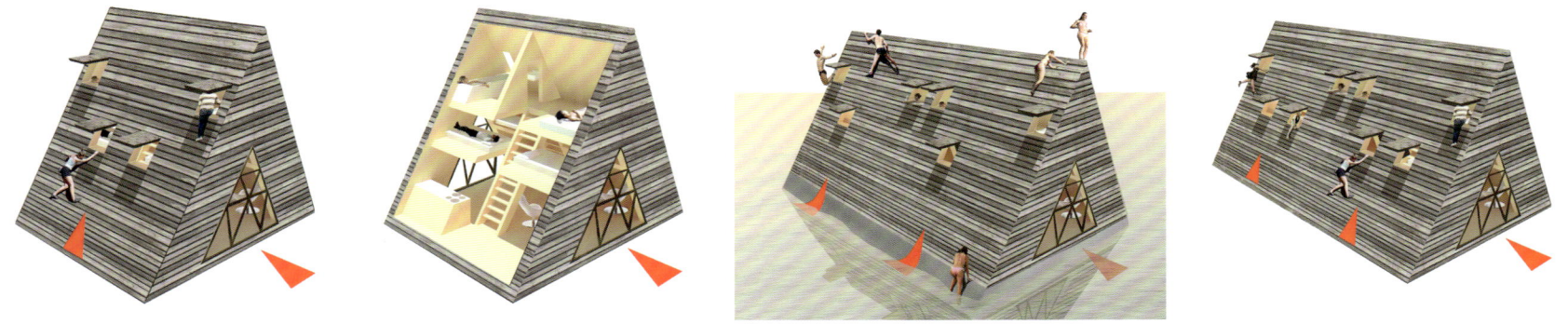

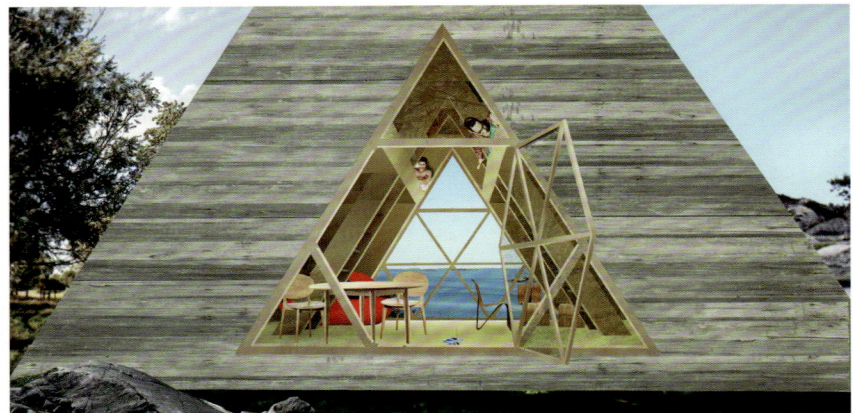

本案是一系列预制的避暑房屋，设计使人们可以在一个相对较小的空间中完成社会生活的很大一部分。这里既是一个家庭的舒适度假屋，也因为有一个睡觉阁楼的存在可以轻松接待一大批客人。房屋是一个锥体的形状，木质立面上的木板带有缝隙，可以供人攀爬。房屋一共被分成三层，中间一层是睡觉阁楼，被夹在一层带有厨房和客厅的开放公共区域以及带有卫生间和主卧的三层之间。每一张床都可以通过一扇与外界连接的小窗口到达，因此给每一位客人专属的私密通道、一个逃生通道和从床上便可看到的美丽风景。房屋稍后在秋天便会在专门从事避暑房屋建造的Sommarnöjen公司出售，该公司一直销售瑞典当代杰出建筑师设计的预制房屋。避暑房屋现在有两个尺寸可供选择，45m²和90m²，但也可以适应不同的要求，做成船只或者更大规模，如一个夏令营。

最小尺寸（45m²）的房屋上层拥有一个主卧，睡觉阁楼可容纳六张床。

90m²的房屋可容纳12张床，在两个主卧之间还有一个大卫生间。

避暑房屋首先会在斯堪的纳维亚地区销售，当然也适用于其他的国家，并可以建造成更大的面积。

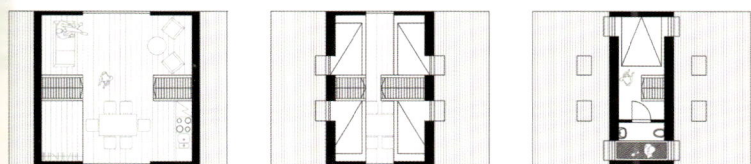

45 m²

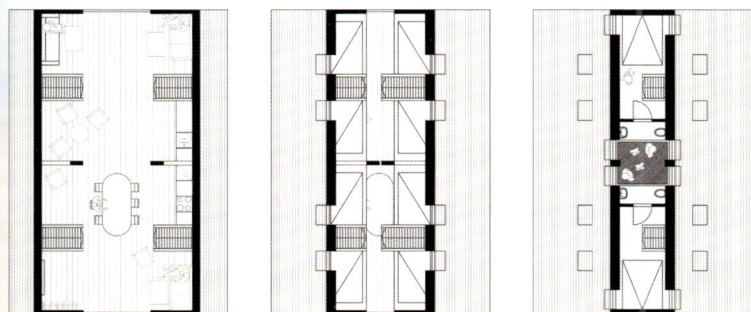

90 m²

INDEX 索引

Ivo Buda Architetto

Ivo Buda architetto is an architectural practice founded by Ivo Buda (Trieste, Italy, 1973) devoted to architecture, art and urbanism, for both public and private sectors.
The practice is driven by a consistent philosophical and holistic approach to the design and technology of each project, with the aim to create buildings with a strong architectural ambition that are intimately connected to context, rationality, function and environment.
The philosophical approach can be explained in following three points:
Art: Without art the architecture risks to become just the sum of the parts, an accumulation of kitchens, bedrooms, offices, gardens and parking lots.
Essence: This is not a question about minimalisms, the essence contains much more than the form itself. It is an endless resource of drama and beauty, is the real nature of the building.
Energy: The practice is taking the challenge of creating durable buildings that respond to a changing environment: the fact is that the biggest part of global energy consumption is produced by buildings and transportation.

AWP

AWP is an award winning interdisciplinary office for territorial reconfiguration and design. It is based in Paris and Basel (AWP-HHF) and develops projects internationally working on a wide variety of programmes : architecture, landscape design, strategic planning, urbanism.
Those projects only differ in terms of context and scale but share the same values and visions: hospitality, beauty, an innovative confrontation between symbol and uses and a renewed relationship between architecture and landscape. Our portfolio of clients includes several European cities, metropolitan and regeneration authorities, cultural institutions and developers. AWP also curates and designs exhibitions for major cultural institutions (such as the GAMC, City of Architecture and Heritage and Pavillon de l'Arsenal, Paris – Fondazione Adriano Olivetti, Rome) and write regularly books and essays. The three partners have exhibited their work and lectured at many architectural venues.

Labics

Founded in 2002 by Maria Claudia Clemente, Francesco Isidori and Marco Sardella. While Labics has been run by Clemente and Isidori since 2005, the three founding members work collaboratively in the architecture department of Rome's "La Sapienza" University in both research and teaching. Labics is responsible for the design of the Obika restaurants in Italy, London, New York and Kuwait, the Citta del Sole development in Rome and the Lausanne Museum of Fine Arts in Switzerland. During this year's interview, Maria Claudia Clemente discussed and showed the design process for the newly finished Fontana square in Quinto de Stampi, Milan; designed over an incredibly intricate grid system, every shape, line and material fits into a impeccably planned whole. This project is a stunning example of how Labics works within a space to create an experience that is not only aesthetically pleasing, but also painstakingly composed with the user in mind.

///Byn

///Byn is created in 2001 in Barcelona by the Architects Nicolas SALTO DEL GIORGIO and Bittor SANCHEZ-MONASTERIO.
///Byn is an investigative architectural office, where the interest in new conceptual research, academics, and technology based design, is applied to projects of every scale, including master plan, urban design, architecture or interior design.
///Byn was understood from the beginning as a nomadic platform that should be flexible enough to travel around the globe, while Bittor & Nicolas were still collaborating with world renowned architects. From 2004 to 2008, the practice moved to New York, and in May 2008 it moved to its current base in Shanghai.
Since May 2009 ///Byn is a full time working office.

Thomas Leeser

2008 Nomination for Cooper-Hewitt National Design Awards
2004 First Prize and commission, American Museum of the Moving Image, NY, invited competition NYC 2012 Olympic Village, invited competition, selected project with MVRDV and StoSS
2003 Three Architectural Record Interiors Awards for Excellence in Design for "Glass," "Bot" and "Pod" New York Foundation for the Arts Architecture and Design Discretionary Grant for "Three Legged Dog Performing Arts Group" New Media Arts Center design
2002 New York Foundation for the Arts, Architecture and Design Comissioning Grant for "Three Legged Dog Performing Arts Group" New Media Arts Center design James Beard Foundation"best design award" nomination for "Glass"
2001 Finalist, EMPAC-Electronic Media and Performing Arts Center, Rensselaer Polytechnic Institute, Troy, New York, invited limited competition
2000 Award winning design for ETH World, two stage international competition for a virtual university

CEBRA

CEBRA is a Danish architectural office with its background in the architectural circles of Aarhus. The office was founded 2001 by the architects Mikkel Frost, Carsten Primdahl and Kolja Nielsen – all graduates from The Aarhus School of Architecture.
In 2008 CEBRA received the most prestigious award in Scandinavia the "Nykredits Architecture Award" and in 2006 CEBRA received the Golden Lion at the Venice Biennale 2006 for the project Co-evolution, which has been exhibited at the Biennale, São Paolo, Beijing, Manchester, Copenhagen and Venice 2006/2007. In 2008 CEBRA participated in the Biennale once again — this time curating the Danish pavilion at the Venice Biennale ECOTOPEDIA — Walk the Talk. The Bakkegaard School was nominated for the Mies van der Rohe award in 2006.
The office has worked with numerous architectural functions and scales but dwellings, sports facilities and school buildings have been predominant.

UNStudio

UNStudio is an international architectural practice, situated in Amsterdam since 1988, with extensive experience in the fields of urbanism, infrastructure, public, private and utility buildings on different scale levels. At the basis of UNStudio are a number of long-term goals, which are intended to define and guide the quality of our performance in the architectural field. We strive to make a significant contribution to the discipline of architecture, to continue to develop our qualities with respect to design, technology, knowledge and management and to be a specialist in public network projects. We see as mutually sustaining the environment, market demands and client wishes that enable our work, and we aim for results in which our goals and our client's goals overlap. In 2009 UNStudio Asia was established, with its first office located in Shanghai, China. UNStudio Asia is a full daughter of UNStudio and is intricately connected to UNStudio Amsterdam. Initially serving to facilitate the design process for the Raffles City project in Hangzhou, it is envisioned that UNStudio Asia will expand to a stable multinational team of all-round and specialist architects.

Jason Zerafa

Jason Zerafa is the founding principal of the Zerafa Architecture Studio. Prior to founding Zerafa Architecture Studio in 2005, Jason Zerafa was a Senior Associate Principal and Senior Designer with Kohn Pedersen Fox Associates PC in New York. Since joining KPF in 1994, Mr. Zerafa was an integral participant in the design and development of a number of significant buildings in the United States, Europe and Asia. His innovative designs for buildings have won major international and national competitions and his design work for KPF has been extensively published.

Mr. Zerafa's recent project for KPF was the Mirae Asset Tower. He also designed the Guohua Smokestack Redevelopment.

As the Senior Designer of competition-winning designs for Canal+Village Louvesciennes, the EMGP Parc du Millenaire in Paris and KPF's designs for the Tour Granite for Societé Generalé and the Generali Headquarters in Paris, Mr. Zerafa helped KPF strengthen their practice in Europe by developing a new domestic client base in France. Here in New York City, Mr. Zerafa designed the recently completed 640 Fifth Avenue, and the New York Jets Training Center.

Other KPF projects on which Mr. Zerafa served as Senior Designer include the Mohegan Sun Casino Phase II expansion and the Queensport Development.

David Serero

David Serero (born 1974, Grenoble, France) received an Architecture Degree from Ecole d'Architecture Paris-Villemin in 1998 and a Master of Architecture and Urban Planning from Columbia University in New York. In 2003, he received the commission of the "Art Arena", an art film museum in London hosting 45 projection spaces for the Roland Collection. The same year, he was selected to conceive the 2009/2011 Memorial in New York City. In 2004, he won the first prize in the International Competition for the Hellenikon Metropolitan Park in Athens, Greece. In 2005, David Serero received the Rome Prize of the French Academy in Rome, where he developed the "Variable Configuration Acoustical Domes", a research on domes, adjustable to modify acoustical behavior and performance of a space.

His work has been widely published and exhibited in shows at the Museum of Modern Art (MOMA) of New York, at the New York Architectural League, the Venice Biennale, and the Mori Museum in Tokyo.

Vlado Valkof

Design Initiatives is innovative, award-winning architecture practice based in Los Angeles, California and Sofia, Bulgaria, EU founded in 2009 by Vlado Valkof.

For two years, after Vlado Valkof [or just Valkof] graduated with Master's degree from UACEG, Sofia in 1998 he had worked as free-lancer. In 2000 he relocated to Southern California first to do his internship in Morphosis and then his post-grad at SCI-ARC. After that he had gained extensive experience in several Los Angeles offices over complex commercial, institutional, mixed-use, residential, and interior projects throughout the US, Europe, Russia and the UAE. In 2009 Valkof had a chance to re-establish his practice under the name of Design Initiatives.

Valkof is a Laureate of The International Academy of Architecture [IAA] and got awarded with The Iakov Chernikhov International Foundation [ICIF] Prize. His work won Second Prize from The United Nations Development Programmed [UNDP] Competition and participated in Beyond Media Festival in Florence, Italy.

Design Initiatives currently is a collaborative environment of three architects but for some projects more architects join remotely from around the world.

3LHD

3LHD is an architectural practice, focused on integrating various disciplines – architecture, urban planning, design and art. 3LHD architects constantly explore new possibilities of interaction between architecture, society and individuals. With contemporary approach, the team of architects resolves all projects in cooperation with many experts from various disciplines. Projects, such as Memorial Bridge in Rijeka, Croatian Pavilion in EXPO 2005 in Japan and EXPO 2008 in Zaragoza, Riva waterfront in Split, Sports Hall Bale in Istria, Centre Zamet in Rijeka, Zagreb Dance Centre in Zagreb and Hotel Lone in Rovinj are some of the important highlights. The work of 3LHD has received important Croatian and international awards, including the award for best building in Sport category on first World Architecture Festival WAF 2008, IOC/IAKS Bronze Medal Award 2009 for best architectural achievement of facilities intended for sports and recreation, AR Emerging Architecture Award (UK), the ID Magazine Award (USA); and Croatian professional awards Drago Galić (2008), Bernando Bernardi (2009; 2005), Viktor Kovačić (2001), and Vladimir Nazor (2009; 1999).

Markus Dochantschi – Founder and Principal StudioMDA

Markus founded StudioMDA in New York 2002. StudioMDA's approach is to creatively engage all parties at the very beginning, forming an inspirational collaboration in defining architecture. Calling a wide range of consultants beyond the architect's typical collaborative, StudioMDA has teamed with artists, video artists, and choreographers, among others to define space generated by various artistic endeavors. Within this context, architectural syntax can become layered into micro and macro scales, balancing the smallest detail with the topographical/ urban fabric.

StudioMDA's awards include: AIA - Oculus Spring, New York Next 10 firms of the future in 2004; in 2007, StudioMDA received an Honorable mention for the New Housing New York Legacy Project organized by the AIA and the City of New York and in 2007 was awarded the First Prize to design and develop the Brooklyn Arts Tower. In 2009, StudioMDA was awarded the First Prize for the Center for Advanced Mobility in Aachen, Germany as Architect and Project Manager.

Cat Huang

Catherine Huang has worked at BIG since 2007 and is currently a Project Architect. She has collaborated with the partners on numerous award winning projects such as the new headquarters for the Icelandic bank Landsbankinn, the Danish Pavilion for the 2010 Shanghai Expo and most recently the Shenzhen Energy Mansion in Shenzhen, China. During her time at Harvard University's Graduate School of Design she conducted research into new models of sustainability for the built environment, with a focus on understanding what buildings can learn from biology.

Catherine has also worked successfully in various international offices such as MoPing Atelier in Beijing, China and the Harvard University Planning Department.

Languages:
English, Chinese (Mandarin) and Spanish
Education:
University of Texas | B.S. Molecular Biology | Austin, TX
Harvard Graduate School of Design | MArch | Cambridge, M

3XN

3XN was founded as Nielsen, Nielsen and Nielsen in Aarhus in 1986 by the architects Kim Herforth Nielsen, Lars Frank Nielsen (partner until 2002) and Hans Peter Svendler Nielsen (partner until 1992). The three Nielsen architects, often referred to as the Nielsens – and today simply as 3XN – quickly became known for two things: their preference for ground-breaking architecture, in defiance of the anti-humanistic modernism, and projects demanding a high level of detail and employing workmanship of the highest quality.

A first breakthrough came with the court house in Holstebro, which was followed by a number of first prizes in architectural competitions, such as the Architects' House in Copenhagen, the Glass Museum in Ebeltoft and the concert hall Muziekgebouw in Amsterdam.

The partnership today consists of Founder Kim Herforth Nielsen; CEO Bo Boje Larsen who became Partner in 2003; and Head of Competition Jan Ammundsen who became Partner in 2007.

modostudio | cibinel laurenti martocchia architetti associati

Located in Rome, modostudio | cibinel laurenti martocchia architetti associati is a multidisciplinary practice of architecture, urban planning and industrial design. Established at the end of 2006 by three principal architects, Fabio Cibinel, Roberto Laurenti, and Giorgio Martocchia, after many years of collaborating with internationally acclaimed architects like Massimiliano Fuksas, Piero Sartogo, Erik Van Egeraat and Kas Oosterhuis, modostudio in a short time was awarded and shortlisted in many international architectural competitions.
Winning prize: New Intecs Spa headquarter invited competition in Rome, (IT),2010
Winning prize: Uhl Foundation international invited competition in Laives, (IT); Ex Fonderie Riunite international competition in Modena, (IT); Los Carmenes shopping mall invited competition in Las Torres de Cotillas, (ES), 2009

Pascal Arquitectos

Pascal Arquitectos was founded in 1979 by Carlos and Gerard Pascal with the main purpose to achieving ultimate and integral development in architecture, interior, landscape, lighting and furniture design. Most of Pascal Arquitectos projects had been published and recognized with national and international awards.
Atelier's production ranges from numerous luxurious residential projects, several residential complexes, institutional and religious buildings, restaurants, to corporate and offices buildings and hotels.
Major restorations of two historic monuments, both recognized by the National Institute of Fine Arts (INBA), had been completed: Edificio Bolivia, an original masterpiece of architect Carlos Villagran Garcia nowadays corporate headquarters of Ford Motor Company de Mexico, and the grand heritage mansion located in Avenida Refoma and Rio Elba street, were restaurant "El Divino" used to be.

Monolab

Monolab is a Rotterdam-based practice for advanced research and design in urbanism and architecture, founded in 1999. Simplicity (mono) is linked to experiment (lab).
Monolab is ambitious and dedicated to advanced and highly integrated projects through simple and crystal clear concepts with high heartbeats. The projects handle complex issues and are fully integrated in their context and our contemporary culture.
Monolab deals with large as well as small-scale projects. The office obtains commissions in a wide field: from furniture design to villas, from residential projects to urban schemes, from stations and city centers to distribution hubs.
ir J.W. (Jan Willem) van Kuilenburg is principal of Monolab and head of final projects at the Fontys Academy of Architecture and Urbanism in Tilburg. He is a member of the advisory committee architecture, research and design at the Dutch Architecture Fund.

Sybarite

Sybarite is an architectural and design practice that aims to tease all of the human senses and to sculpt architecture into the living environment in which it exists, whilst retaining complete functionality.
Torquil McIntosh - Architecte DPLG ARB, Director
Torquil studied at the University of Edinburgh and the Beaux-Arts in Paris, gaining fluency in French and graduating in 1998. His practice experience includes Denis Laming Architectes in Paris and Future Systems in London. He lives in London with his wife and three children.
Simon Mitchell - BA(hons) Dip(hons) Arch RIBA ARB, Director
Simon studied at the University of Greenwich in London, gaining his RIBA Part III qualification in 1998. He has a range of practice experience, most notably with Terry Farrell & Partners and Future Systems where he was Associate Director for several years. He lives in London with his wife and daughter.

J. MAYER H Architects

Founded in 1996 in Berlin, Germany, J. MAYER H Architects' studio, focuses on works at the intersection of architecture, communication and new technology. Recent projects include the Town Hall in Ostfildern, Germany, a student center at Karlsruhe University and the redevelopment of the Plaza de la Encarnacion in Sevilla, Spain. From urban planning schemes and buildings to installation work and objects with new materials, the relationship between the human body, technology and nature form the background for a new production of space.
Jürgen Mayer H. is the founder and principal of this cross-disciplinairy studio. He studied at Stuttgart University, The Cooper Union and Princeton Universtiy.

ADEPT

ADEPT is based in Copenhagen, Denmark and works within the fields of architecture, Urban planning and landscape design. ADEPT is founded and lead by architects: Martin Laursen, Martin Krogh and Anders Lonka.
ADEPT is currently working with projects: "Flintholm Spark" 4,000m² House of Culture and Movement at Frederiksberg (DK), "Village in the Sky" 22,000m² high rise (DK), "Iceland Academy of Arts" 20,000m² educational and cultural institution in Reykjavik (IS), and urban planning projects: "The Tolerant City" 100,000,00m² urban development in Helsingborg (SE), "Køge Coast" 350,000m² urban development project for Køge Municipality, and "The Måløv Axis" 100,000m² urban space in Ballerup (DK).
ADEPT's three partners have all been engaged and involved in teaching, workshops and lectures since they founded their company. Anders Lonka teaches in sustainable urban planning at The School of Architecture at the Royal Danish Academy of Fine Arts, while Martin Laursen and Martin Krogh are external professors at The Faculty of Architecture at AAlborg University.

AART Architects

AART Architects is an international architecture company founded in 2000 with the vision to develop the Scandinavian architectural tradition by creating vibrant, sustainable physical surroundings. AART Architects has offices in Denmark and Norway and is a high-performance team of 50 young designers and technicians that deliver full-service consultants in the field of architecture.
The projects of AART Architects are rooted in the essence of the site and characterized by a strong environmental and social awareness. As architects we believe in the concept of "total design" and collaborate closely with clients, stakeholders and users to produce multi-disciplinary solutions that focus on the environmental, social and economic aspects of every project.
In other words, we believe that long-lasting architecture grows out of a profound understanding of the unique qualities of the specific site, cultural context and client's vision.

Francesco Gatti

Francesco Gatti graduated with honors degree in ROMA TRE University. In 2002 he established the 3GATTI.COM ARCHITECTURE STUDIO. In 2004 he opened a new branch of the office in Shanghai where he completed many projects and won many competition and awards as: the "Buchanan Underground Station" (Glasgow-Scotland), the "Tra la murgia e il mare" urban development (Andria-Italy), the "500m³ design " GBD art residential district (Beijing,China) and the "06 Modern Decoration interior design media price" (Shenzhen,China). His work was published in many international reviews.Gatti participated as jury in Italian and Chinese architecture competitions and taught in ROMA TRE University (Italy) and in Tongji University (China) where he also participated as master of the Archiprix International.

Plasma Studio

Plasma Studio is a leading emergent architecture and design practices with worldwide scope and outlook, engaging seamlessly a wide range of scales and types including furniture design, houses, hotels, cultural projects as well as landscape and urban planning.
Starting with a range of small but challenging refurbishment projects in London between 1999 and 2002, founding partners Eva Castro and Holger Kehne then completed various new buildings in the Italian Dolomites where they opened a studio location with partner Ulla Hell in 2002. Between 2003 and 2005 Plasma worked alongside some of the most famous architects and designers on one of the floors of Hotel Puerta America, Madrid. Plasma's floor was one of the most challenging of the 16 radically different takes and published widely.
Recently Plasma has been involved in several large-scale mixed-use projects in China and is currently lead designer for the International Horticultural Expo in Xi'an with 37 ha and 12,000 sqm of projected buildings.

visiondivision

visiondivision is an architecture firm founded in Stockholm, Sweden by Anders Berensson (b.1980) and Ulf Mejergren (b.1981) in 2005.
Since then, the office has been operating in The Netherlands, Mexico and Argentina with great success.
visiondivision deals with all kinds of architecture and design problems and with any client, always with the promise of making it top shelf.
The office does not have a fixed design idea. Things that make us stand out in this business is our braveness, cleverness and our ability to think inventive.
The office is one of the most published firms in Europe and have already received several prices and mentions even though we only are in our late 20s, the team has experience from the most important contemporary architectural studios in the world like OMA for example, and we have happy clients all over the globe.
visiondivision is good at making conceptual designs and unique solutions for all of our clients.

One By Nine

One By Nine is an architecture studio founded in 2009 by Łukasz Wawrzeńczyk and Konrad Grabczuk. Our portfolio covers architectural projects, commercial and private interior design. We offer broad experience gained in numerous local and international projects, innovative solutions combining aesthetics and functionality, professional consulting, project supervision and most of all passion to create the space in original forms.

Justus Pysall

After completing his diploma in 1989, Pysall went to London and worked three years at Foster Associates. From 1990 until 1992 he lectured at the London Architectural Association as an assistant, and realised at the same time his first project in London's East End. Following this he worked at Jean Nouvel's office in Paris until the foundation of his own office in Berlin in 1993.
Justus Pysall is a member of the Architects' Association Berlin, the AIV, the BDA Berlin — member of the board, the Stiftung Baukultur, the Schinkel Committee of the AIV and the GeSBC (DGNB) — German Sustainable Building Council.
Pysall Architects has been awarded various prizes and mentions, such as the Asakura and the Kajima prize in Japan, the BDA Hamburg prize 2008, the office application award - besto office concept 2010, best architects 2010 and the Gold Certification of the GeSBC (DGNB) 2010.

Studio Nicoletti Associati

Studio Nicoletti Associati, founded in 1957, delivers world-class professional design and management services. Studio Nicoletti Associati offers award winning architectural design, progressive infrastructure engineering, highly ranked project and construction management and are known for quality and professionalism. At over 30 professionals, we are one of the oldest and largest firms in Rome. We can deliver total services to a project or form flexible work teams that can integrate into a larger team in partnering and sub-consulting roles. The studio practice's expanded through all major aspects of urban and building design in Italy, Europe, Africa, USA, Middle and Far East.

David Tajchman

Born in 1977 in Brussels, Belgium, David Tajchman studied architecture at Univerisité Libre de Bruxelles – Institut Victor Horta and at The Bartlett School of Architecture. After having worked as project leader for several architecture and design practices including Dominique Perrault, Jacques Ferrier, Stéphane Maupin and Patrick Jouin, he founded his own architecture agency in 2009 simultaneously in Paris and Brussels.
Teaching and lecturing experiences include a permanent position as Professor at Ecole Spéciale d'Architecture in Paris (2011 to present) and Tutor with Sir Peter Cook (2009 to 2010), guest lecturer at Institut Supérieur d'Architecture Victor Horta in Brussels and The Centre for Contemporary Architecture in Budapest.
David Tajchman's works have been widely published and shown all over the planet. The agency focuses on innovative, original, different and inventive projects. Working at all scales, from urban design to product design, from the city scale to the furniture element, every proposal would be a specific response to the clients needs, site location and cultural background.

OFIS arhitekti

Based in Ljubljana formed by Rok Oman and Spela Videcnik (1998).
rok oman (born 1970)
studied architecture at the ljubljana school of architecture (grad.1998)
and at the architectural association in london (grad.2000).
spela videcnik (born 1971)
studied architecture at the ljubljana school of architecture (grad.1997)
and at the architectural association in london (grad.2000).

Thomas Christoffersen

Thomas Christoffersen began his collaboration with Bjarke Ingels in 2001 when PLOT was first formed and is currently an Associate. Thomas has worked on every notable project from the VM Houses to one of our most recent and global developments, Astana National Library. He is leading a 10-person team while overseeing the detail design and construction of the 33,000m² mobius-shaped structure. Other accomplishments of his include the design for Iceland's National Bank and Stavanger Concert house in Norway.
In addition to his long standing participation in all things BIG, Thomas took a sabbatical year to work in New York City with WORK Architects and has also worked with Stan Allen, David Ling in NYC and Henning Larsen Architects.

James Law Cybertecture International

James Law Cybertecture International is a multidisciplinary practice which pioneered the concept of Cybertecture in 2001. This 21st century discipline weaves the hardware and software of the urban fabric and built reality into a seamless fusion of architecture, technology and engineering, resulting in cyber-based solutions for a lifestyle to Live the Future.
Janmes Law is the visionary founder and Chief Cybertect at James Law Cybertecture International, formed on the first day of the new century in 2001. Dedicated to blend cyber technology within modern architecture, James is paving the way for people to live the future in a seamless and enjoyable manner.

Arch Group Architectural Bureau

The Arch Group Architectural Bureau was established in 2007, when the Arrrh! Bureau (Michael Krymov and Alina Vasilieva) merged with Alexei Goryainov's independent studio. Since then, we completed many architectural and interior design projects. Most of them have been implemented or currently are under construction. Our projects have been distinguished by professional awards and media publications.
The Arch Group Architectural Bureau offers a full range of design works, from initial consultation to concept development to final design to field supervision.
The Arch Group Architectural Bureau is licensed for all major types of design work. Architects M. Krymov and A. Goryainov are members of the Union of Architects of Russia.
Our goal is to identify the maximum aesthetic and functional capacity of the project taking into account the client's wishes. We believe that good architecture should not be a financial burden on the customer's budget. Good architecture is the result of an architect's professional approach and his interest in the outcome. The foundation of quality architecture lies in the attention to details, which is expressed in a well-developed project documentation.

Nuvist

Nuvist is an architecture and design studio which founded in 2006 in Istanbul (Turkey), by Kursad Sekercioglu and Emrah Cetinkaya. Nuvist aims to associate Architectural and Design works with all ranges and methodologies of Art, for supporting and developing each other in possible solutions.
Our design statement starts with "writing"; actually the word of "nuvis" stands for "writing" in the Ottoman language, and in the Ottoman culture the word of "Hoş-Nüvist " is used for the artists of the Ottoman writing art that is called "Hat" or "Hoş-Nüvis " ; those artist are also called as "Hattat". "Hat" and writing is not only expression an idea with symbols but also description of values that enhanced high aesthetic levels. Firstly imagined in mind later happened with sense oriel organs. Sensibility in design brings this imagination and thought approaches which represent and transfer its in foreground.
Consequently our objective is, making environments which help us to explore vary and range design probabilities with using traditional and modern design methods; and also determining and denoting opportunities which 21st century's software and technologies give us how to use information in necessary and suitable places on finding creative & innovative design solutions.

后记

　　本书的编写离不开各位设计师和摄影师的帮助，正是有了他们专业而负责的工作态度，才有了本书的顺利出版。参与本书的编写人员有：Fabio Balducci, Leonardo Consolazione, Manuela Gentile, Domenico Santoro , Vlado Valkof, Anne Valkof, Stanislav Christov, Assen Balkanski, visiondivision, Anders Strange, Anders Tyrrestrup ,Torben Skovbjerg Larsen, Lukasz Wawrzenczyk, Ivo Buda, Manfredo Bianchi, Tomaz Kristof, Martin Laursen, Martin Krogh, Anders Lonka, Jonas Smit Andersen, Kasper Svanberg, David Serero, Yoichi Ozawa, Ran She, Fabrice Zaini,3XN, Paola Bettinsoli, Leonardo Consolazione, Annalucia Scarascia,Francesco Gatti, Carlos Pascal Wolf, Gerard Pascal Wolf, Rok Oman, Spela Videcnik, James Law Cybertecture International, Eva Castro, Holger Kehne, Alfredo Ramirez Galindo, Xiaowei Tong, Mehran Gharleghi, Evan Greenberg, Nicoletta Gerevini, Peter Pichler, Tom Lea, Ying Wang, Katy Barkan, Federico Ruberto, Rui Liu, Danai Sage, Andrej Gregoric, Janez Martincic, Janja del Linz, Katja Aljaz, Robert Janez, SERERO Architects, Nicoletti Associati, Hijjas Kasturi sdn, Thomas Christoffersen, J.W. van Kuilenburg with A. Chlebinska, E. Komarzynska, P. Roger Rzepecki, G. Michaud-Nérard, B. Drogge, M. van Oers, G. Porcu, David Tajchman, Manfredi Nicoletti, Luca Nicoletti, Cat Huang, Gaia Maria Lombardo, Giorgio Pasqualini, Laura Perri, Maria Adele Savioli, Thomas Leeser, CEBRA, AWP (leading consultant) /Atelier Oslo, Sybarite, Timur Bashkaev., Mikhail Krymov, Alexey Goryainov,Modostudio , Cibinel Laurenti Martocchia Architetti Associati, Emrah Cetinkaya, studioMDA, Zerafa Architecture Studio, ///byn , Justus Pysall, Peter Ruge, Bartlomiej Kisielewski, ADEPT (DK), Sou Fujimoto Architects (JP), in collaboration with Rambøll (DK), Topotek1 (DE) and Bosch & Fjord (DK), MVRDV, Rotterdam, and ADEPT, Copenhagen, in a joint effort with SLA landscape architects, Søren Jensen engineers, Imitio, Winnie Ricken, Copenhagen, Max Fordham, 3LHD.